THE

JOURNAL

OF THE

ROYAL GEOGRAPHICAL SOCIETY

OF

LONDON

VOLUME 8

Elibron Classics
www.elibron.com

Elibron Classics series.

ISBN 0-543-78240-9 (paperback)
ISBN 0-543-78239-5 (hardcover)

This Elibron Classics Replica Edition is an unabridged facsimile of the edition published in 1838 by John Murray, London.

THE JOURNAL

OF THE

ROYAL GEOGRAPHICAL SOCIETY

LONDON.

VOLUME THE EIGHTH.

1838.

LONDON:
JOHN MURRAY, ALBEMARLE-STREET.

MDCCCXXXVIII.

LONDON:
Printed by William Clowes and Sons,
Stamford Street.

CONTENTS OF VOLUME VIII.

ANALYSES, &c.

MISCELLANEOUS.

LIST OF ILLUSTRATIONS TO VOL. VIII.

ERRATA.

Vol. VII.

Page 215, in note, *for* Boerecha, *read* Bonechea.
— 216, — *for* India, *read* London.
— 221, line 8*, *for* W. lat., *read* W. long.
— 243-270, *for* Count Gräberg, *read* Count Gråberg, *passim.*
— 244, line 16, *for* 'Abdu, *read* 'Abdi.
— — — 19, *for* ben Ma'súd, &c., *read* 'Abdi-s-salámi-t-Temsemami.
— 246, — 1, *dele* point at end of line, *insert* a comma.
— — — 2, *move down* ر Rá r. to stand alone.
— 247, — 23, *for* 'Aghla, *read* Aghla.
— — — 25, *add* at end of line, Elevated, height.
— 249, — 11, for *Al-mudráwah*, read *Al-madráwah.*
— — — 25, *after* town, *add* and tribe.
— 250, — 14, *for* محمل *read* محمد
— — — 4*, for *Azgun*, read *Azgan.*
— — — *ult.*, read *Zuwaghah.*
— 251, — 5, *for* بحر الكبر *read* بحر الكبير *Add* children of Mohammed.
— — — 14, for *Beïlút*, read *Belút.*
— 252, — 18, add *Táfilelt.*
— — — 5*, add *North Africa.*
— 253, *insert* بوالاوان *Bu-l-awán*, town 32° 43′, 7° 12′; Easy ford, father-passage.
— 254, — 1, *read* دُقّالة *Dukkálah,* the elevated plain.
— 256, — 1, *add* plantation on two heights.—Ar.
— 257, — 13, *add* 'Aïth-Akensús?
— — — *ult.*, *for* excellent, *read* conspicuous.
— 259, — *ult.*, *add* the Luccos of our Maps.
— 261, — 22, *add* and tribe; the city is properly named *Miknas*, population 55,000?
— 263, *for* Cape Fegato, *read* Cape Fegalo.
— 264, — 2, read *Sóma' Hasan.*
— 266, — 12, read *Tarfu-sh-shakkar.*
— — — 17, for *Tawant*, read *Tawúnt.*
— — — 22, *add* filthy.
— — — 26, *add* abounding with plants.
— 267, — 1, *for* west, *read* coast.
— 400, — 7*, *for* alluded, *read* attended.
— 402, in note, *dele* 'which if' to the end of the page.

Vol. VIII.

Page 47, line 18, *for* W.S.W., *read* N.W.
— — — — *for* S.S.E., *read* S.S.W.
— 262, — 16*, *for* all, *read* atl.
— 264, — 13*, *for* altered, *read* allied.

(*) Count from bottom.

Royal Geographical Society.

1838.

AT THE

ANNUAL GENERAL MEETING, May 21, 1838,

The following Report from the Council was read:—

Since our last anniversary, sixty-five new members have been elected, and ten vacancies have occurred in the Society, which now consists of 600 members, exclusive of Foreign Honorary and Corresponding Members.

During the past year we have had to deplore the loss of our late most gracious Sovereign, William the Fourth; a monarch whose name, associated as it is in the annals of our country with the progress of discovery in Africa, Australia, and in both the Arctic and Antarctic Oceans, must ever live in the remembrance of the Geographical Society as its first munificent Patron and Benefactor.

The Council have great gratification in announcing that Her Majesty Queen Victoria has been graciously pleased to become the Patron, and to continue to the Geographical Society the same annual donation of a Royal Premium which had been originally granted to it by her royal predecessor.

Finances.—The state of the finances of the Society, the details of which are annexed, continues satisfactory. The only item of extraordinary expenditure consists in 150*l.* towards the expedition in British Guayana, but this has been paid out of the annual income without touching upon the capital stock invested in the funds.

Publications.—The Journal of the Society for 1837, as in former years, has been published in two parts; and it may be safely asserted that it is not inferior to any of the former volumes, either in the importance of its matter, or in the correctness of its illustrations.

An attempt has also been made to reduce Arabic and Oriental names to one standard of orthography; and on this subject the Council gladly embrace the opportunity afforded them of recording their thanks to their learned Foreign Secretary, to whose able and gratuitous assistance they are indebted for much of the value of the various communications, contained in the Geographical Journal, connected with the East.

The Council have also to acknowledge the continued zealous co-operation of Mr. John Arrowsmith in superintending the drawing and engraving the Illustrations of the volume.

Hitherto the Journal of the Society has been published in two parts, in the months of May and November each year, but the Council, having considered that much of the value of geographical information depends upon its being made public with as little delay as possible, have determined in future to publish the first part of the Journal in February, the second in May, and the third, as at present, in November. Acting upon this resolution, the second part of Vol. VIII. is now laid on the table.

The first edition of Part I. Vol. III. of the Journal having been exhausted, while the demand for it still continues both by the public at large and by new members desirous of procuring complete sets of the Society's Transactions, the Council has directed it to be reprinted, and it is now ready.

The Grammar of the Cree Language, by Mr. Howse, which was undertaken conjointly with the Church Missionary Society, advances towards completion; but it will not be ready for publication before the close of the present year.

Her Majesty's Donation.—The Royal Premium for 1837 has been awarded by the Council to Lieutenant-Colonel CHESNEY, of the Royal Artillery:—for his various travels in the East preparatory to the plan of the Euphrates expedition;—for the energy and perseverance shown by him in its general conduct, and during its progress;—and for the valuable materials both in comparative and physical geography, resulting from the labours of the expedition, in the countries of Northern Syria, Mesopotamia, and the delta of Susiana.

Nor can the Council omit while acknowledging the claims of Colonel Chesney as the original projector and leader of this expedition, to notice how materially he was aided throughout this arduous undertaking by the zealous and cheerful co-operation of the officers and men associated with him; namely, Major Estcourt, 43rd regiment, who was in command during the ascent of the Karún, &c.; the late Lieutenant Murphy, Royal Engineers, astronomer; Lieutenant Lynch, I.N., and after his departure Lieutenant Cleaveland, R.N., who commanded the naval part of the expedition; Mr. Ainsworth, as geologist and naturalist; Mr. Rassam, a native of Mosúl, as interpreter; the late Lieutenant Cockburn and Dr. and Mr. Staunton, of the Royal Artillery; and Messrs. Eden, Charlewood, and Fitz-James, Lieutenants in the Royal Navy.

Other enterprising travellers have also established strong claims to the gratitude of the Society; and among these Captain ALEXANDER, of the 42nd regiment of Royal Highlanders, for his late journey in South Africa from the Cape of Good Hope across the Orange river, and through the country of the Great Namáquas to Walvisch Bay; thence 200 miles due east to the country of the Dámaras, and by a different route back to the Orange river. In the course of this journey several new objects in natural history have been added to our collections—some account of the Dámaras has been obtained—and the features of an extensive part of hitherto unexplored country has been traced on our maps.

Mr. W. I. HAMILTON is especially entitled to honourable mention for the zeal he has shown in the collection of geographical and other information in his various journeys, during the years 1836-7, throughout Asia Minor, from the ruins of Anni on the east to Smyrna on the west: in the course of which he has fixed the sites of several ancient cities—has given an account of the geological formation of the country—and, in comparative and physical geography, has furnished valuable *data* for the improvement of our maps of Asia Minor and part of Armenia.

Messrs. DEASE and SIMPSON claim our warmest applause; these two enterprising officers of the Hudson's Bay Company have successfully traced the hitherto unexplored country between Point Barrow and Franklin's Farthest; and thus has a continuous line of 60° in extent of the northern coast of America been explored by British hardihood and perseverance.

Lastly, the labours of Mr. SCHOMBURGK in British Guayana, which it is hoped are not yet terminated, claim especial notice, for the perseverance he has manifested, in spite of an unhealthy climate, in exploring the rivers Essequibo, Berbice, and Corentyn —in obtaining much topographical information—and for having very materially added to our knowledge of the natural productions of that rich and fertile country.

Auxiliary Association.—The Bombay Geographical Society, originally a branch of our own, has received such encouragement in India as to proceed to the publication of its own Transactions, the two first volumes of which have been transmitted to us; from these, and from the future publications of that Society, now independent, notices will from time to time appear in our Journal.

On this occasion the Council have to direct attention to the great value of the papers on the hydrography of the rivers and coasts of India and Arabia, contributed by the officers of the Indian Navy.

From the Egyptian Society at Cairo no communication has been received.

Original Expeditions.—Of the expeditions more immediately under the control of the Society, the Report of that into the interior of South Africa has been published in the First Part of Vol. VIII. of our Journal.

In British Guayana a detailed Report of Mr. Schomburgk's ascent of the rivers Berbice and Corentyn appeared in Vol. VII.; since which Mr. Schomburgk has again ascended the river Essequibo to its junction with the river Rupunúny: following the upward course of the latter, he ascended its tributaries, the Rewa and Quitaro, and on the 17th November, 1837, the date of his last letters, he had reached the latitude of 2° 31′ N., some miles beyond his extreme point on the former expedition, and was then about to start to explore the mountain range of the Sierra Acaray, and to connect, if possible, his observations with those of Baron Humboldt, on the Orinoco, one of the principal objects originally contemplated by the expedition.

The Council have thought fit to give their countenance and support, in conjunction with the Society for the promotion of Christian Knowledge, to an expedition about to proceed to Kurdistán, its object being to make acquaintance with the Nestorian Christians, and the mountainous country they inhabit in the districts of Hakarí, Julamerik, Amadíyah, &c.

Mr. Ainsworth, well known as naturalist to the Euphrates expedition, and Mr. Rassam, a Nestorian Christian, a native of Mosúl, already mentioned as interpreter to the same expedition, were strongly recommended by Colonel Chesney as fit persons to conduct such a mission; and the Council, considering the exploring Kurdistán a proper object for making use of a part of the funds intrusted to their charge, have appropriated the sum of 500*l.*, to be divided over two years, for this purpose. The Society for promoting Christian Knowledge have contributed a like sum for the furtherance of the objects of this expedition.

In consequence of this decision no time has been lost in making the requisite arrangements for the departure of the expedition; the joint instructions to the travellers have been carefully prepared; they are furnished with good instruments; the necessary passports and firmans are procured; Mr. Rassam has already gone to Malta on his way to Constantinople; Mr. Ainsworth, accompanied by Mr. Russell, as his assistant, will shortly leave London to take up his head-quarters at Mosúl; and the Council confidently anticipate a valuable harvest of geographical and general information, relating to the countries of Mesopotamia and Kurdistán, from the well-tried zeal and energy of the individuals to whom the charge of the expedition is intrusted.

A proposal, originating entirely with Egyptian travellers, for exploring, by means of a native of Dongola, the course and sources of the Bahr el Abiad, or western branch of the river Nile, having been brought before the notice of the Society, the Council have considered it right to vote a sum of 50*l.* for the furtherance of the object of such an expedition.

Foreign and Colonial Correspondence.—Three Corresponding Members have been added to the Society during the past year—Colonel LAPIE of the Dépôt de la Guerre, at Paris; Don José de URCULLU, at Oporto; and Don Pedro de ANGELIS, at Buenos Ayres, author of a valuable collection of documents relative to the Rio de la Plata; and the Council have great pleasure in noticing the gradual and steady increase of the Foreign and Colonial Correspondence of the Society.

Library.—A list of the accessions made to the Library, consisting of about two hundred volumes, is printed with this Report. For many of these works the Society is indebted to the liberality of Foreign Academies; but more particularly to the *Dépôt de la Marine,* at Paris, for several valuable donations.

At a Special Meeting of the Council of the Society, held on July 8, 1837, *the following Address to the Throne was agreed upon.*

TO THE QUEEN'S MOST EXCELLENT MAJESTY.

The humble Address of the President, Vice-Presidents, and Council of the Royal Geographical Society of London.

MOST GRACIOUS SOVEREIGN,

We, your MAJESTY's dutiful and loyal subjects, the President and Council of the Royal Geographical Society, beg leave to approach your Throne, and most respectfully to offer, in the name of the Society, our sincere condolence on the occasion of the death of His late Majesty.

In common with all classes of your MAJESTY's subjects, we have to deplore the loss of a beloved Sovereign; but the Royal Geographical Society has especial cause to revere the memory of William the Fourth as their first gracious and munificent Patron.

Yet whilst expressing our sense of gratitude to our late illustrious Sovereign, we feel that the cheering duty awaits us of respectfully offering our congratulations on the happy accession of your MAJESTY to the Throne of a great empire; and we beg permission to offer our heartfelt thanks for your MAJESTY's gracious condescension and munificence in granting to us the honour of your Royal Patronage and in bestowing upon the Society a Royal Premium for the encouragement of geographical science and discovery.

England has achieved some of her proudest triumphs under the sovereignty of a Queen; the annals of our country record the names of Drake, the celebrated circumnavigator, and of Raleigh, the adventurous discoverer, both distinguished during the reign of Queen Elizabeth; and we confidently anticipate that the reign of your MAJESTY will be equally famed for its glory and prosperity, and for the promotion of geographical knowledge.

That your MAJESTY's reign may be rendered illustrious as the era of important discoveries which may diffuse the blessings of civilization throughout the globe, as well as endeared to the affections of a free and grateful people, is the earnest wish and ardent prayer of your MAJESTY's loyal subjects the President, Vice-Presidents, and Council of the Royal Geographical Society.

(Signed by the President and Council.)

TREASURER'S BALANCE SHEET, 1837.

Dr.	£.	s.	d.	*Cr.*	£.	s.	d.
Balance in hand, January 1st	33	1	7	House Rent	110	0	0
Entrance of 39 Members at 3*l.*	117	0	0	Salaries	300	0	0
Composition of 16 Members at 17*l.*	272	0	0	Printing Journal, Vol. VI.	289	17	0
Subscription of 217 Members at 2*l.*	434	2	0	Illustrations for Journal, Vol. VI.	109	9	0
Arrears paid up	66	0	0	Guayana Expedition	156	4	0
Dividends on Stock (4800*l.* reduced 3½)	168	0	0	Reprinting Journal, Vol. II.	90	0	0
Royal Premium for 1836 and 1837	105	0	0	Royal Premium	52	10	0
Journals sold in 1836	98	13	8	Books and Maps	57	0	0
				Engraving Diploma	17	0	0
				Stationery	10	6	0
				Firing and Lights	6	16	8
				Porterage, Carriage of Parcels, &c.	16	5	0
				Attendance and Wages of Housekeeper	17	0	8
				Postage of Letters	16	19	0
				Expenses of Evening Meetings	9	1	6
				Insurance, Advertisements, and Power of Attorney	4	9	6
				Balance in hand	30	18	11
	£1,293	17	3		£1,293	17	3

The above Accounts for 1837 have been examined by us and found correct.

(Signed) J. TREMAYNE RODD.
W. I. HAMILTON.

(Signed) JOHN BIDDULPH.

ESTIMATE FOR 1838

Dr.

		£	s.	d.
Balance in Hand		30	18	11
Probable Amount of Subscriptions, viz.				
Entrance of 40 Members	£120			
Composition of 20 ditto	34			
Subscription of 210 ditto	420			
		880	0	0
Arrears likely to be recovered		100	0	0
Dividends on Stock		168	0	0
Sale of Journal		100	0	0
Royal Premium		52	10	0
		£1320	8	11

Cr.

	£	s.	d.
House Rent	110	0	0
Salaries	300	0	0
Printing Journal	290	0	0
Illustrations for Journal	110	0	0
Reprinting Journal, Vol. III., Part I.	30	0	0
Guayana Expedition	100	0	0
Kurdistán Expedition	125	0	0
Books and Maps	100	0	0
Office Expenses, including Firing, Lights, Refreshments at Meetings, Attendance, Postage, &c.	100	0	0
Collector	21	0	0
	£1286	0	0

JOHN WASHINGTON.

At a special meeting of the Society, held on Monday, the 14th May, 1838, for the purpose of presenting the Royal Premium to Colonel Chesney, the President addressed him in the following words :—

"Colonel Chesney,—When, on two former occasions, the annual Royal Premiums awarded by the Council of the Royal Geographical Society were presented from this chair, first, to Captain Sir John Ross, and afterwards to Captain Back, they were congratulated on the triumphs they had achieved over the difficulties which had met them in the regions of perpetual snows and ice, amidst inhospitable and untrodden deserts, and on tracts of land which they had brought to light for the first time since the world was tenanted by man.

"It is now my no less pleasing duty to offer to you the warm acknowledgments of this Society, and to present to you this Royal Premium, for the equally brilliant successes which have marked your progress, under the opposite extreme of temperature, in re-opening to the civilised world a large portion of those countries, which, in times of yore, were the seat of extensive monarchies, which abounded in flourishing and populous cities, and whose inhabitants were equally famed for their conquests, their commerce, and their science, but which countries have long been excluded from the pale of polished nations, and plunged in a state of barbarism perhaps more to be deplored than the life of the hunter in the woods, because it presents a combination of the sloth and ignorance of savage life with the vices of degenerate civilization.

"In the year 1830 you were prompted by your own ardent zeal for the honour and welfare of your country, and by your wish to confer upon it a signal service, by extending its commercial resources, with a direct view, too, to the enlargement of geographical knowledge, to visit the south-eastern provinces of the Turkish empire. You spent nearly three years in exploring these regions and in surveying the shores, depths, and shallows of the Euphrates, its resources, difficulties, and facilities, in order to ascertain the feasibility of its navigation with European vessels, properly constructed; and thus making this river, once so celebrated in the annals of antiquity, but which had for twenty centuries rolled along its majestic waters uselessly to man, again available for the noblest purposes of life,—again to be a channel for social and commercial intercourse: but not merely for the benefit of the Assyrian and the Chaldee, but for the great purpose of approximating, as nearly as possible, the shores of Britain, the emporia of Europe, with the great towns of India, now the brightest spot on the continent of Asia.

"This project, the offspring of your enlightened benevolence and patriotism, was followed up by your own characteristic energies; and after many difficulties thrown in your way by the doubts and hesitations natural to those who were to be responsible for the prudence of the undertaking, your views were adopted, and you succeeded in prevailing upon the Government to fit out the Euphrates Expedition.

"On the results of this expedition (whatever differences of opinion may still exist respecting the practicability of adopting the line of the Euphrates as the means of that regular commercial intercourse to which I have alluded), there can be but one opinion on the ability with which it was planned, the foresight with which all the necessary accompaniments of it were provided, the skill by which it was directed, and the cordial co-operation with which it was conducted to its close by all who were assigned to you as assistants in the task. These individuals have received substantial proofs of approbation, whether by immediate promotion, or by equally acceptable testimonies in the shape of employment in the same countries; but it is the peculiar province of this Society, and a most gratifying duty at the same time, to testify their appreciation of the distinguished service you have rendered to the science of geography, as well in your first independent journeys, as afterwards, when as an officer in the public service. You have been the pioneer to point the way, and to open the road into a large range of country hitherto very imperfectly known; you, and those under your command, have really navigated for the first time in modern history two of the most celebrated rivers of the ancient world—the Euphrates and the Tigris: you have added largely to our knowledge of Syria, Mesopotamia, Assyria, and Susiana. You have carried on a line of levels from Iskanderún to Bír, and thence along the whole course of the Euphrates to the Persian Gulf. You have laid down the course of the Orontes from its mouth to Jisr Hadíd; and you have done more than enough to secure to yourself a place amongst the distinguished geographers of our time. These are the grounds upon which the Council of the Royal Geographical Society have judged it right to award to you the first Royal Premium, which the munificence of our illustrious Patroness Queen Victoria has placed at their disposal, for the reward and encouragement of geographical researches; and I have only to add the expression of hearty satisfaction which I feel at being the organ of the Society on this auspicious occasion."

To this address Colonel Chesney replied:—

"Sir,—The honour just conferred by the Royal Geographical Society of Great Britain would be most gratifying to almost any man; and, for many reasons, it must have an increased value in my estimation, if I thought I really had any other claim to it than that of having been at the head of the energetic men who actually performed the work.

"My individual efforts prior to the late expedition have been kindly noticed, but I must hope for permission to pass by altogether the descent of the river Euphrates in 1830, which is now but a matter of imperfect recollection, and the award being in reality for the late expedition, I cannot help feeling some regret that the rules of the Society did not permit me to urge the claims of those which were greater than mine individually; for I, at least, cannot forget that the almost unequalled labour of the transport across Syria was chiefly accomplished whilst I was confined with a brain fever, and that the leading discoveries in geography were made when I was not even present. It is true that every individual performed his part admirably; but the situations of some made them more conspicuous than others: for instance, the discovery of a new river, and the examination of Susiana, took place when I was in India, Major Estcourt being then in command; and, as you all know, the valuable mines, &c. in Kurdistan were met with by Mr. Ainsworth, when coming home from the Euphrates steamer, which had been laid up at Bagdad, after a successful navigation of more than 3000 miles of unknown rivers; and it is sufficient to observe, that this was accomplished under the direction of

Lieutenant Cleaveland, R.N., and his two persevering Lieutenants, Messrs. Charlewood and Fitzjames, without having touched even *once* the dreaded rocks of the rivers Euphrates, Karun, and Tigris, of which we hear so much in this country.

"As, therefore, both the credit of anything like the advancement of geography, as well as the more daring part of the navigation, belongs to others, I should have been still more gratified if the award had passed through me to one of the two gentlemen who remained to the close of the expedition; but as it has been decided otherwise, I accept the medal as their representative, and in doing so, I would offer my warmest thanks to the President and other distinguished individuals composing the Council of this body at large, which, I hope, will add to the distinction they have conferred, by permitting my name to be added to their numerous list of members.

"It may not be quite out of place to say a few words about the late enterprise, which, we all know, was not popular; why, I cannot say, but I am not without the hope that the distinction conferred on the expedition this evening may, in time, lead to a different state of things, as the geography of Asia becomes better known. We have had to contend with prejudice, and ignorance, its handmaid, to an extent which would have been amusing if the effects had not been so fatal.

"Men are ready to encounter the greatest dangers and the most trying difficulties for a brief period, but protracted trials like ours, during a temperature which varied between 108° under a tent, and 8° below Zero in a house, with at least *one* serious attack of illness in every individual case, are seldom met with the same unshaken constancy displayed by the officers and men throughout this gigantic operation, which alternately engaged Lieutenant Lynch, and every other person, myself excepted. This work, which would have been nothing, comparatively, if the Pasha had not broken up the judicious arrangements made as to camels, boats, &c., by Lieutenant Lynch before we arrived, was effected, after the successive illness of Major Estcourt and all the others, by Lieutenant Cleaveland, who was in charge of the operation for the second time, when he brought in the last boiler to Port William, by means of 104 bullocks and fifty-two men, guiding them through the deep mud. Besides the difficulty, there was another marked circumstance attending the transport, viz., the decided honesty of the Arabs, who were intrusted with many valuable things, which were taken, unattended by any European, from the sea to Port William, almost invariably without loss; and subsequently they continued to bring our letters and parcels of money, wherever we happened to be, with equal fidelity; indeed, we never had but one misunderstanding during the two years we were amongst the different tribes of Arabs.

"The descent and survey of the river were commenced 16th March, 1836, by the two vessels, which had been thus carried in frame from the Mediterranean to the Euphrates, and there successfully put together. All went on well up to the 21st of May, when our first and only calamity occurred by the loss of the Tigris, and with her twenty fine fellows, and most of our tools.

"Great as this calamity was, which thus deprived us of half our means, both morally and physically, it had a decisive result on our ulterior proceedings. My original instructions were to break up the expedition on reaching Basrah, which were repeated in a letter I received from the President of the Board of Control about ten days before the storm, the contents of which I had kept to myself, not to discourage the others, and meaning to obey; but when the vessel was carried to the bottom I determined to continue the expedition at my own risk; and I look back with gratitude on the willing support I then received from every one, at a time when a recent calamity of such magnitude was instantly met by an offer to give up their pay, in order to lighten the expense, and lessen my responsibility. The Euphrates steamer having continued the descent to Basrah with the same marked success as before, and there being no means whatever there to replace our tools, or even to obtain plank, we proceeded to Abú-shehr, where she was refitted after much difficulty and great delay.

"On returning to the Euphrates, we examined the River Bahamishír, and Karím—then carried an Indian mail to Bagdad; and after this was done we commenced the ascent of the Euphrates, when its waters were at the lowest; but we found sufficient even for our large vessel, until an accident happened by cracking the cross-head of the air-pump, and she returned with one engine only, without the hope of doing anything more: but as it was opportunely and successfully repaired alongside the Hugh Lindsay, the Euphrates was despatched to continue the examinations of the rivers of Susiana and the Tigris, under Major Estcourt, whilst I proceeded to India to arrange, if possible, for the continuance of the expedition. I succeeded in my object, but it was too late; for the party at Bagdad was already broken up, in consequence of renewed orders from England.

"During an undertaking which is mixed up, like a nursery tale, with rocks, cataracts, deserts, and millions of Arabs living by the plunder of an occasional traveller, we had no accident whatever but the breaking of the cross-head, and our only regret was the loss of many valuable men—amongst whom was a most promising officer of artillery, Lieutenant Robert Cockburn, and another individual, who will long be regretted by this and other societies: Lieutenant Murphy had already completed a series of observations from the Mediterranean Sea to the Persian Gulf, which will transmit his name to posterity as having been the first to lay down a certain base for future exploratory journeys which may be made both to the northward and southward from it as a starting-point. Objects have been obtained during the expedition by the remaining individuals, to whom I hope to do some little justice in the work now in preparation; therefore, I shall now merely beg the President and the rest of his fellow-labourers in the good cause of the advancement of geographical knowledge, to accept the warmest thanks of myself and the other officers, for the marked distinction which has been conferred upon the late expedition—and may every such enterprise be equally honoured on their return by the fostering care of this Society!

"The Euphrates steamer was still at Bagdad without men to navigate her in February, but they were about to be sent from Bombay, and an engineer also. On their arrival, Lieutenant Lynch was to recommence; and if he is supplied with suitable means to return into the Euphrates, it will soon be shown that our opinion as to this river being easy of navigation was well founded, for it will then speak for itself."

At a special meeting of the Society, held on the 8th May, 1837,* for the purpose of presenting the Royal Premium to Captain Fitz Roy, R.N., Mr. Hamilton, one of the Vice-Presidents, said:—

"Captain Fitz Roy,—In consequence of the unavoidable absence of the President of this Society, I have been unexpectedly called upon to announce to you that the Council of the Royal Geographical Society have resolved to award to you the Royal Premium, given by His Majesty, for the year 1836, in testimony of the high sense they entertain of the valuable additions made by you to our knowledge of a large portion of the South American continent and the adjacent islands, whilst you were employed in His Majesty's service, on the late survey of the coasts of Patagonia, Chile, and Peru.

"It was in the autumn of the year 1831, that you were commissioned by the Lords of the Admiralty to proceed in command of His Majesty's ship Beagle, to conduct this important survey, which was to open to the commercial world a new series of approaches into that part of South America, where are daily developing themselves germs of political associations and mercantile prosperity. The eyes of all who were interested in the subject were directed towards your operations, until your return to England in November last, after having surveyed, either in person or at your own expense, seven thousand miles of coast, from the Rio de la Plata on the east, to Guayaquil on the west side of South America, and after having circumnavigated the globe; and they now greet your return with a degree of pleasure and gratification, and with a large portion of national pride, commensurate with the magnitude of the scale on which you have acted, and the unceasing perseverance which you have displayed in conducting to its term the task committed to your guidance.

"During these five years, one of the first of your discoveries laid open to the commercial and scientific world the harbour of Bahia Blanco, in lat. 39° south, the only one, upon the eastern coast, in which a considerable number of line-of-battle ships can lie at anchor. On the same coast, in the face of numerous difficulties, you explored, for the first time, the deep and rapid river of Santa Cruz, from the coast in 50° south lat. to the Cordilleras; the first occasion on which, we believe, this part of the continent of South America has ever been crossed by an European. You surveyed, at your sole expense, that very interesting and important feature in the political and physical geography of those seas, the Falkland Islands. You have given to our maps, besides the great inland waters, called those of Otway and Skyring, a new and important channel through the Tierra del Fuego. On the western side of South America you have, for the first time, laid down the archipelago of islands lying to the south of Chiloe, called Chonos, in lat. 45° south. Amongst various material alterations in the old Spanish surveys of the shores of the Pacific and adjacent islands, may be mentioned, as a sample of the rest,

* Omitted by mistake in last year's Annual Report.

the important fact, that the island of Chiloe was found to be no less than twenty-five miles in error in latitude, as laid down in them. When your term of service was on the point of expiring, and you were about to proceed on your way home, your zeal for science prompted you to engage a vessel, at your own expense, for continuing and completing the survey of the coast of Peru as far as Guayaquil, the result of which has been the examination of a great number of ports and roadsteads, of which many were never before known to be capable of admitting vessels. In circumnavigating the globe, you have for the first time carried a complete chronometric chain of measurement, by twenty-two chronometers, (many of them your own property,) from east to west, round the globe. You have also enabled Mr. Darwin, the well-known naturalist, to add greatly to our knowledge of the natural history of those regions; and you have given us the best account we possess of the earthquake which took place at Concepcion in 1835, a phenomenon which has given rise to much interesting discussion among those who are engaged in physical and geological pursuits.

"Captain Fitz Roy,—those who are best able to form a judgment of the services you have rendered to the science of geography, feel that they give you a fair and just title to be enrolled in the list of the most celebrated navigators whose names adorn the maritime annals of this or any other country. I cannot conclude without the expression of the high gratification which I feel at being thus made the organ of communicating to you this mark of the high admiration and esteem in which you are held by this distinguished Society."

To this address Captain Fitz Roy replied in the following words:—

"Sir,—Before attempting to thank you for the deep gratification which I feel, I must indeed disclaim having such pretensions to the position in which the too favourable opinion of the Royal Geographical Society would place me.

"But, Sir, however little I may really be worthy of much that you have been pleased to say, I feel so deeply that this honourable testimonial of the approbation of my countrymen is a most ample reward for those attempts to serve my country, which I *am* conscious of having made—that I beg you will accept a feebly attempted expression of sincerely grateful feelings, which cannot be shown adequately by words.

"Permit me, Sir, to mention, that the Royal Geographical Society have removed from my breast every painful feeling which had there harboured. I now am rewarded for those services, such as they were, in the execution of which I was encouraged and assisted by the sincere friendship and decided support of the Hydrographer of the Admiralty, and by the most cordial co-operation of Mr. Darwin and the officers of the Beagle. And, Sir, to show my gratitude to the distinguished Society of my countrymen, over whom you are now presiding, I take this opportunity of declaring, that if, at any future time, my humble efforts should be thought likely to be useful, I shall be willing and proud to prefer public duty to private happiness.

"I believe, Sir, there is a gentleman now present whose name I have already mentioned, and of whose claims to your notice I beg you will allow me to say a few words.

"Mr. Charles Darwin embarked in the Beagle in 1831, as a zealous volunteer in the cause of science. At his own expense he passed five years on board the Beagle, or travelling in those countries she visited; and at his return most liberally presented his valuable collections to the public. When it is considered that Mr. Darwin never ceased to be a martyr to sea-sickness, his perseverance may be appreciated. Of the value of his labours I understand you have already been made partially aware; and I believe I am quite correct in saying, that the best judges estimate those labours very highly.

"You have been pleased to mention the beneficial effects which are likely to result from the information obtained during this survey; and I am sure you will be of opinion that they will be much enhanced by that information being made available to the public as speedily as possible. All the charts and plans are finished and deposited in the Hydrographic Office; and the principal written information will be completed during this year.

"You have been pleased, Sir, to notice the chronometric chain of measurement which has been carried round the world from east to west. I have not yet discovered any error in the calculations or observations upon which the results of those measurements depend, and anxiously look forward to the results of some future chain of measurements made from west to east, with a large number of chronometers kept at an uniform temperature.

"Allow me, Sir, to thank yourself in particular for the very kind, and to me so gratifying manner, in which you have communicated the sentiments of this Society, upon whose time and patience I fear I have trespassed too long."

ACCESSIONS TO THE LIBRARY,

21st MAY, 1838.

EUROPE.

Titles of Books.	*Donors.*
British Isles.—Physical and Political Geography of, Nos. I.—X. 1837	
Statisque de la Grande Bretagne et de l'Irlande, par M. A. Moreau de Jonnēs. 2 vols. 8vo. Paris, 1837	M. Moreau de Jonnes.
Londonderry.—Ordnance Survey of the County of; under the direction of Lieut.-Col. Colby, R. E. 4to. Dublin, 1837	The Lord Lieutenant of Ireland.
Sailing Directions from Sunderland Point to Berwick, including the Farn Island, by Capt. Johnson, R.N. 1836	Capt. Beaufort, R.N.
Tide Tables for the English and Irish Channels and the River Thames, for 1838	
The Light Houses of the British Islands, July, 1836	
Belgium.—Flandre Occidentale, Dictionnaire Géographique de la, par Ph. Vandermaelen, 8vo. Bruxelles, 1836	M. Ph. Vandermaelen.
Bruxelles.—Lettre sur l'Etablissement Géographique de, par Ph. Vandermaelen, 12mo. Bruxelles, 1836	
Candia.—Travels in Crete, by Robert Pashley, Esq. 2 vols. 8vo. 1837	
Faroe Islands.—On the Vegetation and Temperature of the, by W. C. Trevelyan, Esq. Pamphlet, 4to. 1837	W. C. Trevelyan, Esq.
France.—Statisque de la France. 4to. Paris, 1837 .	M. Moreau de Jonnes.
Ban de la Roche, Description de, par H. G. Oberlin. 4to. Strasburg, 1806	J. Heisch, Esq.
Etudes Grammaticales sur la Langue Euskarienne ou Basque, par A. T. D'Abbadie et A. Chaho. 8vo. Paris, 1836	M. D'Abbadie.
Italy.—Geographical and Historical Description of Ancient Italy, by the Rev. J. A. Cramer. 2 vols. 8vo. 1826	
Toscana, Dizionario, Geografico, Fisico, Storico della, da Emanuele Repetti. 2 vols. 8vo. Firenze, 1836-8. Letter L.	
——— Statistica del Granducato di, dal Col. Conte Serristori. 5 Nos. Fireuze, 1836-8. .	
Dell' Attuale Condizione della Scienza Statistica in Italia in 1834-6	Count Graberg af Hemsö.
Lanzo—Lettres sur les Vallées de, par Luigi Francesetti, 4to. Turin, 1823	W. Brockedon, Esq.

Titles of Books.	*Donors.*
ITALY—Viaggio alla Meta, al Morone ed alla Maiella, par H. L. E. Capocci. Pamphlet, fol. Napoli, 1835	IL DIRETTORE H. L. E. CAPOCCI.
ANNALI Civili del Regno delle Due Sicilie. 4to. 26 fascicoli. 1833-7	COLONEL VISCONTI.
SICILY, on the Geography, Geology, and Vegetation of, by J. Hogg, Esq. Pamphlet. 8vo.	J. HOGG, ESQ.
POLAND.—Tableau de la Pologne Ancienne et Moderne, par Malte Brun, entièrement refondu, par Leonard Chodzko. 8vo. Bruxelles, 1830	G. H. SMITH, ESQ.
PORTUGAL.—Ensaio Economico sobre o Commercio de Portugal e suas Colonias, por Dom J. J. da C. de Azeredo Coutinho. 8vo. Lisboa, 1816	COUNSELLOR J. J. DA COSTA DE MACEDO.
ESSAI Statisque sur le Royaume de Portugal et de l'Algarve, par Adrian Balbi, 2 vols. 8vo. Paris, 1822	G. H. SMITH, ESQ.
RUSSIA.—Observations Metéorologiques et Magnetiques faites dans l'Empire de Russie, par M. Kupffer. 4to. St. Petersburg, 1837	M. L'ACAD. KUPFFER.
PRECIS du Système de l'Instruction en Russie, par A. Krusenstern. 8vo. St. Petersburg, 1837 . .	J. BARROW, ESQ.
BLACK SEA, an Account of the Russian Ports in the. Pamphlet. 8vo. 1837.	
BLACK SEA, Report on the Commerce of the Ports of, by Julius de Hagemeister, translated by T. F. Triebner. 8vo. London, 1836	G. H. SMITH, ESQ.
SCANDINAVIA.—Några Underrättellser Hörande till kartan ofver södra delen af Skandinavien—af Carl af Forsell. Pamphlet. 4to. Stockholm, 1826	COLONEL C. AF FORSELL.
SPAIN.—Memoria sobre la Posicion Geografica de Sevilla, par Don J. S. Cerquero. Pamphlet. 8vo. . . .	DON M. F. DE NAVARRETE.
ITINERAIRE Descriptif de l'Espagne, par M. le Comte Al. de Laborde. 6 vols. 8vo. Paris, 1827 . .	G. H. SMITH, ESQ.
ETAT Présent de l'Espagne, par M. l'Abbé de Vayrac. 3 vols. 12mo. Amsterdam, 1719	G. H. SMITH, ESQ.
SWITZERLAND.—Ascent to the Summit of Mont Blanc in 1834, by Dr. Barry. 8vo. 1837	DR. BARRY.
TURKEY.—Journey from Constantinople to England, by Dr. Walsh. 12mo. 1831	

ASIA.

ASIA.—Von H. Berghaus. Part 4. In 4to. Gotha, 1837.	
ASIATIC Society of London, Journal of, No. 8, (*in continuation*)	ROYAL ASIATIC SOCIETY.
——— of Bengal, Journal of, to October, 1837 (*in continuation*).	
ARABIA.—Travels in, by J. L. Burckhardt. 2 vols. 8vo. 1829.	
TRAVELS in Omán, the Peninsula of Sinai, &c., by Lieut. J. R. Wellsted, Ind. Navy. 2 vols. 8vo. 1838	LIEUT. WELLSTED, IND. NAVY.
ASIA MINOR.—Missionary Researches in Asia Minor and Armenia, with a visit to the Nestorian and Chaldean Christians, by E. Smith, and H. G. O. Dwight. 8vo. 1833.	
GEOGRAPHICAL Illustrations of Xenophon, by Major Rennell. 4to. London, 1816	

Titles of Books.	*Donors.*
CABUL.—Account of the Kingdom of, and its Dependencies in Persia, Tartary, and India, by the Hon. M. Elphinstone. 4to. 1819	
CAUCASUS.—Der Ugrische Volksstamm oder Untersuchungen über die Landergebiete am Ural und am Kaukasus, von F. H. Müller. 8vo. Berlin, 1837 .	
ALTE Geographie des Kaspischen Meeres des Kaukasus und des südlichen Russlands, von Dr. Eduard Eichwald. 8vo. Berlin, 1838	
VOYAGE dans les Environs du Mont Elbrouz dans le Caucase, par M. Kupffer. 4to. St. Petersburg, 1830.	M. L'ACAD. KUPFFER.
CIRCASSIA.—Travels in Circassia, Krim Tartary, &c., in 1836-7, by Edward Spencer, Esq. 2 vols. 8vo. 1837	MR. COLBURN.
CHINA.—Description Géographique, Historique, et Chronologique de l'Empire de la Chine et de la Tartarie Chinoise, par J. B. du Halde. 4 vols. 4to. à la Haye, 1736	
MEMORIA dos feitos Macaenses contra os piratas da China, por J. J. Andrade. 8vo. Lisboa, 1835 . .	COUNSELLOR J. J. DA COSTA DE MACEDO.
INDIA.—An Account of the Measurement of an arc of the Meridian between the parallels of 18° 3′ and 24° 7′, by Capt. G. Everest, of the Bengal Artillery. 4to. London, 1830	MESSRS. COX.
FOE KOUE KI ou relation des royaumes Bouddhiques:—Voyage dans la Tartarie dans l'Afghanistan et dans l'Inde, executé à la fin du IVme Siècle, par Chi Fa Hian. Traduit du Chinois, et commenté par Messrs. Rémusat, Klaproth, et Landresse. 4to. Paris, 1836	
REPORT on the Eastern Frontier of British India, with an Atlas, by Capt. Pemberton. 8vo. Calcutta, 1835	CAPT. R. BOILEAU PEMBERTON, 44th N. I.
DIE STUPAS oder die Architektonischen Denkmale an der grossen Königstrasse Zwischen Indien, Persien, der Baktrien, von C. Ritter. Pamphlet, 8vo.	PROFESSOR CARL RITTER.
SCENES and Characteristics of Hindostan, with Sketches of Anglo-Indian Society, by Miss E. Roberts. 2 vols. 8vo. 1837	MISS E. ROBERTS.
ANCIENT and Modern Alphabets of the Popular Hindu Languages, by Capt. Henry Harkness. Pamphlet, 8vo. London, 1837	CAPT. HARKNESS.
JAPAN.—Voyages de C. P. Thunberg au Japon. 4 vols. 12mo. Paris, 1796	J. E. GRAY, Esq.
NOVA ZEMBLA.—Ueber das Klima von Nowaja Semlja und die Mittlere temperatur insbesondere, von K. E. F. Baer. Pamphlet	ADMIRAL KRUSENSTERN.
BERICHT über die neuesten Entdechungen an der Küste von Nowaja Semlja, von K. von Baer. Pamphlet, 8vo.	Ditto.
PERSIA.—Narrative of a Journey into Khorasán in 1821-2, by J. B. Fraser, Esq. 4to. London, 1825 .	—
TRAVELS and Adventures in the Persian Provinces on the banks of the Caspian Sea. 4to. London, 1826 .	
RUSSIA.—Reise um die Erde durch Nord Asien und die beiden Oceane in den jahren, 1828-9 and 30, von Adolph Erman. 2d vol. 8vo. Berlin, 1838 .	PROFESSOR ADOLPH. ERMAN.

Titles of Books.	*Donors.*
REISE nach dem Ural, dem Altai und dem Kaspischen Meere Erster Band, von Gustav Rose. 8vo. Berlin, 1837	
SIAM.—Crawfurd's Embassy to Siam. 4to. . .	
SYRIA.—Pococke's Description of Syria and Egypt. 2 vols. folio. London, 1745	
TRAVELS in Palestine and Syria, by George Robinson, Esq. 2 vols. 8vo. 1837	GEO. ROBINSON, Esq
RAISEN durch Syrien und Palastina nach Cpyern und durch Kleinasien in 1766. Von Carsten Niebuhr. 4to. Hamburg, 1837	

AFRICA.

Titles of Books.	*Donors.*
ABYSSINIA, Bruce's Travels in, with Life. 6 vols. 4to. Edinburgh, 1790	
AFRICA.—Erdkunde von Afrika, von Carl Ritter. 3 vols. 8vo.	
———, EAST.—Segunda parte da Memoria Estatistica sobre os dominios Portuguezes na Africa Oriental, por S. X. Botelho. Pamphlet, 8vo. Lisboa, 1837 . .	SENHOR S. X. BOTELHO.
———, WEST.—Expedition into the Interior of Africa, by the river Niger, in 1832—34, by Macgregor Laird and R. A. Oldfield. 2 vols. 8vo.	
ADDRESSES, Petitions, &c. from the Kings and Chiefs of Sudan and the Inhabitants of Sierra Leone, to his late Majesty William IV., and to the late Lieutenant-Governor of that Colony	LIEUT.-COL. CAMPBELL.
MEMORIAS, &c. Contendo a Descripção Geographica e Politica de Angola e de Benguella, por J. C. F. Cardozo de Castellobranco e Torres. 8vo. 1825 .	COUNSELLOR J. J. DA COSTA DE MACEDO.
THE Liberia Farmer, by J. Ashman. 8vo. Philadelphia, 1837	DR. HODGKIN.
———, SOUTH.—Border Colonists, Sketch of the Emigration of, extracted from the unpublished Journal of a Visit to the Chief "Moselekatse," by Capt. Harris. Pamph.	CAPT. HARRIS, E.I.C. Eng.
SUMMARY of the Kaffir War of 1834-5, with notes by the Author of the Zuid Afrikân. Pamphlet, 8vo. 1836. . PROCEEDINGS of the Court of Inquiry on the fate of the Kaffir Chief Hintza. 1 vol. 12mo. Cape-town, 1837	CAPT. ALEXANDER, 42ND R.H.
BARBARIE.—Voyage en, en 1785-6, par M. l'Abbé Poiret. Paris, 1789. 2 vols. 8vo.	J. E. GRAY, ESQ.
CARTHAGE.—Esquisse de la Restauration de la Ville de Carthage, par P. A. Dedreux	SIR GRENVILLE TEMPLE, BART.
EGYPT.—Visit to the Great Oasis, by G. A. Hoskins, Esq. 1 vol. 8vo. 1837	G. A. HOSKINS, ESQ.

AMERICA.

Titles of Books.	*Donors.*
AMERICA, NORTH.—Dr. Richardson on the Zoology of. Pamphlet, 8vo. 1837	DR. RICHARDSON, R.N.
NEW BRUNSWICK.—Statistical and Practical Observations on the Province of, by Alexander Wedderburn. 1 vol. 4to. 1835	LORD GLENELG.
TEXAS.—The History of Texas, with a Map, by David R. Edward. 12mo. Cincinnati, 1836 . . .	G. H. SMITH, ESQ.

Titles of Books.	*Donors.*
UNITED STATES.—Travaux d'Améliorations Intérieures, par le Gouvernement des Etats Unis, par Gme. Tell Poussin. 1 vol. 4to. Paris, 1834	G. H. SMITH, ESQ.
BRAZIL.—Memoria Estatistica, Historica, Geographica da Provincia do Maranhaõ, por Col. A. B. Pereira do Lago. Lisbon, 1822	COUNSELLOR J. J. DA COSTA DE MACEDO.
ROTEIRO da Costa da Provincia do Maranhaõ, por Col. Pereira do Lago. Pamphlet, folio. 1821	
MEMORIAS para a Historia da Capitania de S. Paulo do Estado Brazil, por Gaspar da Madre de Dios. 8vo. Lisbon, 1797	COUNSELLOR J. J. DA COSTA DE MACEDO.
NARRATIVE of a Visit to Brazil, Chile, Peru, &c., by G. T. Mathison. 1 vol. 8vo. 1825	
GUYANA.—Sur quelques Points importans de la Géographie de la Guyane, par Baron A. von Humboldt. Pamphlet, 8vo. Paris, 1837	BARON A. VON HUMBOLDT.
EQUATOR.—A Brief Sketch of the Lands and Mines of the Republic of the Equator, by Colonel R. Wright. Pamphlet	WM. BOLLAERT, ESQ.
PERU.—Pacto y Ley Fundamental de la Confederacion Peru-Boliviana. 12mo. London, 1838	DON VINCENTE PAZOS.
RIO DE LA PLATA.—Coleccion de Obras y Documentos relativos a las Provincias del, por Pedro de Angelis. Vols. 3, 4, and 5 in continuation. Fol. Buenos Ayres, 1836	DON PEDRO DE ANGELIS.

POLYNESIA.

Titles of Books.	*Donors.*
AUSTRALIA.—Das Festland Australien, eine Geographische Monographie, von C. E. Meinicke. 2 vols. 8vo. Breslau, 1837	
NEW SOUTH WALES.—Narrative of a Voyage of Discovery in the Lady Nelson to, in 1800-2, by Lieut. J. Grant. 4to. London, 1803	
TORRES STRAITS.—Narrative of a Voyage to, in Search of the Survivors of the Ship Charles Eaton, by Captain P. P. King, R.N.	CAPTAIN P. P. KING, R.N.
ARAFURA.—Sailing Directions for the Arafura Sea, by Geo. W. Earl, Esq. Pamphlet, 8vo.	G. W. EARL, ESQ.
JAVA.—Uber die Kawi-Sprache auf der Insel Java, von Wilhelm von Humboldt. 4to. Berlin, 1837	BARON ALEXANDER VON HUMBOLDT.
COUP D'ŒIL sur l'Ile de Java et les autres Possessions Néerlandaises dans l'Archipel des Indes, par le Comte C. S. W. de Hoggendorp. 8vo. Bruxelles	
NEW GUINEA.—Verhaal van eene Reize naar de Zuid-West Kuste van Nieuw Guinea, door J. Modera. 8vo. 1830	
REIZE door der weinig bekenden Zuidelijken Molukschen Archipel en Langs de Geheel on Bekende Zuid-West van Nieuw Guinea, in 1825-6, door D. H. Kolff. 1 vol. 8vo. 1820	
PACIFIC OCEAN.—Dalrymple's Collection of Voyages in the. 4to. 1770	
SOUTH SEAS.—Missionary Enterprises in, by the Rev. J. Williams. 8vo. 1837	REV. J. WILLIAMS.

Titles of Books.	*Donors.*
TAHITI, M.S. Diary of Simon Rodriguez, a Spanish Soldier and Interpreter left on the Island for 13 months by the Expedition sent from Lima in 1770 . .	CAPT. FITZROY, R.N.
VAN DIEMEN'S LAND.—Statistical Return of, from 1824 to 1835, by T. J. Ewing, Esq. Folio. Hobart-town, 1836	T. J. EWING, ESQ.

MISCELLANEOUS.

ACADEMIA, Reale das sciencias de Lisboa, Memorias da. 12 vols. 4to. Lisboa, 1796—1837	THE ACADEMY OF SCIENCES AT LISBON.
ACADEMIE des Sciences à Paris, Comptes rendus des Séances de l'. 4to. Paris, May 1837—May 1838 .	ACADEMIE DES SCIENCES A PARIS.
——— Impériale des Sciences à St. Petersburg, Bulletin Scientifique de. 2 vols. 4to. St. Petersburg, 1836-7	L'ACADEMIE DES SCIENCES A ST. PETERSBURG.
ALMANAC, Nautical, for 1840. 8vo. London . . .	CAPT. BEAUFORT, R.N.
——— American, for 1838. 12mo. Philadelphia .	J. E. WORCESTER, ESQ.
——— British, for 1838. 12mo. London . . .	
——— Geographischer für 1838, von H. Berghaus. 12mo., Stuttgart	
AMERICAN Journal of Science and Art, by Dr. Silliman, for 1837	
ANNALEN der Erd-Volker und Staatenkunde, von Dr. H. Berghaus, for 1837 (*in continuation*). 8vo. . . .	
ANNALES des Voyages to April, 1838 (*in continuation*). 8vo., Paris	
ANNALS of Natural History for March, 1838. 8vo. London.	MR. TAYLOR.
ANNUAIRE du Bureau des Longitudes, pour 1837. 12mo. Paris	LE BUREAU DES LONGITUDES.
ASIATIC Society Royal Annual Report and Proceedings of the Committee of Agriculture and Commerce. Pamphlet, 8vo.	ROYAL ASIATIC SOCIETY.
ATHENÆUM Journal to May, 1838. 4to.	THE EDITOR OF THE ATHENÆUM.
ATLANTIC Ocean, Memoirs on the Currents of, by Major Rennell. 8vo. London, 1832. With an Atlas of Charts	LADY RODD.
——— South, New Sailing Directions for, by John Purdy. 8vo. 1837	MR. J. PURDY.
BIBLIOTHECA Marsdeniana, Philologica et Orientalis. 4to. London, 1827	MRS. MARSDEN.
BRITISH Association for the Advancement of Science. 6th vol. 8vo.	THE BRITISH ASSOCIATION.
CLASSICAL Manual of the Greek and Latin Authors, completed to the end of 1836, by Joseph W. Moss. 2 vols. 8vo. London, 1837.	G. W. SMITH, ESQ.
CONNAISSANCE des Tems, pour 1839. 12mo. Paris .	LE BUREAU DES LONGITUDES.
CONSERVATIVES of our Shipping against Enemies. Pamphlet, 8vo.	J. BIRCH, ESQ.

Titles of Books.	*Donors.*
CURIOSITIES of Literature, illustrated by B. Corney, Esq. 8vo. 1837	B. CORNEY, ESQ.
DIRECTIONS for Collecting Specimens in Zoology, Mineralogy, and Botany. Pamphlet, 8vo.	THE BRITISH MUSEUM.
EAST INDIA United Service Journal, Nos. 22—26 and 31—37	MISS E. ROBERTS.
ENGINEERS, Transactions of the Institution of Royal. 4to. 1837	THE INSTITUTION OF ROYAL ENGINEERS.
——— Minutes and Proceedings of the Institution of Civil. 8vo.	THE INSTITUTION OF CIVIL ENGINEERS.
——— LIBRARY, Catalogue and Regulations of. 8vo.	
EDINBURGH New Philosophical Journal. Nos. 1—12. 8vo.	C. DARWIN, ESQ.
EDINBURGH Journal of Natural and Geographical Science. 3rd vol. 1831	W. AINSWORTH, ESQ.
EUPHRATES Expedition, Parliamentary Report of. Folio, 1837.	
FRANKLIN Institute of Pennsylvania, Journal of, for 1837. 8vo. Philadelphia	THE FRANKLIN INSTITUTE.
GEOGRAFICA, Tratado Elementar de, Fisica Antiqua y Moderna, por José de Urcullu. 2 vols. 8vo. Oporto, 1837	DON JOSE DE URCULLU.
GEOGRAPHIE d'Edrisi. Tome V. du Recueil des Voyages et des Mémoires, publié par la Société de Géographie à Paris. 4to. Paris, 1837	LA SOCIETE DE GEOGRAPHIE.
——— des Grecs Analysée, par P. F. I. Gosselin. 4to. Paris, 1790	
——— Systématique et Positive des Anciens, Récherches sur, par P. F. I. Gosselin. 4 vols. 4to. Paris, 1799	
——— Physikalische, von Frederick Hoffman. 8vo. Berlin, 1838	
——— Pflanzen Geographie, Grundriss der, von F. I. F. Meyer. 8vo. Berlin, 1835	DR. F. I. F. MEYER.
GEOGRAPHY Etymological, by J. A. Gibson. 8vo. 1835	
GEOGRAPHICAL Society of London, Journal of the Royal. Vol. vii. Vol. viii., Parts 1 and 2. 8vo.	
——— of Bombay, Proceedings of the. Vols. 1 and 2. 8vo. Bombay	THE GEOGRAPHICAL SOCIETY OF BOMBAY.
GEOLOGICAL Society, Transactions of the, Vol. V. 4to. London, 1837	THE GEOLOGICAL SOCIETY.
——— Proceedings of, for 1837. 8vo.	
President's Address at Anniversary Meeting of, for 1838	
HISTORICAL Eulogium on Don Hipolito Ruiz Lopez, by A. B. Lambert, Esq. Pamphlet, 8vo.	A. B. LAMBERT, ESQ.
HISTORY of Seyd Sayid, Sultan of Muskat, by the Signor Vicenzo. 8vo. London, 1810	MR. J. BOOTH.
LETTER to H. R. H. the Duke of Sussex, with a Plan for the Promotion of Art, Science, and Literature, by T. L. Donaldson. Pamphlet, 8vo. 1837	T. L. DONALDSON, Esq.
LIFE and Correspondence of the Earl of St. Vincent, by Capt. E. P. Brenton, R. N.	MR. COLBURN.
LITERARY Gazette to May, 1838. 4to.	THE EDITOR OF THE LITERARY GAZETTE.
MADRAS Literary and Scientific Society, Journal of, Three Numbers. Madras, 1837	THE LITERARY SOCIETY OF MADRAS.

Titles of Books.	*Donors.*
MADRID; Idea General del Discurso y de las Memorias de la Direccion Hidrografica de. Pamphlet, 8vo. Madrid, 1810	DON M. F. DE NAVARRETE.
MAGNETISM, Report on the Phenomena of Terrestrial, by Capt. Sabine, R. A. Pamphlet, 8vo. 1837 . .	CAPT. BEAUFORT.
MARSDEN'S Miscellaneous Works, 4to. London, 1832 .	MRS. MARSDEN.
MEDICO Botanical Society, Address of the President at Anniversary Meeting for 1838. Pamphlet, 8vo. .	THE MEDICO BOTANICAL SOCIETY.
MEMOIR of Sebastian Cabot, with a Review of the History of Maritime Discovery. 8vo. 1831	
Meteorological Society of London, Laws of. Pamphlet, 8vo.	THE METEOROLOGICAL SOCIETY.
——— Sketches, by W. C. Redfield. Pamphlet, 8vo.	W. C. REDFIELD, ESQ.
MINING Review, by H. English, Esq. 4to. . . .	H. ENGLISH, ESQ.
MISSIONARY Register to May, 1838 (in continuation). 8vo.	
NAUTICAL Magazine to May, 1838 (in continuation). 8vo.	
NUMISMATA Orientalia, Illustrated, by W. Marsden, Esq. 4to. London, 1823	MRS. MARSDEN.
ON THE Relative Quantities of Land and Water on the Surface of the Globe, by S. P. Rigaud. Pamphlet, 8vo. 1837	S. P. RIGAUD.
ORIENTAL Herald, Parbury's, No. I. 8vo. 1838 .	MR. PARBURY.
PACIFIC Ocean, Address on the subject of an Expedition to explore in, by T. N. Reynolds. 8vo. New York, 1837	J. S. POLACK, ESQ.
PENNSYLVANIA Colonization Society, Report of. Pamphlet, 8vo. Philadelphia, 1837	DR. HODGES.
PHENICIAN and Punic Inscriptions, Remarks on, by the Rev. J. Yates. Pamphlet, 8vo. 1837	REV. J. YATES.
PHILOSOPHICAL Transactions of the Royal Society for 1836 7. 4to.	THE ROYAL SOCIETY.
PLUTARCH'S Lives, by Langhorne. 8vo. London, 1828.	
QUARTERLY Review to April, 1838 (in continuation) .	MR. MURRAY.
——— Foreign to April, 1838 (in continuation) .	
Reize om de Werled, in de Jaren, 1823-4, met Z. M. Korvet Lynx, onder bevel van J. P. M. Willinck. 8vo. met Atlas, in 4to., Breda, 1836	
RESEARCHES into the Physical History of Mankind, by Dr. J. C. Prichard. 2 vols. 8vo. London, 1837 .	
ROYAL SOCIETY, Proceedings of the, for 1837, in 8vo.	THE ROYAL SOCIETY.
——— H. R. H. the President's Address at the Anniversary Meeting of, for 1838.	
——— of Literature, Transactions of the, for 1837. 4to.	THE ROYAL SOCIETY OF LITERATURE.
SCALES of equal parts, &c. By C. Holtzapfell. 8vo. London. 1838	MR. C. HOLTZAPFELL.
STATISTICAL Society of London, Proceedings of, for 1837. 8vo.	THE STATISTICAL SOCIETY.
——— Journal of the, 8vo. 1838 . . .	
STATISTICHEN Vereins für das Könegreich Sachsen—Achte Lieferung. Pamphlet, 4to. 1837	THE STATISTICAL SOCIETY OF DRESDEN.
——— Vergleischende Zusammenstellung der Gebornen, Gestorbenen, getrauten und communicanten im Konegreiche Sachsen em Jahre, 1836. Pamphet, 4to.	

Titles of Books.	*Donors.*
STEAM Communication between Great Britain and other parts of the world, by James Macqueen. 8vo. London, 1838	JAMES MACQUEEN, ESQ.
——— Navigation in the Pacific Ocean, by W. Whitelock. Pamphlet, 8vo.	W. WHITELOCK, ESQ.
——— Passage to the East Indies by the Gulf of Mexico and the Pacific, by H. Fairbairn. Pamphlet, 8vo. 1837	MR. FAIRBAIRN.
——— Ship Medea, Memoir of H. M., by Lieut. T. Baldock, R.N. 4to. 1837	CAPT. BEAUFORT, R.N.
——— Vessels, Returns of, in all foreign ports. Folio, 1837.	SIR ROBT. H. INGLIS, BART.
TITHE Commutation Tables, Supplement to, by W. Willich, Esq. Pamphlet, 1838	W. WILLICH, ESQ.
VIAGES y Descubrimientos de los Españoles por mar desde fines del siglo XV., por Don M. F. de Navarrete. Vols. IV. and V. 4to. Madrid, 1837 . .	DON M. F. DE NAVARRETE.
VOYAGE autour du monde, par le Capitaine Jacques Cook, traduit de l'Anglais. 6 vols. 8vo. Paris, 1778 .	B. L. VULLIAMY, ESQ.
VOYAGES and Travels, Phillip's Collection of. 11 vols. 8vo. London, 1821	W. D. COOLEY, ESQ.
———————, Phillip's Collection of. 9 vols. 8vo. London, 1827	
——— Bibliothèque Universelle des, par A. Montémont. Vol. XXI. 8vo. Paris, 1834	MONS. A. MONTEMONT.
XENOPHON's Anabasis, translated by Spelman. 8vo. London, 1776	

Names of individuals to whom the Royal Premium has been awarded:—

1831 Mr. Richard Lander.

1832 Mr. John Biscoe.

1833 Capt. Sir John Ross, R.N.

1834 Lieut. Alexander Burnes, E.I.C.S.

1835 Capt. George Back, R.N.

1836 Capt. Robert Fitz Roy, R.N.

1837 Lieut.-Col. Chesney, R.A.

LIST OF MEMBERS

OF THE

ROYAL GEOGRAPHICAL SOCIETY.

Aberdeen, the Earl of, K.T., F.R.S., L.S., S.A.
Abinger, Right Honourable Lord
Acland, Sir Thomas Dyke, Bart., F.G.S., H.S.
Adam, Lieut.-General Sir Frederic, G.C.B.
Adam, Vice Admiral Sir Chas., K.C.B.
Adamson, John, Esq., F.S.A., L.S., A.S.
Adare, Viscount, M.P.
Ainsworth, William, Esq.
Alcock, Thomas, Esq.
Alderson, Captain, R.E.
Alexander, Captain J. E.
Alexander, James, Esq.
Allen, Captain, R.N.
Alsager, J. M., Esq.
Alsager, Richard, Esq., M.P.
Antrobus, Sir Edmund, Bart.
Archer, Major
Arnold, Rev. Dr., F.R.S.
Arrowsmith, Mr. J.
Arrowsmith, Mr. Samuel
Ashburton, Lord
Askew, Henry William, Esq.
Atkins, John P., Esq.
Auldjo, John, Esq., F.G.S.

B.

Back, Captain, R.N.
Backhouse, John, Esq., F.G.S.
Baily, Francis, Esq., F.R.S., &c.
Baily, Arthur, Esq.
Baillie, David, Esq.
Baker, Colonel
Ball, Major, 49th Foot
Bandinel, James, Esq.
Bankes, W. H., Esq.
Baring, Francis Thornhill, Esq., M.P.
Baring, John, Esq.
Barrow, Sir John, Bart., F.R.S., L.S.
Barrow, John, Esq.
Bate, Mr. R. B.
Bateman, James, Esq.
Batty, Lieutenant-Colonel, F.R.S.
Beaufort, Captain F., R.N., F.R.S., Memb. Inst. France
Beaufoy, J. H., Esq., F.R.S., L.S.
Beckett, the Right Honourable Sir J., Bart., LL.D., M.P., F.R.S.
Bedford, His Grace the Duke of, K.G., F.G.S., &c.
Bedford, Grosvenor Charles, Esq.
Becher, Lieutenant A. B., R.N.
Beechey, Captain Frederick, R.N., F.R.S.
Belcher, Captain Edward, R.N., F.G.S.
Bell, James C. C., Esq.
Bennett, Frederic Debell, Esq.
Bennett, John Joseph, Esq.
Bennett, James, Esq.
Bentham, George, Esq., F.L.S.
Bentinck, Lord William
Bentley, R., Esq.
Berens, Joseph, Esq.
Best, Captain the Hon. T., R.N.
Bexley, Lord, M.A., F.R.S., &c.
Biddulph, John, Esq., F.H.S.
Birch, Jonathan, Esq.
Bird, James, Esq.
Biscoe, John, Esq., R.N.
Blackburne, Colonel William
Blackwood, Capt. the Hon. Price, R.N.
Blackwood, Capt. F. P., R.N.
Blake, William, Esq., M.A., F.R.S., &c.

Blanshard, Henry, Esq.
Blunt, Joseph, Esq.
Boddington, Samuel, Esq.
Borradaile, Abraham, Esq.
Bowles, Captain W., R.N.
Bowles, Colonel
Brandreth, Captain H. Rowland, R.E.
Brayley, E. W., Esq., F.S.A.
Brecknock, Earl of
Brereton, Rev. Dr., F.S.A.
Breton, Lieutenant W. H., R.N.
Brisbane, Sir Thomas M., K.C.B., F.R.S. L. and E., F.L.S.
Broadley, Henry, Esq.
Brockedon, William, Esq., F.R.S.
Brodie, Sir B. Collins, Bart., F.R.S.
Brooke, Sir Arthur de Capell, Bart., M.A. F.R.S., G.S., L.S.
Brooke, James, Esq.
Browne, John, Esq.
Browne, George, Esq.
Brown, Robert, Esq., F.R.S., &c. &c.
Brown, Thomas, Esq., F.R.S., L.S.
Brown, Wade, Esq.
Brumell, George, Esq.
Brunel, M. I., Esq., F.R.S., &c.
Buckland, Professor, D.D., F.R.S., G.S., &c.
Buckle, J. William, Esq.
Buller, Captain Wentworth, R.N.
Bullock, Captain F., R.N.
Burlington, Earl of
Burney, Rev. Chas. P., D.D., F.R.S., G.S., &c. &c.
Burton, Decimus, Esq., F.S.A., G.S.
Burton, Alfred, Esq.
Butler, Archdeacon, F.R.S., S.A.

C.

Cabbell, B. B., Esq., F.S.A.
Caledon, the Earl of, K.P., F.G.S., &c.
Campbell, James, Esq., Jun.
Carey, John, Esq.
Carlisle, the Dean of
Carnarvon, the Earl of
Cartwright, Samuel, Esq., F.G.S.
Cary, Captain the Hon. Plantagenet, R.N.
Castlereagh, Viscount
The Vice-Chancellor, the Right Hon.
Chantrey, Sir Francis, D.C.L., F.R.S.
Chapman, Captain, R.A.
Chapman, Sir M. Lowther, Bart., M.P.
Chapman, Thomas, Esq.
Chatfield, Frederick, Esq.
Chatterton, Sir William, Bart.
Chesney, Lieut.-Col., R.A.
Chichester, the Dean of
Children, J. George, Esq., F.R.S., G.S., &c.
Christophers, John, Esq.
Church, W. H., Esq.
Clarke, Sir Chas. M., Bart., F.R.S.
Clarke, William T., Esq.
Clarke, Captain
Clarke, William Stanley, Esq., F.R.S.
Clarke, W. T., Esq.
Clerke, Major T. H. Shadwell, F.G.S.
Clerk, Sir George, Bart., D.C.L., F.R.S., G.S., &c.
Clinton, General Sir William, G.C.B.
Clive, Viscount, M.A., M.P.
Cockburn, the Right Hon. Sir George, G.C.B., F.R.S.
Cockerell, J. Pepys, Esq.
Coddington, Rev. H., M.A., F.R.S., G.S.
Cogan, Captain R., E.I. Navy
Colby, Lieut.-Colonel, R.E., LL.D., F.R.S.L. and E., G.S., M.R.I.A.
Colchester, Captain Lord, R.N.
Cole, Viscount, M.P., F.R.S., F.G.S., &c.
Colebrooke, Colonel Sir William, R.A.
Coleridge, H. N., Esq.
Collett, William Rickford, Esq.
Collett, C. F., Esq.
Colquhoun, J., Esq.
Colquhoun, Major J. N., R.A., F.R.S.
Colquhoun, Gideon, Esq.
Conybeare, the Rev. W. D., F.R.S.
Cooke, W., Esq.
Cooley, W. D., Esq.
Craik, G. L., Esq.
Craufurd, W. P., Esq., F.G.S.
Craufurd, Captain W., R.N.
Crawfurd, J., Esq., F.R.S., G.S., &c.
Cross, Mr. J.
Cumming, William, Esq.
Cunningham, Allan, Esq.
Curties, John, Esq.
Curtis, Timothy, Esq.
Curzon, the Hon. Robert

D.

Dale, the Rev. T. Aquila
Damer, Colonel the Hon. George L. D., M.P.
Darwin, Charles, Esq.
Dawnay, the Hon. W. H.
Dawson, Captain R. K., R.E.
De Beauvoir, R. Benyon, Esq.
Denison, Lieutenant W. T., R.E.
Denman, Captain the Hon. Thomas, R.N.

Derby, the Earl of, P.L.S., &c., &c.
De Roos, Captain the Hon. J. F. Frederick, R.N.
Dickson, George Frederick, Esq.
Dickenson, Francis H., Esq.
Dickenson, J., Esq., F.R.S.
Dickson, Colonel Sir Alexander, K.C.B., R.A., F.S.A.
Dilke, Charles Wentworth, Esq.
Disney, John, Esq., F.R.S.
Doddington, Samuel, Esq.
Dollond, George, Esq., F.R.S.
Donaldson, John, Esq.
Donkin, Lieut.-Gen. Sir Rufane Shaw, K.C.B., G.C.H., M.P., F.R.S., &c.
Doran, Captain J. G.
Douglas, Edward, Esq.
Doyle, Lieutenant-Colonel Carlo
Drinkwater, Captain Charles, R.N.
Drinkwater, Edward, Esq.
Drummond, Colonel, R.A.
Drummond, Edward, Esq.
Drummond, Captain Thomas, R.E.
Du Cane, Captain, R.N.
Duckett, Sir Geo., Bart., M.A., F.R.S., G.S., &c.
Dundas, the Hon. Captain R., R.N.
Durham, the Earl of, F.G.S., F H.S.

E.

Edwards, Thomas Grove, Esq.
Edwards, Rev. J., M.A.
Egerton, Lord Francis, M.P.
Eliot, Lord
Elliot, Rear-Admiral the Hon. George, F.R.S.
Elphinstone, J. F., Esq.
Elphinstone, the Hon. Mountstuart
Enderby, Charles, Esq.
Enderby, George, Esq.
English, Henry, Esq.
Estcourt, Thomas G. Bucknall, Esq., M.P.
Evans, W., Esq.
Evans, Captain George, R.N.
Everett, Joseph, Esq., F.A.S.

F.

Falconer, Thomas, Esq.
Fanshawe, Lieutenant-Colonel, R.E.
Findlay, Alexander, Esq.
Fisher, Rev. George, M.A., F.R.S.
Fitton, William Henry, Esq., M.D., F.R.S., G.S., &c.
Fitz Roy, Captain, R.N., F.A.S.
Fitzwilliam, the Right Hon. Earl
Forster, Edward, Esq., F.R.S., L.S., &c.
Fowke, Lieutenant Thomas, R.N.
Fox, Lieutenant-Colonel C. R.
Franklin, Captain Sir J., R.N., F.R.S., D.C.L., F.G.S.
Fraser, Colonel John
Fraser, James Baillie, Esq.
Frere, Bartholomew, Esq.
Freshfield, J. W., Esq., F.R.S., F.G.S.
Fyler, George, Esq.

G.

Gardner, Mr. James
Garry, Nicholas, Esq., Dep. Gov. Hudson's Bay Company, F.H.S.
Gascoigne, Capt., Ceylon Rifle Brigade
Gibbs, Charles, Esq.
Gilbert, Davies, Esq., D.C.L., F.R.S., G.S., L.S., S.A., H.S., A.S., Hon. M.R.S.E., M.R.I.A.
Gipps, Sir George
Gladdish, William, Esq.
Goldsmid, Isaac Lyon, Esq., F.R.S., G.S., &c.
Gooden, James, Esq.
Goold, Nathaniel, Esq.
Gordon, Adam, Esq.
Gould, Captain Frederick A.
Gowen, I. R., Esq., F.G.S.
Graham, the Right Hon. Sir James, Bart., M.P., F.R.S., &c.
Gray, John Edw., Esq., F.R.S., F.G.S., H.S.
Graves, Lieutenant, R.N.
Greene, Thomas, Esq., M.P.
Greenough, G. B., Esq., F.R.S., G.S., &c.
Grenville, Right Hon. Thomas, F.S.A.
Gresley, Richard N., Esq.
Grey, Earl de
Grey, Lieutenant George, 83rd Regt.
Grey, Sir Charles, M.P.
Gresswell, Rev. Richard, M.A., F.R.S.
Griffiths, George, Esq.
Griffiths, Richard Clewin, Esq.
Grindlay, Captain, E.I.C.S.
Guillemard, John L., Esq., M.A., F.R.S., G.S., &c.
Gurney, Hudson, Esq., F.R.S., S.A., &c.

H.

Halford, the Rev. Thomas
Halford, Sir Henry, Bart., M.D., G.C.H., F.R.S.
Hall, Captain Basil, R.N., F.R.S., G.S., &c.

Hallam, Henry, Esq., M.A., F.R.S.
Hamilton, J. J. E., Esq.
Hamilton, Terrick, Esq.
Hamilton, William Richard, Esq. F.R.S., &c., &c.
Hamilton, Captain G. H., R.N.
Hamilton, W. J., Esq.
Hammond, Edmond, Esq.
Hammond, George, Esq.
Hammond, William, Esq.
Hammersley, Charles, Esq.
Hanmer, Sir John, Bart.
Harding, Colonel George, R.E.
Hardy, Vice-Admiral Sir T. M. Bart., G.C.B.
Harriott, Captain T. G.
Herries, Right Hon. C. J.
Harrison, Sir George, K.C.H., F.R.S.
Harrison, Benjamin, Esq.
Harrison, William, Esq., F.R.S., G.S., &c.
Harvey, Captain Edward
Hatcher, H., Esq.
Hathorn, George, Esq.
Hatchett, Charles, Esq., F.R.S., R.S.E., &c.
Hawtrey, Rev. Dr.
Hawkins, Dr. Bissett
Hay, Robert William, Esq., F.R.S., &c.
Hay, E. W. A. Drummond, Esq.
Hay, J. Holman, Esq.
Hearne, John, Esq.
Henderson, J., Esq.
Herbert, Captain, R.N.
Herbert, Hon. E.
Herbert, Jacob, Esq.
Hill, Lord Marcus
Hill, Henry, Esq.
Hoare, H. H., Esq.
Hobhouse, H. W., Esq.
Hobhouse, Sir J. C., Bart., M.P. F.R.S.
Hodgkin, Doctor
Hoffman, James, Esq.
Hogg, John, Esq., M.A., F.L.S.
Holland, Doctor Henry, M.D., F.R.S.
Hollier, Richard, Esq.
Holmes, Nathaniel, Esq.
Holmes, James, Esq.
Hooker, Professor, Sir Charles, F.R.S.
Horton, Right Hon. Sir R. W., G.C.H., F.R.S.
Hotham, Admiral Sir William, K.C.B.
Howse, J., Esq.
Hudson, John, Esq.

I.

Inglis, Sir R. H., Bart., M.P., LL.D., F.R.S., &c.
Irby, Captain the Hon. C. Leonard, R.N.

J.

Jackson, Colonel, K.S.W. and S.S.,
James, J. Horton, Esq.
Jenkins, Richard, Esq.
Jerdan, W., Esq.
Jervis, Captain T. B., E.I.C.Eng.
Jones, Charles, Esq.
Jones, Thomas, Esq., F.H.S.
Jones, Colonel Sir J. T., Bart., R.E.
Jones, William H., Esq., F.H.S.
Jones, Captain H. D., R.E.
Jones, Rev. Richard

K.

Kay, Joseph, Esq.
Kelly, William, Esq.
Keppel, Major the Hon. George, F.S.A.
Ker, H. B., Esq.
King, Captain Philip Parker, R.N., F.R.S., F.L.S.
Knapp, Hambly, Esq.
Knight, Charles, Esq., F.L.S.
Knight, H. Gally, Esq., F.L.S., F.H.S.

L.

Lambert, A. Bourke, Esq., F.R.S., G.S., &c.
Lance, John, Esq.
Lander, John, Esq.
Larcom, Lieutenant, R.E.
Law, William J., Esq.
Leake, Colonel W. M., F.R.S., &c.
Lee, John, Esq., LL.D., F.R.S., S.A.
Lee, Rev. James Prince, M.A.
Lefevre, J. G. Shaw, Esq., F.R.S.
Lemon, Sir Charles, Bart., M.P.
Levien, Edward, Esq.
Lewis, Captain Locke, R.E.
Lindley, John, Esq., L.L.D., F.R.S., G.S. H.S.
Lloyd, William Horton, Esq., F.L.S.
Loch, Captain
Long, George, Esq.
Lowe, Henry, Esq.
Lowry, Mr. Joseph Wilson
Ludolf, Count
Lumley, Benjamin, Esq.
Lushington, Lieut. Franklin, 9th Regt.
Lyall, George, Esq.
Lyell, Charles, Esq., F.R.S., F.G.S.

Lynch, Lieut. H. Blosse, Ind. Navy
Lyon, James Wittit, Esq.
Lyon, Captain David

M.

Mac Arthur, John, Esq.
Macculloch, J. R., Esq.
Macfarlane, Major J., E.I.C.S.
Macgillivray, Simon, Esq., F.R.S.
Mackenzie, Alexander, Esq.
Mackenzie, Harry, Esq.
Macdonald, John, Esq.
Macfarlane, Charles, Esq.
Mackillop, James, Esq.
Maclean, George, Esq.
Maconochie, Captain, R.N.
Magrath, Edward, Esq.
Mangles, Captain, R.N., F.R.S.
Marjoribanks, Edward, Esq.
Markham, Edward, Esq.
Marshall, J., Esq.
Martin, Rev. J. W.
Martineau, Joseph, Esq., F.H.S.
Mathison, G. F., Esq.
Maughan, Captain P.
Maxwell, Acheson, Esq.
Mayer, George, Esq.
Meek, James, Esq.
Melville, Lord Viscount, K.T., F.R.S.
Melville, P. C., Esq.
Mercier, Francis, Esq.
Milton, Lord, M.P.
Mitchell, Major
Montefiore, Jacob, Esq.
Montefiore, Sir Moses
Monteith, Lieut.-Colonel, E.I.C.S.
Montgomery, James, Esq.
Moore, G. H., Esq.
Morison, James, Esq.
Morier, James, Esq., F.R.S.
Mornay, Aristides Franklin, Esq., F.L.S.
Motteux, John, Esq.
Moyes, James, Esq.
Mundy, Captain G. R., R.N.
Munster, Right Hon. the Earl of, F.R.S.
Murchison, Roderick Impey, Esq., F.R.S., G.S. and L.S.
Murdoch, Thomas, Esq., F.R.S., S.A.
Murdoch, Thomas Clinton, Esq.
Murray, Lieut.-General the Right Hon. Sir George, G.C.B., G.C.H., F.R.S., &c.
Murray, John, Esq.
Murray, John, Esq., Jun., F.G.S
Murray, T. Laurie, Esq.
Musgrave, T. M., Esq.

N.

Neeld, Joseph, Esq.
Newnham, William, Esq.
Nettleship, Thomas, Esq.
Nicholson, Sir Frederick, Bart.
Nicholson, George T., Esq., B.A.
Nicol, J. D., Esq.
Nott, the Rev. Dr., F.R.S.L.
Norwich, the Right Rev. Edw. Stanley, Lord Bishop of
Nugent, Lord, D.C.L., F.S.A.

O.

O'Brien, Henry, Esq.
Ogle, Vice-Admiral Sir C., Bart.
Oldfield, Richard K., Esq.
Oldrey, Captain William, R.N.
Ommanney, H. M., Esq.
O'Reilly, Colonel
Ouseley, the Right Hon. Sir Gore, Bart., G.C.H., F.R.S., &c.
Outram, Doctor B. F., R.N.
Owen, Vice-Admiral Sir E. W. C. R., K.C.B.
Owen, Captain William Fitzwilliam, R.N.
Oxendon, G., Esq.

P.

Paley, Thomas, Esq.
Palmer, G., Esq., F.G.S., H.S
Palmer, Samuel, Esq.
Parish, Sir Woodbine, K.C.H., F.R.S., G.S.
Parker, Thomas Lister, Esq., F.R.S., &c.
Parry, Captain Sir William Edward, R.N., D.C.L., F.R.S.
Parry, Francis C., Esq., M.D.
Parsons, George S., Esq., R.N.
Pasley, Lieut.-Colonel, R.E., F.R.S.
Pechell, Captain Sir J. S. B., Bt., R.N., K.C.H., F.R.S.
Peel, the Right Hon. Sir Robert, Bart., M.P., D.C.L., F.R.S., S.A., &c.
Pelham, Captain the Hon. Dudley, R.N.
Pelly, J. H., Esq., Governor, Hudson's Bay Company, F.H.S.
Penn, Richard, Esq., F.R.S.
Pepys, W. Hasledine, Esq., F.R.S., L.S., &c.
Petit, Louis Hayes, Esq., M.A., F.R.S., G.S., &c.
Phillimore, Joseph, LL.D.
Phillipps, Sir Thomas, Bart., M.A. F.R.S., &c.

Phillips, Captain C., R.N., F.R.S.
Pigot, Henry, Esq.
Planta, the Right Hon. Joseph
Plowes, John, Esq.
Pocock, J. J., Esq.
Pollington, Viscount, M.P.
Pollock, David, Esq., F.R.S.
Ponsonby, the Hon. W., M.P., F.G.S.,
Ponsonby, Honourable Frederick
Porter, G. R., Esq.
Portlocke, Capt., R.E., F.G.S.
Potter, William S., Esq.
Potts, Charles, Esq.
Prudhoe, Captain the Right Hon. Lord, R.N., F.R.S., S.A., &c.
Powles, John D., Esq.

R.

Radcliffe, John, Esq.
Raine, Crosier, Esq.
Ramage, C. T., Esq.
Rankin, F. Harrison, Esq.
Rawson, W. Rawson, Esq.
Reid, Lieutenant-Colonel
Rennie, George, Esq., F.R.S.
Rennie, Sir John, F.R.S.
Rennie, M. B., Esq.
Renouard, Rev. George C., B.D, M.R.S.A.
Renwick, Lieutenant, R.E.
Rice, the Right Hon. T. Spring, M.P.
Richardson, Dr., R.N., F.R.S., L.S., &c.
Ripon, Right Hon. the Earl of, F.R.S., &c. &c.
Robe, Captain, R.E.
Robe, Captain F. H.
Robinson, Lieutenant C. G., R.N.
Robinson, E. F., Esq.
Rodd, Vice-Admiral Sir J. Tremayne, K.C.B.
Rodd, J. Rennell, Esq.
Rogers, Lieut., 90th Lt. Infantry
Roget, P. M., Esq., M.D., Sec. R.S., F.L.S., F.G.S., M.R.I.A.
Rose, the Hon. Sir George, F.R.S., LL.D.
Ross, Charles, Esq., M.P.
Ross, Major-General Sir Patrick, K.C.B.
Rouse, John, Esq., F.S.A.
Rowlands, Dr. D.
Rudge, Edward, Esq., F.R.S.
Rumboldt, C. E., Esq., F.S.A.
Russell, the Right Hon. Lord John, M.P.
Russell, J. W., Esq., M.P., F.R.S., G.S., &c.
Russell, Captain Robert, R.N.
Ryder, the Hon. F. Dudley

S.

St. Helens, Lord, G.C.H.
Salisbury, the Marquis of
Sandon, Viscount, M.P.
Sandwith, Colonel, E.I.C.S.
Scarlett. Major the Hon. C. Yorke
Schald, Edward Vernon, Esq.
Scott, Claude E., Esq.
Sedgwick, Professor, F.R.S., G.S., &c.
Senior, Nassau William, Esq., M.A.
Sheringham, Lieutenant W. L., R.N.
Sherriff, Francis, Esq.
Shirreff, Captain W. H., R.N.
Shortreed, Lieutenant, E.I.C.S.
Skegg, Edward, Esq.
Skelmersdale, Lord, F.H.S.
Slater, Captain M. A., R.N.
Slater, Joseph, Esq,
Sligo, the Marquess of, K.P., F.H.S., and L.S.
Smirnove, John, Esq., F.R.S., L.S., &c.
Smith, Lieutenant-Colonel Sir C., C.B., R.E.
Smith, Edward Osborne, Esq.
Smith, George Henry, Esq.
Smith, H., Esq.
Smith, George Stavely, Esq.
Smith, George, Esq., F.L.S.
Smith, James, Esq.
Smith, Lieut. Webber, 48th Regt.
Smith, Octavius, Esq.
Smyth, Lieut. Brunswick, 80th Regt.
Smyth, Captain W., R.N.
Smyth, Captain W. H., R.N., K.S.F., F.R.S., Corr. Ins. Fr.
Sotheby, Captain, R.N.
Spencer, Right Honourable the Earl
Spencer, Capt. the Hon. F., C.B., R.N.
Spottiswoode, A., Esq.
Stanley, Sir John T., Bart., F.R.S., S.A.
Stanley, Lieut. Owen, R.N.
Stanley, Lord, M.P.
Starling, Mr. Thomas
Staunton, Sir George T., Bart., F.R.S.
Stavely, Thomas, Esq.
Stephen, Sir George
Stephenson, Daniel, Esq.
Stoddart, Captain C., R. Staff Corps.
Stokes, C., Esq., M.R.A.S., F.R.S., G.S., &c.
Stooks, Thomas, Esq.
Stopford, Admiral the Hon. Sir Robert, G.C.B.
Stuart, Daniel, Esq.

Sturge, J., Esq.
Sturt, Captain Charles, F.L.S.
Surtees, Stevenson V., Esq.
Sutherland, Robert, Esq.
Swinburne, Captain C. H., R.N.
Symonds, Captain Sir William, R.N.
Symonds, W. C., 74th Regiment

T.

Talbot, Earl, K.P., F.R.S. and S.A.
Taylor, Lieutenant-General Sir Herbert, G.C.H.
Taylor, Richard, Esq., F.L.S., G.S., &c.
Taylor, John, Esq., F.R.S., &c.
Temple, Sir Grenville T., Bart.
Templer, J. C., Esq.
Thatcher, Colonel, E.I.C.
Thornton, the Right Hon. Sir Edward, K.C.B.
Tindal, Lord Chief-Justice
Tooke, A. W., Esq., M.A.
Trevelyan, W. C., Esq., M.A., F.G.S., L.S.
Trevor, the Honourable G. R.
Tuckett, Frederick, Esq.
Tuffnell, Henry, Esq., F.R.S., G.S.
Turnbull, Rev. Thomas Smith

V.

Vallé, A. B., Esq.
Vaughan, the Right Hon. Sir Charles, G.C.H.
Verney, Major Sir Harry C., Bart., M.P.
Vetch, Captain, R.E., F.R.S. and G.S.
Vidal, Captain, R.N.
Vigors, N. A., Esq., F.R.S., G.S., M.R.I.A., &c.
Vulliamy, B. L., Esq.
Vyvyan, Sir R. H., Bart., M.P.

W.

Walker, Mr. John
Walker, Mr. Michael
Walker, Lieutenant J. G., R.A.
Ward, H. G., Esq., M.P.
Washington, Captain, R.N.
Washington, Adam, Esq.
Watson, Sir Frederick B., K.C.H., F.R.S.
Weale, Mr. J.
Webb, Thomas, Esq.
Wellington, His Grace the Duke of, K.G., &c.
Wells, the Dean of, F.R.S., F.L.S.
Wells, Major, R.E.
Westall, William, Esq., A.R.A.
Weston, Samuel C., Esq.
Weyland, John, Esq., F.R.S.
Whewell, Rev. W.
Whinyates, Lieutenant-Colonel, R.A.
White, Vice-Admiral J. C.
White, Frederick, Esq.
Wilkinson, William, Esq.
Williams, Rev. Dr.
Willich, Charles M., Esq.
Wills, H. H.
Wilson, John, Esq., F.H.S.
Wilson, L. P., Esq.
Wilson, Thomas, Esq.
Wilson, Dr. Isaac, R.N.
Winterbottom, Edward, Esq.
Wittich, William, Esq.
Wolfe, Lieutenant, R.N.
Wortley, Hon. J. Stuart, F.R.S.
Wray, John, Esq.
Wright, the Rev. George Newnham, M.A.
Wulff, General, R.A.
Wyatville, Sir Jeffrey, R.A., F.R.S., S.A. and G.S.

Y.

Yarborough, the Earl of
Yates, Rev. James, M.A., F.L.S. and G.S.
Yates, John Ashton, Esq.
Yates, Joseph Brookes, Esq.
Young, George F., Esq., M.P.
Young, James, Esq.
Yorke, Lieutenant-Colonel

ROYAL GEOGRAPHICAL SOCIETY.

Patron.

THE QUEEN.

Vice-Patron.

His Royal Highness the Duke of Sussex.

COUNCIL, ELECTED MAY 21, 1838.

President.

W. R. Hamilton, Esq., F.R.S.

Vice-Presidents.

Sir John Barrow, Bart., F.R.S.
The Hon. Mountstuart Elphinstone.
G. B. Greenough, Esq., F.R.S.
Sir Woodbine Parish, F.R.S.

Treasurer.

John Biddulph, Esq.

Trustees.

Sir George T. Staunton, Bart., F.R.S.
Francis Baily, Esq., F.R.S.
John Biddulph, Esq.

Secretaries.

Captain Washington, R.N.
Rev. G. C. Renouard, Foreign and Hon.

Council.

Captain Back, R.N.
John Backhouse, Esq.
Captain Beaufort, R.N., F.R.S.
Captain Bowles, R.N.
William Brockedon, Esq., F.R.S.
Robert Brown, Esq., F.R.S.
Lieut. Col. Chesney, R.A., F.R.S.
Colonel Sir A. Dickson, R.A., K.C.B.
Charles Enderby, Esq.
Colonel Fanshawe, R.E.
Colonel Fox.
J. Baillie Fraser, Esq.
Bartholomew Frere, Esq.
W. I. Hamilton, Esq.
Colonel Jackson.
Captain T. B. Jervis, E. I. C. Eng., F.R.S.
Lieut.-Colonel Leake, F.R.S.
George Long, Esq.
James Morier, Esq., F.R.S.
R. I. Murchison, Esq., F.R.S.
Rear-Adm. Sir J. Tremayne Rodd.

FOREIGN HONORARY MEMBERS

His Imperial Highness the Archduke JOHN OF AUSTRIA
Chevalier BALBI . . Vienna
C. F. BEAUTEMPS-BEAUPRE, Mem. Inst. . . . . . Paris
Leopold VON BUCH, For. M.R.S. L.S. and G.S., Mem. Acad. Berl. . Berlin
Colonel CAMPANA . . . Milan
Chevalier CASSALEGNO . Turin
General CLARKE . United States
Captain DUMONT D'URVILLE . Paris
Captain DUPERREY . . Paris
C. G. EHRENBERG, For. M.R. and L.S., Mem. Acad. Berl. . . Berlin
Carl FALKENSTEIN, Corr. Mem. Acad. Berl. . . . . Dresden
Captain L. DE FREYCINET, Mem. Inst. Fr. Paris
Professor HANSTEEN . Christiania
Baron Alex. VON HUMBOLDT, For. M.R.S., L.S. and G.S., Mem. Inst. Fr., Mem. Acad. Berl., &c. . . Berlin
Padre G. INGHIRAMI . Florence
E. F. JOMARD, Mem. Inst. France, Corr. Mem. Acad. Berl. . . Paris
Admiral VON KRUSENSTERN, Corr. Mem. Acad. Berl. . St. Petersburg
Count Alexandre L. J. LABORDE, Mem. Inst. Fr. . . . Paris
Admiral F. B. LUTKE . St. Petersburg
A. J. LETRONNE, Mem. Inst. Fr. . Paris
Dr. Charles VON MARTIUS, For. M.L.S. Corr. Inst. Fr. and Acad. Berl. Munich
Conrad MANNERT . . Munich
Baron MEYENDORF . St. Petersburg
Professor Carl RITTER, For. M.R.A.S., Mem. Acad. Berl. . . Berlin
Dr. E. RUPPELL, For. M.L.S. Frankfort
G. S. SCHOOLCRAFT . United States
Professor J. F. SCHOUW . Copenhagen
Colonel Frederick VISCONTI . Naples
Dr. George WAHLENBERG, For. M.L.S., Corr. Mem. Acad. Berl. . Upsala
Baron C. A. WALCKENAER, Mem. Inst. Fr. . . . . . Paris
Captain ZAHRTMAN . Copenhagen
Augustus ZEUNE . . Berlin

1835.

Adm. DE HAMELIN, Chef du Dépôt de la Marine . . . . Paris

1836.

Count GRABERG AF HEMSÖ, Corr. Inst. Fr., For. Mem. R.A.S. . Florence
Baron HUGEL . . . Vienna

1837.

Professor Heinrich BERGHAUS . Berlin
Colonel FORSELL . . Stockholm
Professor Finn MAGNUSSEN Copenhagen
General PELET, Chef du Dépôt de la Guerre . . . . Paris
M. Philip VANDERMAELEN . Brussels

CORRESPONDING MEMBERS.

Col. Don José ARENALES Buenos Ayres
Don José Maria CARDANO . Madrid
M. DAUSSY . . . . Paris
M. DUPONCEAU . . Philadelphia
Col. Don Juan GALINDO Costa-rica
William HILHOUSE, Esq. . Demerara
T. A. KJAR, Esq. . St. Thomas's
M. KUPFFER, Mem. Acad. Scien. St. Petersburg
R. H. SCHOMBURGK, Esq. . Demerara
Mr. TANNER . . . Philadelphia
Dr. WOERL . . Freiburg
Don José URCULLU . . Oporto
Mr. WORCESTER . United States

1836.

M. D'AVEZAC . . . Paris
Counsr. J. J. da Costa DE MACEDO, Sec. Roy. Acad. Scien. . . Lisbon
Gen. Don Juan ORBEGOSO . Mexico

1837.

Prof. Adolph ERMAN . . Berlin
Don M. F. DE NAVARRETE . Madrid
Baron WRANGEL . St. Petersburg
Professor Paul CHAIX . Geneva

1838.

Colonel LAPIE . . Paris
Don Pedro DE ANGELIS Buenos Ayres

ADDRESS

TO THE

ROYAL GEOGRAPHICAL SOCIETY OF LONDON;

Delivered at the Anniversary, 21*st May*, 1838,

BY

WILLIAM R. HAMILTON, Esq., F.R.S.
PRESIDENT.

GENTLEMEN,

AFTER the Report of the Council, which has been just read to you, on the Progress of Geography, and the labours of our Society for the promotion of that science during the past year—for which we are indebted, amongst many other most valuable services, to the indefatigable zeal and industry of your Secretary—I fear I may not have it in my power to present to your notice any subject of very new or commanding interest; and I may, perhaps, now and then expose myself to the charge of repetition: but as this is the first occasion on which an anniversary address has been read from the President's Chair to the Royal Geographical Society since its foundation under the auspices of his late Majesty, who on all occasions extended to us his high patronage, and expressed the warmest interest in our prosperity,—and as we have now entered the second septennium of our existence, under the no less fostering hand of our present most gracious Sovereign, the Lady Victoria, I beg to solicit your kind attention to the following remarks on the importance, use, and advance of Geography, and particularly on the benefits which we are fond to believe this branch of knowledge has derived from the institution of this Society.

Science itself is one grand characteristic mark—first, of man as a rational, reflective, and improveable being; and secondly, of man in a state of civilization. Improvement in science is that which, above all things, distinguishes one man from another—one nation from another—

one era from another; and in all the stages of civilized life it has been encouraged in proportion, sometimes to the wealth and extent of a nation, sometimes to the peculiar energies which have been developed in the individual inhabitants of one or other portion of the globe, at other times dependent on the accidental occurrence of taste, magnificence, vanities, and pride of a sovereign, but always in proportion to the liberties and security, whether of property or person, which have been enjoyed by the great mass of the body politic. In this honourable race of improvement, each department of science has had its fair turn, and in some fortunate epochas of the world all have flourished together.

These general observations are equally applicable to the moral, the political, and the æsthetical sciences; and whilst in the infancy of the world, when man could only range in a comparatively narrow sphere, it has naturally occurred, that the energies of the educated classes were limited to one or two branches of inquiry, we have seen, that for a certain period Poetry was the chief object of universal regard; History and the Fine Arts have succeeded, and have had their day of brilliancy and glory, Eloquence has attracted the voice and commanded the applause of literary thousands, the science of War has unfortunately been of all ages, and of all countries, Mathematics and Natural History have been fostered by republics and monarchies; and in addition to these—to come at once to more recent days—Chemistry, Astronomy, Geography, Metallurgy, Statistics, and Geology, have all their ardent votaries, and are rapidly advancing to a degree of perfection commensurate with the capacities of man.

But it is my business, Gentlemen, on the present occasion, to call your attention most particularly to Geography—as an honourable, useful, and highly meritorious pursuit.

Geography, as its name imports, is, in its simple sense, a delineation of the Earth—and would seem at the first blush to be a plain, easy, and, I might almost add, homely acquirement; but the result of experience, and a little reflection, prove to us that it is of a very opposite character: it involves, in the general acceptation of the term, the application of History, Astronomy, Geometry in its most extended sense, Natural Philosophy, Statistics, and to a certain degree, Geology. Before we can satisfactorily delineate and describe any one country, much more the whole globe which we inhabit, we must make ourselves well acquainted with the early inhabitants of its several divisions—the various migrations which have transferred the different hordes and tribes from one place to another, each successively driving out its predecessor—the various colonial establishments which have been planted, for the purposes of commerce or conquest, or for avoiding the evils of a superabundant population—the origin, progress, dispersion, and amalgamation of languages;—and, what is the most essential for the present time, we must know the various forms of government now existing—the different religious sects—the political and moral state of the inhabitants—their commercial and agricultural resources—their manufactures, arts, and usages: and all this from the lowest state of the savage, to the highest refinement of civilized life.

With these details we have to combine the range of coasts—the elevation of the different mountain ranges, from the great stony girdles of the globe to the humble hills and almost insensible undulations—the rocks—the volcanic phenomena, and alluvial districts, whether elevated in large irregular masses, or valleys formed by rivers; we must fix their localities and measure their relative heights. Every river and stream is to be traced from its numerous sources to its outlet in the ocean—each affluent and confluent must be distinctly laid down; the deltas and marshes must be surveyed; the precise site of every village, town, and city must be ascertained, and the whole corrected by trigonometrical admeasurements and celestial observations. But the geographer's task is not terminated with this list of labours on the inanimate world; he must likewise lay before us the different productions of the earth, diversified as they are by soil, climate, and exposure; he must describe to us all the varieties of animal life which that soil supports, from those which are invisible to the eye, to the lords of the forests, and to man himself. Here, indeed, other sciences and other inquirers will come to his assistance; but he must be the pioneer, who is to unlock the treasures which we want to possess; he is to point out the way which others are to go; he is to declare to us, "Here you may search, with the hope that your labours will not be unrewarded; here you will only waste your time, for you will find nothing to repay your toils."

In civilized and well-peopled countries, all this is comparatively easy; and nothing is required but the well-edücated man of science, ardour in the pursuit of knowledge, accuracy of observation, and public or private encouragement. But the real geographer becomes at once an ardent traveller, indifferent whether he plunges into the burning heats of tropical deserts—plains, or swamps—or launches his boat on the unknown stream—or endures the hardship of an arctic climate, amidst perpetual snows and ice—or scales the almost inaccessible heights of the Chimborazo or the Himálaya. Buoyed up in his greatest difficulties by the consciousness that he is labouring for the good of his fellow-creatures, he feels delight in the reflection that he is upon ground hitherto untrodden by man—that every step he makes will serve to enlarge the sphere of human knowledge—and that he is laying up for himself a store of gratitude and of fame.

It would be needless in this Society to enumerate the names of all those whose successful toils in this career, within little more than half a century, have prompted the foregoing observations: those of Cook, Vancouver, Broughton, Grant, and Flinders; Bruce, Browne, Barrow, Ledyard, Parke, Tuckey, Bowdich, Lyon, Denham, Clapperton, Laing, the brothers Lander, Davidson; Valentia, Salt, Wilkinson, Hoskins, Burton, Grenville Temple; Owen, W. H. Smyth, Boteler, Vidal, Allen; Burchell, Campbell, Andrew Smith, and Alexander:—Rennell, Marsden, Horsburgh, Staunton, Raffles, Webb, Moorcroft, Colebrook, Forster, Ouseley, Malcolm, Clarke, Leake, Beaufort, Rich, Kinneir, Ker Porter, Irby and Mangles, Elphinstone, Christie, Pottinger, Morier, Fraser, Basil Hall, Burnes, Chesney, Sadleir, Wellsted, W. Hamilton, and Ainsworth: —Hearne, Mackenzie, Scoresby, Parry, Franklin, Richardson, Back, the two Rosses, Beechey, Elson, Dease, and Simpson:—Mawe, Smyth, Pentland, King, FitzRoy, and Darwin:—Weddell, Biscoe; Bass, Oxley, Cunningham, Sturt, and Mitchell, amongst our own countrymen, will occur to you; and it is highly to the honour of England that such a host of celebrated names, most of them our contemporaries, and almost all within the memory of some of us, have illustrated by their personal labour the geographical annals of our country. Amongst the foreigners of equal merit, we love to do justice to Niebuhr—to Humboldt, in whose name and works alone are comprised all the conceivable elements which make up a scientific traveller, geographer, chemist, naturalist, astronomer, and geologist—Bonpland, his adventurous, but less fortunate companion:—Bougainville, Condamine, La Perouse, D'Entrecasteaux, Baudin, Freycinet, Dumont D'Urville, Duperrey, Roussin, Laplace, Beautemps Beaupré, Bérard, and Blosseville; Beauchamp, Laborde, Jomard, Champollion, Caillaud, Caillié, D'Orbigny, Peytier, Callier, and Texier:—Ulloa, Tofiño, Malaspina, Espinosa, Quadra and Bauzá:—Belzoni, Rossellini, Inghirami; Siebold, Gråberg, Forsell, Hansteen, Graah:—Pallas, Krusenstern, Golownin, Kotzebue, Lütké, Wrangel, Meyendorf, Parrot, Engelhardt, Kupffer, Schubert:—Carver, Lewis, Clarke, Pike, Long, Cass, and Schoolcraft:—Von Buch, Rüppell, Horneman, Spix, Martius, Hügel, Erman, Pœppig, Meyen, Schomburgk, Ehrenberg, Rose, Eichwald, Gutzlaff, Gobat, Dubois de Montpéreux, and lastly Burckhardt, who spent nine years in Syria, Egypt, Nubia, and Arabia, in the employment of the Palestine Association; which Association, now incorporated in this Society, from its early foundation fairly deserves to be called the parent of institutions, having for their object the extension of Geography; and the great liberality of its members has set us a bright and encouraging example.

I would be out of place, on this occasion, to enter more at large on the beneficial influence which the labours of these Navigators and Travellers have had in promoting the wealth and commerce of the world—in extending the sphere of human happiness—in opening the benefits of civilization to myriads of human beings—in bringing us into communication with new worlds, and generally in hastening our approach to a full knowledge of the configuration of the globe. But besides these general, and in a measure speculative advantages from the progress of geography, it has, in connexion with hydrography, an immediate practical result in the preservation of human life, by giving us a more accurate and detailed knowledge of the depths, currents, rocks, straits, and shallows of the ocean, the discovery of new harbours, and a more intelligible and distinct delineation of coasts.

It may be attributed, in great part, to the success of the individuals I have enumerated, and to the reflection of how much more might have been effected by some of them, had they met with more direct support, and had been better instructed and guided in their pursuits—in part, also, to the example I have alluded to, set by the African Association more than fifty years ago, that we have seen during the last ten years the institution of societies similar to our own. I believe the honour of priority of date, in this respect, may be given to the Geographical Society of Paris, to which, in addition to their monthly bulletins, we are indebted for five volumes of highly interesting and valuable memoirs. Our own followed shortly in its track; and this again has given rise to the Geographical Society in Bombay, and to others, founded with similar objects, in Egypt, Berlin, and Frankfort.

To recapitulate the results which have flowed, in the progress of geographical knowledge, from the aid and countenance of the Royal Geographical Society of London, would be to give a catalogue or abstract of the papers published in our Journal. But we may briefly remark—first, That in respect to Europe, the division of the world in which most was already known, and less therefore required to be studied, we have consigned to our pages Colonel Baker's valuable Memoir on the Northern Frontier of Greece, founded on the labours of the Mixed Commission appointed by the Governments of England, Russia, and France, to determine, conjointly with the Turkish plenipotentiary, the limits between the Ottoman and the Greek rule. This memoir is also accompanied by a beautifully executed map; and it connects with the great French survey of the Morea and Northern Greece. A Map of Mount Athos and the adjoining coasts, resulting from the British surveys in the Mediterranean, deserves also our special notice: as well as the learned memoir, by the Dean of Wells, on the Periplus of the Euxine by Arrian, compared with the observations made in the Blonde frigate, in the year 1829–30; and that on the Gulf of Arta by Lieut. Wolfe, in 1830.

A short memoir in the eighth volume of the Journal upon the sites of Minoa and Nisæa, in the territory of Megara, affords a striking instance of the importance of a precise knowledge of the topographical details of a country for a clear understanding of its history; and though all the difficulties attending the question are not yet quite cleared up, yet it is fortunate that the attention of the literary critic and the scientific navigator should have been turned to a spot now so much frequented by our sailors and our travellers, that we may confidently expect that this minute point of comparative geography will not much longer remain a *vexata quæstio*. The difficulty seems first to have been mooted by the learned editor of Thucydides, Dr. Arnold, than whom this country can boast no better authority on such a subject, amongst those who have not had the advantage of bringing familiar acquaintance with the classical writers to bear upon a personal inspection of the locality.

There can be little doubt, however, that the position given by Mr. Spratt to the fortress of Minoa is the correct one, although, according to his own map, the distance between it and Megara is certainly something more than eight stadia; but the authority of Thucydides, which Mr. Spratt quotes for that distance, does not quite bear him out in his assertion; for the text of that historian only says that *Nisæa* (not Minoa) was eight stadia, in the line of the long walls, from Megara; and this accordingly makes Mr. Spratt's hypothesis still more tenable against the conjecture of Dr. Arnold, as his Minoa is to the south of Nisæa, and therefore farther from Megara. Nor need we lay any stress on the different denominations given to Minoa by Thucydides and Strabo, the former calling it an island, and the latter a promontory (ακρα), as the term island (νῆσος) was very frequently applied by the ancients to a promontory or peninsula joined to the main by low banks. There is a remarkable instance of this in the second book of Herodotus, where that historian describes a part of the city of Cnidos as being on the *island* of Triopium, whereas Triopium is, and must always have been, a promontory. Nor need we suppose, as Mr. Bloomfield conjectures, that what in the time of Thucydides was an island became a promontory five centuries after, in the time of Strabo; for Pausanias, who was posterior to Strabo, equally calls it an island.

Then, with regard to the discordance between Thucydides and Strabo, as to the distance between Megara and Nisæa, the former giving eight, and the latter eighteen stadia, it is much more likely that Strabo was in error, if the error be attributable to either, than Thucydides; or we may either suppose that Strabo wrote δέκα ἢ ὄκτω, ten or eight; or with Poppo, that δέκα καὶ was inserted by some copyist; for it is most natural to conceive that the Megarensians would adopt this, the shortest line, for their walls, leading as it did to the nearest point on the coast, where there was a harbour; and Minoa was evidently close to Nisæa, commanding a view of the entrance into the port. Mr. Spratt's localities entirely agree with the whole narrative of Thucydides; and nothing remains to be explained but the five small islands which Strabo says lay between Minoa and Salamis, where there are now but four. This, however, is too trifling a difficulty to throw any doubt upon Mr. Spratt's hypothesis.

The accurate determination, by General Schubert, of 75 points on the shores of the Baltic, in the course of his chronometric expedition in 1833, and noticed in the 6th volume of "The Journal," forms a great addition to mathematical geography. Nor should we pass without notice Colonel Jackson's papers on the Congelation of the Neva at St. Petersburg, and the Temperature of its Waters when covered with Ice, as in some measure connected with geographical inquiry.

But I need not tell you, Gentlemen, that though much has been done, and much is still in progress, and notwithstanding the vast amount of topographical and hydrographical labours, which are enumerated in your Secretary's "Sketch," all is not yet, or likely to be soon accomplished, even in our own quarter of the globe. We have only to cast our eyes over the latest and best maps, executed either at home or abroad, and we shall see, without referring to the vast tracts comprehending the northern portion of Russia in Europe, bordering in part on the White Sea, and between the Gulf of Bothnia and the Ural Mountains, all the northern provinces of Turkey in Europe, comprising what were formerly known under the names of Mœsia, Dacia, Macedonia, Thrace, and Thessaly, are still very imperfectly delineated; and notwithstanding the valuable results of the labours of the Russian engineers in Wallachia, Bulgaria, and Roumelia, the active exertions of some of our own travellers, and the scientific surveys of the late French expedition into the Morea, the details of the interior of a large part of that peninsula still offer an abundant harvest to the geographer; nor will the maps of the interior provinces of Spain, Portugal, or Sicily, deserve a much higher commendation.

Italy still requires the conjoint labours of the antiquary, the historian, and the geographer, to clear up many disputed points of ancient towns, colonial establishments, and territorial divisions. The convulsions of nature have added to the difficulties of inquiries of this description. Much is still to be explored in Etruria, and in Magna Græcia; and it is only within a very few years that the precise line of the march of

Hannibal over the Alps—perhaps the most prominent feature in the ancient geography of this part of the world—has been partially elucidated by the industry of a few inquiring travellers.

What do we know of the immense continent of Africa? Some parts of the coast may have been minutely surveyed, especially those bordering on the Mediterranean ; but if we glance at the inland districts, the whole, with a few exceptions, is (geographically speaking) a dreary and unknown wilderness. The immediate neighbourhood of our own colony of the Cape of Good Hope, some of the more accessible parts of the coast of Barbary, a few points on the western coast of the Red Sea, some tracts of the adjoining desert, portions of the English and French possessions on the Gambia and Senegal, the banks of the Nile below the first or second cataract, Denham and Clapperton's routes to the Lake Chad, that of Park to Sego, and a glimpse of the lately-discovered banks of the Quorra and the Chadda, comprise nearly all we may be said (even imperfectly) to know throughout that portion of the globe. The last named of these events—the tracing of the river Quorra, by the brothers Lander, from Búsah to its outlet in the sea, a range of nearly six hundred miles—has indeed solved one of the most remarkable geographical problems which veil with mystery the interior of this vast continent; and its appearance in the first volume of our Journal gave rise to what may be described as the most masterly discussion by one of the best geographers in this country, that has perhaps appeared in our pages, viz. whether the river so discovered, one of the great arteries of the African continent, was really the Niger of the ancients. In the same year were also printed a brief account of a residence in the city of Marocco, by Captain Washington, as well as the first measurements on record of any of the loftier inland peaks of Mount Atlas. The narrative of the survey of all the northern part of the West Coast of Africa, by Captain Vidal and other officers of his Majesty's navy, is to be found in "The Journal," as well as those of the two recent expeditions of discovery into the interior from the Cape of Good Hope—one under Dr. A. Smith to the north-east, into the countries of the Matabili and the Baquainas; and the other under Captain Alexander, to the north-west, in which were explored the countries of the great Namáquas, and the Boschmans, as far as the river Kiusip, which enters the sea in Walvisch Bay, north of the 23° of south latitude, and thence eastward to the country of the Hill Dámaras.

The description of the African shores of the Red Sea, and the discovery of the ruins of Berenice, by Mr. Wellsted, and Mr. Wilkinson's account of the eastern desert bordering on Upper Egypt (the seat of many of the most valuable mines and quarries worked by the ancient Egyptians),

Lord Prudhoe's memoranda of his journey from Cairo to Sennaar in 1829, have also added materially to our knowledge of these once populous and flourishing districts; and, whilst we are grateful to Sir Grenville Temple for his narrative of the route from Bonah to Konstantinah, and other geographical communications from that neighbourhood, we cannot pass over this part of the subject without a tear of regret at the untimely end of the gallant Davidson, in his unfortunate attempt to penetrate from Marocco over the Atlas to Tumbúktú.

Northern Africa still remains a fertile subject for discovery; and the recent expeditions of the French, which have put them in possession of Algiers and Konstantinah, promise to hasten the period for completing our knowledge of a vast extent of country, which attained a partial civilization at a very early period—where in time we may perhaps find the seeds of much of the historical mythology of the Greeks, which was overrun in various directions by the Roman arms, which was the grave of thousands of their troops before it became the scene of their greatest triumph, and which for centuries afterwards was the granary of Italy.

I cannot forbear availing myself of the occasion to announce to you that an expedition is in progress, and fitting out by the private subscription of individuals interested in African discovery, for penetrating up the Bahr el Abiad, or western and main confluent of the Nile. It is intended to intrust this task to a native of Dongola, who is under suitable instructions for the great object he is to undertake. The many fruitless attempts hitherto made by Europeans to explore this river, almost all of which have proved fatal, have prompted the idea of engaging a native of the country, as less exposed to the evils of the climate and to the hostility of the inhabitants. We have reason also to expect that various discoveries will shortly be made public from this quarter by Mr. Russegger, an Austrian mineralogist, who is at present in Abyssinia, and intended to prosecute his explorations to the banks of the Bahr el Abiad.

With the continent of Asia, the supposed cradle of the human race, but from whence, without entering upon the subject of the early emigrations of man, issued forth, within the historic period, those swarms of barbarians which overwhelmed the Empire of Rome, we are comparatively better acquainted. Our vast possessions in the East, the extensive commerce carried on between the European and the North American States and a large portion of this division of the globe, whether insular or continental, and the long and early connexion between the two continents, have brought us much into contact with the coasts and inhabitants of Asia. But even here, if we require positive data—if we look for anything like mathematical accuracy for the position of cities, the course of rivers, or the heights of mountains, we are sadly at a loss. Much has been done, and is still doing, to bring out a correct map of our own Indian possessions; many interesting portions of this vast continent have lately been laid open by the labours of enterprising travellers, and war, and trade, and politics from time to time go hand in hand in extending our knowledge of the interior. On this subject I cannot avoid pointing your attention to the very able Memoir by Captain Jervis, on the Origin and Progress and Present State of the Surveys in India, published in the seventh volume of our Journal. In the course of these Surveys, the most extensive, and perhaps the most accurate, measure of the earth ever attempted, has been successfully accomplished; Punnae on Cape Comorin, the southern extreme, being in 8° 9′ 38″ north latitude, and Kedar Kanta in the Himálaya Mountains in 31° 2′. The total therefore of the meridional arc thus measured is about one-sixteenth of the entire circumference of our globe.

Captain Johnston's account also of The Himálaya Mountains, and, more recently, Baron C. Hügel's Visit to Kashmír, have added much to our knowledge of that vast range, whose superior elevation above the Andes was first asserted by the lamented Colebrooke.

The mention of this name reminds us of the loss which the scientific world have experienced in the death of this eminent individual. Mr. Colebrooke was not a member of this Society, or his name would not have been passed over in silence amongst those whom we have lost during the last year; but just tributes to his memory have been paid in the Address of the Illustrious Vice Patron of this Society, as President of the Royal Society, as well as by the learned Professor, now presiding over the Geological Society. But Mr. Colebrooke, besides being a most distinguished linguist, particularly in the Oriental languages—and deeply versed in all the learning of the East, whether practical or theoretical, a sound lawyer, an experienced financier, and one of the founders of the Asiatic and Astronomical Societies in this country, added very considerably to our geographical knowledge of the Northern and North-Eastern Provinces of India.

The valuable papers on the Indus and on Sind, the Panjáb and Bucharia, by the well-known traveller Captain Burnes, are too fresh in the memory of those now present, to call for any special notice.

But we have only a very general knowledge of the geographical character of the Birman, Chinese, and Japan empires; the innumerable islands of the latter are still, except occasionally, inaccessible to European navigators. Geographers hardly venture on the most loose description of Thibet, Mongolia, or Chinese Tartary, Siam, and Cochin China. Persia, Turkey in Europe, and Arabia, have indeed been more or less travelled over, and their routes described by Europeans; and we are almost daily gratified with some accessions, characterised by various degrees of accuracy and distinctness, to our knowledge of these regions.

A paper by Mr. Bird, in the fourth volume of the Journal, on the Manners of the natives of the Southern Coast of Arabia and the Shores of the Red Sea, shows much observation united with research in Oriental literature; and we are indebted for the only descriptive accounts hitherto published of the Surveys of the Red Sea and Persian Gulf, executed by the officers of the Indian Navy, to Lieutenant Wellsted, who has also deserved our best thanks for his Memoirs on the Island of Socotra, and on the Oasis of Omán in the Arabian territories of the Imám of Muskat, which had been partially described, though not visited, by Niebuhr.

The Survey of the Province of Azerbijan and of the Shores of the Caspian, by Colonel Monteith—who has also supplied us with the materials for a map of part of the Caucasian district—the accounts of journeys performed by officers attached to the several British missions in Persia, with heights of mountains northward of Tabriz and Tehran, and lately Colonel Shiel's Journey through a part of Kurdistan, are also very useful additions to our stock of geographical knowledge of Asia. But we must bear in mind that it is only within a very few years, within a period almost commensurate with the formation of this Society, that the two most celebrated rivers in Asia, the Indus and the Euphrates, have been navigated, and that by English officers, throughout their whole navigable course; the peninsulas of Arabia and Mount Sinai have only been partially traversed; the same may be said of the great empire of Persia; we only know from occasional routes the interior of that most interesting portion of this continent formerly called Asia Minor; and the great ranges of Imaus and the Hindú Kúsh are still little more than a terra incognita, notwithstanding the successful attempts of some of our most distinguished countrymen to penetrate across them. It is true that these countries attracted the curiosity of many ardent travellers,—I cannot designate them by the name of geographical travellers—in every century since the first dawn of letters and civilization in modern Europe; and they have contributed great accessions to our knowledge of the trade, manufactures, customs, and religious observances of their inhabitants. But it is only within a comparatively very limited period of time that accurate descriptions of the geographical features of the regions thus explored have been attempted, and still more recently have these descriptions been accompanied by their only true tests, geometrical admeasurement and astronomical observations: considerable progress has thus been made in verifying the comparative geography of this part of the globe, illustrious as it has been, more than any other, for revolutions of empires, the construction and demolition of the great monarchies of antiquity, the occasional and temporary introduction of European civilization by colonies from Greece, rapid conquests, and the birth-place of Judaism, Christianity, and Mohammedanism. Much, however, still remains to be accomplished in this respect; and, amongst the prominent events of ancient history, which are still but imperfectly understood, from the superficial description of the countries traversed on these occasions (for geography is one of the eyes of history, as chronology is the other—bereft of either of them, it would become a vast mass of crude heterogeneous matter—"informe, ingens, cui lumen ademptum"), we anxiously look for a more satisfactory illustration of the descent of Cyrus, the founder of the empire of Persia, from his native hills to the plains of Assyria and the shores of the Mediterranean, the military and civil constitution of the empire of Darius and Xerxes, as described by Herodotus, the march of the latter to the shores of the Hellespont, the expedition of Cyrus the younger from Ephesus to Babylon, and the Retreat of the Ten Thousand through the Gordyæan Hills to the Euxine, and thence, whether by sea or land, to the coasts of Bithynia and Thrace, on their return home—the march of Alexander, embracing the provinces of Asia Minor, traversed in various directions, the coast of Syria, Egypt, and Mesopotamia, Assyria, to Persepolis, again in a north-westerly and north-easterly direction through the mountains of Media to the shores of the Caspian, through Parthia, Asia, Bactriana, and Sogdiana, to the Panjáb, his navigation of the Indus, and return with his army and fleet to Babylon. The long and sanguinary wars amongst the successors of Alexander, from the banks of the Nile to the Indus and the Ganges, still want the light of geography, as do also the conquests of the Romans, from their first hostilities with Antiochus of Syria, to the fixing the boundaries of their empire on the Euphrates and Tigris; and particularly the campaign of the Consul Manlius Vulso against the Gallo-Græcians, so fully detailed in Polybius and Livy, and those of Sylla and Lucullus against Mithridates, and of Crassus against the Parthians—the journey of Cicero to assume his proconsulate of Cilicia, the various expeditions of many of the Roman Emperors in this direction, from Trajan to Julian and Heraclius. We have then to illustrate the conquests of the Arabs, when the tide of war again began to flow from an easterly to a westerly direction, aided, as it was, in lowering the pride of Rome, by the successive invasions of Goths, Alani, and Huns, from the frontiers of China to the shores of the Atlantic and the coasts of Barbary. To these succeeded the gradual acquisitions of the Turks, from the Altai to Constantinople and the Danube, and the more or less ephemeral empires of the Tartars, spreading their desolating arms over eastern and western Asia, and the

Crusading expeditions from the Bosphorus to Antioch and Jerusalem. Very few details of any of these important events in the history of the world, and in that of the march, progressive or retrograde, of civilization, are as yet made perfectly clear to the understanding of those, who do not exclusively devote themselves to this study; and it is probably the case with these, in the study of geography, as with all others in the pursuit of every other science, the more they have acquired the more they feel the deficiency of what is yet to be learned. We must not, however, be backward in rendering the tribute of praise and thanks which is justly due, first, to the literary and scientific labours of a D'Anville, a Freret, a Rennell, a Vincent, a Marsden, and others, who in their closets have spent their most valuable hours in elucidating those knotty points of historical geography—nor to the more recent and more active researches of a Niebuhr, a Pococke, an Elphinstone, a Leake, a Moorcroft, a Burnes, a Chandler, a Gell, a W. Hamilton, a Chesney, and an Ainsworth. It is only a few days, since we have been gratified within these walls by the reading of a memoir by the last mentioned of these travellers, on a very minute but very important point in the country I am alluding to, namely, the exact sites of the passes between Asia Minor and Syria, known under the various denominations of the Cilician, the Syrian, and the Armenian Gates—a line of march which, from its being so frequently mentioned by historians, as that which was preferred to all others in the communication between the eastern and western parts of this continent, must have possessed advantages in a military and commercial point of view, which have not yet been sufficiently developed; but resulting as well from the nature of the countries to be traversed, as from the facility of commanding supplies for the support of armies.

I may also be permitted to add a special notice of the memoir in our Journal, of Mr. Brant's journey through parts of Armenia and Asia Minor, in 1835, and more recently of two memoirs, in 1837 and 1838, by Mr. W. Hamilton, which have materially added to our knowledge of the historical geography of Asia Minor. The last-named traveller has ascertained the sites of many ancient cities of much interest, as well by collecting and comparing medals and inscriptions, as by correct admeasurements of distances. This traveller was the first to find the descendants of the ancient Chalybes of Pontus still occupied in the same pursuit, as their ancestors of old, in the same spot, between the Thermodon and Cape Jasonium, and grubbing up the same rugged soil for iron, as Apollonius Rhodius describes them, σιδηροφόρον στυφελὴν χθόνα γατομέοντας. He has also given to us the most satisfactory account we yet possess of the physical geography of this peninsula, by measuring the heights of its table-lands, as well as of many of its mountains.

Much information may also shortly be expected on this part of the world from the forthcoming publications of MM. Callier and Texier, with the details of their discoveries in Asia Minor; and the French government have recently fitted out another expedition, with a view to classical and antiquarian researches on the same continent, under Mr. Raoul Rochette, already well known for his history of the ancient colonies of Greece.

The letters of Lord Lindsay, which I understand will soon make their appearance, will likewise furnish many very interesting notices on the most remarkable localities in Arabia Petræa, in the Hauran, and throughout Syria, illustrated by an observant eye and a well-stocked head. I have, at the same time, much satisfaction in announcing to you that Mr. Ainsworth, already so well known as the surgeon and geologist to the Euphrates expedition, and who has also greatly distinguished himself as a geographer, will shortly proceed, under instructions from the council, on an expedition into Kurdistan; and we may anticipate as the fruits of this journey, which will embrace a large portion of the basins of the Euphrates and the Tigris, most valuable accessions to our knowledge of a district abounding in objects of the highest interest to the historian, the geographer, and the Christian.

I cannot avoid making especial allusion to one subject of inquiry, highly important to a correct knowledge of historical or comparative geography. What volumes have not been written by the learned, who have given their attention to this subject, on the measures of the ancients,—their miles, their stadia, their parasangs, and their plethra; and how much we are still in the dark as to the precise length of any one of these terms? Did the various nations, and various tribes, or subdivisions of nations, as is pretended by some, each use its distinct measure under the same or different appellations? Did all the Greeks use one scale of admeasurement, and all the barbarians another? Can we attain any precise knowledge of these facts by the vague and apparently contradictory accounts of different authors? Were the smaller measures always aliquot parts of the larger? Is the confusion complained of by modern writers to be attributed to one or other of these causes, or to the ignorance and want of information in the ancient historians and geographers, or to the incorrectness of the transcribers of the manuscripts to which we refer? These are questions which require and deserve elucidation, and I trust we may shortly see some new lights thrown upon this difficult subject by a distinguished member of this Society.

In the mean time I have been invited by him to throw out the following hints for your consideration. Although almost all modern geographers—Rennell and Gosselin amongst the rest—have proceeded on the supposition that the Greeks made use of several kinds of stadia, varying from 696 to 750 to a degree, there are still good grounds for believing that this measure was in fact, at all times and in all countries, invariable, having been kept correct by the standard of the Στάδιον, or ground laid out for the foot-races at the solemn and periodical festivals throughout Greece. The extents of all these Σταδιὰ which have hitherto been measured are of one and the same length.

The distance from Olympia to Athens, reported by Herodotus to be 1485 stadia, and that from Olympia to Sparta, stated by Pausanias (vi. 17, 6), five centuries later, at 660 stadia, are shown by the latest maps to have been measured by the same stade.

Plutarch, also, in his life of Caius Gracchus, speaking of the milestones placed on the Roman roads by that public officer, remarks that a mile was equal to something less than eight stades. This is shown to be correct by the true Στάδιον, which is equal to 203 English yards.

The computed stadium in the Roman writers was subject to much variety: this arose partly from the Romans having adopted the stadium as synonymous with 600 feet; but as they of course measured by their own foot, which was one twenty-fifth less than the Greek foot, their stadium became less than the Greek stadium. It may partly also be attributed to negligence, and occasionally the contradictions we meet with may be owing to the corruption of the text; but one main cause of these differences was the ignorance which generally prevailed in ancient times of the real form of mountains, and the exact relative position of particular places.

The evidence of Censorinus as to these different stades with Greek names, opposed as it is to all Greek authorities, is entirely erroneous, or his hypothesis will apply only to Italy.

This is a subject that well deserves to be pursued under the advantages held out by the great improvements in modern science. Geography, as an exact science, must be founded on astronomy; it could therefore make little or no real progress amongst the ancients. Eratosthenes and Hipparchus had indeed improved upon the mathematicians of the time of Aristotle, as appears from the number of stades forming the aliquot parts of the degree of the great circle of the earth; but although Hipparchus had probably some better mode of measuring an angle than that described by Archimedes, yet without instruments of more delicate construction it was impossible to make much further advances either in geography or in astronomy, or these would inevitably have led to the discovery that trigonometry was the true basis of geography. When we consider how recent is our own knowledge of the true form of many—I may indeed say of *all*—countries, as far as we yet *do* know them, how continually we are even now misstating the real as well as the relative position of places, we need not wonder at the want of precision in the writings of the ancients on these matters. Nevertheless, after the labours of Hipparchus, and of those who attended to such subjects up to the third century of the Christian era, with the help, too, of a more extended navigation, and the practice introduced by the Romans of traversing in various directions the countries they conquered, by military roads, which exercised in every part of the empire the skill of the architect and engineer, they did obtain a certain degree of practical knowledge of relative sites; but the only real result to geography, as a science, was the compilation of Itineraries and Peripli, abounding in false computations and textual errors, the greater part of them being composed apparently by men of very limited acquirements even for those days.

Geography is, in truth, so peculiarly a progressive science that it may be described as a continual correction of errors; and approximations are always useful in degree; but as in astronomy these errors are, or at least must be considered, as the truth, until they are corrected, and as they furnish the only materials, and they are the only mode to further discovery, nothing can be more unphilosophical scientifically, or more unsound critically, though nothing is more common, than for geographers to condemn the labours of their predecessors, without adverting to the *history* of the geography of each country, which is scarcely ever attended to.

On the subject of the importance of geography to history we may mention also, as one object of praiseworthy study, the ascertaining the exact site of all the cities founded by Alexander during his conquest of Asia. This would materially tend to illustrate the moral and political effects of those memorable campaigns. A learned Dane, the Chevalier Bronsted, has undertaken this task, and he has very appropriately called in to his aid the science of Numismatics. He has collected for the Royal Numismatic Cabinet at Copenhagen, in the first place, all the original medals of Alexander which the pecuniary means at his disposal enabled him to purchase, and, where these were deficient, the casts of all others contained in the several public and private collections in this country and on the continent. A careful comparison of these most authentic documents, most of them varying as they do in the types and other marks, whether of time or place, imprinted upon them, will enable him to effect this service for historical geography in a manner which cannot fail to secure to this classical traveller the warm thanks of every lover of ancient art and ancient history.

The Society have already heard that the Council have awarded her Majesty's annual premium for Geographical Discovery to Col. Chesney of the Royal Artillery. Amidst several other distinguished travellers, whose names and merits in the cause were brought before them on this occasion, they thought that the claims of Col. Chesney were the most conspicuous: in the first place, and most especially, for the spirit with which, unassisted by any other encouragement than his own ardent mind, he accomplished his first survey of the Euphrates, and formed the project of the practicability of its navigation for commercial and political purposes, and at the same time added materially to our knowledge of the countries bordering on its banks; secondly, for the skill and perseverance with which, after much opposition, he effected this splendid enterprise, in which the discretion and prudence of the commander were as conspicuous as the zeal, energy, and ability, by which he was seconded in those who were placed under his direction; and, thirdly, for the great mass of new and valuable geographical and scientific information which he has brought home.

The details of this expedition are as yet only to be found in an abridged memoir of his proceedings, by Col. Chesney, inserted in the miscellaneous portion of our Journal for last year, and in the official papers laid upon the table of the two Houses of Parliament, besides a partial notice by Mr. Ainsworth, the naturalist appointed to accompany the expedition, which I have just alluded to.

I shall therefore only repeat to you on this occasion, in the words of your Secretary, that "Materials for a correct map of a large portion of Northern Syria have been collected; a line of levels was carried across from Iskanderún on the Mediterranean to Birehjík on the Euphrates, and thence to the Persian Gulf; Northern Mesopotamia was explored, and the grand survey of the river commenced with the descent of the two steamers, one of which was lost in the hurricane near 'Anah, and continued by the other to the embouchure of the river in the Persian Gulf. At a subsequent period two different ascents were made of the Karún, and two descents of the Bahamishír; while the country intervening between the Jeráhí and the Euphrates, the great delta of Susiana, concerning which so much that is incorrect is still in our maps, was examined. The river Tigris was twice ascended to upwards of 400 miles beyond its junction with the Euphrates; a second line of levels was carried between the Euphrates and the Tigris; new positions obtained by the journeys across the deserts; and a geological section of the Taurus, of several hundred miles in extent, were amongst the last of the successful labours of the expedition."

After a perusal of this very interesting memoir, the Council felt that there was no necessity (and I am confident that the Society will coincide in that conviction) to wait for the publication of the full details of the expedition, till they should give to Col. Chesney the stamp of their approbation, and, I may add, their admiration, of his conduct throughout, as commander of the expedition, by awarding to him the royal premium for the year 1837.

They were anxious, also, to seize the first opportunity in their power to show how fully they joined in the expressions used by one of their body, who, when noticing some years ago Colonel Chesney's first reports on the navigation of the Euphrates, commenced his observations by saying, that "Whilst new parts of the world are daily becoming better known through European enterprise, many most interesting portions of the Ancient World would seem likely to be forgotten or neglected, though, both for their physical character and the historical recollections attached to them, they furnish materials as ample for the inquirer into natural phenomena, and infinitely more abundant for him who studies the history of man, and the revolutions of political societies."

But whilst we are thus collecting materials, and encouraging farther researches towards a more correct knowledge of the most remarkable periods in the history of the Ancient World, we are not therefore to shut our eyes to objects of more immediate and daily interest. A still wider field here opens to our view. Instead of unlocking the lost or hidden treasures of antiquity, and only now and then obtaining a result which may compensate for our labours, we have the sure prospect of a rich harvest. The vast untrodden steppes in the interior of Africa may be pregnant with objects of interest in the vegetable or mineral kingdom, which, with the progress of discovery, may open out new sources of wealth to our commercial adventurers, whilst the inhabitants, we may hope, will in the fulness of time be ready to receive the blessings of liberty, civilization, and religion, which we may impart to them; and I have already alluded to the first navigation of the Niger or Quorra, as being one of the most remarkable conquests, which have been gained in our time by the arts of peace and honourable enterprise.

But it is in the New World or Worlds that the great field of geographical discovery is still to be cultivated. And here also the pages of our Journal may be consulted with advantage. The setting on foot of the Arctic Land Expedition in 1832, in search of Captain Sir John Ross and his adventurous companions, may be noted as one of the useful abours of the Society; and the Journal contains the first account of the results of that expedition in the discovery of the great river which flows from the Great Slave Lake, in a north-east direction, into the open sea, nearly ninety miles south of the Boothia of Captain Ross, with the current flowing into it from the westward, and which river now justly bears the name of "Back."

The opinions also of some of our most celebrated navigators, urging another expedition of discovery, in which the Society took an active part, will be found in the annals of the year 1836. An account also of the monuments and relics of the ancient inhabitants of New Spain will fully repay the time bestowed upon it. I may observe in passing, that these monuments offer a very remarkable coincidence with phenomena long since observed in the Ancient World, denoting the existence of a race of men who seem to have chiefly occupied themselves (in a state of civilization of which we have no records, and of which we can form no plausible conjecture) in raising edifices of enormous grandeur, and almost imperishable strength: such, I mean, as the ancient Treasuries of Mycenæ, and Erchomenos in Greece, the Cyclopean Walls of Tyrins, Stonehenge and Abury in our own island, the Cromlech Circles at Carrowmore in Ireland, and the innumerable pillars of Carnac on the coast of Brittany in France.

But to return from this digression. In South America, King's and Fitz-Roy's Account of the Ten Years' Survey of the Shores and Rivers of Patagonia, Chile, and Peru, stands prominently in relief: the brief but comprehensive Summary of the Voyage of the Beagle, which contains every result that can be useful to the seaman, and was published only a few days after the arrival of the ship in England, may be held up as a model, and as an example well worthy of imitation by all voyagers and travellers.

A brief account of the Bolivian Andes, by Mr. Pentland; of the Affluents of the Amazons, and of the descent of this river from Peru to Para, by Captain Smyth; of the exploration of the rivers Santa Cruz and Negro, in Patagonia, by Captain Fitz-Roy and Villariño; an abstract of the documents now publishing by Don Pedro de Angelis, at Buenos Ayres, and prepared for the Society by Sir W. Parish, and the very full reports of Mr. Schomburgk of his discoveries in British Guayana, by which very large and important additions have been made to the available resources of that flourishing colony, are amongst the chief contents of the Society's Journal relating to this quarter of the world. I may here add that the last accounts received from Mr. Schomburgk, whose scientific labours have received the warm applause of Alexander von Humboldt, hold out the promise of his being able during this year to effectuate the passage of the mountains, which separate British Guayana from the Basin of the Amazons, and eventually to reach Esmeralda on the Oronoco, and thus unite his labours with those of Humboldt, who has lately given, in the "Annales Maritimes" for the last year, some very valuable notices on the geography of Guayana, in the fullest acceptation of that term.

The only complete account of the progress of inland discovery in Australia, during the later times, is to be found in the Society's Journal. A general view is also there given of the state of discovery in Australia in the year 1833, drawn up by Mr. A. Cunningham, at the desire of his late Majesty's Principal Secretary of State for the Colonies; and has been succeeded by Major Mitchell's narratives of his expeditions into the interior, that of the establishments at Melville Island and Port Essington, and some account of the settlement at Swan River.

An attempt has also been made, in which the Council depend on the Society's warm concurrence, to introduce into this large portion of our colonial possessions a more accurate and appropriate nomenclature, applicable to its principal divisions.

We look with great anxiety for the results of the exploring expedition from the north-west coast of Australia, under Lieutenants Grey and Lushington.

Whilst upon the late discoveries in this part of the world, I have much pleasure in announcing to you, that information was received only two days ago, that the river Murray, on the south coast of Australia, has been found to have an open navigable mouth, with four fathoms water at its entrance, flowing into the sea at Encounter Bay, and joining the eastern side of Lake Alexandrina. Mr. Sturt, the first discoverer of this river and lake, in 1833, appears to have only missed this outlet by keeping to the western side of the lake.

In the Antarctic Ocean, the discovery of two masses of land, in long. 47° E. and 67° W., in the year 1833, by Mr. Biscoe, to whom one of our gold medals has been awarded, was first made known in the pages of our Journal; as were also the narratives of two visits to the islands of the Pacific Ocean—viz., one by Captain Waldegrave in 1833, and that by Mr. Bennett in 1836.

As intimately connected with some of the higher departments of geographical science, I must not omit to mention how much we are indebted to the learned investigations of the late Baron William von Humboldt, the worthy rival of his distinguished brother, on lingual and ethnological affinities, particularly in reference to the Malay languages, used more or less over the Polynesian Islands, which were noticed in the seventh volume of our Journal. Admiral Krusenstern also has conferred an essential benefit on the nautical world, by his corrections of our charts of the Pacific Ocean, where numerous islands have been discovered, and many important points fixed by astronomical observations.

Having taken this cursory and very imperfect review of what has been done, is doing, and may be expected to be done, in promoting geographical knowledge—either by sending out expeditions, or by proffering the means of spreading the results of expeditions, under the auspices of the Society—it can scarcely be said that the Society has not, in part at least, fulfilled the promises held out when it was first formed, or that it has not honourably exerted itself in the task which it undertook. Nor can it be maintained that a pursuit which has so immediately for its object the extension of knowledge to ourselves, and the communication of the blessings which, as Europeans, we enjoy, to a large portion of the human race—now in a state, more or less, of comparative barbarism—is without its charms. The world, we may truly say, is all before us;—we have only to go a few steps from our own doors, and we shall never fail to find objects of research calculated to excite our curiosity, and reward our endeavours to benefit mankind. The calculating utilitarian, and the enthusiastic philanthropist,—the exact mathematician, and the practical statesman,—here meet upon the same ground. The pursuits of geography may not, perhaps, be so tangible or so seductive as some others—as, for example, those of the geologist; but whilst they have no prejudice to encounter, no jealousy of the subversion of supposed truths to fight against, no arbitrary systems to build up, or to dislodge, they contribute more immediately to enlarge the sphere of our observations;—they tend more directly to increase the happiness, and improve the moral welfare of our fellow-creatures;—they give us a deeper and more accurate insight into the history of man—of man, too, in all the diversified stages of civilization;—they teach us to be contented with our own lot, and to bless Providence for the blessings with which we are surrounded;—they excite all the best sensibilities of our nature, by bringing us into contact with savages who can scarcely count upon their fingers, or who with difficulty find wherewithal to supply the merest necessities of life from the weeds of the field, or from the inhabitants of the air, the forest, or the deep; we are gratified by the reflection, that, whilst we introduce our own better habits and better principles to some nations, we are weaning others, if such there be, from the horrors of cannibalism; and whilst we are pointing out to our merchants new points of the globe, to which they may steer, we are the pioneers to guide them in safety through the shoals, by which they were endangered on the shores which they have been in the habit of frequenting.

In an age, too, when the exact sciences are more than ever cultivated—when in all our pursuits, moral, political, statistical, and purely scientific, we are more than ever led to insist upon facts and the most rigid experiment, Geography lays in its claim with confidence to its full share of public attention and support. It is that science which, more than all others, is founded upon facts, and upon facts alone: it admits of no theory, no hypothesis, no analogy, no metaphysical deductions, on why or wherefore. Analogies may indeed be discovered between one continent and another—one system of mountains and another—one range of steppes and another—one line of coast and another—one great fluvial basin and another—one cluster of islands and another—one great continent and another; but none of those can be counted upon *à priori*. We must suppose nothing; we must prophesy nothing; nothing must be taken for granted. Every individual spot must be subjected to the test of local observation—must be measured upon the earth—must be measured in the sky, by night and by day: its relative and its absolute position must be fixed upon unerring data. And these are the principles which mainly distinguish the geographical pursuits of the present age from those of the critical and learned geographers of the last.

But, besides the actual fixing of places on the earth's surface, and the consequent progress towards perfection in the construction of our maps and charts—or, more strictly speaking, in aid of this the great end and aim of Geography—there are various collateral objects of research, of greater or less importance, to which the attention of this Society has been called, to which it has already been given in part, and on some of which we may hope that it will hereafter be more effectually bestowed, when our means are increased, and when the accommodations we can offer to our members shall be improved.

Amongst these succedanea to the improvement of geographical knowledge may be enumerated—First and foremost,

1. The increase of our Geographical Library.

2. The acquisition of Portulani, or the maritime surveys effected in the middle ages, and at the era of the revival of literature.

3. The completion of our collection of Maps, of this and all other countries: of those older maps, too, which, though no longer of use for practical purposes, are of the highest interest in tracing the history of geographical knowledge.

4. The possession of the most useful instruments for measuring distances and angles, and for making astronomical observations, with a view to point out to travellers those which they should take with them, and to instruct them, where it may be required, in the best mode of making use of them.

5. A room in the apartments of the Society where these instruments may be carefully lodged and kept open for public inspection by the members of the society.

6. A larger and more convenient apartment; which, though absolutely indispensable for the securing many of the purposes for which the Society was formed, the funds of the Society are as yet quite unequal to procure.

7. The collecting for preservation, or for collation, of original MS.

documents in various languages, relating to subjects of geographical inquiry; this, it has been said, has hitherto been too much neglected, and it cannot be doubted, that many valuable documents of this description have fallen the prey, like others of an analogous nature, to revolutions, civil tumults, or hostile invasions.

8. The introduction of a more definite and exact nomenclature in geography, which might be admitted, as far as the difference of language would allow, into the vocabulary of other nations, so that all might use an universal geographical language.

9. To see whether some approximation might not be made to more appropriate names for newly discovered lands, and their several divisions and subdivisions.

10. To examine whether it might not be possible to apply generally the decimal divisions of numbers, so as to supersede the specific denominations of measures now peculiar to each country.

11. The introduction into the mechanical process of constructing maps and charts, of some more intelligible and easy mode of denoting elevations, peculiarities of the surface of lands, the velocities and falls of rivers.

These and various other suggestions have been warmly urged upon the consideration of the Council for extending the sphere of the usefulness of the Society; but they are all more or less dependent on our attaining one or two other objects of a more homely description; namely, a very considerable increase to the number of our subscribing Members, or the acquisition of an apartment adequate to our wants: this last is only feasible by the accomplishment of the former, or by a grant from Government. One of the difficulties in the way of our obtaining sufficient accommodation, is the paramount necessity of our retaining a central situation in the metropolis—first, for the convenience of Members, and again, for the advantages of vicinity to the public offices, with which we are in constant, almost daily communication. The Council have long been alive to the deficiency of the rooms we now occupy; but we have as yet no prospect of improving ourselves in this respect without such a large outlay of money, both in premium and in annual rent, as to be entirely beyond our present means: the increase of our numbers is indeed progressive and highly satisfactory, particularly when we take into consideration the increase in the number of other Societies, every year giving birth to them, which have pursuits more or less analogous to our own; and thereby diverting the attention, and drawing largely on the pockets of the public: but, until this difficulty is somehow or other got over, and till we can ensure them more comfortable accommodation than we can at present, we cannot expect the Members of the Society, even those the most qualified and disposed, to give us the benefit of their time and labour.

With this inducement, we might fairly hope in a few years to effect some also of the other results which have been proposed to us by one of our most intelligent and active contributors, and thus farther to redeem the pledges offered in the name of the Society at its foundation; as, for example,—An alphabetical list of all rivers, as far as they are known, together with their tributaries, arranged in a tabular form, comprising the longitude and latitude and elevation of their sources, longitude and latitude of the confluence of their principal affluents, and of their embouchures, length of course, breadth, depth, rapidity of currents, length of navigable course, quantity of water furnished: also of lakes, their elevation above the sea, length, breadth, and surface. A synonymy of the names of places in alphabetical order, giving the position of the places, and the authorities for the different names by which they have been in former times, or are at present, designated. A list of desiderata, or things necessary to be known regarding the geography of any part of the world, of any country or division of a country. A similar list of desiderata for the hydrography of the earth, comprising tides, currents, soundings, temperature of the water, liability to periodical tempests, and regular winds. A critical notice or catalogue raisonné, as far as possible, of all maps and charts extant, specifying the degree of reliance that may be placed upon them, the defects observable, and any details of execution, such as comprehensive arbitrary signs, &c., which it would be advisable to imitate; the scale, the place from which the first meridian is counted, &c. A similar catalogue of Geographical works, from the most ancient down to the present time, indicating, as far as possible, in what public libraries or private collections the rarer works may be consulted.

These and other similar objects are certainly worthy the attention of our Members generally, and more especially of the Council; and whenever our funds shall be materially improved, or the other facilities to which I have alluded shall be placed at our disposal, I have no doubt that some of them, if not all, may be effected: but in the mean time I cordially avail myself of this opportunity to invite, in the name of the Council, any contributions towards them, which our Members may have it in their power to make, assuring them that their labours will be duly appreciated, and the utmost exertions will be made, consistent with our means, to make them available for the extension of the pursuits in which we are engaged.

In the mean time we shall steadily pursue the way in which we have begun; and we shall think we have not laboured in vain, as long as we can accumulate and lay before the public, what is, after all, the end of geographical study, *precise geographical data.*

I cannot conclude this Address without inviting your attention to the liberal and unvarying encouragement which the Royal Geographical Society have received, since their first institution, from Her Majesty's Government. Some of the most valuable papers read at our Meetings and published in our Journals have been communicated from the offices of the Secretaries of State for Foreign Affairs and for the Colonies, and from the Admiralty; frequent mention has been made in our Annual Reports of the large pecuniary assistance given by the Government to the two Expeditions sent into Southern Africa and to British Guayana, under Captain Alexander and Mr. Schomburgk, and I have no doubt that the same liberal confidence will continue to be extended to us, and that the Government and the Society will respectively promote the same harmonious intercourse, equally conducive to the benefit of the public and the extension of geographical knowledge. We are bound also to make similar acknowledgments for the assistance we have received from the Directors of the East India Company, to whom we are indebted for access to the papers of Messrs. Moorcroft and Trebeck, the first accounts of whose travels in the Himálaya and the adjoining countries appeared in our Journal, and who have lately granted to us unlimited access to all their geographical documents, with liberty to give them to the public in our Journal: we with pleasure also acknowledge the uniform co-operation we have at all times received, when occasion offered, from the Governor and Deputy Governor of the Hudson's Bay Company. This Company has indeed so recently come before us, as the projectors, and, through their own immediate servants, as the conductors of one of the most successful expeditions for exploring the Arctic regions, that I cannot better relieve you from the kind attention with which you have received my Address, than by the expression of our earnest hopes, that, under this auspicious commencement, we may continue to be fortunate fellow-labourers in the same vineyard, that the farther attempts which this Company are making to complete the survey of the North American coast may be equally crowned with success, and that the solution of this great problem may be speedily added to the many triumphs over difficulties and dangers, which form the wreath of glory around the brow of the British enterprise.

ROYAL GEOGRAPHICAL SOCIETY.

Patron.

THE QUEEN.

Vice-Patron.

His Royal Highness the Duke of Sussex.

COUNCIL, ELECTED MAY 21, 1838.

President.

W. R. Hamilton, Esq., F.R.S.

Vice-Presidents.

Sir John Barrow, Bart., F.R.S.
The Hon. Mountstuart Elphinstone
G. B. Greenough, Esq., F.R.S.
Sir Woodbine Parish, F.R.S.

Treasurer.

John Biddulph, Esq.

Trustees.

Sir George T. Staunton, Bart., F.R.S.
Francis Baily, Esq., F.R.S.
John Biddulph, Esq.

Secretaries.

Captain Washington, R.N.
Rev. G. C. Renouard, Foreign and Hon.

Council.

Captain Back, R.N.
John Backhouse, Esq.
Captain Beaufort, R.N., F.R.S.
Captain Bowles, R.N.
William Brockedon, Esq., F.R.S.
Robert Brown, Esq., F.R.S.
Lieut. Col. Chesney, R.A., F.R.S.
Colonel Sir A. Dickson, R.A., K.C.B.
Charles Enderby, Esq.
Colonel Fanshawe, R.E.
Colonel Fox.
J. Baillie Fraser, Esq.
Bartholomew Frere, Esq.
W. I. Hamilton, Esq.
Colonel Jackson.
Captain T. B. Jervis, E. I. C. Eng., F.R.S
Lieut.-Colonel Leake, F.R.S.
George Long, Esq.
James Meek, Esq.
James Morier, Esq., F.R.S.
R. I. Murchison, Esq., F.R.S.

FOREIGN HONORARY MEMBERS

His Imperial Highness the Archduke John of Austria
Chevalier Balbi . . Vienna
C. F. Beautemps-Beaupré, Mem. Inst. . . . . . Paris
Leopold von Buch, For. M.R.S. L.S. and G.S., Mem. Acad. Berl. . Berlin
Colonel Campana . . . Milan
Chevalier Cassalegno . Turin
General Clarke . United States
Captain Dumont D'Urville . Paris
Captain Duperrey . . Paris
C. G. Ehrenberg, For. M.R. and L.S., Mem. Acad. Berl. . . Berlin
Carl Falkenstein, Corr. Mem. Acad. Berl. . . . . Dresden
Captain L. De Freycinet, Mem. Inst. Fr. Paris
Professor Hansteen . Christiania
Baron Alex. von Humboldt, For. M.R.S., L.S. and G.S., Mem. Inst. Fr., Mem. Acad. Berl., &c. . . Berlin
Padre G. Inghirami . Florence
E. F. Jomard, Mem. Inst. France, Corr. Mem. Acad. Berl. . . Paris
Admiral von Krusenstern, Corr. Mem. Acad. Berl. . St. Petersburg
Count Alexandre L. J. Laborde, Mem. Inst. Fr. . . . Paris
Admiral F. B. Lutke . St. Petersburg
A. J. Letronne, Mem. Inst. Fr. . Paris
Dr. Charles von Martius, For. M.L.S. Corr. Inst. Fr. and Acad. Berl. Munich
Conrad Mannert . . Munich
Baron Meyendorf . St. Petersburg
Professor Carl Ritter, For. M.R.A.S. Mem. Acad. Berl. . . Berlin
Dr. E. Ruppell, For. M.L.S. Frankfort
G. S. Schoolcraft . United States
Professor J. F. Schouw . Copenhagen
Colonel Frederick Visconti . Naples
Dr. George Wahlenberg, For. M.L.S., Corr. Mem. Acad. Berl. . Upsala
Baron C. A. Walckenaer, Mem. Inst. Fr. . . . . . Paris
Captain Zahrtman . Copenhagen
Augustus Zeune . . Berlin

1835.

Adm. de Hamelin, Chef du Dépôt de la Marine . . . . Paris

1836.

Count Gråberg af Hemsö, Corr. Inst. Fr., For. Mem. R.A.S. . Florence
Baron Hugel . . . Vienna

1837.

Professor Heinrich Berghaus . Berlin
Colonel Forsell . . Stockholm
Professor Finn Magnussen Copenhagen
General Pelet, Chef du Dépôt de la Guerre . . . . Paris
M. Philip Vandermaelen . Brussels

CORRESPONDING MEMBERS.

Col. Don José Arenales Buenos Ayres
Don José Maria Cardano . Madrid
M. Daussy . . . . . Paris
Dr. Duponceau . . Philadelphia
Col. Don Juan Galindo Costa-rica
William Hilhouse, Esq. . Demerara
T. A. Kjar, Esq. . St. Thomas's
M. Kupffer, Mem. Acad. Scien. St. Petersburg
R. H. Schomburgk, Esq. . Demerara
Mr. Tanner . . . Philadelphia
Dr. Woerl . . Freiburg
Don José Urcullu . . Oporto
Mr. Worcester . United States

1836.

M. D'Avezac . . . Paris
Counsr. J. J. da Costa de Macedo, Sec. Roy. Acad. Scien. . . Lisbon
Gen. Don Juan Orbegoso . Mexico

1837.

Prof. Adolph Erman . . Berlin
Don M. F. de Navarrete . Madrid
Baron Wrangel . St. Petersburg
Professor Paul Chaix . Geneva

1838.

Colonel Lapie . . Paris
Don Pedro de Angelis Buenos Ayres
Colonel Oberreit . . Berlin
Colonel Skribaneck . . Vienna

PAPERS READ

BEFORE THE

ROYAL GEOGRAPHICAL SOCIETY.

I.—*Report of an Expedition of Discovery, through the countries of the Great Namáquas, Boschmans, and the Hill Dámaras, in South Africa.* By Captain J. E. ALEXANDER, 42nd Royal Highlanders, K.L.S. Read 22nd January, 1838.

IT is remarkable that, during the three centuries and a half which have elapsed since the celebrated Portuguese navigator, Bartolomeo Diaz, first doubled the "Stormy Cape," the progress of discovery should have advanced so slowly, that up to this day the whole of the western region of Southern Africa, to the north of the Orange River, has hitherto remained a blank on our maps;—the Great Fish River, supposed to extend upwards of 300 hundred miles in length from north to south, and said to receive, both from the eastward and from the westward, more than twenty tributaries, is there only indicated by a dotted line;—of the range and height of its mountains and elevated plains no trace exists;—and of its geological structure and general features we are utterly ignorant. Gordon in 1777, Paterson in 1778, Le Vailant in 1781, and Thompson in 1827, may have reached, *in this direction,* the southern bank of the Gariep; but neither in the last or present century is it recorded that any European traveller has crossed, within 400 miles of its mouth, to the northern bank of the Orange River:*—but where the spirit of enterprise has not yet reached, a nobler spirit has directed the steps of other wanderers; and, for nearly a quarter of a century, a few missionaries have from time to time endeavoured to spread the truths of the gospel in the district of the Orange River, and one, the Rev. Mr. Schmelin, performed a journey of some extent, several years ago, through a part of Great Namáqua land.

To penetrate farther to the north, then, in this direction, and to endeavour to become acquainted with the Dámaras, a nation

---

* Barrow in 1797, Truter and Somerville in 1801, Lichtenstein in 1805, Burchell in 1809, Campbell in 1813, Thompson in 1827, Hume in 1834, with many others, and lastly, Dr. A. Smith in 1835, have crossed the Orange River, but that was between 23° and 24° east longitude, near its mid-course, or about 400 miles from its mouth. W. Van Reenen also, in 1791, ascended the Great Fish River, but his account is not published.

inhabiting between the 21st and 24th parallels, and only known to us by report, seemed to be now the chief object of geographical research, since the recent journey of Dr. A. Smith had rendered it needless to proceed with the Delagoa expedition. Accordingly, after mature deliberation, with the authority of His Excellency Sir Benjamin D'Urban, and by the advice of those best qualified to give an opinion in the colony, I determined to explore the country to the north of the Orange River, west coast, as the best means of promoting the great object for which I had left England, and carrying out the views of Her Majesty's Government, and of the Geographical Society.

Having made the necessary preparations, as detailed in former letters already published, I quitted Cape Town on the 10th September, 1836, and travelled to the northward towards Clanwilliam, with considerable interruption from rain. My party consisted of seven attendants—four Europeans, one Bengalee, and two stout Bastaards (as the Cape term is for any dark man, with an admixture, however small, of European blood); two horses, a waggon, and span of fourteen oxen; and half a dozen dogs, which are the best sentries in the wilds:—my waggon was freighted with arms and ammunition, besides beads, cutlery, and, above all, tobacco, to purchase provisions in the interior, and to make my way among the natives.

The country, at this season of the year, was exceedingly beautiful—wild-flowers, seen only in conservatories in England, appeared on every side; there was nothing arid or barren in the landscape, but a verdant carpet, variegated with gay colours, was spread before us; in the distance, and on the right, were the lofty peaks of the primitive range of the mountains of Drakenstein, averaging in height more than 2000 feet above the sea.

The Dutch farmers seemed to be generally amusing themselves in following stein-boks, with horse and hound, and shooting them with their long roërs or rifles.

Our travelling dress consisted of grey broad-brimmed hats, covered with ostrich feathers on account of the heat; grey or blue smock-frocks, over cotton shirts; leather trousers, and buskins of untanned hides: we had neither stocks, socks, nor gloves; and the only preparation for going to sleep in our sheep-skins, during the year we were out, was taking off our shoes, shaking the sand out of them, and putting them on again, for fear of hungry dogs or hyenas eating them.

On our right we next saw the district of the Twenty-four Rivers, described in such glowing colours by Vailant, and then crossed the Berg River by a punt. After toiling through the heavy sands under the Picket Berg (which it is supposed would be a favourable situation for the growth of coffee), I left the waggon at Uitkomst, the residence of a worthy Dutch farmer of the name of Vanzyl, and rode about twenty miles to the east to Clanwilliam.

From this last village, seated in a sort of basin, about half a mile from the eastern bank of the Olifant River, I made two excursions to the Cedar Mountains, ten miles farther to the east. They are a fine primitive range, the peaks of which rise from 1600 to 5000 feet above the sea, and have a very picturesque outline. The heights given to me, and said to have been measured by Baron von Wurmb, were, Sneeuw-Berg, 5000 feet; Groene-Berg, 4860; Rondeberg, 2990, &c.

The principal rock of the higher parts of the Cedar Mountains appears to be an ash-coloured quartzose sandstone; the secondary range contains many marine petrifactions, shells and fish, at a height of 2000 feet above the sea. There are also black, red, and striped jaspers, hornblende, garnets imbedded in argillaceous schist, numerous ores of iron, much quartz, and a great variety of beautiful agates.

The valleys between the hills are rich in a dark-coloured vegetable mould, which is exceedingly productive. Corn, tobacco, and some wine are here produced; whilst there is a constant and abundant supply of water, which is more or less chalybeate.

I was particularly interested about the cedar-trees, "the glories of Lebanon," which formerly covered this beautiful range of mountains, and which in part still do so. They occupy ravines in the higher parts of the range, perhaps as high as 3000 feet, and one cut down in 1836 measured thirty-six feet in girth, whilst 1000 feet of plank were sawn out of its giant arms.

No care has hitherto been taken of these valuable trees; the farmers, the Bastaards, and Hottentots, living in the neighbourhood, cut them down without leave or license, and burn the grass, to improve the pasture, by which many old trees, and thousands of young plants, are annually consumed.

As the cedar-trees might, if preserved, become of great advantage to the colony generally, I represented, in the proper quarter, the manner in which they are constantly and wantonly destroyed; and it is to be hoped that means will be taken for preventing the future waste of that most valuable and imperishable timber, with which the temple of Solomon was built.

There are many Boschman caves in the Cedar Mountains: they are generally at some height, varying from 300 to 1000 feet above the valley, and are not of any great depth, say 30 or 40 feet, but they are very interesting, as containing the drawings in red ochre of a wild people who have for some years disappeared from this locality. In one cave there is a spirited representation of a combat with bows and arrows; in another, a flock of large-tailed sheep and lambs are accurately delineated: I was really surprised at the fidelity of the outlines.

From Clanwilliam I passed on to a place called Heerelodgement, where, under a huge overhanging rock, are many names of old travellers and hunters, from 1714 downwards, among others, F. VAILANT, 1783, is carved. Swimming my waggon over the Olifant River (which once in four or five years overflows its banks, when the increase is a hundred fold), I passed by the Rhenish Mission Institution of Ebenezer, and with two of the worthy brethren rode about twenty miles in a W.S.W. direction, to the mouth of the Olifant River.

Rocks abound at its mouth, and it would be difficult to make it accessible for vessels without a great deal of cutting. Plans are continually brought forward to enter the barred mouths of the South African rivers, but if pains were taken to improve the landing at Table Bay and Algoa Bay, by means of good piers, that would be a sufficient improvement for some years to come.

Continuing northwards, with a range of mountains on our right, and passing onward towards the Groene River, we found the country less fertile than before, covered with shrubby plants, and with much brackish-water, whilst the Boors, from July to October, inclusive, live in mat-houses for the purpose of changing their pasture-grounds.

On the 10th of October, I ascended the Kamies or Lion Mountain, a continuation of the Cedar Mountain range, and took up my residence for a time at the Wesleyan Mission Station of Lily Fountain, containing about 800 persons, to make arrangements, by despatching messages to native chiefs about the Great River, for my further progress.

A remnant of the nation of Little Namáquas has been saved by missionary exertion, and settled on the table-land and in the valleys of the Kamies Mountain, the highest peak of which, about five miles south of Lily Fountain, rises about 5000 feet above the sea, which would seem to be the highest point of the range; to the eastward the view is bounded by near hills, but to the west it extends over gradually-decreasing ranges to the Atlantic Ocean. Under the old Dutch rule the free natives were almost driven entirely out of the colony, with the exception of those who remained as slaves to severe task-masters.

Having been requested by the governor of the Cape to inspect the mouth of the Orange River, I left my people to collect objects of natural history on the Kamies Mountain, and to prepare packs for bullocks when we might be forced to abandon the waggon, and rode in a north-western direction about fifty miles, to the house of Mr. Schmelin at the London Mission Station of Komákas (signifying red clay). Descending from the Kamiesberg, we rode over a grassy plain; at 25 miles the country became hilly, and at Keerom, or 'turnabout,' is a remarkable smooth rounded hill of granite, about 300 feet high, round which the road wound and gradually inclined to the northward to Komákas; where are five fountains, and five or six huts, containing about thirty Namáquas, but at times as many as twenty huts, and upwards of one hundred people, are collected together here. On to the north-west the ground rises to 1000 feet above the sea. Here I ascertained the existence of a new bay called Rooé Wall, or Red Wall Bay, forty miles south of the Zwarlintjies River, and at the mouth of the Spook River. There is good anchorage for schooners and small craft in this bay, which it is to be hoped will be turned to account to open a market for this part of the colony.

Quitting Komákas, we continued to the northward over a sandy plain covered with shrubs, having a range of mountains on our right, and hills to the left: I saw many ostriches and stein-boks on the road, with the prints of zebras. The worthy old German missionary Mr. Schmelin, and two Rhenish missionaries, accompanied me to ascertain if the mouth of the Orange River was fit for a new station.

The labours of the missionaries in this part of South Africa have been attended with much success, and it is to be hoped that they will be encouraged to extend their field of usefulness further into the benighted interior, so that wars between the tribes may cease, and that trade, civilization, and Christianity may be promoted. At 25 miles crossed the dry bed of the Kowsie River about 20 miles from its mouth, and travelled to the northward over sandy plains with bushes; to the eastward, a range of mountains about 1000 feet high; and to the west, undulating hills of 300 feet. At 20 miles from the Kowsie, reached Ukribip, where are some Namáqua huts; a few miles beyond is Kama, a village of ten huts; the range of the Kamiesberg appears to extend in a north-north-west direction, about 20 miles on our right, reaching to the banks of, and perhaps 30 miles beyond, the Orange River. After 60 miles of travelling over the same description of country, we reached the village of Aris, on the banks of the river, at 25 miles from its outlet into the Atlantic.

Where I first saw the Great or Orange River, it was 500 yards across, and the banks were lined with black ebony, willows, mimosas, &c. The river was at this time shallow enough to be easily forded, but we saw marks on the trees and banks indicating an occasional rise to a very great height.

The Little Namáquas, and some Bastaards, living here, have flocks and herds, but subsist principally by the chase, and by the produce of a seal island, which I found to lie between the mouths of the Orange and Kowsie Rivers; the latter is the boundary of the colony. The natives are clothed in sheep-skins, some of which are made into trousers and petticoats; and they would willingly exchange their stock for cloth and cutlery. They sell valuable seal-skins for five shillings each.

The land is flat about the mouth of the Orange River, and on the north side is good pasture. The breadth of the entrance at the beginning of November was about 170 yards, whilst inside was a lake about four miles across, and abounding in wild fowl. The air was darkened and filled with the cries of wild geese and ducks, flamingoes, pelicans, gulls, &c. Here is most excellent shooting-ground for the sportsman, and 'harder' and 'springer' fish abound for the seine.

I saw no rocks at the mouth of the Orange River; there is probably, however, a shoal of sand outside, but with care it seems that the mouth of the river could be entered by a schooner. I never heard that any soundings had been taken at the mouth of the Orange; indeed, the increasing wants of the colony demand a far more detailed survey of the south, south-eastern, and south-western coast of Africa, than any that exists at present. About 30 miles east-south-east of Aris, at a distance of four days' journey, with a waggon, from the mouth of the Orange River, and near the south bank, I made, by means of a Bastaard, the discovery of a great mass of copper ore, of so rich a quality, that, by analysis of Sir John Herschel, at the Cape, to whom a specimen was sent, the ore yielded 65 per cent. of metal.*

This copper is quite accessible, and it might be either smelted on the spot with Orange River wood, of which there is plenty, or it might be floated down to the mouth on rafts, which rafts might then be sawn up and sent to the Cape, where wood is always in great demand for building.

Honey is found in such quantities about the Orange River, particularly in the months of December and January, that a Namáqua, who had a waggon, assured me, that in two days, whilst on a honey hunt, he loaded his waggon with skin-bags of honey above the side planks; a trade in bees'-wax therefore might also be carried on here; whilst on the coast there are extensive beds of muscles for shell lime. The climate is very healthy for Europeans.

I found iron also not far off, but on the north side of the river, and as the site of these valuable ores is far beyond the colony, no expense would be incurred in purchasing a right to work them. The natives also would be pleased to see strangers among them,

* Another specimen from the same place, assayed in London, yielded 28 per cent.

if they were kept under proper control. Of course careful surveys ought first to be made of the region of the ores, before any speculation is entered into, to avoid unnecessary risk or loss.

After a few days we retraced our steps to the Kamiesberg, and on crossing the Kowsie River, on the 6th of November, the thermometer was 103° Fahr. in the shade. This was now the begin-ging of the hot season.

I found at Lily Fountain Mr. Edwards, the Wesleyan missionary, just returned from the Cape, from whom I received every civility and assistance.

Nov. 16.—Having completed our arrangements we left Lily Fountain on our journey to the north, escorted, by the Governor's order, by a Field Cornet and twelve mounted Boors, armed with long guns, as far as the warm bath, to produce a proper effect on the Namáquas, and to show them that the expedition was a government one. None of the Boors had been at the Orange River before; and, what with the fear of lions, Boschmans, and their old runaway slaves, they did not at all relish the expedition; and I believe, from the crying we saw at leave-taking, that their wives and children never expected to see them again.

Descending from the elevated station of Lily Fountain into the plain, we crossed the dry bed of a tributary to the river Kowsie, and continued to the northward; hills about 300 feet high on each side of the road; no cultivation, and hardly an inhabitant to be seen. About twenty miles we halted at *Silver Fountain*, where are a few huts; ten miles further north reached the Copper Berg, a range of granitic mountains, without wood, about 1000 feet in height, two miles to the east of the road. Nearly ten miles to the south-west of this spot is a conspicuous mountain, called the *Vogel Klip*, or Bird Stone, about 1200 feet in height. Proceeding to the northward we passed through a stony valley, with bushes of mimosa; and about twenty miles farther reached Bezondermeid (Strange woman), a London Missionary station, under the charge of the veteran Mr. Wimmer; here are about thirty Namáquas. Beyond this, to the N.N.E., black hills, about 500 feet high, are scattered over the plain on the right, till we reach a range of sand-hills, within a few miles of the Orange River; here, turning due east, we travelled along their southern foot for about fifteen miles, to *Kárahas*, or Koran-ford.

The country, on both sides of the Orange River, is exceedingly barren and dreary for many miles from the river. Bare and black hills are surrounded by plains of yellow sand, with very scanty vegetation. One of my horses being totally unable to walk, from sheer starvation, I bought another from a Dutchman, which was better able to live on bushes.

Nov. 25.—Arrived at the Great River, we found Mr. Jackson, a Wesleyan missionary, with the chief Abram, and sixteen swimmers, on the opposite bank ready to assist us over; but we forded the Orange River, which was only waist-deep, without much difficulty, and I looked with wonder on the wild and grotesque forms of the granite hills which enclosed its banks. The valley through which it here flows, in a due west direction, is about half a mile broad, the bed of the river from 400 to 500 yards; its banks are well wooded with willow, mimosa, &c.; the hills, both north and south, rise upwards of 500 feet above the bed of the stream.

Travelling north-east for two days, over the same description of country, I arrived, on the 27th of November, at Nabís, or the Warm Bath, the most northerly mission station, and 450 miles from Cape Town. Here I intended to halt till the thunder-rains of the beginning of the year should enable me to advance.

The station of Nabís, or Nisbett's Bath (as it is now called), is among rocks, and on both banks of the 'Hoom River, a branch of the Orange. A great plain spreads around, occasionally visited by lions, whilst spring-boks, ostriches, and zebras, are not unfrequently met with. Black conical hills of 200 and 300 feet elevation are seen here and there on the plain.

The tribe of Great Namáquas, who reside principally at the Bath, live in fifty or sixty circular huts, composed of arched boughs covered with rush mats; they sleep on skins; these and the wooden cylindrical vessels for their milk are the chief articles of furniture they possess.

The men and women of the Great Namáquas are taller than the Namáquas we had formerly seen; but they have the same high cheek-bones, small eyes and noses, and yellow (Malay) complexions. Men and women generally wear skin dresses, some mantles or karosses, others cut their clothes after the European fashion. The tribe of Abram (the Bondelzwart) is too remote from the colony, and the tribe bears too indifferent a character as yet, to be visited by traders.

One of the first things I did was to clear out the sand and stones from the Warm Bath, which is generally at 103° of Fahr., with a stream six inches broad by one and a half deep. Myself and people set the example of bathing, which was soon so perseveringly followed by the Namáquas, that it was only at night we could get near the bath, at the risk of meeting with a lion there.

I employed my people in shooting, in preparing the packs, and in training pack bullocks, which were exceedingly difficult to be procured ready trained. I bought as many sheep, too, as I could, for beads and handkerchiefs, as a stand by, in case of our being occasionally without game.

What facilities would not camels afford to travellers in this land!

—patient, enduring, eating any shrubby plant, and carrying the load of three bullocks—by means of camels the Equator might be reached without great difficulty, and yet no one at the Cape has had the enterprise to introduce a few from Bombay even by way of experiment.

I made two excursions from the Bath—first to Africaner's kraal, about 80 miles to the eastward: our road lay over the same barren plains in a south-east direction for about 40 miles, during which we crossed one small stream, the Kururú, or Noisy, flowing towards the south; at 50 miles we reached Naros, within five miles of the bank of the Orange River, and then turned north-east to the village of Africaner, the well-known robber-chief. We found a few of his descendants, altogether about 30 persons, living beside a fruitful garden, in which tobacco, melons, and calabashes, were raised, and to which the water of a tolerably abundant spring was led out. We returned by a more direct route in two days and a half to Nabís. The other excursion was down the 'Hoom River, to the range of hills called Twanos, about 500 feet high, 12 miles in a south-south-east direction from the Bath.

In the beginning of 1837, the heat at the Bath was very oppressive, daily it was 100, and sometimes 110, in the shade. We saw rain-clouds occasionally in the distance, but none came to us. The stream of water at the Bath, too, became more feeble, the grass turned black with the heat, and, tired of waiting for rain, I determined to risk a move, and accordingly, with a lightened waggon, some pack-oxen, and a flock of sheep, I journeyed northward.

Jan. 18, 1837.—Abram and sixteen of his people, armed with guns, reluctantly accompanied me, and ascending along the banks of the 'Hoom River, for six days, we passed near the spot where Trelfall the missionary was murdered a few years before, whilst attempting to travel towards the Dámaras with only two men. At Kanús we left the waggon on the banks of the 'Hoom, with half my party, and on the 24th proceeded to the north-east, for about 40 miles, into the recesses of the 'Karas (or sharp) Mountains, to endeavour to recover by negotiation some cattle and sheep, of which my old interpreter, Chaubep, had been deprived by a robber-chief. On the evening of the second day we reached the banks of the Keikap, or Witch River, which flows to the south-east, and continued to the northward along its western bank, crossing two of its small tributaries, to Henrick's Place. The 'Karas Mountains vary in height, the most elevated reaching 3000 feet above the plain.

We found the Rob Roy of these wilds in a mountain glen: at first he made a show of resistance to us, by drawing out about 30 musketeers, but eventually he was compelled to surrender the cattle; he threatened the waggon, too, with an attack, but surrounding it with an abbattis of bushes we defied him.

We returned from the Karas to Kanús, and in crossing the plain saw many spring-boks, zebras, a cameleopard, and the *spoor* of a lion. On reaching the camp I dismissed Abram and his people, whom I found to be great cowards.

Jan. 29.—We continued our journey to the north-north-west, quitting the banks of the 'Hoom, and travelling over grassy plains; on our right were mountains about 15 miles distant. At 26 miles crossed three streamlets flowing to the south-west; 10 miles beyond passed through a *poort*, or pass, between hills 500 feet high, and crossing a grassy plain, with mimosa shrubs, reached the banks of the Kaap, flowing to the north-west; thence across undulating stony plains, 40 miles, brought us to Nanebís on the banks of the Kamop, or Lion River, flowing west to the Fish River.

Feb. 22.—Left Nanebís, and inclining to the north-west for twenty miles, we came to the 'Oup, or Fish River, flowing to the southward, through sandy banks; stream about 80 yards broad; mimosa, *dubbee*, and ebony-trees scattered along its banks, and crossed it on the afternoon of the 24th.

Feb. 26.—After much trouble with our guides, we continued our journey to the westward, but finding no water we turned more to the west-south-west, where a line of trees indicated a stream, and at 15 miles crossed the dry bed of the river Nukaníp (Black bulb), and found bad water by digging, and suffered much for want of it. Continuing our journey to the north-west, across plains of Koran (bustards), we arrived at the deserted mission station of Bethany on the 6th March. Here, on a rising ground, are the roofless remains of a church and some houses, with a fine spring of water; it is now tenanted by 50 Namáquas. Kúisip, a Namáqua chief, and three men, now offered to accompany me to the sea, for presents of muskets, ammunition, &c., which offer I was obliged to accept, although they brought with them between thirty and forty hungry followers, men who could devour 10 lbs. of wild flesh per day. Most of them wore the kaross or skin-mantle, also part of a jackal's skin in front attached to a waist-belt, whilst behind dangled a square-cut piece of leather; a few of this escort had guns, the rest carried bows and arrows, javelins, and clubs.

Quitting Bethany we travelled to the north-north-west 20 miles, when we crossed the Koan-quip River flowing south, and continuing on the same course, leaving the Tamuhap, or Dropping Hills, above 400 feet above the plain, on the left, we passed over fine grassy plains, with mimosa bushes, and saw herds of zebras, &c.

March 13.—A descent to Tuais, or Muddy Spring, on the bank of the Gnuanuip, a tributary of the Koan-quip, to the east of which extend the range of Unuma, or Bulb Mountains, in a north and south direction for upwards of 50 miles.

March 18.—Again crossed the Koan-quip, coming from the north-west; noticed here a conspicuous table-mountain in the Unuma range, bearing north-east, which I called D'Urban Mount, rising about 2000 feet above the plain. Proceeding northward, we crossed several beds of streams, which appeared to flow in a south-easterly direction, towards the 'Oup, or Fish River, the principal drain of this part of Africa. In the 'Oup there is always abundance of water; and fish, chiefly of the *silurus* tribe, in the deep pools.

We had already experienced many of the inconveniences attending African travelling — such as the bullocks straying for hours, and sometimes for a day or two, in search of grass and water—for at night we could not do otherwise than drive them as near our bivouac as possible, and see them lie down, before we betook ourselves to our sheep-skin coverings and stony or sandy bed.

We were also occasionally much distressed for water, had long and painful searches after it—dug with our hands, and with a spade and pickaxe, many times in vain, at the site of old pools; at other times we smelt the fetid water ten yards off, the only supply for ourselves and for the cattle. It surprised me often that we did not get seriously ill from the thick and polluted water we were forced to put up with, occasionally more like chocolate than the pure element.

There being of course no roads, the waggon gave us occasionally much annoyance and occasioned great delay; it would at one time stick fast in heavy sand, whilst half-a-dozen hands lifted the wheels, and the driver plied, with great shouting, his immense bamboo whip—then a precipice was to be descended, the stones were removed and the waggon carefully conducted with hide ropes down the steep. It is difficult to conceive greater trials of patience than what occur on a South African journey, through a new country.

March 21.—We reached Aanabip, a chain of ponds; to the west was the range of the Tarup Hills; we now entered on the great plain of Kei-kaap, a dreary barren flat, which we crossed in a north direction for a few days' journey; here we saw the mirage two or three times, whilst we were in much distress and trouble for water. We were now in the country of the Boschmans.

One day Kúisip, the chief, after looking at the ground, set off at a gallop on his ox, followed by two or three of his men; I followed, expecting to fall in with some large game, as we had just before surrounded and shot some zebras. We pushed on rapidly for a mile and a half, and then came to a hollow place, where, among the shrubs, were three hovels, composed of stakes and bushes; by one of these sat an old man—dark, dried up, and covered in part with a kaross of spring-bok skin—his bow and arrows lay beside him; he was engaged cooking some leaves in a small earthen pot, and two or three naked children crawled about on the ground beside him. This was the first family of Boschmans we saw.

Subsequent to this we saw many Boschmans, and as we were now (on the 21st of March) in the country of rhinoceroses, we saw hunters following on the spoor or track of these huge animals, to destroy them with their javelins, which they do from behind circular enclosures of stone, breast high, darting their light weapons into the tough hide as the rhinoceros passes, the monster having been previously scared from his lair by stones thrown at him.

I saw the manner the Boschmans frighten the lion from its dead prey, by shouting and springing about, in order that the "children of the Desert" may have a share; their manner of making fire with two sticks; of creeping on game; and remarked that they had no scruple to give up their women to the Namáquas for a piece of meat or tobacco.

March 24.—Still traversing the Keikaap, its surface covered with thorny bushes. In the east, about two miles distant, were flat-topped mountains, rising more than 1500 feet above the plain, not improbably the step of the table-land, or terrace, which we afterwards crossed on our homeward route; to the west were hills between us and the sea. The three or four streams which cross the plain in a south-easterly direction are bordered, as usual, with thorn-trees.

March 30.—An extensive plain lay before us, covered with grass and scattered bushes; bounding the view to the north was a range of flat-topped mountains, of 2000 feet elevation; their sides were bare and scarped, and broken in one place by a deep ravine; this was the Kópum'naas, or Bull's Mouth pass. Proceeding from it, and winding through the plain in a south-westerly direction, was a line of high trees, marking the course of the Chúntop River, the favourite resort of large game. The landscape was very beautiful, and we were highly excited with the prospect of much sport. Twelve rhinoceroses were seen in one day; two of which charged me, but I avoided them among bushes, and got a shot at the last as he passed me. We had plenty of wild flesh at the Kópum'naas, and all night long the Namáquas "stamped the marrow-bones."

April 3.—The Bull's Mouth pass winds through mountains of

between 2000 and 3000 feet elevation, for a distance of forty miles nearly. The first part of the valley of the Chúntop River was rather narrow, but beautiful with the red and grey colours of the rocks which hung above it, and the bottom covered with the fantastic forms of various species of mimosa; under these the tracks of the huge rhinoceros was everywhere seen under his favourite trees. The valley opened out and the mountains receded on each side of us, leaving a plain between them of a few miles in breadth. We travelled through the pass first north, then west, and then north again, descending the course of the Chúntop till we reached Ababíes, or Calabash Kraal.

With considerable labour we cleared a road through the Bull's Mouth, whilst baboons cried angrily round us, and for the first time a waggon was dragged through it, with difficulty. Beyond the pass we came to a region of fig-trees, each of which was about 60 feet high, and covered with ripe fruit. A dozen Boschmans (men, women, and children) were in a hole in the bank of the Chúntop River, and lived on the fruit of one of these fine trees.

April 6.—We next reached the dreary Desert of Tans, which is an immense plain of grey sand, undiversified by a single bush or shrub, and with only single blades of white and sun-burnt grass waving here and there; heat, and glare, and an awful silence were round us in this horrid waste. In the far distance, to the north-east, full forty miles, were black mountains; amongst which, most conspicuous, was a huge flat-topped eminence, rising 4000 feet above the plain, (from which I gave the Desert its name,) called the Tans or Screen Mountain, for it shuts in all the neighbouring heights. On our left were ridges of sand.

Here we nearly all perished for want of water; the thermometer was at 100°: we came to no water for two days; many of the people threw themselves down on the sand, and desired to be buried, and others were forced to resort to the last means to endeavour to allay their raging thirst. For three days the poor oxen had had nothing to drink, and could not bellow, though they tried to do so. I thought, though I did not say so, that it was impossible we could escape; but at length, on the 8th of April, with great suffering, we reached the bed of the Kúisip or Root River, flowing to the west, but with the loss of both my horses, some oxen, sheep, and dogs. Water was immediately carried back, in the stomachs of wild animals, to the more feeble of the party who had resigned themselves to their fate, and happily all the people were saved.

Where we first saw the Kúisip, its bed was between black and frowning cliffs; those on the north bank were part of a great and nameless mountain, whilst our precipitous descent on the south was, at least, 600 feet to the water, which lay in long pools in the dark channel, 100 yards wide.

As far as we could see up the river, which was no great distance from its sudden windings, impending and perpendicular sandy crags enclosed it, whilst down the river its bed is similarly enclosed, for a distance of several days' journey. My people named the spot where we first reached the Kúisip the Devil's Den, from its frightful appearance, though here we had been mercifully preserved, by a most seasonable supply of water, from the jaws of death.

Further down the Kúisip high trees began to appear in its bed, under which we found rich grass, on which our poor cattle luxuriated.

I had abandoned the waggon to the tender mercies of the Boschmans (who were prowling about us with their poisoned arrows), for the sand-hills on the south bank of the Kúisip opposed an impassable barrier, when one of the head men of the Namáquas offered to take it back towards the Orange River. He did so, with ten men and twenty-five oxen.

I went on with the remainder of the party down the Kúisip, with riding and pack-oxen, but we again suffered much for want of water, even in the bed of the river: on one occasion we saw no water, even with digging, for a distance of thirty miles. It was often difficult to rouse the people from their despondency; but finding one day the traces of men, we hunted after them, and caught two of a tribe of Great Namáquas, belonging to a large tribe on the river. These men, being well used by us, showed us water, and conducted us to the sea at Walvisch Bay, which we arrived at on the 19th of April, being the first Europeans who had reached it by land from the Cape of Good Hope.

Before our arrival at the coast, our provisions, too, had failed us; we could get no game. Hides (reserved for making shoes) were therefore roasted, beaten between two stones, and eaten. At the sea we dug with our hands in the sand for clams, and searched the shore for fish, and fortunately found stranded *cabaljaos* of fifty pounds weight, which we ate with great relish.

Walvisch Bay extends along-shore, as it were, for twenty miles. A long spit of sand, terminating in Pelican Point, in lat. 22° 52½′ forms its south-western shore, and offers an effectual barrier to the roll of the South Atlantic. Vast flocks of pelicans, flamingos, geese, gulls, &c., frequent this sheltered bay.

To the east of the Bay there are long ridges of sand, covered here and there with dark bushes; behind these ridges, and in the valleys, brackish water is everywhere to be found by digging. Between the ridges and the sea, where we halted at a pool, there was a plain covered with reeds and grass, whilst the heavy roar of breakers broke on our ear continually from the west. Morning and evening the atmosphere was obscured by a grey fog, confining our view of the landscape, which, in consequence of the mist and the dimly-seen sand-hills, had a peculiarly wild and dreary appearance.

We looked in vain for the ship-of-war which Admiral Sir Patrick Campbell had kindly promised to send round to help us. But two American whalers (the Commodore Perry and Pocahontas) came in to remain for four months, to catch whales: with the captains I exchanged rope, knives, fishing-lines, zebra head-skins, &c., for biscuit. The natives of the Bay, also, at length gaining confidence, brought us some lean sheep and goats to barter for knives and tinder-boxes. On the beach we saw the bones of several sailors who had been murdered by them, probably for interfering with their women. Interference of this sort I, of course, always tried to prevent as much as possible among my people, for it is the most usual cause of disaster on an expedition of discovery.

I now saw the mouth of the *Swakop*, or Bowel River, about fifteen miles farther north, flowing (like the Kúisip) into Walvisch Bay, in 22° 38′ south lat. The Swakop is the Somerset of Mr. John Arrowsmith's map of Africa: near its mouth, and extending as far as the eye could reach to the northward, are lofty mountains called the *Quanwas* or Clay-trap Mountains; they reach from 2,500 to 3000 feet above the sea, and must form a conspicuous land-mark on making this coast. They are known to sailors as the Blue Mountains. The Dámara negroes live in the upper parts of the Swakop, and I tried hard to get beyond it to visit a nation of red men, who, strange to say, live beyond the negroes; that is, in about 20° of south lat. probably. This nation is called the Nubbís, or many people, and has been visited by the chief of Walvisch Bay; but he unfortunately was absent when I was there, and I could persuade none of his people to guide me to them, for there was an exterminating war raging between two tribes of Dámaras on the route. The negroes were on the move, and, consequently, dangerous.

One of the American captains offered me a passage to Saint Helena, seven days' sail, for myself and my seven attendants, for 1000 rix-dollars, or 75*l.*—the price of a whale—but I thought I had not done enough yet. I now resolved, seeing that I could not get farther north, to go east as far as I could.

May 3.—Accordingly, after a fortnight spent here, we left the Bay, and retraced our steps for eight days up the Kúisip. We subsisted in part on an admirable new fruit called náras, the size of a shaddock, about five inches in diameter, covered with prickles, containing inside, pulp, and seeds like those of a melon, and growing on a bush of three or four feet high, the branches of which have opposite thorns but no leaves. The náras forms the entire subsistence of the people of the Kúisip, and about Walvisch Bay, for two or three months in the year, and they require no other food, and not even water, when it is in season.*

May 11.—To avoid a considerable southern bend of the Kúisip, we crossed, in an east direction, a hard and gritty plain, without bush or grass, between it and the Swakop; and the following day saw cameleons, said to be poisonous, on the open plain, and many miles from either bushes or water. We also passed, just to the northward, a strange and fantastic pile of rocks called Einhíras, or the Hill of the Laughing Hyena, which two Boschman guides told us was inhabited by snakes with red heads and prominent brows. As we were pushing on at the rate of thirty miles in the day and night for water, we could not stop to endeavour to catch one of these extraordinary reptiles.

May 12.—On our left was *Tarahap* (Quiver) Mountain, and *Hokap* (Spotted Body) Mountain, about 1200 feet above the plain. Before us a long range of mountains extending N.W. and S.E., and called 'Túmas, or the Mountains of the Wilderness, rising about 2000 feet.

May 13.—Reached the Humaris, or Rolling River, a northern affluent of the Kúisip, and followed its course in a S.S.E. direction. Its banks were enclosed by high cliffs, and we descended as by steps 400 feet to the bed of its recipient.

We found the Humaris full of rhinoceroses. We also ate plenty of zebra's flesh here, but both are rank and disagreeable; we found, however, the soup made from the rhinoceros particularly strengthening for the chase, which, having now no horses, we followed on foot.

On two or three occasions the people got a supply of locusts, which they roasted and ate, pounded like snuff. I found them not unpalateable, and certainly better than shoe-leather.

We again reached our old acquaintance the Kúisip; and here, for the first time for three days, the oxen luxuriated in plenty of grass and water. The black rhinoceros of the Humaris lives on mimosa bushes; it is the white rhinoceros (which we had not yet seen, and which is a more timid animal), that lives on grass. The Kúisip here does not flow in so deep a channel as lower down in its course; still it was enclosed by rocks and hills, though not so impending as at the spot we first saw it.

May 14 (Sunday).—In the bed of the Kúisip we halted, and assembled the people as usual for divine service.

---

* I have given some seed of the náras, and a description of it, to Dr. Lindley.

May 15.—We crossed the Kúisip, which here comes from the N.N.E., and steering an easterly course we entered Dámaraland. Under a hill, and on an elevated plateau, about a mile from a pool of water, we saw the first Dámara village; it consisted of eighteen conical huts, constructed of stakes driven into the ground, brought to a point at the top, and covered with bushes. Some of the houses had a rude sort of porch, also composed of stakes and bushes, and the whole were arranged in a circular form.

Long lines of thorny bushes, miles of them in fact, were observed on the slopes near; these were placed to direct rhinoceroses or zebras to pit-falls, placed at intervals in the lines. We as yet saw no people, for there had been a scarcity of water during the last season, and we understood from our Boschman guides that the Dámaras had moved further east, and we followed after them.

Of Dámaras there are two great nations—the Dámaras of the Plains, who are very rich in cattle (and between whom and the Namáquas are frequent wars in consequence of cattle)—and the Dámaras of the Hills, who have no cattle, but subsist by hunting and on roots.

The Dámaras of the Plains extend from the Swakop, north and east; whilst the Hill Dámaras are found from the Kúisip, for a considerable distance, south and east: both tribes are negroes; black, with woolly hair, small round noses, and thickish lips. The Dámaras of the Plains, from their superior living, are much more robust than those of the hills; I saw many of the former living as slaves among the Namáquas; and, observing one day a fine Dámara boy in a state of starvation, under a Namáqua mistress, (being obliged to search for gum and lizards for food,) I offered to buy him, that I might eventually emancipate him in England, and I obtained my young shepherd for two cotton handkerchiefs, and two strings of glass beads, in value about four shillings!

The huts of the Dámaras of the Plains are also conical, but the stakes are wattled, plastered with clay, and covered with hides; the bed consists of thorn bushes, on which skins are spread. In 1834 the Dámaras of the Plains occupied the country south of the Kúisip till driven back by the Namáquas, headed by a warlike chief, to their present ground about the Swakop. Some of the huts of the Dámaras of the Plains were left standing about the Kúisip.

The Dámaras of the Plains are circumcised, and extract the two front teeth of the lower jaw; they are almost entirely naked, men and women wearing only a sort of skin kilt about the waist, and reaching to the knees. In war the men have a plume of ostrich feathers on the head, a leopard or lion skin thrown over the shoulders, the waist bound round with thongs, besides the usual kilt and sandals; whilst their arms are bows and arrows, a club, and a stabbing assegai, or javelin, composed entirely of iron, shaft and blade. The Dámaras of the Plains speak a language peculiar to themselves, of which I have got a short vocabulary. The Hill Dámaras speak the clicking Namáqua language.*

The Dámaras of the Hills have bows and arrows, and an assegai with a wooden shaft. They generally wear a short mantle of deer-skin on their shoulders, and about their waist many flaps of leather behind and before; that behind contains a large pocket for roots.

May 16.—We crossed a great plain on which wild horses or zebras, in groups of five or six; each group under a fine-looking leader, were browsing on the abundant grass; before us were lofty mountains, and in the distance, right and left of us, they were also seen.

This part of Dámara-land is very beautiful; consisting of wide plains covered with herbage, broad and grassy slopes at the foot of hills, occasionally thorn bushes, (among which is frequently seen the black back of a rhinoceros,) and mountains rising to an elevation of some thousand feet—some of a tabular shape, and others with the most sharp and rugged profiles it is possible to conceive.

Directing our course to the east-south-east we reached the great mountain of Tans, or the Screen, before mentioned; its height is apparently greater than that of Table Mountain, which it resembles with its flattened top. Tans may rise probably 4000 feet above the plain.

With great labour, arising from the steepness of the ascent, and the packs of the bullocks continually shifting, we ascended some of the off-sets of the mountain, and found ourselves on an elevated table-land; here the thermometer was 65° at noon. In the rocky recesses of this plateau dwelt small communities of Hill Dámaras. These people had never before seen a white man, thought us very ugly, and that we had no skin; but it was said that they had a tradition among them, that white men should pass through their country, and they had been anxious to see them, as they expected to get something from them. I asked these rude people, among other questions, how they purchased their wives from their parents; they said that they collected as many onions and striped mice as they could, and gave these, as an acceptable present of food, to the father of the young woman they wished to obtain.

About this time many lions prowled about us; nor was it advisable to sit up and watch for them, as a person sitting up by a fire, or even lying on the ground with his knees up, is the most likely to be carried off by them, or shot at with poisoned arrows

* The Namáquas call the whole nation of Dámaras, *Dam-áp*; and distinguish the *Hill* or south-eastern Dámaras by the name of *Húmi* (hill) or *Háu* (dung or vile) *Dam-áp*. The Plain or Cattle Dámaras call themselves *Om-oto-ronto-rondú* or *Oke-temba kachi-híqui*.

by creeping savages. Myself and people, therefore, lay on the open plains in our sheep-skin coverings, having first arranged a few bushes (when we had not rocks to screen us) in a semicircular form to give us shelter to windward, whilst at our feet was a fire; the dogs and the bullocks lay around us; and we got up and shouted, or fired off a shot, or rang a bell, when the lions came too near. I, however, lost three head of cattle by them.

East of Tans are beautiful plains of grass and trees. We here saw brindled gnus for the first time; these the Boschmans destroyed, disguised as ostriches. We also killed and ate several zebras and rhinoceroses, white and black; but we felt the pangs of hunger severely at times, and ox-hide was roasted and eaten, as before. The water was in some places black, in others green, or impure with the visits of wild animals.

May 22.—At starting in the morning the thermometer was at 40°. Of course my poor people felt this temperature very severely, and at night we could not sleep for the cold. At noon the thermometer rose to 60° generally. We were now traversing table-land, and passed the various small rivers, and saw the mountains indicated on the sketch map.

May 24.—We came to the Kei-kúrúp, or "first Ugly River," flowing through beautiful plains of grass and trees, and abounding in game. Here, also, we saw many picturesque and detached mountains. Under one of these the plain was covered with mat-houses of Namáquas, and bush-huts of Hill Dámaras; in short, there was here a town of about 1200 persons, called Ni-ais, or Black; while to the eastward from north-east to south-east a range of mountains, some reaching 2000 feet in height, bounded the horizon in that direction.

A party, mounted on galloping bullocks, came out to meet us, and I was conducted with every courtesy to the hut of the chief, Aaramap. He had lately fought some bloody battles with the Dámaras of the Plains, endeavouring to regain their old ground, where we now were, and the possession of some hot and cold springs, beside a small lake, three days to the north-east of us.

Our wants were now abundantly supplied; we got sheep to purchase for knives and cotton handkerchiefs, and bullocks for shawls and axes. We also drank milk till we were tired, purchasing it with needles. Honey-beer was drank; and the reed, pot, and Dámara dances were performed to welcome us.

The first of these, the reed-dance, is performed by men standing in a circle, facing inwards, and blowing into reeds of various lengths, and stamping on the ground, whilst the women, clapping their hands and singing, dance round them. The pot-dance consists of one man dancing in the midst of a number of women, who sit on the ground singing and clapping their hands, whilst one beats with her fingers on a wooden vessel covered with skin; and the Dámara dance consisted of half a dozen women standing in a line, clapping their hands and singing, whilst a man, with a pair of spring-bok horns on his head and rattling seeds on his ankles, stamped before them.

From information obtained while here, we learnt that to the north-east, and three days' journey from Ni-ais, are very abundant supplies of water and grass; hot and cold springs issue from both sides of a ridge of hills; to the north they unite their waters, and form a stream which falls into a lake, of several miles in extent, its shores in many places covered with reeds. Wild-fowl and fish are to be got here. The stream from the lake then forms one of the sources of the Swakop. On the south side of the ridge the hot and cold springs lose themselves in a plain covered with rich grass. This altogether is a most highly-favoured and beautiful part of Dámara-land, which, however, is said continually to increase in fertility to the north. We were also told that about ten days, or 150 miles from Ni-ais, E.S.E., is the Nósop, or Kaffir River, full of trees and elephants. The chief of Ni-ais shot fifteen elephants at the Nósop last year. It is said to flow from near the source of the Swakop and Kúisip, with an easterly bend, to the Orange River. We were also told that the rainy season at Ni-ais is in the month of August.

We had now got about 200 miles into the interior, from the coast, and had reached the fine cattle country indicated (from native report) on Arrowsmith's map. We were at a short distance too from the sources of the Kúisip and Swakop, and had crossed beyond the principal sources of the 'Oúp, or Fish River. We desired much to go further north or east, but we could get no guides to show us the waters to the north, for the Dámaras of the Plains lay there ready for war, and no one would, for any consideration, convey a message to them; whilst to the east there was said to be an impassable desert, which no one at Ni-ais had ever ventured to cross. In fact, the chief told me, that without going first south, nearly to the Orange River, it was impossible to go easterly towards Latakú, which lay south-east of us.

I had now, therefore, got to the utmost limit of my range, north and east—and I saw that, in the present state of affairs between the Dámaras of the Plains, those of the Hills, and the Namáquas, that it was impossible to pass through the country of the former from the south, and that the best chance for future travellers, to visit the interior about 21° of south latitude, is to come down from the Portuguese settlements, extending to 17° S. on the west coast, and whither the Dámaras (by reports of prisoners) resort to barter cattle for iron, &c. I therefore turned my face to the south, having first had a grand lion hunt, on foot, at a run for eight miles, at the end of which the monster turned, charged us, and fell within four yards of the muzzles of the guns. As soon as we had got the skin off, the Hill Dámaras cut up the flesh of the lion, and carried it off to eat.

On the 31st of May then, accompanied by the Chief Aaramap, and many of his people, we travelled to the S.S.W., along the banks of the Kei-kurúp—thermometer at noon 65° in the shade. At thirty miles from Ni-ais we came to a beautiful valley, enclosed with low hills, and commanding a view to the N.E. of the lofty mountains before mentioned. There was quite a forest of thorn-trees, of several miles in extent, in this valley; the grass stood in it like corn, and amongst it were flocks of pheasants and guinea-fowl.

In the centre of the valley, among rocks of granite, rises a warm spring, at 126° of Fahr.; the waters flowed freely from it, and were lost in the plain. Moreover, some attempts had been made to cultivate calabashes and tobacco by means of it, and I gave some melon and cucumber seed, which I had with me, to the chief, to extend his garden.

We shot some new and rare birds here; and as the valley is fertile, wood and water abundant, and as it is in the midst of game, and of a numerous population of Namáquas and Hill Dámaras, this spot seems well adapted for the establishment of a mission station, and accordingly I have indicated its existence and its advantages to the Wesleyan Missionary Society, as, although a distant point, it might yet be connected with their present stations at Kamiesberg and Nisbett's Bath.

The people are anxious for missionaries; the women in particular said, "send us teachers for us and for our children;" we therefore ought speedily to respond to their appeal.

We saw indications of the Dámaras of the Plains at this new Warm Bath, such as the head of an ox placed on a tree to mark the grave of a warrior below. Here, some time before, a fearful tragedy had been enacted. The Namáquas, living at the Bath, were aroused one night by the howling, as it seemed, of wolves, when there was a fearful rush of negro savages, who destroyed the women and children, whilst the men, less encumbered, escaped to a neighbouring hill. The wife of the chief was secured by the Dámaras, and had her hands cut off by the wrists next morning, before her own castle, previous to her being put to death.

We saw here, and at other places on the journey, occasional traces of Haiji Aibib, or of the Namáqua deity. A long pile of stones was pointed out, on which lay a few branches, the offerings of wayfarers. Under the pile dwells, say the Namáquas, a person they call Haiji Aibib, of whom they stand in some fear, but of whose appearance and power they entertain no distinct ideas.

The only trace of religion I found among the Boschmans was

at a pool of water commonly choked up with sand, near the Tans Mountain, where, before they dug for water, they presented an arrow, or a piece of skin or flesh, to a large red man with a white head, who is supposed to inhabit the place; at the same time they repeated a prayer for success in hunting. To dig for water here without this ceremony, they say, occasions sickness and death.

From the Hill Dámaras I could make nothing out to show they had any, the most imperfect, religious impressions. "Who made the sun?" I asked them. "We don't know; we are a stupid people, we don't know anything—only let us get plenty to eat, that is all we care for,"—was the common answer I got from this benighted people.

By a present of tobacco I found I could always get the people to speak; by means of my medicine-chest a certain influence was obtained over them; and by a violin, tambourine, &c., they were kept in good humour.

June 1.—We crossed the *Oanop* (Tell-tale) River, running to the south-east, and which is lost in the sands of the desert, in that direction. Yet the Nósop, or Copper River, was said to be ten days from us in that direction, and full of elephants. After a march of ten miles we came to a place called *Gnu-tuais*, or Black Mud, where, on the side of a hill, was pointed out to me a trench of some yards in length, where the Dámaras are in the habit of digging for copper, of which they manufacture rings. I collected some of the ore near the surface of the ground; it seemed rich and abundant; but this of course is not so valuable as the copper-mine formerly mentioned, from its remoteness and difficulty of access.

Meeting with various hunting adventures, and killing kudús, &c., we travelled steadily on to the S.S.W., crossing several beds of streams flowing to the S.E., through plains on which were white rhinoceroses: we walked in the morning and evening, and rode our oxen in the middle of the day.

June 6.—We were out of Dámara-land, and found ourselves in a flat and uninteresting country, though everywhere to the east of the Fish River we found grassy plains and a good cattle country.

After crossing the Kukama (Brown) River, we turned to the S.S.E., towards the Cho-unp (Calf) River, which is a tributary to the Great Fish River; we crossed the Cho-unp on the 8th of June, and afterwards recrossed it three times. We now began to see many cameleopards; they were commonly in herds of a dozen, with two videttes on an eminence looking out over the bushes. We ate the flesh of the giraffe, which we liked better than that of any of the other wild animals we had tasted, and we had partaken of everything from a lion to a locust.

I need hardly say that flesh every day, and often of the most disgusting description, and eaten without bread, salt, or vegetables, becomes tiresome enough—such we felt it.

The appearance of the country throughout this tract was undulating, without any marked features—grass, bushes, and occasional pools of water, besides the streams we crossed, but we did not see a single inhabitant: we knew that there were Dámaras on the hills to our right, but they kept out of sight—our only associates were wild beasts.

On the 10th of June we recrossed the Great Fish River, just below its junction with the Cho-unp, and by means of a small net, of thirty-five feet long, and swimming and wading with the people, we got out of deep pools in the river many dozen fish of the genus *silurus,* and also others resembling mullet. On several previous occasions we got good meals by means of the net. The bed of the Fish River was here about 150 yards broad; the banks low, and well wooded.

We now found the vast and silent plains to the west of the Fish River most monotonous to traverse, and sufficient to take the spirit of wandering out of the most determined traveller—that is, for some time.

On the 16th of June I parted with the Chief Kúisip, who wished to go home by a short route, to fight a man who threatened him. I had found him very faithful, and of great assistance, and he and Choubip went away pleased with a handsome present of a new musket each, ammunition, cloth, cutlery, beads, and buttons.

Quitting the banks of the Fish River, we proceeded to the S.S.W. for about fifty miles, over dry, stony, and dreary plains, till, on the 18th of June, we reached the banks of the beautiful *Húntop* River, which we had before crossed on our outward journey, about fifty miles farther to the north-west.

June 20.—We reached the Kei-sú, whence we turned much more to the westward, and winding among hills we ascended to the summit of the *Unuma* mountains, the great table range we had before seen from the west, and passed along them with a small escort, headed by Henrick Boys, our chief hunter, who was fleet enough to catch and kill a zebra with a knife. We suffered much from cold; the thermometer was at freezing point, and we rose every morning, from our sandy or stony bed, with the kaross covered with hoar frost. We had also one or two showers of rain; and were in some danger from pools poisoned by the Boschmans for zebras. After a few days, descending into the plain of the Koan-quip, I recovered my waggon on the 30th of June.

The Namáquas of the Koan-quip are in the yearly habit of resorting to Angra Pequena Bay, on the west coast, to barter their cattle for powder and lead with whalers; thus they receive only two bottles-full of coarse powder, or forty balls, for a good ox. Sometimes, however, they are made drunk, and get nothing at all.

I asked the Namáquas why they did not come into the colony to exchange their cattle with the colonists for cloth, &c.? They answered, "We are afraid to do so." But it is evident that, if they were to be protected by a magistrate, living near the frontier, they would have no cause for apprehension; and I therefore represented the case strongly to the Government at the Cape, that a trade may be commenced in Namáqua-land, the beneficial influence of which may eventually be greatly extended.

July 13.—Having retraced our steps down the Koan-quip we again reached Bethany, and after two days' halt, we proceeded on our journey, in a more south-westerly direction than our outward route had been. The country at this season of the year was parched and burnt up till we got as far as Herís, or wet ground. We travelled over broad plains, with mountains to the left, but without a single inhabitant, from seven miles to the south of Bethany, (where there were a few Namáquā families on the Quahanap, or Javelin River,) to the Gariep: during the first part of the distance the country abounded in lions, which, as we advanced, were succeeded by leopards.

July 25.—Crossed the Heineip, or Thick Bush River. Here we saw iron occurring in veins on the side of a hill. There is also said to be to the N.E. of Bethany, near the Great Fish River, a plain covered with large masses of iron, some of which require several men to lift them. I have got specimens of this iron. It is not unlikely, too, from the indications we saw, that coal may be found to the south of Bethany.

July 27.—At Herís the first signs of the South African spring were seen; the plains were covered with dark-green shrubby plants, relieved by the bright hues of flowers; and as we walked after our cattle in the morning and evening march, the sense of smelling was regaled with the aromatic scent of desert plants, which burst, in a variety of forms, from the soil which had recently been moistened with showers. About this time we subsisted for two or three weeks on ostrich eggs.

July 30.—We found the region of the Orange River wild and barren as before—the hills scattered in great confusion. We crossed the river at the ford called Kunarúsip, or that of the Ebony Black Sheep, distant about forty miles from the sea; the stream was about 450 yards wide, and flowing briskly, yet I forded it with water only up to my waist. The banks were low: looking towards the east mountains closed in the view, affording a wild and romantic scene. To the westward high hills intervened between us and the sea.

August 1.—After a day's halt, we left the banks of the Great River, and though it was the beginning of the South African spring, the poor cattle had about this time no grass, only the tops of bushes, and the leaves of the ebony-tree, for a fortnight. My cattle, therefore, were fast knocking up; we had already left at intervals on the road a span of fourteen, and had eaten another span. I therefore sent into the colony to Komákas for assistance, which was immediately rendered by the excellent Bastaards under the care of the Rev. Mr. Schmelin. At fifty miles farther south we regained the route we had followed on our way to Aris; and, travelling southwards, on

August 8, I again gladly crossed the Kowsie River, the boundary of the colony, and tarrying a short time with the hospitable old missionary, Mr. Schmelin, who has translated the four Evangelists into the Namáqua language, I journeyed to the Olifant River, and by the beautiful district of the Twenty-four Rivers, by the Paarl and Stellenbosch, and during the finest season of the year, when the face of nature was fresh and green with verdure and brilliant with wild-flowers, I reached Cape Town on the 21st of September in health and strength, feeling very grateful for the manner in which myself and people had been mercifully preserved during the year of our journeyings in the bush and on the road.

In justice to my people, I must say that they were animated by a good spirit throughout the expedition (certainly the roughest I had undertaken during an experience of sixteen years of voyages and travels); they were respectful and obedient, and showed no want of courage; and, above all, they severally submitted with patience to the privations and hardships to which such an expedition was always exposed.

During the whole journey the rocks were found to be chiefly of primitive formation; granite, old red sandstone, mica slate, (particularly about the Kúisip,) quartz with imbedded crystals of feldspar, hornblend, &c. I brought away many specimens of copper and some of iron.

I made several drawings of the natives and of remarkable scenery, such as the Table Mountain of Tans, the Bull's Mouth Pass, and of the mountains to the eastward of Ni-ais.*

The accompanying map has been compiled from my original daily tracks, showing the course by compass, and the distances calculated from the rate of a waggon drawn by bullocks, travelling three British miles an hour; when we rode on pack-oxen, four miles an-hour have been allowed. At starting I was well provided with instruments, but being without any companion or

* These, with my extended Notes, I hope to be able shortly to lay before the public.

assistant, added to the difficulties attendant on travelling in Africa —the necessity for making personally every arrangement for our march and for our night's halt—the difficulty of providing food—of dealing with the natives—of preventing drunkenness and quarrels among a party at times fifty in number—and the constant exertion required to keep all in good humour, occupied my whole time, and did not allow me leisure to make as much use of the instruments as I could have wished; yet as the route chiefly lies between known points, and as the daily tracks have been revised by that careful geographer Mr. John Arrowsmith, there can be little doubt that the present map of the western region is a very near approximation to the truth, and offers a route of upwards of 1500 miles which has never appeared in any former map of Southern Africa.

With respect to the objects in natural history collected during my journey, I have the gratification to subjoin the notes of Mr. W. Ogilby on the mammalia; of Mr. John Gould on the birds; and of Professor Lindley on the plants. Their names alone will be sufficient guarantee for the accuracy of the descriptions; and I may be here permitted to offer them my sincere thanks for their kindness in so promptly acceding to my request.

List of new and rare Mammalia collected by Captain Alexander during his recent expedition into the country of the Dámaras:—

I.—QUADRUMANA.

1. *Cynocephalus Porcarius* (Boddeart).

II.—CHEIROPTERA.

2. *Nycteris Affinis* (Dr. Smith).

III.—INSECTIVORA.

3. *Chrysochloris Dámarensis* (Ogilby), new species.
4. *Macroscelides Alexandri* (Ogilby), new species.
5. *Macroscelides Melanotis* (Ogilby), new species.

IV.—CARNIVORA.

6. *Gulo Capensis* (Schreber).
7. *Mustela Zorilla* (Desmarest).
8. *Viverra Felina?* (Thunberg).
9. *Herpestes Melanurus* (Dr. Smith).
10. *Cynictis Ogilbii* (Dr. Smith).
11. *Proteles Cristata* (Penny Cyclopædia, i. 2).
12. *Canis Megalotis* (Cuvier).
13. *Canis Mesomelas* (Erxleben).
14. *Felis Leo* (Linnæus).
15. *Felis Nigripes* (Burchell).

V.—Rodentia.

16. *Bathyergus Dámarensis* (Ogilby), new species.
17. *Graphyurus Elegans* (Ogilby), new species.
18. *Geosciurus Capensis* (Dr. Smith).
19. *Lepus Rupestris* (Dr. Smith).

VI.—Pachydermata.

20. *Equus Zebra* (Linnæus).
21. *Rhinoceros Africanus* (Desmarest).
22. *Rhinoceros Simus?* (Burchell), an imperfect skull.
23. *Hyrax Capensis* (Schreber).

VII.—Ruminantia.

24. *Antilope Eucore* (Forster).
25. *Antilope Tragulus* (Forster).
26. *Antilope Traguloides* (Ogilby), new species.

It will be observed from the preceding catalogue that nearly a fourth of the mammals collected by Captain Alexander are new species, that is to say, six out of twenty-six; and of the remaining twenty, seven, viz., Nos. 2, 9, 10, 11, 12, 20, and 21, are still rare in cabinets of natural history. This is a very large proportion of new and rare specimens from a country so repeatedly and so carefully explored as South Africa; and I heartily congratulate Captain Alexander upon the success which has attended this part of his mission. I have endeavoured to acknowledge the obligations which Captain Alexander has conferred upon the cultivators of natural science, by connecting his name with one of the new species which he has discovered.

W. O.

Upon examining the interesting collection of Birds brought home by Captain Alexander from the interior of South Africa, I find many rare species, and several which appear to me to be new to science. The collection consists of 320 specimens and 125 species. The following is a slight enumeration of them:—

Of the Raptorial order, or birds of prey, there are thirty-four specimens and sixteen species, viz., two vultures, *Neophron Peronopterus* and *N. Monachus;* twenty-two falcons (ten species), among the more remarkable of which is a very beautiful eagle, with a red breast and white tail, and a very diminutive true falcon, half as large again as a sparrow, with a red back and spotted tail; eleven owls (four species); one of the genus *Surnia,* two of *Scops,* and one of *Athene.*

Of the Insessorial order, or perching birds, there are 235 specimens and seventy species. Of these seventeen specimens (six species) belong to the Fissirostral tribe, viz., 1 *Caprimulgus;* 2

species of *Coracias;* one of which, the natives say, alights on the horn of the rhinoceros. These examples are highly interesting, as showing the southern limit of the range of this beautiful tropical form. Two species of *Merops* and one *Alcedo.* Of the Dentirostral tribe there are sixty specimens (twenty-two species), among which occur examples of the following genera:—*Lanius* (two species), *Crateropus bicolor, Petrocincla, Saxicola, Ixos.* Of the Conirostres eighty-six specimens (twenty-five species), comprising examples of the genera *Euplectes, Estrilda, Amadina, Ploceus, Pastor, Lamprotornis,* and *Corvus.* Of the Scansores there are forty-two specimens (twelve species), of the genera *Colius, Bucco, Picus,* a very rare *Corythaix* described by Dr. Smith, and an apparently-new *Agapornis* or small parrot. Of the Tenuirostres thirty specimens (six species), of the genera *Upupa, Rhinopomastus, Cinnyris,* &c. Of the Rasorial order there are eight or ten species, among which are two or three species of pigeons and examples of the genera *Otis* (three species), *Cursorius bisinctus, Pterocles, Francolinus, Struthio, Charadrius,* &c.

The collection is not so rich in birds of the Grallatorial and Natatorial orders as in those of the preceding. Among the Grallatores are examples of the genera *Numenius, Ardea, Tringa,* and *Umbretta,* and in the Natatores, *Anser, Anas, Gallinula,* and *Podiceps.*

J. G.

Among the Plants, collected beyond the country usually visited by strangers, occur two or three species of *Pappophorum;* a curious spiny plant with hoary leaves and large flowers, somewhat resembling an *Escobedia* in size and appearance, but belonging to the natural order *Solanaceæ;* several Acanthaceous plants, particularly one with bright blue flowers and spiny leaves, allied to *Barleria* and *Acanthodium;* the rare *Otoptera Burchellii;* a fine plant related to *Sesamum;* two trees belonging to different species of the genus *Ficus;* several *Amarantaceæ,* and an apparently-new species of *Aptosimum.* The most curious plant is, however, what is called the *Naras,* bearing a spiny fruit double the size of an orange; of this the single specimen is so imperfect that even with the assistance of the seeds it is impossible as yet to form any decided opinion concerning it. The specimen resembles *Schepperia juncea* so much, that it would be mistaken for it if it were not for the seeds, which are extremely like those of some cucurbitaceous plant.

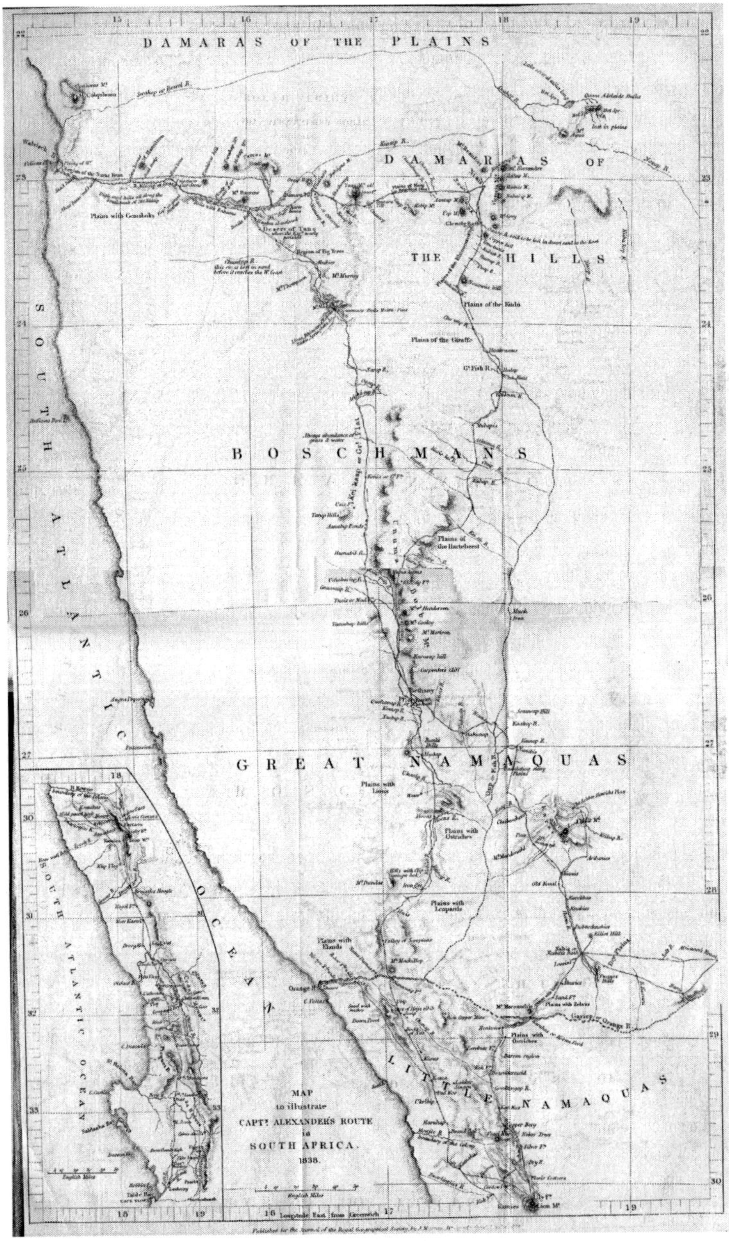
DAMARAS OF THE PLAINS
DAMARAS OF THE HILLS
BOSCHMANS
GREAT NAMAQUAS
LITTLE NAMAQUAS
SOUTH ATLANTIC OCEAN
Plains with Gemsboks
Plains of the Koodo
Plains of the Giraffe
Plains of the Hartebeest
Plains with Lions
Plains with Ostriches
Plains with Leopards
Plains with Elands
Plains with Zebras
Orange R.
English Miles
Longitude East from Greenwich
MAP
to illustrate
CAPT. ALEXANDER'S ROUTE
in
SOUTH AFRICA.
1838.

II.—*Itinerary from Tabríz to Tehrán,** via *Ahar, Mishkín, Ardabíl, Tálish, Gílán, and Kazvín, in* 1837. By E. D'Arcy Todd, Bengal Artillery, Major, serving in Persia. Communicated by John Backhouse, Esq.

[The following Itinerary was accompanied by a map on the scale of *six* British statute miles to an inch, or $\frac{1}{380160}$ of the natural scale: this has been reduced, and made use of in the annexed map of the north-western portion of Persia. The bearings in the Itinerary are all *magnetic*, and were taken with a Kater's compass; the distances are in British statute miles, and estimated from the pace of a horse walking on level ground at the rate of four miles an hour. The positions of latitude and longitude depend upon the town of *Resht*, which, according to the observations of Mr. Baillie Fraser, is in 37° 17′ 30″ N. latitude, 49° 43′ longitude E. of Greenwich. Part of the route was examined by Colonel Monteith in 1827, but in this Itinerary many more names are given, with the advantage of being written according to one standard of orthography, for which see vol. vii. p. 245.—Ed.]

Dec. 27*th*, 1836.—To Khojeh,† N. E. 20 m. On quitting Tabríz, the road for about a mile runs along the right bank of the Bosmiyeh river, and then ascending, approaches the hills to the north-east of the city. At 6 m. reach the summit of the ascent, which is stony, and in some places exceedingly steep. The descent is shorter and less steep than the ascent, and leads into a fine cultivated valley. The road level and free from stones. The village of Khójeh contains about sixty or seventy houses, and yields to Naṣru-llah Khán, Seri Kesеklí Báshí, to whom it belongs, 300 tómáns,‡ and 140 hulwáns of grain yearly.

Dec. 28*th*.—To Ḳizil Ḳapán (Red Barn. T.) N. E. 20 m. After leaving Khójeh, the country is covered with small isolated green and red hills, varying from 30 to 100 feet in height. At 8 m. passed the village of Shebrek, three or four miles to the left of the road. At 16 m. crossed a richly-cultivated valley running N.W. and S.E., with a small hamlet at some distance to the right. Afterwards gently-undulating ground. A range of snowy mountains bearing S.S.E. The road level and passable for carriages. The village of Ḳizil Ḳapán, half a mile to the left of the road, has suffered dreadfully from a late earthquake, which has shaken every house in the village.

Dec. 29*th*.—To Ahar N. E. 8 m. East 12 m. The road gradually descends from Ḳizil Ḳapán into a valley. At 3 m. a

---

* Properly Tihrán (the pure), *i* being commonly pronounced *e*. See Uylenbroek's Iracæ Persicæ Descriptio. Lugd. Bat. 1822. 4to. pp. 13, 18.—F. S.

† Khwájah (*w* not sounded), pronounced Khójah, 'a husband, old man, preceptor,' by the Turks, with the stress on the last syllable.

‡ A tómán is worth about 16 shillings. A hulwán is an ass-load.

ruined káravánseráï marks the commencement of a gentle ascent, about a mile in length, by the side of a small stream; the descent for three miles is more abrupt. A ruined káravánseráï stands at the foot of the pass. One mile beyond, the road turns to the eastward, and follows the course of the Ahar river on its right bank. A range of hills to the north sprinkled with snow. The road, after clearing the pass, is excellent, being hard, level, and free from stones. Much cultivation; wheat and barley on both sides of the river.

The town of Ahar is the capital of Ḳará-dágh,* and the residence of the governor, Mírzá Moḥammed Raḥím Khán; it contains about 700 houses, and has been surrounded by a wall, now in ruins. Mount Savalán† bears S. 71° E. distant 50 m. from Ahar. The range of mountains to the south of Ahar is called Kashka‡ by the people of the district.

From Ahar to Ardabíl, E.S.E. nearly, distance eighty-two miles, by the road.

Dec. 30*th.*—To Ḳishláḳ,§ S. 80° E. 10 m. At 4 m. the road crosses the stream of Ahar, and continues on its southern bank. From 4 m. to 10 m., a rich tract of country occurs, called Savalán Shámí, about six miles in length and four in breadth, covered with cultivation, except in the vicinity of the river, which is reserved for pasturage. A range of mountains called Shaïvek, distant about eight miles to the north of the road. At 9 m., passed the village of Kázeh Kond, containing twenty or thirty houses, situated on the left bank of the Ahar stream; after crossing which, the road passes over a succession of ridges and ravines which descend to the river from the Kashká range, distant about four miles on the right, and crosses several small streams. Ḳishláḳ is a village containing about thirty houses.

To Seyyidler‖ S. 85° E. 2 m. A small village half a mile to the left of road.

Dec. 31*st.*—To Ḳullah Báshí¶ S. 60° E. 2 m. A small village to the left of the road, situated on the pinnacle of an isolated hill, at the foot of which flows the Ahar stream. It is surrounded by a wall, and contains about twenty poor huts.

To Yengíjeh** S. 80° E. 9 m. For the first four miles the road is uneven and stony, over hills on the right bank of the stream. At 4 m. the road quits the river, which flows to the N.E. to join the Ḳará Ṣú. At 7 m. passed the village of Mízár, containing about twenty houses, situated on the spur of a hill to the right. After

---

* Kará-tágh (T.), 'black mountain.'
† Sayalán, or Saïlán in Jihán-numá, p. 384.
‡ Kashaghá, or Ḳashaghí (T.), 'a curry-comb.' § (T.) 'Winter quarters.'
‖ (T.) 'The Seyyids,' *i. e.*, Descendants of Moḥammed.—F. S.
¶ (T.) 'Chief Tower,' ** (T.) Little-New-Town.

leaving the river, the road leads over undulating ground, sloping to the north, and richly cultivated with wheat and rice.

To Ṣúj Búláḳ N. 75° E. 3 m. A small village to the right, and another half a mile on the left. Half a mile beyond is the small village of Hawashí on the right, and half a mile further the village of Khanchel. Rich cultivation on both sides of the road.

Jan. 1st, 1837.—To Aḥmed Beglú N. 50° E. 4 m. A considerable village, situated on the side of a deep ravine running from the hills to the right in the direction of the Ahar river.

To Karramaú S. 85° E. 3½ m. Large village, with a fine stream. At 2 m. the large village of Mírkend half a mile on the right, and the small village of Zerdán Zamín at the same distance on the left. The village of Ush Tepah,* about a mile to the north of Karramaú, and 1 m. beyond, the village of Iyák to the right.

To Naṣr-ábád† E. 2½ m. Large village with abundance of water. A range of mountains sprinkled with snow about twenty miles distant, extending N.W. and S.E.

Jan. 2nd.—To Barzil N. 65° E. 1½ m. Large village with abundance of water. The residence of Rustam Khán, chief of Mishkín. Rich cultivation of wheat, barley, and rice, on both sides of the road.

To 'Alí S. 80° E. 2½ m. Large village, with fine stream. One mile beyond Barzil, the village of Khiyú to the right of the road.

Jan 3rd.—To O'nár N. 60° E. 8 m. Large village surrounded by gardens and orchards in a valley running down from Mount Savalán. Abundance of water. The valley richly cultivated. Between 'Alí and O'nár is the vilage of Ḳará Ḳaya, about four miles to the left of the road, on the left or southern bank of the Ḳará-ṣú.‡ Within two miles of O'nár lies the village of Lóró, about a mile to the left of the road.

Twenty-two villages are said to be within a short distance of O'nár, namely, Lóró, O'rdú-lí, Jálir, Kangerlú, Dádíbeglí, Sheïkhlú, Kapakhah, Ḳádirlí, Lumber, Lahak, Takhtábád, Argá, 'Alí-ábád, § Kaviyet, Dashkesán, Noghdí 'Ulyá, || Noghdí Ṣuflá, ¶ Berazmiyán, Arbáb, Lájawah, Babiyán, and Shaáb-u-d-dín.

O'nár yields 400 tómáns revenue. Mount Savalán bears from O'nár S. 7° W. about 17 m.

To 'Alí-ábád N. 40° E. 1½ m. Small village.

The Ḳará Sú about three miles on the left of the road. From 6 m. to 10 m. beyond O'nár, steep ascents and descents occur. At 9 m. is the ravine of Dújakh, which marks the boundary between Mishkín and Ardabíl.

---

* Uch tepeh, or tapah, (T.) Three Hills?
† Abode of Victory. (P.)
‡ 'Black Water.' (T.)
§ ''Alí's Abode.' (P.)
|| Upper Noghdí. (A.)
¶ Lower Noghdí. (A.)

Jan. 4th.—To Samarín S. 60° E. 17 m. Large village with abundance of water. Small village of Kurd Ḳishlák* on the left of the road.

To Sháhsewár† S. 35° E. 4 m. A considerable village. The village of Bárí 2 m. on the right of the road. The Ḳará-ṣú is about 2 m. on the left. Passed the two small villages of Gerján and Sumah.

Jan. 5th.—To Ardabíl S. 45° E. 10 m. Rich cultivation of wheat, barley, and rice, between Samarín and Ardabíl; road excellent; gradual slope of country from the skirts of Savalán to the Ḳará-ṣú.

The town of Ardabíl has suffered much within the last two years from visitations of the plague, the population has much decreased, and the town presents the appearance of a mass of ruins; the fortress lies 500 yards to the E.S.E. of the town.

There is a great deal of waste land in the vicinity of Ardabíl, but much of this might be brought under cultivation at a trifling expense. The waters of Nóhúr, about twenty miles S.E. of Ardabíl, are described as being surrounded by hills and having only one outlet; but as there is no dam across this opening, they find egress at one season, and are thus lost. These waters are said to be sufficient to irrigate a tract of country that might contain a hundred villages, but which at present lies waste. The expense of a dam, according to Persian calculation, would be about 5000 tómáns (4000*l.*)

Mount Savalán‡ bears from Ardabíl N. 85° W. nearly 25 miles.

The plain of Ardabíl is about 5000 feet above the level of the sea, from a rough experiment made with a thermometer, water boiling at 202° Fahr.§

From Ardabíl to Germí in U'jarúd, N.N.W. nearly; distance seventy-three miles by the road.

At 3 m. passed the village of Chenzenek, with the village of 'Alí-búláḳ adjoining on the right. Abundance of water; road perfectly level. A small tributary of the Ḳará-ṣú passes close to these villages from the S.E.

Cultivation of wheat and barley on both sides of the road.

Jan. 14th.—Noadeh N. E. 10 m. A large village on the right bank of the Ḳará-ṣú, which we cross. The river forms the boundary between Ardabíl and Velkhí, but contains very little water at this season, while in the spring it is a considerable stream.

---

* Kurd's winter-quarters (T.)—F. S. † Royal-rider. (P.)

‡ Mount Savalán was ascended by Captain Shee of the Madras Infantry and some other Englishmen in 1827, who state that water boiled at its summit at 188° of Fahr., which would give, says Colonel Monteith, a height of nearly 13,000 feet above the sea. See Journal, Vol. iii. p. 28. This requires confirmation, and it may be hoped that some of our countrymen wandering in Persia will obtain barometrical measurements for the height of this mountain.—Ed.

§ Ardabil contained about 3000 inhabitants in 1822.—Ed.

Námín N. 35° E. 5 m. Village, the residence of Mír Kázim Khán, chief of Velkhí, Astarah, and U'jarúd. Abundance of water; road excellent. Namín is situated in a ravine descending from the mountains to the N.E.* Between Nowahdeh and Námín, four villages on the right and right front, the road turns to the westward.

Amberán N. 30° W. 8 m. A large and populous village on the right; village of Ammigán on the left. Abundance of water; road passes over hills which slope to the west.

At 2 m. road enters and ascends for 2 m. the ravine of Shattehderra, stony and bad; the hill of Gerhlar, a holy place, to the left, and 1 m. beyond Kishlák. Descend 1 m. to Pilá-Cháï, or Pilá-rúd, a village of seventy houses, scattered along the banks of the stream which flows from the N.E., and gives its name to the village. The hamlet of Ushná lies 1 m. higher up the stream. Much cultivation near the banks of the river.

From Pilá-Cháï the road turns due west, and at 5 m. enters a ravine. Sheríf Beglú at 9 m. a large village on the right bank of a ravine.

The range of hills which marks the Russian frontier, about eight miles on the right, extends N.W. and S.E.

At 2 m. Rezí, a village of about twenty-five houses, on the right bank of a stream which has its source in the hills to the right, and flows towards the S.W. The village of Rahím Beglú 2 m., and Khalífahlú 4 m. higher up the stream; road excellent.

At 6 m. cross in a north direction the ravine and village of Bak Cháï, and gradually approach within 2 m. the frontier range of hills.

At 9 m. cross the dry bed of the Cholpah Cháï. The peak of Histír, the highest point of the frontier range, bearing E. 3 m.

At 13 m. a ruined caravanserai; the range of hills 3 m. on the right, called Dáyimán. Low rounded hills to the left of the road.

At 15 m. an opening in the hills, marking the northern extremity of the Dáyimán range, and the commencement of the mountains of Kullah Básh; road ascends.

At 17 m. reach the summit of the pass called Agzí Gadúk, which marks the commencement of the district of U'jarúd; an abrupt descent of two miles leads into a valley.

At 20 m. the road turns N.E. and crosses some hills to the right.

A large black stone called Beldásh, to the right of the road, marks the summer encamping place of a portion of the Persian tribe of Peránbilí.

---

* That is, from Mount Shindán, distant only four miles to the N.E.; and, according to Colonel Monteith, reaching the height of nearly 7,000 feet above the sea. See Journal, Vol. iii. p. 26.—Ed.

At 23 m. enter a defile, and cross the Dizdeh Cháï, a small stream running from the eastward.

At 26 m. the pass of Sukkál Lishlí in a north direction: the descent steep and very stony; the road turns to the eastward after clearing the pass, and crosses over some hillocks affording fine pasturage.

Jan. 23rd.—At 30 m., Germí, chief place of the district of U'jarúd, a large and populous village on the bank of a ravine; the inhabitants chiefly pastoral, possessing large flocks of sheep and droves of cattle.

From Germí to Permír, and thence, *via* Adíná Bázár, return to Germí, distance thirty-three miles.

At 5 m. Sengír, winter encampment of the Peránbilís; ravine running N. and S. from the hills called Jekír, with the village of Alazar (upper) on the east side. Much pasturage.

Immediately after crossing the ravine, the road turns abruptly to the S.E. for 5 m., leaving the village of Tapah half a mile on the right, and crosses an offset of the Jekír hills by a steep ascent; a gradual descent of 1 m. in a south direction, by a sheep track, rather than a road, with a deep valley immediately on the left, leads to Permír, a village of fifteen or twenty houses, situated in a deep glen; abundance of water. The frontier said to be 5 m. distant to the south.

Directly after leaving Permír, turned N.N.E., and continued along the left bank of the Permír Cháï, which flows to join the Waïsel Cháï at 10 m. distance.

At 2 m., Demir-chelú, a village of the tribe of Míránkúh.

At 6 m., Duninkend, a kishlák of the Dilloghardalús.

At ½ m., on the right, passed the site of O'rtákend, a small village; 1 m. beyond, the site of A'tálú; and 1 m. further, the site of Hasan-ábád, all three villages destroyed during the last summer by the Russians.

At 10 m. reach the Adíná Cazur river, or Waïsel Cháï, and continue N.W. along its left bank; it here marks the Russian frontier, and at 8 m. below falls into the Balahrúd, which flows to the N.E.

After two m. along its bank, quit the stream, and ascend the hills in a S.W. direction; pass a kishlák of three or four tents, and at 8 m. reach the village of Alazar (lower), rejoin the road to Germí, and retrace our steps to Sheríf Beglú, where we turn off to the S.S.E. for 7 m. over a good and level road to the village of Tagh-dizdeh Cháï, containing twenty-five houses, built within the last four years. The ground gently sloping towards the Kará-sú; and at 12 m. beyond, by an excellent road, again reach the village of Námín.

From Námín to Enzelí, distance 114 miles.

On quitting Námín, 1 m. to the S.E., is the village of Dáderán, whence the road turns E. by N., leaving Khán Aghá (old), on the right, and crossing the brow of a hill sprinkled with underwood, passes within about half a mile of Mount Shindán,* a perpendicular rock, which marks the south-western point of Russian Ṭálish.

From this point the road descends through thick wood, scarcely passable at this season from mud and melting snow, leaving the village of Venabín on the brow of a hill, a quarter of a mile on the right; 7 m. beyond is the hamlet of Ḥájí Amír. An abrupt descent through thick underwood, till we reach the stream of Ḳalá Ḳúshí, a feeder of the Astarah river, which we cross, and continue along its left bank over the projecting offsets of thickly-wooded hills. The Ḳalá Ḳúshí is said to mark the frontier from its source near the mountain of Shindán to the point where it joins the Astarah river, which thence becomes the boundary between Russia and Persia, to its embouchure in the Caspian.

At 28 m., having crossed and recrossed the river eight times, reach the village of Astarah, seated on each side the stream, at a quarter of a mile from its outlet into the Caspian.

Feb. 1*st.*—To Heví, or Hevír, south, 20 m. Fine, hard, level road along the sea-beach. A range of hills, three or four miles to the right, covered with a dense forest which stretches down to the water's edge. At half a mile cross the embouchure of the Khójeh Kerrí stream, which is said to contain a considerable volume of water after the melting of the snow; at present it is fordable; water up to horses' girths. At 7½ m. forded the stream of Búz Kúsham, or Khán-rúd, a village of the same name, said to be about two miles up the stream. Very little water at this season. At 9 m. cross the stream of Levandavil. At 10 m. that of Chilivand. At 16 m. that of Lemír, which separates Astarah from Kerganah-rúd; at 18 m. that of Chúbár.

Hevír, a small village half a mile from the sea, on the banks of a stream;† the wood partially cleared, the houses scattered in groups, surrounded by rice-fields.

At 6 m. from Hevír, cross the stream of Khutwah Seráï, fordable at this season. At 11 m. a Russian fishing station at the embouchure of a small stream. Passed several other small streams, the names of which I could not ascertain. The coast forms a succession of fine bays, with forest down to the beach. Several Russian vessels at anchor two or three miles off shore.

---

* The abandoned fortress of Shindán stands on the summit of a bare, isolated, and rugged rock, nearly 7,000 feet in height, and forms the leading feature of this range of mountains. See Journal, Vol. iii. p. 26.—Ed.

† This stream will admit a vessel of fifty tons, says Colonel Monteith. Journal, Vol. iii. p. 26.—Ed.

At 23 m. is another Russian fishing station of five or six huts, immediately after which the road turns to the west, and ascends by its left bank the Kerganah-rúd (Kerganah river, P.); the pathway through thick underwood.

The village of Kerganah-rúd lies on both sides of the stream; it is the winter residence of Bálá Khán,* the chief of the district. The houses are scattered amongst half-cleared thick wood and rice cultivation.

Returned by a narrow pathway through the forest to the seashore. The district of Kerganah-rúd ends, and that of Asalim begins, about a farsakh to the south of the Kerganah-rúd.

Seven miles beyond is a small village in the district of Asalim, under Muṣṭafá Khán, about a mile up a stream of the same name.

Road similar to that of yesterday. Hills about two miles distant from the shore, their offsets projecting into the sea, with fine bays between the points.

Hills and forest gradually recede from the shore. Road continues along the sea-beach.

At 14 m. from Ḳaláh-seráï, is the village of Cháï-biják, nearly depopulated by the plague. Forest gives place to swamps and brushwood. Soon after leaving Chái-biják, the road enters upon a tongue of land, 13 m. long, and varying from 300 to 600 yards in breadth, extending in an E.S.E direction, which separates the Caspian from the Múdáb,† or backwater.

Road excellent; a sandy ridge hides the Caspian from view.

Enzelí, or Inzilí, is a town of 200 houses,‡ situated at the eastern extremity of the sandy spit; the houses are scattered over a space of half a mile in length, and separated from each other by reed enclosures. Here is a tolerable bázár, the population chiefly Russian. A narrow strait, about 500 yards broad, divides the tongue of land upon which Enzelí is built from a similar strip of sand called Kázerán. There is not depth of water in the entrance to admit vessels of any great burthen into the bay, as it seldom exceeds five feet. At some seasons the water of the bay is so fresh as to be drinkable, there being, it is said, about seventy streams that discharge themselves into it.

From Enzelí to Ḳazvín, distance 114 miles.

S. 35° E. 11 m. leads across the Múdáb, or backwater, to the southern shore, which is covered with thick underwood, and ascend the Yóderúd, a small winding stream not more than fifteen yards broad, to the landing-place at Pírí-bázár (old man of the market, P.), so named after Pír Ḥasan, a saint who lived here in the reign of Sháh Ismá'íl.

---

* Upper Inn. (P.)

† 'Amúd-áb?

‡ Two thousand five hundred inhabitants, says Colonel Monteith, in 1827. See Journal, Vol. iii. p. 18.—ED.

The road for the next $5\frac{1}{2}$ m. is through a quagmire, the mud of which is in some places up to the horse's girths. Dense wood on either side. Such is the communication between its sea-port and Resht, the capital town of the province of Gílán. The present town of Resht* is comparatively modern; it was built near the site of a more ancient city about 350 years ago. The date of its foundation after the Hijrah being preserved in its component letters رشت by their numerical value.†

On quitting Resht in a S.S.E. direction, forded a small stream, and continued along a made road in tolerable repair, although in some places it has been cut through by streams, and rendered nearly impassable for laden mules.

Dú Shambeh-bázár (Monday market, P.), at 8 m., is a small hamlet. Diverged to the left at Dú Shambeh-bázár leaving the high road which leads through Jehennam (Hell), a quagmire said to be utterly impassable at this season, and followed a circuitous pathway over rice-fields, ditches, and marshes, through a wilderness of mulberry-gardens, interspersed with detached huts and small villages.

At 6 m. reach the village of 'Aïn-yeh Ver, or 'Aïn-weh Vezán, on the north-western bank of the Safíd-rúd,‡ and follow its course upwards in a S.W. direction; the road excellent, hard sand and gravel along the bed of the river, which is here about 400 yards broad; the water at this season flows in numberless channels, which cross and recross the road at every half-mile; all these are fordable. The principal stream a quarter of a mile to the left of the road.

At 8 m. beyond, are Imám-Zádeh Háshim,§ and Caravanserai, the former perched on the knoll of a wooded hill to the right of the road, the latter is the work of the Mo'temidu-d-dauleh, late governor of the province. It is as yet unfinished, but workmen are employed upon it, and it promises to be a substantial and useful building.

The road continues along the left bank of the Safíd-rúd, and round the slopes of wooded hills on the right. On either side of the stream are hills covered with forest down to the water's edge. On the left, the mountain of Derfek-dágh, distant about seven miles, covered with snow.

At 16 m. Ráhdárí,‖ or Custom-house station, in a narrow pass

* Resht contains a population of 50,000, and is the most flourishing place in Persia, says Colonel Monteith, in 1827. Mr. Baillie Fraser states its population from 60,000 to 80,000, in 1834.—Ed.

† Rá=200—Shín=300—Tá=400=900; but A. H. 900=A.D. 1495.

‡ White River. (P.) § Imán's Son Háshim. (P.)

‖ Ráh-dár, "road-keeper," "guardian of the road;" whence ráh-dárí, "the office of guarding the roads," the toll levied for their protection, and the place where the toll is collected.

below the ráhdárí; to the left, the Safíd-rúd runs through a narrow channel with abrupt rocks on either side.

At 20 m. a village surrounded by rice-fields, one of a group consisting of six or seven villages, called Rustam-ábád.*

At 22 m. the road winds round hills; the Safíd-rúd flows in an undivided stream to the left at some distance below the road. Forest gradually disappears; the hills on both sides of the stream sprinkled with juniper and willow.

Rúd-bár (stream, P.) at 28 m. is a large village, the houses and bazars scattered amongst a forest of willows. Shortly after a difficult pass occurs, bluff rocks on either side the stream reach down to the water's edge; road stony. At 4 m. cross the river by a bridge of brick and stone, of seven arches, called the bridge of Menjil, built by the Sefevís, partially destroyed in the time of Nádir Sháh, and rebuilt by Suleïmán Khán-kájár about forty years ago. The two centre arches have been lately carried away by floods; the piers, which are still standing, are connected by beams and planks. The junction of the Kizil U'zán flowing from the west, and the Sháh-rúd from the S.E., takes place about a mile above the bridge. The united streams take the name of Safíd-rúd. The road, after crossing the bridge, turns abruptly to the S.E., and leaves the latter river.

Menjil, P. (for Menzil, *i. e.* Station), is a village of sixty houses, about 1½ m. from the bridge, and ½ m. from the right bank of the Sháh-rúd. Menjil† is the last village of Gílán, being on the boundary-line between that province and 'Irák 'Ajemí.

The road leads over low hills, occasionally descending to the right bank of the Sháh-rúd. Willows and rushes to the water's edge.

Much cultivation. A range of snowy mountains in the S.W., eight or ten miles distant.

At 12 m. from Menjil, cross to the left bank of the Sháh-rúd by a bridge of masonry. Shortly after the road quits the river and ascends for one mile the left bank of a small tributary, called Varalú-cháï, to a caravanserai, called Páï-chinár, or Plane-tree-foot. (P.) Thence cross the Varalú-cháï and ascend the face of hills on the right by a steep ascent.

Kharzán, at 8 m. beyond, is a village and caravanserai, the latter a fine building erected by the Mo'temidu-d-dauleh; gradual ascent; the road in some parts is steep and covered with snow, and for 8 m. farther continues ascending mountains, covered with deep snow.

---

* Rustam's abode.

† Menzil is about 800 feet above the level of the sea, or 900 feet above the Caspian, and here and in the pass of Rúd-bár are the only places in which the olive is indigenous. Monteith in Journal, Vol. iii. p. 14.—Ed.

At Mazráh, a village, end of ascent. The plain of Ḳazvín visible from the summit of pass. Descent gradual; snow diminishing.

Three miles beyond is a deserted village on right; road leads over low hills; gradual descent for two miles farther, when the road leaves the hills and enters the plain of Ḳazvín. Pass the hamlet of Aghá Bábá on a small stream flowing from the N.E., and an hour after leave the village of Niẓám-ábád half a mile on the right. Six miles farther, by a level road, the latter part through vineyards, we approach a large town surrounded by a wall and ditch, and enter the city of Ḳazvín. From Ḳazvín to Ṭehrán the distance is about ninety miles, in an E.S.E. direction, the high road leading through a long valley better cultivated than is usual in Persia, and bounded to the north by the lofty range of El-búrz.

III.—*Extracts from Notes, made during the Campaign to Ḳostanṭinah, in September,* 1837. By Major Sir GRENVILLE T. TEMPLE, Bart., M.R.G.S., and Member of the Scientific Commission attached to the French Army in Africa. Read February 12th, 1838.

Kostanṭinah, October 20th, 1837.

September 19*th,* 1837.—I landed at Bónah: on approaching the town and off a point of land lie two rocks called "the Lion," from correctly representing, when seen from a particular spot, a lion couchant. Bónah,* the ancient *Aphrodisium,* is called in Arabic 'Annábah,† "the place of jujubes." The town was destroyed in 1832, but is now rebuilt; the streets have in several instances been widened and carried in straight lines. Many good houses have been constructed, and good shops, reading-rooms, coffee-houses, restaurants, and even a theatre established. The walls by which it is surrounded were erected in 450 of the Hijrah [A.D. 1058.] The ruins of *Hippona*‡ are distant a mile, standing on *mamelons* between the waters of the Sebús§ and the Bejímah. The cisterns are very considerable. I made a plan and took drawings of them. I also observed the site of a theatre, which has never been noticed by other travellers. The bridge over the Bejímah is of Roman construction, and lately repaired.

---

* Bónah is an ancient Arab name, Idrísí (Jaubert) 246, &c.—F. S.

† From 'Aneb; Zizyphus jujuba.—F. S.

‡ The ancient Hippo.

§ The MS. of Sir G. Temple contains many names of places in Arabic, but as the words have been strictly rendered in the Roman character according to the standard adopted by the Society, as given in Vol. VII., p. 243, it has not been thought requisite to print them also in Arabic;—the MS. is preserved for reference in the library of the Society.—ED.

This river is but four feet deep at Hippona, and a much smaller stream than the Sebús. Two streams called Míyáh-báridah, "cold waters," and Wad-edlís, "the river of the rush" (a particular sort so called in the Berber dialect) and both rising about 36° 36′ N., unite soon after, and form a stream, which at first bears the name of the Abú Ufrah, then that of Mab'újah, and finally of Bejímah. The Sebús, which during the time of the Romans was navigable, and on whose banks they had established a dock-yard, might still be rendered so for some distance without difficulty, by removing the bar at its mouth, in which are now only from three to four feet water; inside this, however, it deepens to twelve and thirteen. It runs up S. to 36° 30′ N., and then turns to the W. I shall hereafter speak more of this river. To the east of Bónah and Hippona extend the level plains of the Bení Werjín and the Merdás, uncultivated save where exist three or four oases. To the W. are the mountains of Yedúgh, among which we distinguish the Jebel Esh-hebah,* "grey mountain." To the S.W. are the mountains of Belílyetah, "the weathercock," the highest of whose summits is called Abú-fernún. Between these two ranges runs the valley of the Kharesas, extending for ten miles S.W. to the lake of Fezárah. This lake is said to be ten geographical miles in length, by six miles and a half in breadth: it is very shallow, and abounds with flamingos and wild fowl, and its shores with snipe and wild boar. The French have not commenced the colonization of the environs of Bónah.

September 26th.—I left Bónah and marched with a division of the French army to join the head-quarters at Mejáz 'Amár. Crossed the Bejímah, leaving Hippona on the right: after passing over a few elevations of ground, we entered the great and level plain through which flows the Sebús; this river was on our left. Three block-houses are established between Bónah and the fortified camp of Ed-dere'án; at the second is a Roman bridge of two arches over the Bejímah, or Mab'újah, for I do not exactly know where the stream changes its name: the road which passes over this bridge is the shorter, but we left it for the longer one of Ed-dere'án more to the southward, in order to obtain supplies, as Neshmáyah, the second camp, was too distant to reach in one day with the infantry, artillery, and baggage. At the third block-house, called Sídí Maklúf, from the adjoining sanctuary, is a well; to the right is a large tree, which being seen from a great distance, forms a good land-mark: it is called Shajeratu-s-selám, "tree of peace." The fortified camp of Ed-dere'án, eleven geographical miles S. by W. from Bónah, stands on a low ridge of

* Rising 3189 feet above the sea. Bérard.—ED.

heights. The country about it is inhabited by the Aulád Abú 'Azíz. The derivation of this name (Ed-dere'án) I could not ascertain correctly; some say, from Edrár, the Berber word for a hill, others, from Dhirá', "the arm," "the cubit measure." My own idea is that it is connected with the name of Adrian, who may here have erected a camp or station, as the French have done at the present period. From the Roman bridge at the second blockhouse the Romans had two roads, one passing by Ed-dere'án, on its way to the *Aquæ Tibilitanæ*, another, which crossed the bridge to *Russicada* (near Stora), passing along the level ground on the south of the Fezárah lake. Our horses were watered in the stream of Witsibah, which flows into the Sebús opposite Sídí Denden.

September 27th.—Leaving Ed-dere'-án, we soon after passed on our right an isolated hill studded with fantastically grouped rocks: it is called by the Arabs Ḥajaru-n-naḥal, "the rock of the bees;" but the French give it the more formidable appellation of Montagne des Lions, from the number of those animals they have there met with. On our left and left-front were the hills of El Wust [the middle], and in the latter direction the Marábuṭ of Sídí 'Omar. Six miles from Ed-dere'án we reached the Miyáh Báridah. The Roman road from hence ascends this stream to Askúr the ancient *Ascurus*. We continued our march more to the westward, ascending a steep hill. On our right was the chain of Dra' el 'Arsh,* and on the left the steep peak of Bír-urla, and behind it, at the distance of one mile and three quarters, is Askúr, which I was prevented visiting, as an escort could not then be given. Two leagues and three quarters from Miyáh Báridah we reached Neshmáyah; but the elm trees, which gave it its name, no longer exist, at least I did not perceive them. The fortified camp of Neshmáyah is smaller than that of Ed-dere'án: it is placed in a basin or hollow surrounded by heights, from which it might comfortably be cannonaded. Close by the camp flow four small streams, the Wádí Sehúgah, the Wádí Kúrs, and two others, whose names I did not learn: these united form the Wad-edlís, which united with the Miyáh Báridah, constitute eventually the Bejímah. The Bení Fúghál occupy the country on the west, and the Bení Káïd on the east.

September 28th.—Immediately on leaving the camp we commenced ascending on the south, and by the pass called El-mulfah, the range of the Fejúj, which notwithstanding the road made last year by the French, occasioned much delay to the artillery. On the summit of this hill are the ruins of a *vigie* erected by the French. From this point the eye embraces a great extent of

---

* Dhirá'u-l-'arsh *i.e.* Throne's arm.—F. S.

country, the Kaṣbah of Bónah, Ed-dere'án, the sea, the lake of Fezárah, Hammám el Berda'ah,* Ḳalemah,† the Serj el 'Aúdah, and many mountains, rocky peaks, and winding valleys. The scenery around is fine, bold, and picturesque, but at present, tenantless and treeless. Descending the hill, we find, where we again join the Roman road, the ruins of a small Roman town, perhaps of *Tibilis,* built on a low *mamelon,* and embosomed in olive and other trees. Here we also strike the Wád el ḥammám, "river of the bath," running south to join the Sebús, and bordered by oleanders. Eight miles from Neshmáyah we reached the camp called Ḥammám el Berda'áh, "the bath of the *bât,*" which name it derives from the adjoining tepid springs, that part at least of it which relates to baths; respecting the *bât* I could learn nothing. These waters are the *Aquæ Tibilitanæ,* that is, unless we must place the latter at Ḥammám Meskhúṭín. At all events, they were known and used by the Romans, for we here see considerable remains of their baths, and some slight ones also of their houses. With the stones of these ruins the present fort has been constructed. The temperature of the water varies in parts of the baths from 29 to 30 centigrade (84° to 86° Fahr.). These baths are now surrounded by a thick and luxuriant grove of vines, olives, figs, and oleanders. The troops fully appreciated the luxury of these waters.

September 29th.—We descended the course of the Wád el Ḥammám for two miles and a half, when it falls into the Sebús, which here flows from west to east. On our right were the lower ranges of the Jebel Debbágh, "hill of the currier;" and on the left those of the Jebel Awara, the latter covered with a thick wood of olive trees. Through the vista of the valley formed by the heights, we beheld to the south the towers and walls of Ḳalemah.

The Sebús, where we crossed it, was then only two feet deep, but during the rainy season it sweeps furiously along its course in heavy masses. The French army last year lost many men, horses, and carriages in its passage. It is rendered still more dangerous by its bed being filled with large, loose, and rolling blocks of stone. Ascending its right bank, we reached, after one third of a league, the camp of Ḳalemah. As I before observed, I shall not touch upon the subject of antiquities; suffice it, therefore, to say, that this camp occupies the site and the remains of *Suthul,* which also bore the name of *Calama,* one of the principal and richest towns of Numidia. The ruins of the Numidic city stretch over a great surface; what is surrounded by the still existing stone walls and their thirteen towers, is the space occupied by the much

* Packsaddle-bath.—F. S. † Pronounced Gelmah.

more circumscribed town of the Lower Empire. Kalemah is situated on the northern declivity of a branch of the Jebel Mahónah, one of whose peaks, called Serj-el-'Audah, "the mare's saddle," forms a very striking feature of the landscape. Kalemah is in the territory of the Aulád 'Aríd: the country on the left bank of the Sebús, in the direction of Mejáz 'Amár, belongs to the Werz-ed-dín. The troops did not pass through Kalemah, but pursued their march from the confluence of the Wád el Hammám with the Sebús, along the left bank of the latter. Kalemah lies nearly south from Hammám el berda'ah, and from it to Mejáz 'Amár, the course is a little S. of W. After riding seven miles from Kalemah, we reached the camp of Mejáz 'Amár, "the occupied passage;" the whole of the valley between which and Kalemah, and through which flows the Sebús, is not only beautiful by comparison with the country we had hitherto traversed, but would also be considered so under all circumstances. This valley is embellished by the rushing waters of the river, by a thick growth of trees and shrubs, and by verdant and elastic turf. It is bordered on the S. by the range of Mahónah, and on the N. by those of Bení 'Addah and Debbágh; the outlines of these hills are bold and picturesque, and their flanks are clothed with trees and shrubs, among which predominate the olive, and several varieties of the lentisk. The camp of Mejáz 'Amár, at which the expeditionary army was assembled previous to its advance upon Kostantinah, is situated at the confluence of the Wád-esh-Shárif and the Wád-es-sĕdam, "river of the irruption;" these two waters united, bear the name of Sebús. The former is much more muddy and slower in its motion than the latter, and its bulk of water is greater; it comes from the S., the latter from the W. Two wooden bridges have been constructed by the French over the latter, one for the passage of the heaviest artillery, the other solely for infantry. The right bank in some parts rises perpendiculary to a considerable height. The entrenched camp occupies both banks of the Sĕdam. Ahmed Bey, with the *élite* of his forces, bravely and vigorously attacked the part on the right bank, but was repulsed. The beauty of the country, the positions occupied by the combatants, the variety of their costumes, the fineness of the weather, and the animation and spirit of the affair, served to form one of the prettiest pictures it is possible to imagine. Two and a quarter miles to the W., are the famous hot-springs, called Hammám Meskhútín, "the enchanted baths," which I was prevented then visiting, but hope to see on my return, when we shall have more leisure time. Some excavations were made at Mejáz 'Amár, by order of General Damrémont, and the remains of a Roman villa discovered. I cannot, however, say what name this place bore under the Romans.

October 1st.—The army commenced its march, taking the road by Rás el 'Akabah and the course of the Wád Zenátí, which promised us a greater supply of water than the much shorter way by Alígah. We immediately commenced ascending the pass of Rás el 'Akabah, "the summit of the acclivity," situated between the Jebel 'Anúnah on the left, and the Jebel Sadá on the right. The Roman road to 'Anúnah keeps to the left of the present French one. The country about 'Anúnah is extremely picturesque, possessing a happy combination of mountains, rocks, water, and trees. We bivouacked on Rás-el 'Aḳabah; here rain fell, which continued till late on the following morning. Our horses went to a considerable distance to drink; the artillery found it impossible to reach the summit of the pass this day. The French military road ends at Rás el 'Aḳabah. The distance from Mejáz 'Amár is 7½ miles.

October 2nd.—As soon as the artillery joined us we resumed our route. The country on that side of the mountains where we were now marching presented a totally different aspect from the other: the hills which now bordered our route had not the bold and picturesque forms of the others; not a tree or a shrub was to be seen—not a man or an animal was visible; all was silent, desolate, and dreary. The land had, however, all been under cultivation, which was not the case on the other side, though the ground appeared to be equally fertile. Judging from the stubble, this year's crop was a plentiful one. During the day I observed many remains of Roman posts and stations. We crossed two small streams flowing from left to right; the last is called 'Ain-et-toráb, "the spring of earth;" they both join the Zenátí.* We shortly after crossed this latter river, which here flows a little to the W. of N., and ascending its current, bivouacked round the Marabuṭ of Sídí Tamtám, between the Jebel Insel† on the S. and the Kesen 'Asel, "honey," on the N., distance 6½ miles from Rás el 'Aḳabah. Although we saw no Arabs during the day, we frequently heard and felt them at night, when they permitted themselves to fire upon us.

October 3rd.—Crossed the dry bed of a stream, and afterwards the Wádí 'Asel, with water, which it discharges into the Zenátí on the left. On the bank of the Wádí 'Asel is a small Roman fort of stone. We continued to ascend the course of the Zenátí along its left bank, and towards the end of the march, crossed it three

* When at Kostantínah I was told by some Arabs that the Wádí Zenátí does not join or form the Sebús; but that from Sídí Tamtám it flows north and joins the other waters which discharge themselves into the gulf of Stora. Researches respecting rivers in Africa are not easily pursued, from the variety of names which the Arabs give to them in different parts of their course, thereby creating much confusion.

† Inzil. Dismount ?—F. S.

times near its source, called Rás-ez-Zenátí. This is the spot about which the waters of seven small streams, Seba' 'Ayún, unite to form it. From the heights* beyond the head waters of the Zenátí, we beheld thick and lofty columns of smoke rising from the plain. The Arabs on our approach fired their *dowárs* † and their stacks of straw; this conduct they pursued all the way to Ḳosṭanṭinah. Our cavalry threw itself forward to rescue the forage, and in part succeeded, as it also did in discovering some maṭamórs [subterraneous granaries], in which the Arabs conceal their wheat and barley. We bivouacked on the Wád el Baḳarah, "river of the cow." The country about the sources of the Zenátí is very much broken and tormented, forming numerous cones or mamelons, and has the appearance of being volcanic.

October 4th.—During the night and early in the morning, we experienced considerable cold, the thermometer marking only five centigrades above 0. (41° Fahr.) Soon after leaving the Wád el Baḳarah, we passed on our left the ruins of a Roman *statio,* or small town, and beyond it reached the most southern point attained on our march to Ḳosṭanṭinah, viz., 36° 12′ N. We observed many foot-prints of camels; crossed a stream with no running water, but containing some in the holes or deeper parts of its bed. Crossed the Wádí Muheïrís (the pounding mortar), and bivouacked on its right bank; this river flows to the westward, and under the name of Abú Merzúḳ, joins the Rummel, or Rumlí, "the sandy," close to Ḳosṭanṭinah.

October 5th.—We continued our course, descending along the right bank of the Muheïrís, leaving the Jebel Abú Ghareb on the right, and crossing a stream of the same name which flows into the Abú-merzúḳ, and then ascended the heights of Ṣúma'ah,‡ "the tower, or minaret," so called from a Roman mausoleum which crowns its summit, and which is distinguished from a considerable distance, and bears in fact the appearance of a tower. From this point we enjoyed a magnificent view, rendered extremely interesting to us by embracing in its range the battlements of Ḳosṭanṭinah, the absorbing object of our thoughts. We also beheld the Bey's camp on our left front, the Setáḥ el Manṣúrah, "terrace of victory," the red Jebel-esh-shaṭábah, the perpendicular rocky side of the Jebel Sídí Súleïmán, both being portions of the range of Zuwíyah, the Kudyat 'A'tí,§ &c. Crossing a small stream, and halting for the night on the banks of another, called Wad el Akmimín,|| we were employed for an hour, before making our

* The plateau which divides the basins of the Zenátí and Muheïrís, and separates their head waters, is 2,824 feet above Bónah.—A.

† Dowár, or rather dwár, is a colloquial abbreviation of adwár, (*pl.* of daur,) 'circles,' because the Arab tents are pitched in a circle.—F. S.

‡ Ṣauma'ah, vulgo Ṣma'h, or Súm'ah.—F.S.

§ Coming hill.

|| Distance from Wad el Baḳarah sixteen miles and a quarter.

soup, in a very pretty little affair with the Arab cavalry. Our progress from Rás el 'Akabah had been slow, as the engineers were in many places obliged to make a road for the passage of the artillery and waggons.

October 6th.—Crossing the Wadí Gumas,* we had another little affair with the Arabs, and then ascended to the Marábuṭ of Sídí Mabrúk, 4½ miles from Wád el Akmímín on the Seṭáh, el Manṣúrah, which is a branch of the Jebb-el-Waḥsh.† At this Marábuṭ were established the head-quarters of the army: we galloped on to the edge of the terrace, when suddenly burst upon our sight the whole town of Ḳosṭanṭinah lying at our feet, and separated from us by the deep, perpendicular, and rocky ravine through which rush the waters of the Rumlí. As soon as we were seen from the town, its batteries opened upon us, and every window and rock became animated by the fire of musketry. The women at the same time raised their voices in the usual sounds of *lú-lú-lú,* and the men shouted curses against us. All these sounds were loudly and often repeatedly reproduced by the prolonged and many-toned echoes of the ravine. The siege lasted till the 13th; batteries had been established at Manṣúrah, and on the other side of the Rumlí on the Kudyah 'A'tí, "hill of 'A'tí." The weather during all this period was rainy and cold; there was no shelter, and no fuel to make fires; no forage was to be obtained, so that the horses had only half rations of barley, which we had brought with us; fevers and dysentery raged throughout the camp, to which was subsequently added the cholera; men and horses died rapidly. At last a breach having been made, the town was on the 13th most gallantly carried by assault, and the tri-colour floated from the minaret of the Ḳaṣbah.

Ḳosṭanṭinah or Ḳosanṭínah, قسنطينة, stands on a peninsula formed by the Rumlí, "sandy," or Wád er-rummel, "the river of sand." The part on which the town is built was at one time certainly connected with the heights of Seṭáh-el-Manṣurah, and of Sídí Meshid, and was separated from them, not, as is always stated, by the hands of men, but by an earthquake or some convulsion of nature, though at what period this took place we cannot ascertain, no tradition whatever of such an event existing. Both sides of the ravine are of calcareous rock, covered with a very shallow coat of vegetable earth. We here discovered four strata: the lower one is black, pure schist; the second is calcareous, black, and, on exposure, separates in laminæ; the third calcareous, dark-grey, compact, and very hard; the upper one calcareous, warm yellow, or pale orange, spongy, and filled with fossil remains. These strata are nearly horizontal; the black rocks of

* Ḳummás, or kammás, diver's or leaper's river?—F. S. † Wild well.

Kostantinah, and the black stones with which its principal edifices were formerly constructed, are not, as commonly stated, either of lava or basalt; the stones used in construction were taken from the third stratum. In more modern times the upper stratum has been used as being much softer and more easily worked. The Rumlí, which at or near the town turns twenty mills, enters from the S.E. the deep fissure or ravine called El Huwah, existing between the heights of Mansurah and the town. The entrance is extremely narrow, the breadth there from side to side not being more than four yards. The rocks rise perpendicularly on each flank, but there exist narrow ledges at different elevations, which enable a pedestrian to follow the whole of its course to El Kantarah, from which there exists no difficulty in continuing it to the cascade where the waters debouche from the ravine. A part of the works below the cascade are of marble, and thence called Dár-er-rukhám. El Kantarah, as its name indicates, is a bridge across this ravine, placed at the angle which it here makes; the entrance bearing from it about S.S.E., the cascade W.S.W. From the summit of this bridge to the water of the river the height is 114 yards. The bridge itself, which rests on a natural one called Gorra,* is fifty-two yards. The open arch of the natural bridge, called Dholmah, "darkness," is of considerable height; but I was unable correctly to ascertain what part of the remaining sixty-two yards it occupied; the extreme length of the bridge on its summit from the Báb el Kantarah to its commencement on the opposite side, and following its curve, is 113 yards; its breadth eight yards. Higher up the ravine are either the commencement or the remains of two other bridges or aqueducts, also of Roman construction. The modern part of El Kantarah was built by Mahonese, about forty-five years ago, with, it is said, materials prepared at Mahon! From El Kantarah to the cascade are four natural bridges; the arch of the one nearest the cascade so perfectly resembles a Gothic arch, that at first it appears to be the work of man. The cascade is divided into three falls, which together may be from forty-five to fifty yards, but I did not measure them. The rock which overhangs it on the town side, or left bank, is in perpendicular height exactly m. 109·3 yards, to which must be added a slope formed by the fall of stones and earth, which measures thirty-three yards more; total above the summit of the cascade 142·5 yards: from the top of this rock, named Keff Shakará, women guilty of adultery are precipitated. Kostantinah, before it received its present name from Constantine its restorer, was called by the Romans *Cirta;* but under the Numidic dominion it must certainly have had another name, for *Cirta* corrupted from קרתא *Carta,* the

* Korrah, pupil of the eye?—F. S.

Punic for "a city," was only applied to it by the natives, as we employ the word "town" for London—the city—the town *par excellence.* I entertain not the least doubt of the Numidic city having been of far more considerable extent than the town of Constantina, which was what it still is at present. The Numidic city extended not only over Kudyat 'A'tí, but also on the right of the ravine, and at the base of the Seṭáḥ el Manṣúrah. I will not here bring forward the proofs of my assertions, but for the rest they are numerous and convincing. Ḳosanṭínah still retains a great portion of the walls of Constantina, which extended from where the Rumlí or Er-Rumel enters the ravine, across the neck of land, and, when it was necessary, along the summit of the rocky precipice which forms the W. and N.W. boundary of the town. From the Ḳaṣbah, which almost overhangs the cascade, we see no traces of it along the edge of the ravine; nature had here rendered them superfluous. The Mohammedans have, however, erected batteries near El Ḳanṭarah to defend it. Ḳosanṭínah* (I call it by its Arab name to distinguish when speaking of it under the dominion of the Romans,) has four gates—Báb el Ḳanṭarah; Báb el Jabiyah, nearest to the entrance of the Rumlí into the ravine; Báb el Wád, now closed; and Báb Jadíd, nearest to the river, after it has made the circuit of the town. All these are of Arab construction, built, however, in great part with materials of Roman edifices; the superb gates, with columns of red marble, mentioned by several travellers, do not now exist; and I may here observe, that the ancient edifices have of late years suffered much, and in many instances have been entirely destroyed, in order to obtain materials for the fortifications of the town: especially this was the case with Aḥmed Bey, when preparing to defend himself against the French forces. The town walls on the land side, five feet thick, and in many parts with casemates behind them, are built of Roman wrought stones. Kosanṭínah measures nearly 2700 yards, or $1\frac{1}{3}$ mile geographical, in circumference; the accounts which state its population at between 25,000 to 30,000 are probably correct. The period, however, at which I arrived in the town was not a favourable one for ascertaining its correctness, for, with the exception of dead bodies, we scarcely saw more than 200 or 300 of its inhabitants; the rest all fled when they perceived the certainty of our taking their town. Judging from the size and decorations of the houses, and rich furniture and stores we found in them, a considerable portion of the inhabitants appear to have been very wealthy, and to have indulged in much luxury. Neither did I see any indications of extreme poverty in any of the habitations; there appeared to exist a general degree

* Ḳosanṭínah, a corruption of Ḳosṭanṭínah, has long been in use. See Idrísí, Africa (Ed. Hartmann), pp. 123, 203, &c.

of comfort which is seldom found in any large towns, even in Europe. The greater part of the houses are built from two to five feet above the ground, on large square-cut blocks of the dark-grey calcareous stones, the remains of ancient buildings. Kosṭanṭínah from a distance has not the gay and white appearance of the towns of the East, or even of other parts of Barbary: this is owing to the peculiarity of the houses, not being covered with white-washed terraces, but with tiled roofs, *à dos d'âne*. During wet weather, Kosṭanṭínah, as seen from Manṣúrah, or any of the other commanding elevations, presents itself, from this circumstance, in a most gloomy and dull aspect. None of the mosques, public buildings, or houses, are remarkable for any beauty or elegance of architectural design. Judging from the size and height of the minarets, or rather towers, (for they resemble not the graceful ones of Turkey,) and not from their fame or sanctity, there are nine principal mosques in the town; but since this, the names of thirteen have been given me, besides several chapels. The Bey's new palace, built about eight years ago, is a large edifice, and in its interior very handsome; white marble courts, galleries, fountains, and columns; bright and gaily painted walls; vivid and glossy *azulejos*, with Arabesque patterns; orange, citron, and pomegranate trees; mirrors, and numerous glass lamps suspended in all directions; with a due mixture of rich carpets, cushions, lion and leopard skins, form on the whole a pleasing *ensemble*.

The inhabitants state that Kosṭanṭínah contains 9000 houses and 40,000 persons, but perhaps both these statements exceed the truth; however, from the reasons before given, I can form no decided opinion, and the exact number of houses had not, when I left the town, been correctly ascertained. About 17,000 soldiers might be quartered in the barracks, fonduks, palaces, and the large houses of the wealthier inhabitants, without having recourse to billets on all the houses in general.

Before our arrival provisions were very abundant and cheap; wheat, per sack, 8*s.* 6*d.*; barley, per sack, 4*s.* 3*d*; beef, per lb. ½*d.*; mutton, per quarter, 1*s.*; fowls, each, 5*d.* to 6*d.*

The inhabitants chiefly bake their bread at home; and the few public ovens which existed before those constructed by the French could only bake about 3000 to 3500 rations daily.

The land round the town belongs for the most part to the community in general, and is let out to a few of the principal families; these let it again to the actual cultivators, who receive one-fifth of the produce. The land is fertile and produces generally 30 per cent.

The principal manufactures in the town are those of saddles, bridles, boots, slippers and gaiters. The leather for these objects is dyed of a dingy red colour, with the bark of the delbragh. A few coarse blankets are also made. About twenty-five men were employed by the Bey in the manufacture of gun-powder. All the arms are made in the hilly districts of the Benî 'Abbás. A considerable trade was carried on with the south, from which, in return for corn, saddlery, and objects of European manufactures, the Kosṭanṭinians received gold dust, ostrich feathers, slaves, and the finer sort of ḥáïks, both of wool and silk. From 1200 to 1500 mule-loads of corn, &c., were yearly sent to Tunis, from which European goods were brought in return.

The valley to the N.W. and W., through which flows the Rumlí, after it has disengaged itself from the ravine, is of great beauty; the river winds much in its course, and is bordered by a few villas and numerous gardens, rich in every variety of vegetable and fruit trees, with extensive groves of pomegranate, olive, fig, orange, and citron; the view in this direction is bounded in the distance by a bold range of mountains. North, or little westerly of Kosṭanṭínah is the hill of Sídí Meshiḍ and the Jebel el Waḥsh [the wild mountain], to the east Manṣúrah, to the south Kudyaṭ 'A'tí, and to the west of south, Jebel-esh-Shaṭábah [mountain of heather], the rocky Sídí Suleïmán, and the range of Zawáwiyah. In the Jebel Jebbás (plaister) a part of the last-named range, gypsum is found, and the stones are burnt in the ravine Shabt-er-rúsas. The ancient remains in and around Kosanṭínah, I shall merely enumerate, without entering into descriptions. Where now stands the Ḳaṣbah was the Numidic citadel, some parts of the walls of which are still seen, as well as the remains of a large edifice, with the bases and pedestals of very large columns of the dark-grey stone; these bases measure seven feet square. This must either have been a double temple, or a palace (of Syphax?). It was also the Roman citadel, and called by them, as an inscription informs us, the Capitol; great portions of the Roman walls exist. Here are also large cisterns, divided into twenty-one compartments; and there is also a church, of perhaps the time of Constantine. In the town is a Tetrapylon, or rather was, for one of its gates or arches has been destroyed. Near it, but unconnected, are two other arches. A great part of El Ḳanṭarah is Roman; the bas-relief on it, of a woman and two elephants, was incorrectly copied by Shaw. On the east, or right of the ravine, is a circus, of which the arch, called Ḳaṣr Gulah, (now destroyed,) formed the entrance; near it are two sets of cisterns. At the entrance of the river into the ravine are the remains of a large stone dam extending across its breadth. On the right bank of the Rumlí, and close to its junction with the Abú-Merzúḳ, is a fine aqueduct; six arches of the lower tier only remain, but over the river it must have had three tiers of arches to have carried the water to the elevation at which, on Kudyaṭ

'A'tí, we afterwards find parts of its channel. A short way higher up the stream, are the vestiges of a Roman bridge, and the traces of the road from Cirta to Carthage. On the different heights which, though incorrectly, are generally by Europeans included in the name of Kudyaṭ 'A'tí, are many remains, cisterns, channels of aqueducts, two paved Roman roads, houses, &c. At the foot of the precipice, on the west of the town, are the tepid Roman baths, now called of Sídí Meïmún, the waters mark 31 centr. [102° Fahr.], and abound with tortoises. Lower down are the ruins of a Numidian mausoleum, which, like those of Kubr-er-rúmiyah [Roman graves] and Medrashem, rose by steps to a point; it is however square, and not round. On the ridge between Báb Jedíd and Kudyaṭ 'A'tí was the theatre, another of which existed not far from the Bardo. Many inscriptions exist in and near Kosantínah.

Abú 'Obeïd el Bekrí* states that Kosṭanṭínah is bathed and surrounded by three great and navigable rivers, which rise from the sources called Inghál, or black, and then lose themselves in the deep ravine close to the town. We cannot reconcile this passage with what at present exists, as there are only two rivers, the Rumlí and the Abú Merzúk, for we cannot count the salt stream of Wád el Melaḥ, both from its insignificance and its not entering the ravine; the only manner of explaining the passage is by counting Wád el Hammám and the Wádí Sigan, which uniting a few leagues to the south of Kosṭanṭínah form the Rumlí. But none of these four streams are navigable for even the smallest boats, both from want of water and the rocky and broken nature of the beds. I have before stated that at the entrance of the ravine are the remains of a stone barrier across it; this of course increased the bulk of water above, but was probably constructed more for the sake of irrigation, or defence, than navigation. I may here observe, that if the Rumel did not always flow through the ravine, its former bed was in a valley near that of the Wád el Melaḥ, and that it joined its present bed below the town, at the point where the latter stream now discharges its waters into it: however, these are useless conjectures, for a convulsion which could have formed so great a chasm may well have caused water to spring from a place where it had not before flowed.

My observations on the part of the country I traversed must necessarily, for evident reasons, be extremely few, and those very slight. With regard, for example, to its inhabitants—of their manners, customs, and character, I can form no judgment, as I never met them except in the field of arms, or a few in the town, vanquished, submissive, and at one's feet. In dress and

* Notices et Extraits des MSS. de la Bibliothèque du Roi, xii. p. 516.

personal appearance I have observed no peculiarities to distinguish them from the people of Algiers or Tunis. From Bónah to Ḳosṭantínah I did not see above six Arabs, besides those fighting against us. Ḳosṭantínah contains many Jews, who appeared to maintain the character which their brethren have gained throughout the world. Many Turks were in the town; and they fought admirably; so also did the Kabáïl,* and the Arabs. Many French persons have accused the latter of not displaying much courage, saying that they never stand to receive a charge, even from an inferior force:—this is generally true—but is solely to be attributed to their peculiar style of warfare, and not to deficiency of courage. They would act unwisely to change this system, for it is good, and well adapted to the country; and so has it, as in former days with their ancestors, proved to their enemies. Independently of the women in the Bey's ḥarem I saw but few, and none of any great beauty.

The country from Bónah to Rás el 'Aḳabah is little cultivated, scarcely at all, but it supports numerous herds and flocks. There are in parts woods of considerable extent, but they are composed chiefly of large shrubs, and with the exception of the olive, very few trees—none adapted for ship-building. From Rás el 'Aḳabah to Ḳosṭantínah nearly the whole country is cultivated, and produces much corn and barley, but not a tree or even a shrub, with the exception of a few asparagus bushes, and oleanders on the banks of the streams. To the west of Ḳosṭantínah fruit trees are reared, among which are found a very few date trees, but their fruit does not ripen; the olives are among the largest, and the finest I have ever seen; the cultivated pomegranates are large and of exquisite flavour: this tree, as well as the fig, grow also in a wild state; the wild lotus, in Ar. Sidrah,† is found in abundance, especially to the north. The melons of Ḳosṭantínah are the most delicious that exist—they are called Merḥúm,‡ and are not found at Bonáh; they are green, with yellow spots; the flesh is of a beautiful green colour, and may all be eaten close to the skin, which in thickness does not exceed that of a dollar.

During our advance, the country was parched and burnt up, but after the heavy and continued rains which had so much annoyed us, it almost instantly clothed itself with brilliant verdure, studded with the purple flowers of numberless iris. The Sebús is by far the largest of the rivers we saw; the Sherf, and the Rumel, are the next; then follow the Abú-merzúḳ, the Zenátí, and the Bejímah; some of these would, however, in Europe, never

* Plural of Kabíleh, a tribe.—F. S.

† Zizyphus Lotus.

‡ 'Blessed.' These melons closely resemble those of Ḳaṣabah, or Ṭurghúd-lí, 40 m. east of Smyrna, on the road to Ephesus, which are an incomparable fruit.—F. S.

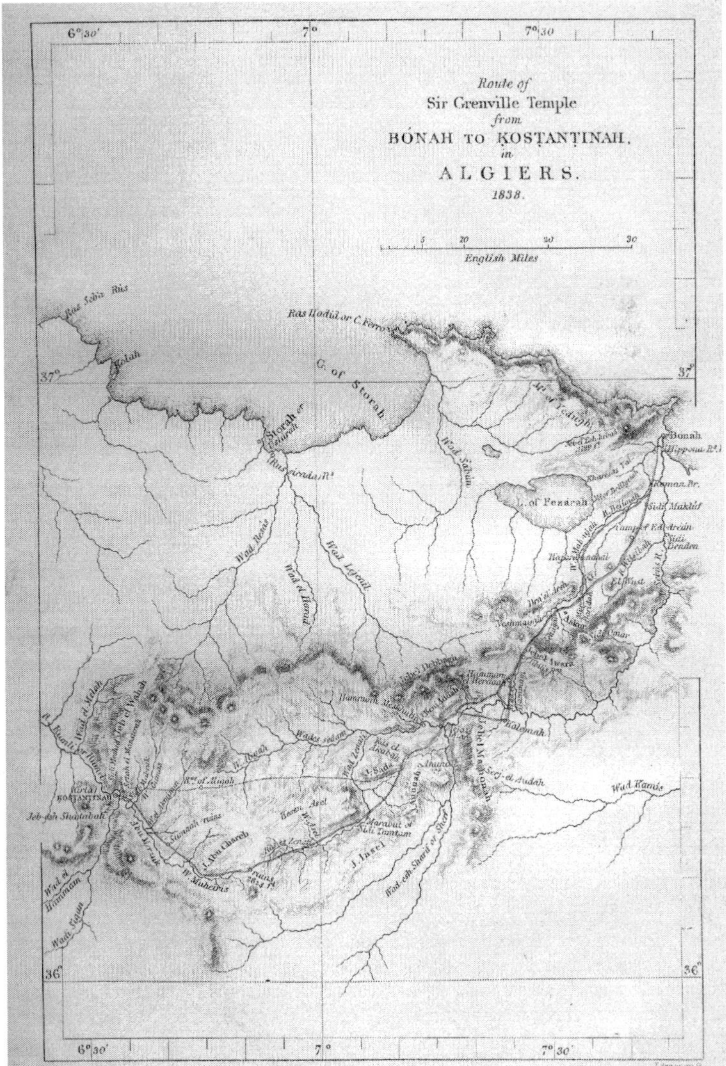

Published for the Journal of the Royal Geographical Society by J. Murray, Albemarle Street, London, 1838.

be dignified by the name of rivers, and many others would never be noticed, so very small are they: most of them, in summer, have no flowing water, retaining only in the holes, or deeper parts of their beds, the remains of what had been furnished by the rains. Independently of the domestic animals, I *saw* but few others, namely, camels, jackals, hares, tortoises, hyænas, and leopards, and only very few of these; there are, however, many lions and wild boars; partridges, snipe, and wild-fowl, are found in great numbers, and in still greater numbers did we find vultures and crows; but the presence of these latter was accounted for by the great number of horses that died during the campaign. Of insects, the most common were scorpions, mosquitos, and a very large and black species of earwig which I had never seen before. The thermometer in my house at Kostantínah ranged from 10 to 15 centigrades (50° to 60° Fahr.), but seldom rose higher than 12 (53·3 Fahr.)

In returning to Bónah, I followed, with trifling exceptions, the same route. From Rás-el-'Akabah, I visited the ruins of *Annona;* from Mejáz 'Amár, I went direct to Hammám Berda'ah, thus cutting off a great angle; and from Neshmáyah, I rode along the course of the Mab'újah, to the Roman bridge near the second blockhouse, thus avoiding Ed-dere'án. Our étapes during the return were as follows:—Wádí Muheïrís, Rás-ez-Zenátí, 'Aïn-et-toráb, Mejáz 'Amar, Neshmáyah, Bónah.

My researches were chiefly directed to the subject of antiquities, and the illustration of ancient history and geography, as well as to the obtaining Arabic MSS. relative to Arab history. I had intended giving the latitudes, longitudes, and elevations of different parts of the Beylik, from the observations made by M. Falbe; but after due consideration, I do not think I should be justified in doing so, as he may wish to publish them first himself in his own country. On the subject of zoology, I may mention, that while at Kostantínah, Ferhád ben Sa'íd, chief of the powerful tribe of Auléd Sa'íd, came from the Great Desert with nearly 1000 cavalry, to offer his allegiance to the French, and on this occasion, I observed that their horses were generally much superior to those we had hitherto met.

The numerous ruins of Roman posts and stations that we saw, though in no way whatever remarkable either for their size, architectural design, or preservation, were extremely interesting, as clearly showing the excellent plan adopted by that nation when colonising a conquered country. These posts were of two sorts, those which secured the roads, and others which guarded the estates at some distance from them; but I have not leisure at present to enter into the details of their colonization system.

IV.—*Notes on a Journey from Tabríz, through Kurdistán,* via *Ván, Bitlís, Se'ert and Erbíl, to Suleïmániyeh, in July and August,* 1836. By Lieut.-Col. J. Shiel. Communicated by the Hon. W. Fox Strangways. Read February, 1838.

At Tabríz two routes were proposed to me for reaching the Turkish camp, which was supposed to be situated to the north-east of Mósul. One was by proceeding to Júlámerik,* an independent Chiefship in the mountains of Kurdistán, bordering on Persia, and from thence through Tiyárí, the territory of the Kaldání (Chaldæan) or Nestorian Christians, whose almost impracticable country joined to their own warlike character, enables them to avoid rendering obedience or tribute to Turk, Kurd, or Persian, and still corresponds with Xenophon's † character of the Chaldæans: "they are said to be a free people and warlike." The objections to this route were the necessity of assuming the character of a Dervísh, that is of travelling under the appearance of great poverty, as my informant said that otherwise there could be no security, and of performing a part of the journey on foot, a portion of the road being totally impracticable for cattle. I therefore selected the route by Ván, although more distant.

July 15th.—We left Tabríz and proceeding in a western direction arrived on the evening of the 17th at the town of Dilmán, situated in the large and fertile valley of Selmás, which is bounded by the Lake of Urumiyah,‡ on the east, and by the mountains of Kurdistán on the west. In this district Armenian Christians are very numerous, and exclusively occupy whole villages; there are also some villages inhabited by Roman Catholics, under the spiritual guidance of a Khalífah or Superintendent, appointed by the Vicar General in Baghdád. Like the Nestorians, these last call themselves Kaldánís, which is probably a national, not a religious designation, while they also style themselves Kátólíks, a name by which they are known all over Kurdistán. A strong mutual antipathy exists between them and the Nestorians. I remember a bishop of the latter church once telling me, that the Armenians were not very good,—the Musselmáns were much worse, but that the vilest of all mankind were the Kátólíks. The contentions of the fifth century are not yet forgotten, and they still retain the doctrine that separated them from the Roman Catholic Church.§ Yet the Nestorians are generally tolerant; they are

* Júlámerk, according to Father Garzoni, who lived eighteen years among the Kúrds. (Gramm. Kurda, p. 18.)

† Anabasis IV. iii. 4. VII. viii. 14.

‡ Urmiyah, according to Hájí Khalífeh, and in Armenian, Urmi (Jihán-numá, p. 385); therefore Urumiyah is a corruption.—F. S.

§ See Asseman. Bibl. Orient. Vol. III. part ii. p. 67 and 199. La Croze, Doucieu, &c. Mosheim, Eccles. Hist. IV. 264, &c.

anxious for instruction, and they have given a welcome reception to some American missionaries who have established themselves in Urumiyah.* The Roman Catholic bishops of Selmás are generally educated at Rome. The present bishop informed me that he had spent fifteen years in the College of the Propaganda, and added to my surprise, knowing the pertinacity with which the Latin language is adhered to in Europe, that the mass and other services were celebrated in the Chaldæan † tongue, which was the only one employed in their books. When I alluded to the contrary practice in Europe, he said that his flock would not submit to any deviation from the customs of their forefathers.

Selmás is also a home for many of the Russian deserters in the service of Persia. Here they marry and settle when they are worn out and unfit for service and form a sort of colony. There are also many Leks ‡ established in Selmás. This is a tribe of the ancient Persian race which dwells chiefly in the south of Persia. They were brought here by Nádir Sháh, but they have forgotten their language, and now speak only Turkish.

Dilmán, commonly called "the City," is a new town built by Amír Khán Kájár, a relation of Fath 'Alí Sháh. There is an old city of the same name placed one farsakh§ to the west, now almost in ruins; the only reason I could discover for removal to the present site, is its greater distance from the Kurds, and therefore greater security. The new town is of considerable extent, and is said to contain 15,000 inhabitants. Like almost all the villages in the district of Selmás, Dilmán is surrounded by gardens; the streets are clean, but the bázárs are poor and ill supplied. Caravans are sent from Dilmán to Ván, Júlámerik, Tiflís, and Erz-Rúm.

Our quarters were in the Mehmán-Khánch, or Guest-house, a substitute for an inn. This, though in Persia not a very honourable mode of entertainment, is far more convenient, and much less expensive than the usual mode of lodging in the house of a private individual.

July 18*th*.—We resumed our journey, and were joined on the march by four Kurd horsemen, who were directed to accompany me to Ván by their chief Yehyá ‖ Khán, to whose sister the present King of Persia is married. He is the head of a small tribe called Chehrí, a branch of the powerful tribe of Hekkárí which rules over a large portion of the central part of Kurdistán. The chief of Júlámerik is the head of the Hekkárís, but his authority

* See Missionary Researches in Armenia, &c. in 1830-31, by Messrs. Smith and Dwight, with an excellent introductory memoir by Mr. Josiah Conder.—Ed.

† That is the Syriac, called Chaldæan by the natives of Kurdistán.—F. S.

‡ A Kúrd tribe. See Journal, Vol. VII. p. 232.

§ Or farsang. Four British miles; for this must be the Tabrízi farsakh, (Ouseley's Travels, iii, 376.)—F. S.

‖ John.

is only nominal, the tribe being divided among a number of Begs,* who pay him no obedience, and they have rendered themselves almost independent of the Turkish Empire, to which they properly are subject.

The escort was well mounted on prancing Kurd horses: this breed is usually small, but strong, active, and hardy, and derived from the Arab stock: the chief of these Kurds was dressed in the usual gay costume of the better class of his countrymen; short yellow boots, large cloth trousers, two or three jackets of striped cotton, a mantle, and an enormous turban of striped red and white silk and cotton, or of handkerchiefs of a variety of colours. Their arms are, a lance carried on the shoulder, and a large pair of pistols in the girdle. The Kurds are said to excel the Persians in courage and in horsemanship, but the fusil of the latter gives a great advantage, for however prodigal a Kúrd may be in exposing his own person, his anxiety for the safety of his steed rarely allows him to encounter the Parthian warfare of the natives of Persia with success.

We passed through the old town of Dilmán, a large portion of which is in ruins. From the number of mounds in the neighbourhood it has the appearance of having been once of considerable extent, and it is described by the Orientalist, St. Martin, as being a very ancient Armenian city. Near the town we saw two very high minarets standing alone in the plain, the mosques to which they were probably attached having fallen to decay; they were said to have been built by the 'Osmánlís, who possessed this part of Persia upwards of a century ago.

My inquiries respecting some ancient sculpture on the face of a mountain near the old town of Dilmán were answered with difficulty: after a long search I discovered that it was half a farsakh to the S.E.; our road was west, and my party having preceded me, together with its being near sunset, obliged me to abandon my intention of visiting it. I learnt that, in the range of hills separating Selmás from the district of Urumiyah, there was some sculpture of a similar description, but whether it has been examined by any European I know not.

Our course was westerly, and two miles after leaving the old town of Dilmán, we entered the hills of Kurdistán: they were small, but of curious shape, as crowns, pyramids, &c. We then turned north-north-west, and ascended the banks of a small stream which runs under Dilmán, and at 9 P.M. the Kurds informed us that we were close to a small deserted Kurd hamlet, and that there was no other habitation for five farsakhs; here we halted for three hours.

July 19th.—We set off at 2 A.M., and ascended through a nar-

* Pronounced by the Turks Bey.

row defile twelve miles in length with low hills on both sides. The road was often excessively bad, which in the dark caused great confusion, and the continued ascent rendered the air so cold, that towards morning a cloak was an insufficient covering. We travelled on the banks of the same stream, each side of which was covered with a profusion of herbs and weeds, and abundance of poppies. At the close of the defile there was an abrupt ascent, and we then entered a valley or rather table-land of considerable extent. After travelling two miles we were opposite to Heráwel-dágh,* which lay about the same distance from us to the west. This mountain had been visible twenty-five miles on the other side of Dilmán, and had then the appearance of great height, but on near approach, it lost much of its apparent altitude.

We soon after passed two Kurd villages, in the neighbourhood of which there was much cultivation, but the habitations in them were of the most wretched description.

We continued our journey through the same undulating valley, and at about the twentieth mile, we were in view of the valley of Elbáḳ,† the entrance of which bore west-south-west ten miles distant; it was said to be five miles wide, and twenty miles in length. It was situated on our left-hand, at the very foot of the black, steep, snow-capped range which rose behind Elbáḳ. The soil we travelled over to-day is Persian, but Elbáḳ belongs to the Hekkárí tribe, and is nominally subject to Júlámerik.

This valley was once rich and fertile, though now its wealth is much diminished, which is ascribed to the emigration of many of its Armenian inhabitants after the last war between Russia and Persia. The Kurds highly value the Armenians, whose industry is a source of profit; they treat them well too, better, at all events, than the Persians, among whom it is not uncommon to carry off their daughters, and force them to turn Mohammedans. A Kurd, the chief of a village, once boasted to me, that he had just enticed an Armenian priest to settle in his village; "for now," said he, "when I invite Christians to establish themselves here, and they inquire about a priest, I am able to say to them, here you have him."

After a tedious march of nine hours, in a north-north-west direction, and various ascents and descents among hills, we arrived at the district and village of Ḳotúr, which are attached to the governorship of Khóï in Persia. Overlooking this village of about sixty houses, is a tolerable mud-fort on a mound, yet it was plun-

---

* Herawel-dágh is probably the Mount Akroual of Colonel Monteith's map, and may be estimated at 9000 feet above the sea.—Ed.

† This must be the valley through which the river Záb flows, in a S.W. direction, as, if our maps are correct, here is the source of that river at about 7500 feet above the sea; the valley is the Ali Baugh of Monteith's map.—Ed.

dered in October, 1835, and held possession of during a month, till all the grain was carried off by a branch of the tribe of Hekkárí, nominally under the Páshálik of Ván, ruled by two brothers named Khán Mehmúd and Khán Abdál.

We were well received at the village, which is chiefly inhabited by Armenians; the Kedkhodá,* or chief, was extremely civil, and supplied us with the best house, which belonged to an Armenian. A more unclean place I have seldom seen than his dwelling, which brought strongly to my recollection the habits of my dirty old acquaintances in Tibet. "The dingy denizens are reared in dirt" most conspicuously. Every body was filthy and in rags; but their poverty was more apparent than real, for the house we lodged in was crowded, like the generality of the other houses, with cows, calves, buffaloes, &c. The houses were not much higher than four feet outside, but as there was a considerable descent at the entrance, the height was much greater within. These habitations did not differ materially from those described by Xenophon after his entrance into Armenia, and the practice he mentions of the inhabitants and the cattle being lodged in the same house is still preserved. There was no symptom of the beer in jars, to which he alludes in the same passage.

The landlord complained of the oppressions which Armenians living at a distance from large towns endure. He stated that each family paid the priest of the village 12*s.* and fifty pounds of bread yearly, and a very trifling sum at marriages and funerals; he considered the pay inadequate.

Though the village was small, the cultivation was extensive; oats were growing wild among the barley, but I do not think they are cultivated in any part of Persia.

The village is situated at the foot of a black, bluff, high range, called Haleb dágh† of which the direction is about east and west: a stream, called here, Ḳotúr Cháï,‡ passes close to the village, and after flowing to the eastward, under the city of Khóï, falls into the Aras.§ The distance we travelled to-day was about twenty-eight miles.

July 20th.—We left Ḳotúr at half-past one, A.M., and crossed to the left bank of the Ḳotúr Cháï, which flows with a very rapid stream. We then entered a wide defile, named Bálánjik,‖ twelve or fourteen miles in length, with the above stream flowing through it. Our general course was west; at about the tenth mile we ascended to the right from the defile, leaving the stream to the left. At about the fifteenth mile, we entered an immense chemen,

* Pronounced at Constantinople, Kyahyá, and spelt Kya by Sir John Macdonald Kinneir. It signifies properly, 'agent, homme d'affaires.'—F. S.

† Apparently Mount Erlau, of Colonel Monteith's map of Armenia.—ED.

‡ Kotur, river. § Araxes. ‖ Bâlânjik, valley.

or meadow, and at twenty-five miles halted at the ruined village of Múllá Hasan, on the bank of a small stream flowing from the south. The horses were let loose to graze in the meadow, but saddled to be ready at a moment's notice. This is the most unsafe part of the road between Khóï and Ván, from the vicinity of the tribe of Khán Mahmúd, who has taken possession of this meadow. During the day, a Kurd of most villanous and wild aspect rode up to our ground: his countrymen of our escort declared that his object was to examine whether we were merchants or Kháns; the latter is the title with which the Kurds have dubbed me at the villages, probably for the purpose of increasing their importance, but the rank is an inconvenient one for my purse. Another Kurd, a very old man, rode up soon after, and complained bitterly of the loss of his cows, which had been stolen last night, and of which he was now in search. This old Kúrd said the meadow was the boundary between I'rán and Islám, at which the Persians with me laughed, and it put me in mind of the Spanish question—Were the English Christians? Neither Turk, nor Persian regards the other as a Musselmán.

We mounted again at 3 P.M., and crossed a stream which rises at the eastern extremity of the meadow, and is called the Mehmúdjik Cháï.* We then entered a defile of the same name, which gradually opened into a valley, and then into a wide plain. The road, as it had been during the entire day's march, was very good. At the thirty-fifth mile from Kotur, we recrossed to the left bank of the Mehmúdjik Cháï, which flows into the Erchekjún lake.

We met a small caravan of bullocks carrying wheat from Ván to Khóï. The Kurds are the only people in Persia by whom I have seen bullocks used for carrying burdens.

At about the fortieth mile, the noble mountain of Sípán dágh was visible; it bore N.W.†

After travelling forty-four miles, we halted at sunset at the Armenian village of Erchek: it was only after much altercation that we were able to procure a habitation. A swaggering Turk, dressed in plaid trousers, with an enormous pair of pistols in his belt, acted as master of the ceremonies, and a very uncourteous one he proved. The servants, with the usual craft of Persians, called me a messenger, if not an ambassador, from the king of the Inglís; even the obtuse Turk felt surprised at this announcement, but it altered his demeanor. Although we had been the whole day exposed to the sun, we felt no inconvenience from the heat. We

* Little Mahmúd's river.

† Situated to the north of the Lake of Ván, distant more than sixty miles.—ED.

must, therefore, be considerably elevated above Tabríz,* where mid-day travelling would be now extremely inconvenient.

Erchek contains one hundred houses, which resemble those at Ḳotúr. The village is about two miles distant from the east side of the Lake Erchekjún. This is a fine sheet of water of an oval shape, its largest diameter being north and south, and perhaps about twelve or fifteen miles in length by eight or nine in width, although the villagers declared the length was at least forty. The water is brackish, and contains small fish of good quality. The lake is bounded by mountains on all sides, excepting the east, where the shores are flat.

July 21st.—We left Erchek at half-past three A.M., and proceeded nearly south along the shore of the lake for nearly three miles, when we were close to its southern extremity. We then passed through some valleys partially cultivated, and over two small passes, the general direction being W.S.W. At about the thirteenth mile after crossing a low pass, we came in view of very fine scenery; the high scraggy naked rock of Warak dágh was six miles distant to the S.E., while to the south lay the plain of Ván covered with villages, gardens, and cultivation. The black rock of Warak† is a remarkable object; the direction of it is from N.E. to S.W., but the mountain is of no great extent, probably fifteen miles in length; high among the crags there is a solitary Armenian church, held in deep veneration; the reason given for placing it there is, that Moses was in the habit of praying upon that spot; but M. St. Martin‡ says that it owes its sanctity to a cross which was erected on the site of the church by a celebrated female Armenian saint, named Hrhip'hsimé, (Rípsimé,) who suffered martyrdom soon after Christianity was introduced into Armenia. A mile farther on, the lake of Ván was visible, and soon after we beheld the rock of Ván and Sípán dágh, while the snowy mountains of Erdóz, which bound the southern extremity of the lake, closed the view. Sípán dágh§ is a splendid mountain; inferior to Aghrí dágh‖ or Ararat, it perhaps nearly equals Demávend and Savelán in Persia: in appearance it exceeds them, but this may be owing to the absence of other mountains of great height.¶ In shape it resembles

* Tabríz must be from 4,500 to 5,000 feet above the sea, as the mean height of the barometer there in March, 1818, was 25,100 inches.—ED.

† Varak tágh. St. Martin, Mém. i. 54.—F. S.

‡ Ibid. ii. 427.

§ Seïbán of St. Martin (i. 52), but Sípán is, doubtless, right.—F. S.

‖ Ibid. i. 42. Jihán-numá, p. 408.

¶ In Colonel Monteith's map of Armenia, Sípán dágh is marked at 7,000 feet, but it must be nearer 11,000; we know that Demávend is 14,500 feet above the sea, or 10,500 above the plain of Tehrán (see p. 72); now the level of the lake of Ván is certainly not less than that of Tabríz, probably it may be assumed at 5,000 feet; and if Sípán dágh looks equally lofty with Demávend, to an eye accustomed to look at the latter from Tehrán, we may venture to place it at 11,000 feet without much a truncated cone: it bore from us about N.N.W., and much snow lay on the summit. Macdonald Kinneir places this mountain at the N.W. angle of the lake, which it almost overhangs; but I feel persuaded that it is much nearer to the N.E. corner. The Kurds say that, before resting on Ararat, the ark touched this mountain, upon which Noah exclaimed, "Ṣubḥanú-llah!" (Praise be to God!) which expression has been converted into Sípán.

We found the villagers ploughing in the plain of Ván. The plough was formed entirely of wood, and, like that used at Urumiyah, had two wheels, which I had not observed in any other part of Persia, one of the wheels being much larger than the other; the object of this arrangement was unknown, excepting that their fathers followed the same practice. The plough was drawn by eight bullocks.

Is-ḥaḳ Páshá, the governor of Ván, received me with civility, and assigned me a dwelling in a very pleasant garden-house attached to his own residence. The Persians accuse the Turks of luxury, and they certainly have some idea of comfort in their sitting-rooms. Instead of the nakedness of a Persian apartment, where a carpet is the only furniture, the room was well supplied with comfortable sofas and cushions.

The Páshá was so impatient to see me, that he sent a message that if I did not visit him, he would come to my quarters, which I of course would not allow. Before I could reach him, I was obliged to pass through an array of at least forty servants, attired in all sorts of garments, Turkish, Kurd, Persian, and Arab, besides others of a very fantastic description. He received me standing, pretending to read a note. The Sunnís, or, at all events, the 'Os̈mánlís, dislike rising to a Christian, and when they wish to be civil, they generally contrive not to be seated when he enters the room. Each nation has its prejudices. A Persian rises, but it is an abomination to him to use the same bath as a Christian, while to this a Turk is indifferent.

The Páshá, who is an 'Os̈mánlí, not a Kurd, was a very gentlemanlike old man, and so like an European in his dress, that, were it not for his Turkish red cap, he might have passed for an Italian or a Spaniard. We drank sherbet, coffee, and tea, while he read the letters of which I was the bearer. He strongly expressed his desire to be of use to any one employed in the business of the King of England, between whom (praise be to God!) and the Ali 'Othmán, the house of 'Os̈mán, there was now, as there always had been, an intimate alliance. The Páshá then sent for

chance of error. It would be an invaluable service rendered to physical geography if some traveller would carry a good mountain barometer throughout this elevated land of Armenia.—Ed.

his Visier,* to consult him regarding the best road for proceeding to Reshíd Páshá's camp. It appeared that they were in complete ignorance of the present position of Reshíd Páshá, and could not determine whether he was at Dyár-bekir, Jezíreh-ibn-'Omar,† or Mósul. The plan he recommended was to proceed to Jezíreh, and there endeavour to discover Reshíd Páshá's movements. To that city there are three roads,—one by Bitlís‡ and Se'rd,§ and the others directly south through the centre of Kurdistán. The first was the least difficult and safest, though the longest; the other routes were very difficult from mountainous and bad roads; and, in addition to this, the Páshá was apprehensive that the Hekkárí and other Kurd chiefs, through whose territories I must pass, would not, now that they were relieved from alarm by the distance of Reshíd Páshá, allow me to proceed on my journey, which their suspicions would induce them to consider as relating to themselves. The Páshá said that any accident that might occur to me would be a discredit to him, and that, therefore, he would send messengers to those chiefs, whose country was not more than thirty miles south of Ván, so that I should not be detained more than two days.

The Páshá visited me in the evening, accompanied by his son, a handsome boy of ten years of age. So great is the deference towards parents among Oriental nations, that the boy did not sit down in the presence of his father, who ordered him out of the room while he drank tea. During three days I was the guest of the Páshá. Early each morning the Ḳehveh Altí‖ was sent to me; this consisted of bread, cream, honey, curds, and then coffee; hence the name Ḳehveh Altí (under the coffee). Some fifteen or twenty dishes were brought for my solitary breakfast and dinner. Contrary to the inconvenient custom of the Persians, where the dinner is placed on the ground, and all eat by stooping down to the dishes, here an immense pewter tray five feet in diameter was placed on a low stool, and covered with food; but Turkish cookery is an abomination. Each dish, whether of meat or vegetables, was filled with grease and curds, and the favourite sherbet consisted of milk, sugar, and garlic, several bulbs of the latter being in the bowl.

Is-ḥaḳ Páshá spoke with great enthusiasm of the improving condition of Turkey, and particularly of the army; he described the redíf,¶ or militia, to be established all over European Turkey,

* Vezír.

† Properly Jezíreh Bení 'Omar, *i. e.* the Island of the Children of 'Omar. 'Omar, being a proper name, cannot take the article. It is more commonly called Jezíreh alone.—F. S.

‡ Bidlís. Jihan, p. 415.

§ Se'ert.

‖ 'The sub-coffee,' or substratum of the coffee drinking; an ante-breakfast.—F. S.

¶ Redíf, a pursuivant in Arabic.—F. S.

and in some parts of Anáṭólí, or Asiatic Turkey. His ignorance on certain points was strange: he inquired if England and India were east or west from Constantinople, and where Austria lay.

The scenery at Ván is the most beautiful I have seen in Asia: the town is situated in a large plain, said to be twelve farsakhs * in circuit, studded with villages and gardens. The imposing mountains of Warak, Sípán, and Erdóz, are in full view, bounding the plain on the N.N.W. and S.E., while to the west lies the beautiful lake of Ván, distant one mile and a half. The rock of Ván is a most striking object. It is shaped somewhat like a camel's back, rising in the centre and falling at both ends. The ridge runs east and west, and is about 600 yards long, divided into three parts, each of which is about 200 yards in length. The rock stands alone, without any other hills in the vicinity, and is therefore more remarkable in its appearance. The middle and highest part is separated from the two ends by dikes, which are cut through the solid rock, so that each part is a separate fortification, the capture of either extremity by no means ensuring that of any other portion. The middle division is about 120 feet in height, and perpendicular on the south side; on the north it is formed in part of a very abrupt rock and in part of an earthy slope, but very steep and strong; the height of the two other positions at the lower part is twenty feet, and the whole rock is encompassed with a wall of stones and earth, with bastions, some of which are square, and the others round. On that part of the north face, where earth takes the place of rock, there are no less than five successive tiers of walls and bastions. The town is placed under the southern face of the rock, and is enclosed with a wall of mud and stone, having large round and small square bastions, protected, though not on all sides, by a ditch. The population, including the suburbs, which are placed in the gardens outside the walls, is said to consist of 12,000 people, of whom 2,000 are Armenians, who are very numerous in this Páshálik. The other inhabitants of the town are chiefly Turks, the Kurds being few. As usual in Turkey, the little trade the town possesses is in the hands of Armenians; the town contains two large churches, four large mosques, two baths, and two caravanserais; though the streets are narrow, the town is tolerably clean; the houses are built of mud and bricks, and, contrary to the practice of Persia, where nothing but a gloomy wall meets the eye, every dwelling has latticed windows to the street, and many have wooden rooms at the top, overhanging the street, where the 'Osmánlís sit and smoke. Over every door the words " Allahu Akber" (God

---

* Forty-three English miles.—F. S.

is great) are inscribed. The bázárs are few, and chiefly inhabited by Armenian weavers and mercers. The manufactures are the coarse cotton chintzes worn by the Kurds and Turks; cotton and corn are imported from Persia, for which money is paid.

The lake is described by Macdonald Kinneir to be twenty-five or thirty miles in length, and from nine to twelve in breadth, yet it has the appearance of being double that extent, and in fact a much greater size is attributed to it by the inhabitants: the water is brackish, but drinkable; a few boats are employed on it in trading between Ván, Akhlát, and Tedván,* on the west side; yet, though we marched several days on the shores of the lake, not one was to be seen. At a village near Ván I saw a boat on the stocks; it was formed of planks six feet in length, six inches in breadth, and fastened with iron nails; the length of the boat was about forty feet, the bottom was flat, and eight feet in breadth, while the top was about twenty.

Ván, from its strength and favourable position near the lake, was probably a place of importance in very remote antiquity, and this is in some degree confirmed by various inscriptions; on the south face of the highest part of the rock, sixty or seventy feet from the ground, there is an inscription about five feet by four in size. Intervening houses prevented a nearer inspection than 150 yards, and even with a good telescope, I was only able to conjecture that it was in the arrow-headed character. I have since learned what I was then ignorant of, that this inscription had been examined and copied by Dr. Schultz, the German traveller, who was murdered near Júlámerik, in 1829,† and who found it to be in the arrow-headed character. This gentleman is reported to have suspended himself from the top by ropes, in which position he copied the inscription. The interior of the fort is said to contain other inscriptions of the same kind, but the Páshá declined giving me permission to view it, alleging that he himself was not allowed to enter the fort without the sanction of the Ser-'asker Páshá of Erz-Rúm, under whose government he is placed.

The climate of Ván is extremely severe; none of the fruit was yet ripe: snow falls about the 20th of November, and sometimes remains six months; a portion of the lake is frozen in very severe winters.

---

* Tátván, remarkable on account of Khosrev Páshá's caravanserai, chapel, mosque, and baths, built A. H. 980=A.D. 1573. Jihán-numá, p. 415.—F. S.

† Professor Schultz, of the University of Giessen, undertook a journey of literary research in Asiatic Turkey and Persia, under the auspices of Baron Damas, in 1826. For some account of his discoveries at Ván, and its environs, where he copied forty-two inscriptions in the cuneiform character, see his letter to M. St. Martin, in the Nouveau Journal Asiat. for 1828, Vol. ii. p. 160—188. This enterprising and much-to-be lamented traveller, on his second journey from Constantinople in 1829, was murdered, it is said, near Júlamerk, or near Dereh, in the valley of the Záb.—ED.

As the Páshá's servants had seen me use a pocket sextant, he sent me a request in the evening to be allowed to examine the moon with it: soon after he returned it, with the remark, that he was unable to distinguish in the least degree better what was passing either in the moon or in any of the stars.

July 22.—The Páshá's messenger not having returned on the 22nd, I resolved to continue my journey by the road of Bitlís: for this purpose, it was necessary to proceed through the territory of Khán Mehmúd, a Kurd chief, nominally under the control of the Páshá of Ván, but who not only paid him no submission, but had lately seized a district called Khavasúr, which had previously been under the authority of the Páshá. This Khán Mehmúd was evidently considered as a very formidable person, and it was obvious that the Páshá did not feel assured in what light he would regard my visit to Reshíd Páshá. His minister, therefore, invited me to dinner, and I there met an agent of Khán Mehmúd, under whose protection I was formally placed. He assured me that his chief would have much pleasure in seeing me, and recommended me to quit the main road and visit him in his castle of Pasvákh. To this arrangement I feigned to acquiesce, though I must confess, without having the least intention of fulfilling it. This agent was a very pleasant, chatty, ignorant fellow, with that sharp knavish look which distinguishes Kurds from the heavy solemn aspect of the 'Osmánlís. He had never before heard of the Yengí Dunyá,* (America,) and was thunderstruck on learning that some of the inhabitants never used clothes: he inquired if they could speak.

During the time we were at Ván, three deserters from the Russian army made their appearance; one was a German, the others Poles. They had deserted from Eriván, but were unable to explain by what road they had contrived to reach Ván. All they knew was, that they had passed through the country of the Kurds, by whom they had been plundered of their clothes: their wish was to reach Constantinople by accompanying an Armenian caravan.

St. Martin, the historian of Armenia, says, that according to the traditions of the Armenians, Ván is a very ancient city, having been founded by Semiramis, and called by her Shemiramgerd: this account appears to be confirmed by the researches of Professor Schultz, who is said to have conceived that he deciphered the word Shemiram in one of the arrow-headed inscriptions which he copied. So late as the fourteenth century there existed buildings which the inhabitants attributed to the ancient sovereigns of Asia, and which were of such ponderous construction, that they resisted the efforts of the soldiers of Tímúr Leng for their destruction. Ruined by the course of time, the city of Semiramis was rebuilt

* New World.

by King Ván, who lived a short time previously to the expedition of Alexander the Great, and bestowed his own name on it; but, having again fallen into decay, it was restored by Vagh-Arshag,* brother to Arsaces, the first king of Armenia of the race of the Arsacidæ, about one hundred and fifty years before Christ. The city fell successively under the domination of the Seljúkís, of Tímúr Leng, of the Turkománs, and finally, it was captured by the 'Osmánlís in 1533, and has remained in their possession ever since that period.

The Greek name given to the Lake of Ván, or at least that ascribed to it by Ptolemy, according to St. Martin, is Arsissa, which is supposed to be derived from Arjísh, a town on the northern side of the lake under Sípán-dágh; by the Armenians the lake is called Akhtamár,† which is the name of an island west of Vastán, and also of a large village on the shores of the lake seven miles south-west of the city: this village is by the 'Osmanlís called Artemíd. In some old maps I have seen Ván called Artemita.‡

The Páshálik of Ván is nominally of large extent, but the authority of the Páshá reaches but little beyond the plain in which the city is placed, excepting to the north, where the towns on the shores of the lake, and those under Sípán-dagh, Akhlát, Arjísh, Eljaras,§ Bègerí,‖ and Albák, are under his control. Several tribes of Cháder-Nishín (tent-dwelling) Kurds live in the northern part of the Páshálik of Ván, which reaches to the territory of Báyazíd. These are the Haïderánlús of 1500 tents, the Sipkí of 1000 tents, the Shúlú of 200 tents, Hamzeh-begí 200 tents, who have the reputation of being excellent cavalry. On the west, Ván is bounded by the territory of Bitlís,¶ but this part of the Páshálik, as well as that on the south, is almost entirely in the hands of Khán Mehmúd.

Five miles south of Ván, is a low range of hills, on the south side of which is the valley of Khavasúr, inhabited by Armenians, which has been seized by Khán Mehmúd. Khavasúr reaches from Warak-dágh on the east, to the valley of Vastán, which lies between the lake at its southern extremity and the Erdóz range. To these mountains Macdonald Kinneir gives the names of Haterash and Hertowshee, but neither of these names is employed by the inhabitants.** South of Khavasúr, and bounded on the south by a high range called Ṣú-suzán-dágh,†† is the large

---

* Valarsaces.

† In Turkish and Persian, Aghtamar in Armenia, in which *gh* is substituted for *l*, Boghos being equivalent to Paulos.—F. S.

‡ Artemita, a Parthian city, is mentioned by Tacitus, vi. 41.

§ 'Adu-ljeváz.—J. N. p. 411. ‖ Bárgírí.—F. S. ¶ Properly Bidlís.—F. S.

** He probably misunderstood his guides who spoke in Turkish-Persian.—F. S.

†† Waterless Mountains, the án is a Persian termination added to the Turkish phrase sú suz, 'waterless.'—F. S.

district of Meḥmúdíyah, of which Khúsh-áb* is the capital. It is the hereditary territory of Khán Meḥmúd, and reaches from Elbág† on the east, to the western extremity of the Erdóz range, which bounds lake Ván on the south. South of Sú-suzán-dágh lies the mountain of Sháh-dágh,‡ with a town of the same name said to be sixteen hours distant from Ván. The town is the capital of the large district and tribe called Hertaushí, nominally under Ván, but perfectly independent. All these valleys are narrow; south of Hertaushí is 'Amádíyaḥ.

Between Ván and Vastán lies the valley of Kavásh, placed between the lake and Erdóz range. The capital of this district is Pasvakh, a strong fort, now the residence of Khán Meḥmud, in one of the crags of Erdóz.

South of Erdóz is the chiefship of Ispert, between which and Buhtán§ lies the district of Mukusú: the Beys of these districts are under no subjection; besides which, almost every hill possesses a fort where the owner resides in independence. Buhtán is a large district stretching on the south to Jezíreh-ibn 'Omer, which until it was taken possession of by Reshíd Páshá, was the capital. Amádíyah seems to bound it on the east and Se'ert on the west.

The above is the information I was able to collect during my stay of two days at Ván, from the few people with whom I was in communication; and must be received with caution. The Páshá had given orders, under the pretence of preventing me from being disturbed, and I am unable to assign the real motive, that no one should have access to me. I may here notice the difficulty of obtaining information in Kurdistán. An 'Osmánlí is generally in profound ignorance of everything not passing before him. A Kurd is more intelligent, but too commonly answers to any question regarding mountains, rivers, or roads—"God knows! how should I know?" Besides this, except in large towns, the Kurds rarely understand any language but their own; but the greatest impediment to information is their extreme suspiciousness. My usual introduction to each village, was the report that there was not a mountain, village, stream, or road which I did not write down; and when making use of a pocket compass, the guides often taxed me with endeavouring to find a road for cannon to conduct the Persians to the aid of Reshíd Páshá to subdue the Kurds. Between Salmás and Ván the road is in general good.

July 23rd.—The messenger whom the Páshá had despatched

* Khúsh-áb—sweet-water.
† Elbáḳ (J. N. p. 420). In Armenian Agpag. St. Martin, i. 177.—F. S.
‡ King's Hill.
§ Bahdínán (the people of good faith).

not having returned, we left Ván in the evening accompanied by a single Kurd, who was to proceed with us to Bitlís. The Vizier explained that it would make no difference if we were accompanied by one or one hundred men, as no one would dare to molest us without the orders of Khán Mehmúd, and if he was averse to our proceeding, the presence of a large or small guard would make no difference in the impediments by which he could oppose our progress.

In return for the civilities of the Páshá, I sent him a watch, with which I heard he was greatly delighted. He soon afterwards sent me a pony which, being perfectly useless to me at the time, I begged he would keep until my return. His attendants, and perhaps he himself, were much offended at my declining his present: they said, that if I did not choose to accept it, I might at least have had the civility to examine the Páshá's horse, praise his points, swear he was of high value and breed, and then request the Páshá's master of the horse to take charge of him until my return. It was impossible not to feel that the 'Osmánlí ideas of courtesy were more refined than those of the Feringí.

We skirted the lake for four miles, and having reached the southern extremity, turned to the west and reached Artemíd three miles farther on. This is a large Armenian village of about 350 houses, placed on some heights above the shores of the lake, and completely buried in orchards, throughout which the houses are dispersed. It has no appearance of antiquity, nor were there any mounds or traces of ruins in the vicinity. The civilities of the Páshá were still continued; an excellent dinner was in readiness on our arrival, and food was supplied for our horses.

July 24th.—We mounted at half-past three A.M. The road was exceedingly bad in some places, and at times over a perfectly smooth rock, where it would seem impossible for a horse to travel; but a Persian horse will go anywhere. Occasionally the road led us close to the lake, which was clear and blue like the sea.* Three miles after leaving Artemíd we were in the territories of Khán Mehmúd. At six miles we crossed the mouth of the Khavasúr valley, and at seven miles the Khúsháb river flowing from the east nearly. Our course was now about west; the lake was close to us on the right, while the black mountains of Erdóz were about two miles distant on the left: these mountains are very high, and at a distance have a particularly rugged aspect. We were now in Vastán, an undulating valley of three miles in breadth. Ten miles from Artemid we reached the fort of Vastán,† a royal residence in the eleventh century, but now without a

* So the lake of Urmíyah, which is thence called Kabudán (blue) by the Armenians. St. Martin, Mém. i. 59.

† Or Usdán, in the territory of Rhesh-duni.—St. Martin, i, 141.

trace of greatness. The fort, a small mud building, is placed on a hill overhanging the lake, while the village and gardens are under Erdóz. Our Kurd attendant insisted that we should ascend to the fort to eat the Ḳehweh-Altí. The 'Os̀mánlís appear unable to travel without resting frequently. Subsequently when I had 'Os̀manlí guides, we never passed a fort or village without their proposing, "Let us enter and rest ourselves: let us eat: let us take some coffee; and let us have a Chibúḳ." The governor, a very ferocious-looking Kurd, with only half an upper-lip, and his garrison of half-a-dozen men received me very civilly. It had been thought necessary to send a messenger to Khán Meḥmúd to inform him of our approach, but I found that he had only just left Vasṭán. I became a good deal incensed, and expressed my anger in warm terms. The garrison understood very little Turkish, and took high offence at what they imagined to be an invective directed against Khán Meḥmúd, but I soon pacified them by a small present in return for the Ḳehweh-Altí.

Leaving Vasṭán, we continued our journey by the side of the lake, and after travelling a few miles we entered the valley of Kavásh, which is a continuation of that of Vasṭán, and presents the same appearance, Erdóz being upon the left four or five miles distant, and the lake on the right hand. About the seventeenth mile from Artemíd, we were opposite to the rocky island of Akhtamár, which St. Martin supposes to have given its name to the lake. On the island, which, however, seems really to be two islands close together, are a much venerated church and a monastery, both visible from the shore: they were once affluent, but were despoiled of their wealth some years ago, by Khán Meḥmúd, the powerful Rob Roy of this part of Kurdistán, or rather Armenia, for such it may more truly be called, almost all the villages being inhabited by that people. A few miles further on, we were opposite to the small island of Limn,* which was said to be five miles distant from us, and to be uninhabited. At about the twenty-fourth mile from Artemíd we were opposite to Pasvákh, the residence of Khán Meḥmúd. We could distinguish the castle among the rocks of Erdóz. At about the thirty-fourth mile we reached the Armenian village of Núr Kúh, distant about one mile from the lake, and situated in a plain, much of which is in cultivation. The valley of Kavásh contains many villages.

The dames of Kurdistán are not held in the same restraint as their neighbours in Persia and Turkey. Among the Khánehnishín, (the dwellers in houses,) only women of high rank conceal their faces; but among the dwellers in tents all exhibit their

---

* Limn opposite to Amig.—St. Martin, i. 137.

features without reserve. The women, among the latter, acquire great control in their families, and have considerable intercourse with the men of their encampment. It is very common for the young men to run away with the young women of another tribe or encampment, which produces violent quarrels, for a Kurd resents an affront of this nature with almost the same vindictiveness as a blood-feud. The Kurd who accompanied me from Ván, in speaking of his countrymen, said, that the dwellers in houses were a bad race, but that the dwellers in tents were beasts, and not to be included among mankind. It is certain that they are, if possible, greater liars than the inhabitants of Persia.

July 25th.—We left Núr Kúh at half-past two A.M., and soon after entered a wide defile, eight miles in length, which was closed by a steep ascent, a short distance beyond which we passed under the high peak of 'Akad,* which lay between the road and the lake: here Demir dágh, the Iron Mountain, a branch of Nemrúd-dágh, was visible. We then travelled ten miles through a mountainous country, ascending and descending alternately. The mountains, which had hitherto been naked rocks or bare earth, now began to assume a more pleasing appearance; they were here tolerably covered with verdure, and occasionally with open woods of stunted oaks. At about the eighteenth mile we descended a steep hill to the valley of Pasvákh, where we were again near the lake which we had quitted after leaving Núr Kúh. Here we halted two hours at the village of Gúleh. After this, we again left the lake, which was hid by intervening hills, and travelled eight or nine miles to the village of Surp. This is a large village pleasantly situated close to a small bay in the lake, with high mountains behind, thickly covered with shrubby oaks. This was the termination of Khán Mehmúd's country. We then travelled along the face of hills, over a very narrow, bad road, overhanging the lake, and passed close to the pretty village of Harzúk, situated in a nook which defends it from the cold in winter. In Kurdistán, near the villages, there is generally a considerable quantity of cultivation, but all the rest is waste. We finished our journey of about twelve hours on horseback, or thirty-five miles, at the large village of Almaliyah, belonging to Bitlís. It is placed near the lake, and is surrounded by orchards. As usual, we underwent the process of being guests. This practice is so habitual in these countries, though not in Persia, that it is usual to count journeys by the same word; thus the distance from Ván to Bitlís would be described by saying, it is four Gonakhs * or guests. The solitude of our march was very striking; no caravans, no travellers, not

* 'Aḳd-ṭágh. Knot-Mount?—F. S.

† The author means Ḳónáḳs; but that word signifies "resting-places," "sleeping-places," not "guests."

even a Dervísh, who is seen almost everywhere, interrupts the silence of these mountains.

The practice of medicine seems to be of a violent description in Kurdistán; suited, perhaps, to the rough, uncouth manners of the inhabitants. The Kurd who was with me, happening to have a fit of the ague, stripped himself stark naked, which is considered highly indecent in all Mohammedan countries; but, according to his own account, being only a tent-dweller, he was a beast, and he then jumped into the lake. His ablution was so little beneficial, that he was soon unable to ride, and found it necessary to mount one of the baggage-ponies.

July 26th.—We mounted at five A.M., and reached Bitlís at ten, about twenty miles. After travelling about three miles through the valley of Almaliyah, we ascended a height and left the lake, having reached its western shore. We proceeded through a plain or valley which was said to extend to Músh: at the eighth mile we were opposite to Demir-dágh and Nemrúd, which were distant about two and five miles respectively, and bore about north. At a distance they both have the appearance of high mountains, which they probably are, but a nearer approach destroys that impression. They rise abruptly from the plain over which we were travelling. We then turned nearly south, having been previously marching about west, and entered a wide defile, with a stream flowing through it, called here Bitlís Ṣúï,* which is said to rise in Demir-dágh. We descended almost imperceptibly for nine or ten miles, and then reached the city of Bitlís.

We were lodged in the Governor's house, a large stone square building inclosing a wide court, and placed on the top of a high hill, where it stood alone, overhanging a part of the city. The Governor, a Kurd Beg, named Sherif Beg, was absent in Reshíd Páshá's camp; but his wife sent his two young sons to congratulate me on my arrival, which they did with the graceful manners one usually finds in Asiatic children of high rank. The Beg's house, in point of furniture, was in a wretched condition; but it possessed strength, which, probably, was the only thing he cared for. A balcony, or verandah,† fifteen feet high, surrounding the court, leads into a number of rooms, few of which had even a ragged carpet to conceal the stone floor. The Anderún or haram ‡ is, however, the place to see the comfort of a Mohammedan, for in that sanctuary he can enjoy it unmolested, and without its being made a plea for extortion. We were, of course, the lady's guests,

* River of Bidlís.

† Varanda, a word borrowed by the Portuguese from the Hindús. To it our Anglo-Indians commonly append the letter *h*, which they as commonly omit in Arabic words, where it ought to be written.—F. S.

‡ Interior.

but her cookery was abominable; the usual curds and garlic were predominant, together with apricots stewed in grease.

The court contained a number of horses ready saddled, in which state they continue the whole day, but at night they are unsaddled: the answer to my inquiries was, that it was prudent to be prepared for every event, for no one could tell what might happen in Kurdistán. I found this to be a general practice.

The city of Bitlís* has a very remarkable appearance: it is placed in a wide ravine, which is open to the east but closed by high mountains to the west; the houses are dispersed over the sides of the steep banks of the stream which runs through it, and on several neighbouring hills. The form of the town is, therefore, most irregular; the houses are built of red stone, which is cut into square blocks, and the generality are of two stories, with grated windows to the street, which produces more resemblance to the towns of Europe than to those of Persia. Like Ván, the streets are paved with round stones. From the irregular manner in which the houses are scattered over the hills, intermingled with gardens, the town covers a considerable extent of ground: it is not inclosed by a wall; but this is scarcely necessary, each house being in fact a fortress, and a strong one too. The town is said to contain 1500 houses, of which 500 are occupied by Armenians.† To this class belong the bakers, butchers, grocers, &c. of the city, they being considered *pure*, in a religious sense, by the Sunnís; while in Persia it has sometimes happened, that they are not even allowed to purchase bread at the same shops as the Mohammedans: on the other hand, a Turk will on no account use the salutation "Selàmun 'Aleïkum" to a Christian, which a Persian does not scruple to do.

Bitlís contains four caravanserais, three large and twelve small mosques,‡ three baths, eight Armenian churches, and one Nestorian: the large mosques have each one very tall minaret, which has a pleasing effect, and they are said to be very ancient Mohammedan buildings. Of butchers, bakers, gun-smiths, and silver-smiths, the number is very considerable, there being nearly twenty of each trade. The principal manufacture is coarse striped cotton cloth, and the chief export is tobacco. Pears, apples, plums,

---

* Bitlís stands in lat. 38° 40′ N., long. 41° 57′ E., according to Colonel Monteith's *Map of Armenia;* and in lat. 38° 34′ long. 42° 30′ E., according to M. Lapie in the Notes to *M. Jaubert's Voyage en Arménie*, p. 475. Macdonald Kinneir in 1814 made its lat. 38° 35′, long. 42° 50′ E. Astronomical observations are much wanted here as well as at Ván.—ED.

† Kinneir, in 1814, says, 12,000 persons, one-half of which are Christians, *Journey through Armenia*, &c. p. 394; and 26,000 in *Geographical Survey of Persia*, p. 331, including the neighbouring villages.—ED.

‡ Kinneir says, about thirty mosques and eight churches, p. 394; but if half the population be Armenian, this number of mosques is probably inaccurate.—ED.

apricots, grapes, melons, cucumbers, lettuces, cabbages, and other vegetables, come to perfection. The climate is cooler than at Tabríz, though much warmer than in the country we had lately travelled through.

The most remarkable object in Bitlís is the old castle, which is placed in the centre of the town, on a rock thirty feet in height, and built up with stone to the elevation of about one hundred feet; the walls are extremely thick, and a single gate leads through the narrow passage which gives admission to the fort. The extent of the inside may be 120 yards; it is now in ruins and filled with old houses. The wall is strengthened by several square bastions on the outside: at the height of sixty feet there is an inscription in Arabic, cut in stone. An old man informed me that, within his own remembrance, there was an inscription on the wall which stated that the castle was built 300 years before Mohammed.

The women at Bitlís walk about with very little concealment of their faces, and display that very ugly Asiatic ornament, common in India, the nose-ring; this appendage is not worn in Persia.

A very reverend-looking Múllá paid me a visit during breakfast: he drank a tumbler of undiluted rum, talked of Columbus, and Napoleon, and of an Englishman he had seen in Bitlís twenty years ago, Macdonald Kinneir no doubt: he took his departure, praying me to come and drink with him at Músh, where he resided.

The territory of Bitlís extends twelve hours towards Se'rt, twelve towards Músh, four towards Diyárbekir, and four towards Ván. The only notice which St. Martin takes of this city is, that it has been almost always governed by Kurd Begs, whose subjects are the most civilised of their race.

July 27.—Accompanied by two guides on foot, we left Bitlís at eight A.M., and descended to the Bitlís Cháï, which is the only name by which the river is here known. We proceeded at first down the left bank, and then passed over to the other side, by one of the numerous neat stone bridges by which it is crossed: the road lay through the narrow defile which is formed by the stream, and was the worst, if not the only really bad one we encountered. The path was high above the river, and sometimes so narrow as scarcely to give room for the horses' hoofs, and at other times so impeded by large stones, that our progress did not exceed one mile an hour. At about the fifth mile the road was crossed by a high ridge of rock which reached to the stream. Instead of carrying the path over it, an opening was cut through the solid rock of fifteen feet in width and height, and twenty feet in length. There was no inscription on it, and the answer of the guides to my inquiry was the usual one of "God knows." At the tenth mile, fatigued by the heat which our descent down the stream had increased considerably, and by the badness of the road, we halted a short time under some walnut-trees. The Persian muleteer cursed himself, his luck, and avarice, that brought him to such a country; and he subsequently had ample reason for repentance, for long before we finished our journey four of his cattle died, besides two of my own horses. The ravine, particularly on the left bank, was enclosed by very high mountains, which probably were offsets from Erdóz: our guides' ignorance of the Turkish language rendered it impossible to obtain any information from them: the mountains were clothed with woods of oak, and the banks of the stream were covered with walnut-trees, mulberries, raspberries, vines, and abundance of herbs, but all enjoyment was destroyed by the heat and fatigue. During the march we changed our guides for two others. Unlike the guides of Macdonald Kinneir, who complains bitterly of their violence and rudeness, we found them alert, civil, and good-humoured. These men were armed with a rifle-firelock, swords, daggers, and a basket-shield studded with brass knobs: their rifles have a piece of iron two inches high, near the lock, with five or six small holes which serve for sights, and excellent shots are said to be made with them. From the mountainous nature of the country, there is very little cavalry in this and the adjacent part of Kurdistán, but every man has a rifle, and therefore the irregular infantry is numerous.

We met a caravan of mules proceeding to Bitlís from Shírwán, said to be eight hours south-east, with salt. They regarded us with astonishment, few Persians ever coming to this part of Kurdistán. My own dress being almost entirely Persian, I was always taken for an inhabitant of that country, and was constantly addressed with the words, "O son of 'Ajem, where goest thou?" At about the twelfth mile from Bitlís we passed close to an old castle, with a ruined caravanserai, on the opposite side of the road: the castle was a square fort with bastions, and was built of large uncut stones and mortar: it looked old, but the only attainable information regarding it was, that it was built by a person named Kaï Fendúk, but who he was, or when he existed, was unknown.

At the sixteenth mile we crossed over a stone bridge to the left bank of the stream, which was rapid and deep, though not more than twelve yards in width: at about the twentieth mile we ascended a high mountain in a south direction, and left the river, which flowed to the west. The ascent was most fatiguing to our exhausted cattle: we then descended to the district of Várkhán, and at eight in the evening, completely worn out, reached the large village of the same name. Although we had been ten hours on horseback I do not think we travelled a greater distance than twenty-six miles, yet so great was the fatigue, that I doubt whether the horses ever recovered from the effect of it.

So great was the change of climate, that we found Indian corn growing here: the night being too hot to sleep in a house, we spread our carpet under some trees. Close to us was Sheríf Beg, the Governor of Bitlís, who had just returned from the camp of Reshíd Páshá. He sent me a dinner of curds, grapes, milk, and that most indigestible of dishes, a wheat piláú. Soon afterwards he came to see me, that he might request me to tell Reshíd Páshá that his country was in excellent order, and that I had been well treated. He was a dashing Kurd of twenty-five years of age, and chiefly remarkable for his dress. It consisted of short yellow boots, blue cloth trousers of prodigious dimensions, three jackets of silk and cloth of different colours, and one of them with sleeves two yards in length; a wide silk sash round his waist, and an enormous turban of silk of every colour: a white Arab cloak was thrown round him, and a dagger, long pistols in his belt, and a sword completed his equipment. In Kurdistán, the sword is worn with the edge to the rear, which the Kurds contend is the best method for drawing the weapon. This dress is very fantastic, but very gay, and is imitated by every one, more or less, according to his means. The lower classes wear the coarse woollen manufactures of their villages, made into a short jacket and trousers. The Armenian villagers can scarcely be said to wear any dress at all: it is generally made of shreds and patches, and the marvel is, how the man gets in and out of it, if he ever takes that trouble. Instead of trousers, the Armenian women wear what the Persian women call trousers of one leg, by which expression a petticoat is meant; they wear, as is customary among the Armenians, a large white cotton veil, and the unbecoming slip of white cloth with which they partially conceal the mouth.

July 28.—We left Várkhán at seven A.M., and soon after arrived near a small stream, down the left bank of which we travelled: the only name I could obtain for it was the Se'rt Sú or river. We crossed it several times, and at the eleventh mile it was twenty-five yards wide, but shallow and not rapid, contrary to the appearance of the Bitlís Chái, or river. We passed a caravan travelling to that city with tobacco and wheat. We saw a number of cotton-fields, which, as well as the hot wind which blew the greater part of the day, showed that we had got into a very different climate from that of Ván. We continued to descend, though not abruptly, by a tolerably good road through an open country; the woods became more scanty, and the mountains decreased much in size. At about the twentieth mile we saw a high range straight and steep, running from about north-north-east, to south-south-west, which was perhaps thirty miles from us to the south-east. Nothing was known of them excepting that they were in Búhtán. After being eleven hours on horseback we arrived at Se'rt,* but I do not think the distance was more than twenty-four miles: the fatigue of yesterday's march had nearly disabled our horses.

Se'rt, or properly Ise'rd,† is placed in the midst of a large undulating plain without a single tree, surrounded at a considerable distance by high mountains. The quantity of cultivation in the vicinity of the city is great, particularly of melons and cucumbers: in the midst of each field there is a small stone house, well loop-holed, for the protection of the property. There are at least twenty of these edifices in the neighbourhood of Se'rt, which give it the appearance of being surrounded by a number of small forts. The town is about two miles and a half in circuit, inclosed by a wall of stone and lime, with round and square bastions, but destroyed in many places, and without any ditch. A great part of the space inside the wall has no buildings, and the city is said not to contain more than 1000 ‡ houses of Kurds, Armenians, and Nestorians. There are three large mosques, and several small ones, two churches, five baths, and one caravanserai. The Governor's house is a large building sunk in a deep moat, which can be filled with water: this castle has bastions and loop-holes in abundance. The houses are all arched and built of stone, with very thick walls; but this does not diminish the heat of the interior. My title of Elchí § was of very little service to me at Se'rt; I was put into a house so insufferably offensive and hot, that it was impossible to sleep in it. The Governor's hospitality induced him to send me a solitary bowl of milk for dinner, and the habit of being well entertained in the large towns made me fancy that he had behaved ill; in fact he had done so according to the notions of the country.

The nearest road to Jezíreh-ibn 'Omer was on the left bank of the Bitlís Chaï (?), and through Búhtán, which it was my intention to pursue, if possible. The distance was only sixteen hours, but I was informed that it was utterly impracticable, as the Beg of Búhtán was in rebellion, and had killed some people belonging to Reshíd Páshá. The impracticability of the road was pointed out to me by the fact that three hundred men were about to proceed to join Reshíd Páshá by another road of thirty-eight hours: they were to start at midnight, and I was recommended to accompany them, which I intended to do, notwithstanding that

---

* Kinneir makes the distance between Bitlís and Se'rt sixty-three miles; he probably travelled by a different road.—ED.

† Se'rd and Se'rt in the Jihán-Numá, pp. 436, 439.

‡ Kinneir says about 3000 persons in 1814.—ED.

§ Or, I'lchí, i. e. Ambassador.

we had only arrived at sunset: but they departed without me. My horses and cattle being extremely fatigued, I hired mules to relieve them for three marches, but they proved a source of great vexation; their slow pace, and the muleteers' habit of halting every farsakh, led to hourly disputes with them.

The following information was given to me at Se'rt: the city of 'Amádyíah is said to be forty-eight hours south-east from Se'rt, but it seems unlikely to be at so great a distance.* From Se'rt to Bitlís there are three roads of sixteen, eighteen, and twenty-two hours respectively: we travelled by the road said to be eighteen hours. Besides this, there is a road of thirty-eight hours to Múshdirect, which does not pass through Bitlís: this must be the road which Macdonald Kinneir supposes the ten thousand to have taken after they crossed the river which *he* calls the Khábúr† at Se'rt. Búhtán is a large district, which extends from Jezíreh twenty-four hours to the north. Diyárbekir,‡ to which Se'rt is now attached, is twenty-four hours from the latter city. A little to the north of west of Se'rt, a high peak is visible, said to be distant twenty miles. It lies in Hhazán, a mountainous ridge which has not been subdued by the Turks, and where a number of Kurd Begs live in disorder and violence. North of Hhazán there is another district in the same state, called Motkah.

Macdonald Kinneir, and I believe D'Anville also, says, that Se'rt represents the ancient Tigranocerta, and was founded by Tigranes the Great, a descendant of Arsaces, the conqueror of Armenia. No traces of ruins were seen by me in the very cursory inspection I was able to bestow between sunset and dark, but I was informed that some do exist. St. Martin says that D'Anville's conjecture is founded only on the apparent resemblance of the last part of the word Tigranocerta to Se'rt: he contends that Amida on the east bank of the Tigris, now Diyárbekir on the west bank, occupies the site of Tigranocerta. The Armenians, he says, call the city of Tigranes, Tigranogerd, and all their writers consider it to be the same as Amida. Tigranocerta, St. Martin states to have been thirty-seven hours from Nisibin, which is certainly less than the distance between Se'rt and that city; neither is Se'rt near Mount Masius, which separated the two cities. Tigranes the Second, or the Great, was defeated by Lu-

* In Colonel Monteith's map of Armenia, the town of Amádíyah lies 57 geographical miles south-east, and Diyár-bekr 70 geographical miles west-south-west of Se'ert: this does not at all agree with the information obtained here, but the true position of any of these places, and especially of Amádíyah, is unknown.—Ed.

† Kinneir must be mistaken in calling this river the Khabúr; it is only known here as the Bitlís Cháï or Se'rt Sú—and there can be little doubt it flows about south to the Tigris: the Khabúr must rise in Amadíyah, as has been already pointed out in Rich's Kurdistán. All our maps are wrong in this point.—Ed.

‡ Diyár Bekr, the tents or houses of Bekr. The *i* in Bekir is a Turkish addition.—F. S.

cullus near Tigranocerta, about sixty-seven years before Christ. He was the descendant of Arsaces the Great, or Mithridates, who conquered Armenia one hundred years after the Parthian conquest of Persia by Arsaces the First (about two hundred and fifty years B. C.). Tigranes the First lived in the reign of Cyrus: he was a descendant of Haig, who 2200 B. C. left Babylon, and founded the kingdom of Armenia.

July 29.—After much trouble we discovered the barn in which the Kurd muleteers had hid themselves to avoid a night-march. We left Se'rt at four A.M., attended by an 'Osmánlí horseman belonging to the governor. After travelling two miles, we saw the Bitlís Cháï, two or three miles distant on the right, and at the third mile passed through the flourishing village of Shírván, of about 200 houses, lying on both sides of the road. At about the sixth mile, over a tolerably good road, we reached the Bitlís Cháï, now swelled into a moderately large river. We proceeded three miles down its left bank, and then forded it a short distance above a small fort on a hill called Gardil. The breadth of the river was about fifty yards, the current rapid, and the depth less than knee-deep. If Macdonald Kinneir's opinion that the Ten Thousand crossed the river which he calls the Khábúr in the vicinity of Se'rt be correct, this ought to be the spot, supposing the information that I had received at Se'rt of there being only one ford in that neighbourhood, to be true.

In passing close to the small fort on the hill, the 'Osmánlís proposed that we should "enter and rest, and eat, and smoke, and drink coffee." This, as the stage was distant, we resisted, upon which they all took their departure, and left us to pursue the journey without them, telling us they would rejoin us afterwards. The road lay at a short distance from the river: at the tenth mile the latter ran towards the east, while the road was nearly south; at the eleventh mile we were again near the river, and at the fifteenth mile we left it altogether, our road being nearly in the same direction: the road was good, and lay, after passing the Bitlís Cháï, through the valley in which the river flows.

The Kurd muleteers had been constantly protesting most bitterly and loudly against the rapid pace at which we travelled. To rest our fatigued horses we were all mounted on mules, and these we never urged beyond a moderate walk. They frequently pressed us to dismount and halt, but we always refused compliance. At last, when we arrived at a most inviting spot for grazing, they lost all patience, and one, who had a gun in his hand, came up and said, in the most positive language his Turkish would admit of, that we must halt. Perceiving that I disregarded him, he took hold of the bridle and endeavoured to turn the mule's head. I hit him a blow on the arm with my stick, on which he released the mule and put himself into a violent passion and flourished his gun with great energy. As I perceived that he had not removed the piece of cotton cloth which is placed in the pan to preserve the powder, and that he therefore could not fire, I took no notice of his antics, but one of my servants dismounted and collared him: they both began to struggle so violently that it became necessary for myself and the other Kurds to interfere and separate them. Both were armed, but neither for a moment thought of using his weapons; a blood-feud is too serious a thing to be readily incurred even in Kurdistán. Seeing that we were determined not to be forced into compliance, they discontinued their opposition and remained behind, protesting that their wives and children would be ruined, as the mules would surely die. This little altercation produced a very advantageous change in the conduct of these Kurds, who during the two subsequent days that they remained with us were very civil and tractable.

Soon after this we saw a river on the left, and asking the 'Osmánlí the name of it, he replied that it was the Búhtán Cháï: supposing this to mean the Khábúr of Kinnier, the river I have called Bitlís Cháï, I inquired no further, as the name of Khabúr is not known here. At about the eighteenth mile we arrived at the village of Til,* belonging to Se'rt, with a small fort on a hill near the Tigris, which the 'Osmánlí guide and several of the Kurds persisted, in spite of my explanations, in calling the Murád. They agreed that it was the river which flowed by Músul and Jezíreh-ibn 'Omar, and I therefore conclude it was the Tigris.

We halted three hours under a tree, where we were joined by the chief of the village and half the inhabitants, who seated themselves about me without the slightest ceremony. An 'Osmánlí or a Kurd considers himself entitled to take a great deal of liberty with every one not a Muselmán, and that he himself, however humble his rank, is superior to every Frank; yet to their superiors of their own religion they show an abjectness resembling that of India, and far exceeding anything practised in Persia. Not a syllable of Turkish was understood by these Kurds, who appeared in complete ignorance of everything beyond their own village.

The chief had ordered breakfast, and when it was ready a long narrow table-cloth of decayed leather, or rather a hide, full of dirt and patches, was spread on the ground: on it were laid four bowls of boiled lamb and broth; the host gave the signal in the name of God, and the havoc began. They washed their hands neither before nor after eating, and seizing the meat with both

* Tell in Arabic, a hill or mound.—F. S.

hands, tore it with their teeth: not a morsel was left undevoured. They then began to discuss what was to be done with me. I was like a bale of goods; each chief had charge of me when I was in his district, sent me out of it in safety, and got a receipt when there was any one that could write.

While we were at breakfast several shots were fired; but of these neither I, nor apparently any one else, took the least notice. I afterwards learnt that they proceeded from the people of the village and those of Búhtán, who lived on the other side of the river called Búhtán Cháï: they were skirmishing from opposite sides of the river.

We proceeded a mile up the left bank of the Tigris, which we forded; it was at least 150 yards in breadth, nearly waist deep, and very rapid. We were now in the Páshálik of Diyárbekr. We then proceeded nearly a mile down the right bank, and reached the village of Móyen close to the river: there was an abundance of vineyards about this village: the vines were not planted in ridges as in Persia, nor trained on stakes as in Europe, but were dispersed irregularly over the fields. To each house there was attached a high platform erected on poles and covered with twigs and leaves; here the people slept on account of the heat.

From this point I saw a river, which since reading Rich's Kurdistán, I believe to be the Bitlís or Se'rt river, falling into the Tigris from the north, at a short distance below the village of Til.

At about the twenty-first mile we proceeded nearly south, leaving the Tigris flowing to the east, and after four miles of very fatiguing ascent and descent the road again approached close to the river, which was now flowing a little to the eastward of south: the Tigris was not above ninety yards in breadth, but very rapid, and said to be very deep: high mountains lay on both sides of the river, and Búhtán was on the left bank.

Late in the evening, at about the thirtieth mile, we reached the large village of Chelek, of four or five hundred houses, inhabited by Kurds and Ya'ḳúbí* Christians, and defended by a strong handsome castle on a rock, in which the chief resides. We were treated here with much inhospitality; the chief neglected to furnish us with any food or corn, and the inhabitants refused to sell anything, on the plea that they possessed only a sufficiency for their own wants: this is a frequent inconvenience in Kurdistán, where the people are unwilling to sell anything; and on the present occasion neither the entreaties nor the remonstrances of the guide were attended with the least effect. He at last contrived to procure our most urgent want, corn, probably by bribing the servants of the chief, but I did not consider any scrupulous inquiry at all necessary.

* 'Jacobite.'—F. S.

Opposite to the village, on the other side of the river, in Búhtán, there was much cultivation, with the owners of which, the villagers told me they skirmish daily. The heat is so great here, that the greater part of the inhabitants had spread their bedding close to the river on the damp sand; they said that it would be otherwise impossible to obtain any sleep. We had latterly experienced considerable inconvenience from the necessity of travelling all day in the sun, but we in general found abundance of fountains in the hills near the road: this water is extremely cold, yet we drank it with impunity in the greatest heat; a hot wind blew during the entire day and night.

July 30.—We left Chelek at 4 A.M., accompanied by a guard of four Kurds on foot, besides the 'Osmánlí. It was a pity they had no language, as the Persians say of a man who cannot talk Persian, for more civil, attentive fellows, I never met. They were employed the entire march in helping me over streams, removing bushes, branches, and stones, which lay on the road, and in stealing cucumbers from the fields we passed through; the latter vegetable is so good and wholesome, that people eat a dozen of them without suffering any injury.

We travelled two miles near the bank of the river, and then left it flowing about S.E. while we proceeded to the south, nor did we see it again until we reached Jezíreh-ibn Omar. On leaving the Tigris, we ascended some hills, and soon after crossed a wide torrent flowing from nearly north, called the Só-úḳ Ṣú.* We then ascended a hill, near the top of which we passed through a tunnel of rock twenty yards in length and about twenty feet in height and breadth: the Kurds could give no account of this excavation, excepting that they believed it to be the work of men: it had the appearance of being artificial, but the object for which it was made was not very apparent, as it was near the top of the hill; the road through it was incumbered with rocks, which had either fallen from the sides, or been left there when the work was discontinued. Below the hill lay the village of Hesáv, in a narrow valley, encompassed by wheat and barley fields, still green, and by extensive vineyards.

About the ninth mile, we passed over a rocky mountain, in part of which steps had been cut, together with holes for horses' feet; beneath, on our right, lay the village and long valley of Derije,† where rice, cotton, melon, cucumber, and pumpkin grounds were cultivated to a large extent. The road, though in general tolerably good, was sometimes exceedingly bad; the mountains had lost much of the magnificence of those in the neighbourhood of Se'rt, and on the left bank of the Tigris; but they were still high and

* Cold water.—F. S.

† Derehjí?—F. S.

well covered with wood; these chiefly consist of stunted oak, oak bushes, fir, holly, and a few elms, together with raspberries, blackberries, barberries, and a profusion of smaller plants. After passing near two other villages, and through a great deal more cultivated ground than usual, we arrived at the Ya'ḳúbí [Jacobite], village of Kermó. We had been eight hours on horseback, but the road being very bad, and there being besides much ascent and descent, the distance did not probably exceed twenty miles.

At Kermó, the inhabitants were most civil and hospitable; they lodged me in a hut made of leaves and branches of trees, which in summer is a far more agreeable residence than a house: the women are dressed in a red cotton petticoat, red jacket, and a red veil or sheet, which reaches to the ground. On their heads they wear a string of silver coins. Kermó contains about a hundred houses.

A large concourse of the villagers assembled round me; they kissed my hand and said, that like all Franks, I was their brother: they called themselves Ya'ḳúbís, or Suryánís,* enemies, as they express it, of the Nestorians, but friends with the Kátólíḳs, the Armenians, and Kupts † of Egypt: their language, they said, was Syrian, but that they also spoke Arabic; and they added, that throughout the Páshálíks of Diyárbekr and Múṣul, their church was the most powerful. Their belief seems to be, that Jesus Christ possessed two natures, divine and human, and one will, while the Nestorian creed ascribes only one nature to him.‡

July 31.—We mounted at midnight, and travelled for eight miles by a continued, but almost imperceptible descent: we then descended a hill, and arrived in an immense rocky plain, with low hills dispersed over it, of which none of the bounds were visible, excepting towards the north: the plain is called by the same name as the small town we were marching to, Mediyád. It is crowded with villages, and in spite of the stones which filled every field, covered with unirrigated cultivation, but of poor and scanty crops. In some of the fields the stones were piled up to the height of fifteen feet. The only water procurable in this plain is from wells of great depth, covered at the top with a large stone perforated to admit a bucket. The inhabitants are Ya'ḳúbís and Kurds, with a few Yezídís; the Christians do not, like other Kurds, in general carry arms, excepting on a journey, but all possess rifles: in their bearing and appearance, they resemble the manly and independent Nestorians of Áẕerbaïján, without any of

---

* Syrian.—F. S. † Ḳubṭs.

‡ This is not quite correct; the Jacobites are not strictly monophysites; *i. e.*, they acknowledge only one nature in Christ, the two being amalgamated into one. (Mosheim, iv. 259.) These Jacobites were probably converts to the Church of Rome.

the meanness which a long course of servitude has imprinted on the countenance and character of the Armenians.

We arrived at Mediyád after ten hours' travelling, or twenty-eight miles: both our baggage and riding horses were so exhausted that a slow walk is the utmost they could be urged to do; the village is large, and inhabited chiefly by Ya'kúbís, and here the governor of the district, an 'Osmánlí, resides. We proceeded to his dwelling to obtain quarters, but he was absent: this house was a long narrow building twenty-five yards in length, the rooms being in an upper story with a wide gallery in front. This was filled with a crowd of 'Osmánlí servants, who scarcely condescended to answer our questions; it was declared impossible to obtain a house, but we were offered the use of the guest's room. I found this to be an extremely comfortable apartment, well supplied with couches, cushions and carpets; but being already occupied by several 'Osmánlís, I preferred an uncarpeted room which was offered in its place. No breakfast was forthcoming after our long march, which was a very unusual piece of inhospitality. I visited the Jacobite church, where I met one of the priests of the village; the burial-ground surrounded the church, and from it issued the abominable smell of bodies in a state of corruption. The church was as bare as the most earnest lover of simplicity could desire, but this probably arose more from poverty than from any other cause. The priest showed me two books in Syriac characters, which he said were the Bible and Prayer-book: he declined selling them, on the ground that they would be replaced with difficulty. Besides the church which I saw, there is a large new one at a short distance, of the magnificence of which the people boasted highly: the priest deplored their poverty; he said that a man possessing two bullocks was rich, and that if he displayed a greater number the Kurds, or Yezídís would certainly steal them; the latter were represented as being much worse than the other Kurds.

Mediyád is placed on the direct road between Jezíreh-ibn 'Omar, and Diyárbekr.

In several of the villages in the plain of Mediyád almost every house was surmounted by a loop-holed turret, reminding one of what Xenophon says, after crossing the river Centrites, on entering Armenia, "upon most of the houses there were turrets."

When we were about to depart from Mediyád, the whole of the 'Osmánlí servants, forgetting their previous neglect, assembled round me, demanding "bakhshísh," *i. e.* a present. The Turks have the reputation of being insatiable on this point: it never occurred during this journey; but it is said, that whenever a visit is paid to a man of high rank, his servants* insist that the honour shall not be enjoyed gratis, and demand payment forthwith: this disagreeable custom has no existence in Persia.

August 1.—We left Mediyád at 9 P.M. yesterday. The road was bad and stony, and soon led us through the defiles formed by low hills. At about the twentieth mile we reached the Jacobite village of Eḥbáb, where we halted for an hour; the general direction of the road had been south and south-east-by-south. After leaving Eḥbáb we ascended some steep hills by an exceedingly bad road, so obstructed by large stones that our tired horses could scarcely make any progress. We then proceeded through a defile three miles in length, called the Pass of Eḥbáb, after which we descended full 1500 feet into an immense plain, which, for want of another name, I shall call the Plain of Jezíreh, although it reaches, at least on the right bank of the Tigris, as far as Múṣul, and perhaps Baghdád. After the fatigue and vexation of travelling so long among mountains, it was now delightful to behold the prospect of performing a part of our journey on level ground.

After travelling three miles over the plain, in a south direction, we arrived at a mud fort on a mound, with a ruined village beneath, called Ernúz. I entered the fort, and was introduced into the chief's apartment, an almost naked room, covered in part with a tattered carpet. He was a wild-looking, dissipated little 'Oṡmánlí, who had been placed there by Reshíd Páshá, in his recent progress from Diyárbekr to Jezíreh. After the Khúshgeldiniz, "you are welcome," and coffee, he proceeded to display his qualifications as a judge, in a case which my appearance had interrupted. A Jew and a Christian accused a Muselmán of theft: before the complaint had been concluded, and without the presence of the accused, this lover of good order swore, by all the prophets he could muster, that he would fine him 200 krúsh,† or 2*l*., and he kept his promise. The thief and a witness were brought in; the judge enjoined the latter to tell the truth,—"Pesawink, kúpek oghlí"—"Sir Pandarus, you son of a dog, if you tell a lie I will cut your throat;" and he looked as if he would do it. After hearing the evidence, he ordered the fine to be paid. "Listen to my witnesses," said the accused; "You son of a dog's dog, pay the money, or I will rip open your belly," shouted this mad 'Oṡmánlí. He kept half the fine, and gave the other to the complainants. He then proposed to me to drink arrack with him, but fatigued with the long ride of twelve hours, although the distance was only thirty miles, and with the clamour in his room, I left his house, and passed the day under a tree,

* It is not usual in Turkey to visit a great man without making a present to him and his servants; but this is not absolutely demanded.—F. S.

† Or Ghurúsh, *i. e.* Piastres.

exposed to a strong hot wind. He was extremely civil in supplying us with food and corn, which was fortunate, as none would have been procurable in the village. He visited me in the evening, and brought with him a bottle of arrack, which he protested was excellent, and insisted on my tasting it. He informed me that the villages in his neighbourhood were inhabited by Kurds, Jezídís, Ya'kúbís, and a few Jews; and that the languages they used were Kurd, Arabic, and Syriac. A vast number of mounds were scattered over this plain; some were bare, others had forts on their summits, and villages below.

August 2.—We mounted last night at ten P.M., the extreme heat rendering it difficult to travel during the day. We were accompanied by an 'Osmánlí courier, and four Kurds on foot, whom the chief of Ernúz considered it necessary to send with us, from apprehension of the Yezídís, from whose incursions, he said, the road was not perfectly safe. These Kurds were themselves Yezídís, but nothing could persuade the 'Osmánlí courier to question them regarding their religion. He declared that any allusion to it would give them deep offence; and besides, they knew nothing whatever about it. Our general course was east and east-by-north. The road for fifteen miles was level, and extremely good, after which it again became covered with large stones. We passed through two villages, called Seseyán and Khánek. At the latter place we were detained two hours in changing the Kurds, without whose protection the 'Osmánlí refused to move. At ten A.M. we arrived at the almost ruined village of 'Aïn-ser, where we fortunately found a few trees to pass the day under, but where no food of any description was to be got at any price.

The plain abounded in villages, but many had been destroyed by Reshíd Páshá three months before, in his passage from Diyárbekr to Jezíreh, in consequence of opposition to his troops by the Amír, or Chief of Búhtán, who had crossed the Tigris, and taken possession of this plain. We are approaching a lofty range of mountains, which seem to run north and south, and which, we are informed, are the mountains of Búhtán,* close to Jezíreh, on the left bank of the Tigris. The mountains called Zákhú Bíng ṭághlar, "the thousand mountains of Zákhú," are also visible, bearing east, and extending east and west. In the evening we travelled five miles south, to the village of Tiláberí. The Múllá, or priest of the village, said that if it had not been for the recent presence of the army of Reshíd Páshá here, it would be impossible for me to travel accompanied by only five or six servants. A large quantity of tobacco was growing at this

* Bahdínán?

village. We were alarmed here by tidings that the plague was raging at Jezíreh, whither we were cautioned not to proceed: this news happily proved false, the disease being only an intermittent fever and ague, which probably always exists during summer in this hot climate. The plain we are in is an enormous meadow, covered with thick high grass.

August 3.—We left Tiláberí at midnight, accompanied only by the 'Osmánlí courier. We travelled by a good road through the same grassy, undulating plain, and reached Jezíreh-ibn 'Omar in six hours. About daylight, in passing through a village where there was some difficulty in finding the road, the courier seized a villager as a guide; but we had not proceeded above two hundred yards when twenty or thirty of the villagers sallied out to rescue their companion: the ground was unfortunately covered with large stones, with which they gave us a most unmerciful pelting, and bruised the courier, and one or two of the horses: he tried to explain that we were not what they probably mistook us for, a party of plunderers; but his Kurdish appeared not to be understood: the servants unslung their arms, with the apparent intention of making use of them, but I loudly urged them to retire and refrain from firing, knowing well that the destruction of our entire party would be the consequence of the death of any of the Kurds. One carbine was fired, but luckily without effect. The guide made his escape in the row, and we, having certainly the worst of the conflict, withdrew and were not pursued.

Two miles from Jezíreh we crossed a stream (running from the Tigris, I think, and rejoining it a few miles below), near the ruins of a bridge, of which no part of the arches is left. It was very massive, built of mortar and large stones, and faced with black stones one foot and a half square. We then descended about 300 feet the low hills which form the bank of the river, and crossed the small arm of the Tigris, which forms the island [Jezíreh], on which the town of Jezíreh is built.* It was only a few yards in breadth, and ancle deep at this season: near the part that we crossed, there was another bridge similar to the former, of which five arches are still remaining. The town occupies nearly the entire island, which is about two miles and a half in circumference: the town is of an oval shape, and is surrounded by a low wall, in ruins in many places, and without a ditch; the wall is faced like the bridge, with the same black square stones. From the low situation of the

* The town of Jezíreh, we learn from Mr. Ainsworth, who carried a mountain-barometer there in the spring of 1837, is about 900 feet above the level of the sea, and the extensive plain over which Colonel Shiel travelled to the south-west of the Tigris is at least 1500 feet in height. Mósúl is about 350, Mardin about 3000, and and Nisibin 1300 feet above the sea.—ED.

town in the bed of the river, and from the height of the banks, the heat is extreme, and not a single tree is to be seen on the island or its vicinity. The most complete desolation existed in the town; it was almost in ruins, and it was only after a long search that we were able to find a wretched hovel to pass the day in: no inhabitants were to be seen; it absolutely contained none, excepting a few hundred sickly miserable soldiers. Plague, cholera, and war had ruined this city; neither barley, nor straw, nor grass, could be got for our horses; no bread, no firewood, nor anything else whatever for ourselves, either from the governor or the bázár; the former was busy in preparing some soldiers for a marauding party across the river into Búhtán. Jezíreh had been the capital of Búhtán; for several years the Amír had refused to pay tribute or acknowledge subjection to the Sultan; Reshíd Páshá had seized his capital, and he was now in rebellion in his own mountains, and had become so formidable that they were not free from apprehension at Jezíreh. It was, therefore, no matter of regret that I resolved while at Ván not to travel through his territories. When we were approaching the town, we heard three cannon shot, and thought they were actually engaged: the firing, however, was in celebration of the capture of Erbíl, by 'Alí Páshá of Baghdád. It was at Jezíreh ibn 'Omar that Macdonald Kinneir was imprisoned and heavily fined by the Kurd Beg. The large palace described by him is now completely in ruins: it is close to the eastern and larger branch of the Tigris; and opposite to it, on the other side of the river, are two pretty white forts, which completely overlook the town. St. Martin says, that Jezíreh ibn 'Omar, the island of the sons of 'Omar, is called in Syriac, Zozartá Zabelítá, and Bazebda, and that it was situated in a country called Zabdicēné, conquered by Diocletian and Galerianus from Nársí, and restored to the Persians after the death of Julian. The Tigris appears to flow from the N.E., before it reaches the island, it then takes a bend and flows from the north: the mountains of Búhtán* are close to the left bank; they look very dark and magnificent, and are so high as to have, even now, some snow on the peaks: they come from N.N.E., but at a short distance above Jezíreh they turn to about east, and slightly south, having an opening of two or three miles at the angle: the river follows the course of this range only a few miles below the town, when the mountains stretch towards the east and the river towards the south: where the river leaves the mountains there is a plain of fifteen miles, bounded on the south by the thousand mountains of Zákhú: the Búhtán range, as well as the Zákhú, which is visible, looks very rugged and formidable; in the former there are said to be nu-

* Júdí Ṭágh?

merous villages of Nestorians and Yezídís. It appearing hopeless to expect any food at Jezíreh, we left it two hours before sunset, in company with a ḳevvás, or courier, which was the only escort the governor had it in his power to grant. Our course for five miles was south, among low hills and near the bank of the river; at the second mile we crossed the stream, over which the first bridge we saw before our arrival at Jezíreh was built. The Tigris was about 100 yards wide, not rapid, but apparently deep. After the fifth mile, our direction changed to S.E. by S., the river being a mile off, and the road lay through the same immense meadow through which we had been travelling, but it was totally without villages or inhabitants, notwithstanding that it was watered by many small streams. Many of the Kurd I'ls,* it is said, who in summer live in Búhtán, pitch their tents here in winter. At about the twelfth mile, when our unfortunate baggage-horses were almost incapable of moving, we arrived some time after dark at a Kurd O'bá, or encampment, about a mile and a half from the Tigris: it consisted of some fifty tents of the tribe of Ḥesenánlí: the camp was pitched in a circle, near the centre of which the chief's tent was placed: the tents were made of coarse black woollen cloths for the walls, with neat roofs of wicker work, and round each tent the sheep, lambs, bullocks, and horses, which were all, excepting the latter, very numerous, and made a terrible din, were assembled. The ḳevvás took us, as a matter of course, and without any introduction, to the chief's, or guest's tent; this was twenty yards in length, divided into two parts by a slight wicker-work, in one of which lived the Amír and the guests, in the other, the women of the chief: by mistake I entered the latter, and advanced to seat myself near four persons whom I supposed to be young men, the sons of the chief: they looked at me with surprise, but did not utter a word, or answer my "selámun 'aleïkum;" at length a Kurd approached and requested me, though without any signs of displeasure, to withdraw. In the men's apartment there were fifteen or twenty people smoking their pipes, who took not the least notice of me; one of them, however, shouted the word, guest, I suppose to the cook, for soon afterwards a dinner of bread, cold cabbage, curds, and honey, was brought in, upon which I made a most hearty meal. From the O'bá, the Zákhú range (western end of it), lay S.E.

August 4.—We left the O'bá accompanied only by the ḳevvás, nor had we a larger escort till we reached the encampment of 'Alí Páshá; but it is probable that the present security can only be ascribed to the vicinity of Reshíd Páshá's camp, and that on other occasions greater precautions in travelling would be

* Hordes, or clans.—F. S.

necessary. We travelled along the banks of the Tigris through well-watered meadows nearly south, and at the fourth mile passed a stream from the north-west, which we were told was the river of Diyárbekr (?); at the fifth mile we reached the ferry of the Tigris; it was the same spot at which Reshíd Páshá crossed the river with his guns and troops: the ferryman told us that he had used floats of inflated sheep-skins for transporting the former. We were delayed two hours from the want of boats: there were but two, and one was so large that the boatmen could not manage it; besides which, there was a body of Kurd horsemen who were crossing from the opposite side. It was very amusing to observe the mode of crossing the horses: they were all collected together and urged by loud shouts into the water, while some twenty naked Kurds seized the manes of the leaders and swam over with them. These horses seemed to understand the affair completely; but not so our Persian steeds, whose sires had never seen a river: they could only be induced to enter the water by pelting them with stones, and when they had swam half way across, they all suddenly wheeled round and returned: this manœuvre they repeated two or three times, but so tractable is the Persian horse, that some days afterwards, when we crossed the Záb, they swam over without the least hesitation. The stream which I was here told was the Diyárbekr river, enters the Tigris a few yards above the ferry, with a rapid current twenty-five yards in breadth. The Tigris is about 220 yards in breadth, with a rapid stream; it comes from the N.W., and above the ferry bends and flows from the N.E.; the west extremity of the Zákhú range bore from this E.S.E.; the town of Zákhú is said to be E. by N. six hours, and placed in an island of the Khábúr, a little to the north of the Zákhú range: the city of 'Amádíyah is said to be N.E. We were informed that boats do not go up and down the river at this part. Keleks, or floats of inflated sheep-skins, descend from Diyárbekr to Músul, where the skins are sold. Having crossed to the left bank of the Tigris, we were in the meadow plain about six miles wide, which separates the Búhtán from the Zákhú range; the former of these rises at least 3000, and the latter about 2000 feet above the plain. We marched to the south-east, while the Tigris took a bend to the south-west, and after two miles we reached the right bank of the Khábúr, flowing from the east and falling into the Tigris two miles to the right: we forded it knee deep, the stream being extremely rapid and fifty yards wide, but from the width of the banks it would appear that after spring it must be 300 or 400 yards in breadth. We then proceeded through well-watered uncultivated meadows, while immense plains of the same description stretched to the west and south on the right bank of the Tigris; at about the twelfth mile a road branched off to Músul, said to be two days' journey distant. We then ascended a low ridge, the tail of the western extremity of the Zákhú range, which lay east from us three miles, and which was probably also the distance of the Tigris from the same spot: the ground between us and the Tigris was undulating. At about the fifteenth mile, the heat becoming extreme, we halted at an uninhabited fort on a hill, and were now a short distance to the south of the thousand mountains of Zákhú, the direction of which was as nearly as possible east and west. Reshíd Páshá had carried his guns from the ferry to the town of Zákhú, which was in the possession of the Amír of Ráwanduz. After capturing that place it was requisite to carry his artillery across the Zákhú range; but the badness of the road obliged him to send it down the banks of the Khábúr, and he transported it over the ridge we had just passed, by a road nearer to the Tigris.

While we were in this fort we were joined by a party of twenty horsemen from Akháltzikheh,* in the province of Ḳárṣ, who were proceeding to join Reshíd Páshá; their chief, a young man, called himself a Georgian, although a native of Akháltzikheh. The inside of the fort not being very clean, he asked permission to sit on my carpet, which was spread in the only shady place under the wall; neither of us had anything to eat, but he drank five cups of tea, and was inclined to continue his libations had I not requested him to desist. He spoke very little Turkish, but he described himself to be an adventurer proceeding to Reshíd Páshá's camp in search of service. He seemed to have succeeded in his pursuit, for I afterwards saw him in the camp in attendance on the Páshá. The Zákhú range is that which Macdonald Kinneir considers as having opposed the progress of the Greeks before they reached the Carduchi. Xenophon describes it to be a place where on one side of the Greek army there were exceedingly high hills, and on the other a deep river; and again, it is noticed as a spot "where the Tigris is, from its breadth and depth, absolutely impassable, no road appearing, and the craggy mountains of the Carduchi hanging over the river." The Zákhú range does not, as Macdonald Kinneir asserts, extend to the bank of the Tigris; on the contrary, it is six miles from it, and the intermediate country is far from being impassable: the Zákhú range is sufficiently rugged, but it does not appear to be impassable to infantry: the range seems to be extremely narrow, and to consist of a single ridge; yet after all, there are no other mountains that appear to represent, so well as those, the spot where the Greeks met with this impediment; for though the Búhtán range

* Or Akhiskah.

represents far more accurately the locality described by Xenophon, yet as in reaching it he must have crossed the Khábúr, it is difficult to suppose he would omit allusion to it; and it is strange that on crossing the Zákhú range he makes no reference to that river, as he must certainly have been close to it. If the Búhtán range were to take the place of the Zákhú range it would be necessary to convert the Erzen branch of the Tigris into the Centrites, which perhaps would place geographers in some difficulty. From the Zákhú range to Se'rt, through 'Amádíyah, is a long distance to march in seven days, as conjectured by Macdonald Kinneir. In the evening we continued our march, and after three hours, or nine miles, reached the village of Kherbenuz. The high road to Múṣul from the uninhabited fort was in the direction of S.S.E., but we proceeded nearly due east, close under the Zákhú range: the road occasionally was rather hilly, but more to the right it seemed very level; we still travelled through the same meadow. At night the grass is set on fire, and immense fires are to be seen in every direction: the black patches thus formed look like woods during the day, but not a single tree is to be seen. About the fourth mile from the fort we passed through a large grove without any village near it, dedicated to a saint; at the fifth we passed through the village of Merjsúr. From Kherbenuz Zákhú city is said to be three hours E.N.E., and the pass leading to it is called Pavishmú; a large portion of the inhabitants is stated to be composed of Jews. We found excellent grapes at this village, but it was only by repeated remonstrances that we were able to induce the inhabitants to sell to us either food or forage: a furious hot wind blew the whole night.

August 5.—We left Kherbenuz at three A. M., and travelled about fourteen miles in five hours to a tree, near a stream, where we halted: for six miles we travelled E.S.E., and then turned into the Múṣul road from Zákhú at S.S.E., which of course separated us from the Zákhú range; the Ḳevvás was most anxious to proceed under the range where the villages are numerous, and where he probably expected good fare: he alleged that by approaching the Tigris, there was danger from the Arabs, who were in the habit of swimming over on their horses; but having heard that the lower was a better and nearer road, I forced him to proceed by it. The Zákhú range had diminished in height, but it had become steeper; its distance from the place where we halted might be about six miles. 'Amádíyah is said to be sixteen hours E. by N. The road was extremely good, and lay through an uncultivated, though sufficiently watered country. At two A.M. we mounted, and travelled four miles S.S.E., three miles S.E., and five miles S.E. by E. to the village of Dúlóm: the heat and glare were excessive, and were accompanied by a strong hot wind.

We were warned against the Bádi Shám, "the wind from Damascus,"* which comes suddenly with a very bad odour, and kills those on whom it takes effect. Garlic roots are said to be a sovereign defence against these gales, but it seems to be a love for the vegetable that has converted it into a medicine. On a very hot day a person brought me what seemed to be a bowl of milk; on tasting it, I found it was full of garlic: I protested against such a compound. "There is no God but the Great God," said he; "what can be better on a day as hot as hell than milk and garlic?"

This part of the eastern bank of the Tigris had been in the possession of the Amír of Ráwanduz, whose followers had destroyed this village on the approach of Reshíd Páshá. The inhabitants praised the discipline of the army of the latter. Múṣul is said to be twelve hours E.S.E. from this; the Zákhú range appears to be about twenty-five miles in length; the whole of it is not exactly E. and W.; six miles before it reaches its eastern extremity, it runs from W.N.W. to E.S.E. and sinks into low hills; it then rises again into steep, bare mountains, under the name of the Sótí range, which stretch east to the Záb (and perhaps further), under the name of the 'Amádíyah and Zebárí Mountains.

August 6.—Having heard that Reshíd Páshá had crossed the Záb, I determined to endeavour to overtake him, and left Dúlóm at midnight, with one pony lightly laden: the course for eighteen miles was about S.E. by E., and then two miles E.N.E. close under the Sótí range; the road was good, and lay among low stony hills. At the third mile we passed close to the village of Gapán, at the ninth that of Dákah, and at the sixteenth that of Búndanah. Under the Sótí range there were many villages, which seemed to be placed in that position for the purpose of security. Our distance from those hills varied from a half mile to two miles: they are still low though steep, but further east they rise into high mountains: their direction is nearly E. and W.; at the twentieth mile we reached the town of Elkósh. Elkósh is a large Kaṣabah,† very strong, and built on the side of a low rocky hill; it is surrounded by a strong stone wall, and the houses are built in the most substantial manner of the same materials; all are arched at the lower story, and being built on a declivity, they rise above each other, which, together with all the houses being loop-holed and very strong, make it a place of great strength. Four years ago the Amír of Ráwanduz cut off the heads of three hundred people for resisting him in this town. The inhabitants are said to amount to two or three thousand people,

* Or from Syria?—F. S. † Market-town.—F. S.

who are all Roman Catholics, and speak nothing but Arabic. The men dress like Kurds of the poorer class; the women wear blue trousers, and over them a large blue shift: they wear no veils; the hair hangs down behind in two plaits or tails; round the head a garland of silver coins is placed. I often examined these coins, which the women allowed me to do without reluctance, in the expectation of finding ancient ones, but they generally consisted of old European coins.

Reshíd Páshá's agent here was most civil: he got ready what I considered an excellent breakfast, but two hours afterwards another made its appearance of a most substantial description; the first was only the Ḳehwah-Altí the host declared, and he refused to partake until I pressed, which in an 'Osmánlí was an extraordinary piece of delicacy. After breakfast, in spite of fatigue and the intense heat, I visited one of the churches; it was a very extensive building, with walls of immense thickness, and was said to have been built by the Jews, to whom the town belonged, but at what time no one knew. The inhabitants themselves looked like Jews or Arabs, but with coarser features; their manner and air were perfectly independent, without the least approach to the servility of the oppressed Armenians. The altar resembled those used in Catholic churches in Ireland; candlesticks were placed on it and a covered chalice; nothing could exceed the simplicity of the whole church, apparently arising more from choice than poverty: there was nothing to disturb this effect except a small frame of woodcuts of French saints, of horrible execution, and among whom St. Louis took the lead. The priest of the church presented himself; he said they were Kátólíḳs, and obedient to the great Pápás in Italy; that they abstain from flesh on Wednesdays and Fridays; that the mass was in Kaldaní* (not Syriac), in which language the Scriptures were written; and that the priests marry once: there was an immense Bible in the church, written in large Chaldee† characters. I then proceeded to view the synagogue. No Jews now reside in Elḳósh, but it is much frequented by them as a place of pilgrimage. The synagogue is a large building, quite as substantial as the church, and, like all synagogues which I have seen, perfectly plain. No Jews are now attached to it; but at certain seasons they assemble from the neighbouring districts to visit the tomb of Náhúm Peïghember, "the prophet Nahum," whose tomb is in the synagogue. The walls are covered with small Hebrew inscriptions on paper: there was a large Bible on two rolls of parchment, inclosed in a

* This is probably a mistake, as the Chaldee has been long extinct; perhaps there may be some difference of dialect between these Mesopotamians and the Western Syrians.

† Syriac; the Chaldaïc character is the same as the common Hebrew.—F. S.

wooden cylindrical case, which opened in the middle; the writing was beautiful, or the printing, for I could not distinguish which it was;* there was an abundance of books in the synagogue in beautiful Hebrew characters. Two miles east of Elkósh, placed high among the crags of Sótí, there is a large and extraordinary-looking church, which heat, fatigue, and the lateness of the hour prevented me from visiting: the Catholic guide attached the highest importance to the edifice, though he did not seem to know exactly why, or perhaps we did not understand him, as he spoke nothing but Arabic. We quitted Elkósh at one o'clock, and travelled nearly east in a blazing sun, with an intolerable glare, the road being close to the Sótí range, which was low, perhaps 1000 feet above the plain, but rugged and steep; at the twenty-second mile we passed the village of Borjá; at the twenty-fourth mile the Yezídí villages of Hespedún and Taïtah; and at the twenty-eighth mile the Yezídí village of Bádú. All these were on the left, close under the Sótí range; while on the right were the villages of Bóbán Teseher, and several whose names I could not learn. All these villages have forts attached to them; there was one very remarkable-looking fort and village perched on a high mountain, with a rampart at the top. At the thirty-second mile, when we were passing through the Yezídí village of 'Alì Aghà, close to the Sótí range, and about a mile from the village where we intended to make our halt, several of the attendants of the chief rushed down from an eminence they were standing on, and seizing my horse's bridle, insisted upon my becoming their master's guest that night. Hoping that this would give me an opportunity of inquiring into their religion, I willingly complied. They conducted us to the roof of a house, where the chief, accompanied by a number of his relations, paid me a visit. Not one of them could speak anything but Kurdish. Coffee was brought, and soon after a large substantial dinner, with wheat instead of rice for piláú; every one ate in the strictest silence, and every one of the twenty who formed the party despatched his dinner before we had time to begin. This was the third good meal we had had that day, and formed a strong contrast with our previous abstinence, for after the first march from Se'rt, I had eaten little but bread, which we had brought from Persia, and tea; this sometimes arose from there being nothing procurable, and more frequently from the feeling of fatigue being greater than that of hunger. The whole party quitted me at the same moment, and gave me no opportunity of making inquiries; I urged the kevvás to attempt it, but he declined to have anything to say

* Manuscript, no doubt, as rolls are never printed, and such would be an abomination to the Jews, whose synagogue rolls are written according to prescribed rules with great care.—F. S.

about it. We thus found these far-famed Yezídís the most civil people in Kurdistán; they are reported to be the greatest robbers among their countrymen, and their civility may perhaps be ascribed to the vicinity of Reshíd Páshá. The common reason assigned for their being called Yezídís is their approval of the murder of Husëin, grandson of Mohammed, by Yezíd: they are said to worship, or at least to deprecate the Devil, calling him God: this would appear to suggest the opinion that the old religion of Persia is not eradicated from among those people, and that the Devil is the Principle of Evil, which, under the name of Ahrimán, the Gebrs worshipped as the equal of Hormúzd, the Author of Good. 'Amádíyah is said to be eight hours from 'Alí Aghá, and Músul to lie S.S.W.

August 7.—We left 'Alí Aghá at three A.M., and reached the village of Házirjút at ten, about twenty miles distant, having travelled over an excellent road and through a flat country: for eight miles we went E. then E.S.E. to Házirjút; at the ninth mile passed near the village of Jewán, and soon after near the Yezídí village of Memvesh at the twelfth mile the high single mountain of Meklúb, rising abruptly from the plain about 1200 feet, was on our right, two miles distant. At the third or fourth mile the Sóti range disappeared, but it was soon succeeded by another remarkable, black, rugged, and bleak high range, perhaps 3000 feet in height, running in nearly the same direction E. and W.; the name of this ridge is El Khaïr, or the Mountains of 'Amádíyah, which lies on the other side of it: further east, this ridge bends a little to the south; at the foot of it numbers of Yezídís dwell. Fourteen miles from 'Alí Aghá, we crossed a stream flowing from the N.E. called the Gómel. Most probably the battle of Arbēla took place somewhere in this neighbourhood. The ground is level as described by the historian, yet it is strange how he could avoid allusion to the remarkable range of El Khaïr. From Házirjút nearly in the direction of east, a very high mountain was visible, which we were told was the mountain of Ráwanduz, distant twenty-six hours: this district is called Kalkúsh: it is level, and contains many villages, and a considerable quantity of cultivated ground. We left Házirjút at four P.M., and travelled in an east and east-by-north direction eight miles to Kelí, having passed near the villages of Ashtín and Khenáb Kendig. The El Khaïr range here bends slightly towards the south: from Kelí another excessively bad pass into 'Amádíyah lay north-north-east, distant three miles. From Kelí 'Amádíyah is said to be twelve hours; Tiyárí, the district of the Nestorians, fourteen hours, to reach which are passes through 'Amádíyah. Ráwanduz is twenty-four hours east by south, and Múṣul eleven hours west-south-west. At eight P.M. we left Kelí, and travelled eight hours on a very bad and hilly road, but not impracticable for cannon, Reshíd Páshá having conveyed his guns hither. After having travelled that day not less than thirty-five miles, we halted at a short distance from a village. The heat both night and day was very great.

August 8.—We marched at daybreak, and after two miles to the east and two miles to the north, over hilly ground, we arrived near the town, or rather large village, of Akereh, of about five hundred houses, surrounded by fine gardens. We learned that Reshíd Páshá had left Akereh the day before, and had proceeded farther eastward: he had besieged, or rather surrounded the fort for twenty-two days. It is a very strong castle, placed on a rock projecting from El Khaïr, the town being at the foot of the rock. He contrived to bring a couple of guns to bear on the masonry of the wall above the rock, upon which the garrison surrendered, though without the least necessity; for even if he had succeeded in breaching it, he had no troops that would storm such a breach. None of the gardens about Akereh were destroyed or plundered, which shows the progress the Turks have made in this important point of discipline; the same remark is applicable wherever they marched. After leaving Akereh, we went twelve miles in an easterly direction, and then three miles to Reshíd Páshá's camp: we travelled at the foot of the El Khaïr range, crossing the hills at its base, from which many torrents gushed: several villages were placed high in the inmost recesses of the mountain, and being in general surrounded by trees and some cultivation, presented a delightful contrast to the adjacent sterility. The peasants who passed us spoke no language but Arabic, so that, having no interpreter, we could not make any inquiries: we were even deprived of our very useless kevvás, who vanished most unaccountably in the morning; nor did he ever appear to claim his fee. At the tenth mile we passed through the large village of Zin-jí, inhabited by Arabs: close to it there is a violent torrent with a stone bridge over it, and a path leading to Ad-desht,* which appears to be a level tract lying between 'Amádíyah and Zebárí. The torrent seems to have cut down the rock three or four hundred feet, and the pass is only practicable to men on foot. It is certainly a strange and gloomy defile: rice, cotton, pomegranates, figs, &c. grow here in abundance. The El Khaïr range soon afterwards loses its steepness and sterility, and becomes lower and more verdant; the mountains extend as nearly as possible from east to west; the Záb is said to flow on the other or northern side. At the fifteenth mile we reached the camp, which was pitched on a mountain at some distance from any village.

---

* The plain.

August 9th.—On the 9th the extreme heat and burning wind rendered me exceedingly ill; I was attacked by a violent pain in the side, and complete loss of strength and energy, but no fever; it appeared to be a sort of stroke of the sun: my arm and shoulder became red and painful; I was covered with blisters. In consequence of this, I removed on the 10th to a deserted village, one mile and a half from the camp; in the afternoon, having learned that the Páshá was about to march, I prepared to follow him. We marched north into the district of Zebárí, crossing the range, which is a continuation of the Mountains of 'Amádíyah, but, though very high, not having the same rugged, barren aspect: the ascent was extremely fatiguing, and the descent in the dark nearly as much so. We travelled about ten miles to the camp, which was placed in a fine valley with several villages in it, but uninhabited; grapes, figs, and walnuts grew wild: no one knew the name of this valley, but the Záb was said to be three hours to the N.E. There were no Kurds in this camp, which was a small one of 2000 men, the greater part of the army, 7000 regular infantry, being in advance some miles. One Kurd whom I met told me we were in the district of Zebárí, which had been attached to 'Amádíyah before the Amír of Rawandúz seized the latter Páshálik. Tiyárí, he said, lay to the N.W., 'Amádíyah W.N.W., and Rawandúz E.S.E. He described the district of Zebárí as being one day's journey and a half in length, from north to south, and one in breadth, and as consisting entirely of mountains. That of 'Amádíyah, he stated to be eight days' journey in length and four in breadth, which I cannot avoid thinking is an exaggeration.

August 13.—On the 13th I paid my farewell visit to the Páshá.

On descending the Zebárí range, we could see the Záb flowing from the E.N.E. Several strings of camels passed us laden with grain for Reshíd Páshá's camp. This animal seems able to travel in all situations; mountains and plains, blazing sun, frost and snow, seem alike to him. These were beautiful animals of their kind, unlike the awkward heavy camels of Persia and India; they were slender, active creatures, and nearly white; they were Arab camels, and came from the plains on the west side of the Tigris; but every animal thrives in Arabia; man, (in form at least,) the horse, camel, ass, goat, are all excellent.

We left the village in the afternoon of the 13th, in company with a ḳevvás, who was ignorant of the road: we went five miles S.E. by S. over low hills, and then S.S.W. four miles, through a woody ravine filled with vines and blackberries, to the small village of Jelam, which formerly belonged to the Amír of Rawandúz. The people fearing, no doubt, that there would be no payment, were most anxious to induce us to proceed to another village, which they pretended was close; but when they found that we resolved to stay, they were extremely civil: the old chief of the village complained that his son was forcibly detained at Rawandúz as a soldier; it seemed to be the Amír's plan to retain his subjects in good order by taking a male from each family into his service.

August 14.—We left Jelam at day-break, and travelling in the general direction of S.E. by S. for ten miles, over low grassy hills, we reached Dáb, a small village on the right bank of the Záb, which flowed from nearly north; after great detention in procuring skins to make a kelek, or float, we crossed to the left bank: the stream is rapid and about 100 yards in breadth, and said to be deep. The kelek is formed by tying a number of inflated skins under an open frame and covering the whole with branches of trees; it makes a very commodious conveyance. This was guided by two men, each of whom had a large hollow calabash under his arm: one of these watermen took a string in his hand and swam across, dragging the kelek with him, and putting one in mind of the horses applied to a similar purpose on the Oxus; the other pushed it from behind. The Záb is supposed to be the Zabatus of Xenophon and the Lycus of Arrian; the latter mentions, that immediately after the battle of Arbēla, Alexander crossed the Lycus with his troops, and proceeded to Arbēla by midnight. After crossing the Záb, we travelled among hills in almost every direction, though the general course was S.S.E. for four hours, or twelve miles. We passed at a distance several villages, which were almost all deserted, the inhabitants having built upon some neighbouring cool spot a new village of huts formed of leaves and branches. On the approach of winter, they return to their far less agreeable, permanent residences. We then travelled eastward for two hours, and came in sight of 'Alí Páshá's camp. It was now dark, and after scrambling for another hour through a very bad road among hills, we were at length unable to proceed, and halted near a stream, where we passed the night without any food for ourselves or our horses.

August 15.—We marched three miles eastwards to an eminence, on which a portion of the army of 'Alí Páshá of Baghdád was encamped, under Mustafá Páshá; at the summit of the eminence was the small fort of Darvín, which had been taken two days before.

August 16.—We left the camp at day-break, and proceeded in the direction of S.S.W. over low hills for eighteen miles, and then travelled for four miles to the village of Beherkah, across the plain of the Tigris. In the evening we travelled ten miles S.S.W. to Arbēla:* the road was excellent and level, and far to the left

* Erbil in Arabic.

lay the mountains of Kurdistán: the town is placed on a large mound sixty or seventy feet in height, and 300 yards in length by 200 in breadth; it is inclosed at the summit, with a brick wall having bastions, with a few small guns in them: at the foot of the mound there is another town, inclosed by a mud wall, a great part of it being in ruins, in which respect it resembles both the upper and lower town; the latter especially is almost desolate. There are no ruins or remarkable buildings. A short distance to the west of the town, there is an immense brick pillar standing by itself in the plain; it looks old, but seems to be a Mohammedan building; nothing is known of it excepting that it once was the minaret of a mosque. I saw no river near Erbil, and the people declared that there is none. The troops of the Amír of Rawandúz made a short resistance at this place, but a small and ineffectual mine having been exploded in the mound, they were alarmed, and surrendered. Erbil contains 6,000 people, three large mosques, and two baths. The next march being a long one, and the heat being very great, we halted at Erbil until the evening of the 17th; we then marched in a S.S.E. direction across a flat plain, and passed two villages at a short distance. After a fatiguing march of about forty miles, we reached Altún Kúprí at sunrise. This town is placed on an island in the Altún Ṣú,* (which I suppose is the little Záb,) which we crossed into the town by a bridge, whence the town is said to have been named on account of the lucrative toll formerly levied, Altún Kúprí meaning 'gold bridge.' The river was shallow, but is deep at other seasons, and about fifty yards wide before it divides: it was flowing from the N.E., and rises in U'shneh, a district of Persia, near the Urumiyah lake. The chief of this town was extremely civil; he gave me a room overhanging the river, and thirty feet above it, commanding a fine view of the country. This town is said to have formed the boundary of the acquisitions of the Amír of Rawandúz to the south. Altún Kúprí, according to the statement of the chief, once contained 8000 people, but plague and famine had greatly thinned it.

August 18.—We left Altún Kúprí in the evening, and travelled S.S.E., and after halting four hours during the night, reached Kerkúk in the morning. The distance was said to be twenty-five miles; we were still accompanied by our very useless guard of Arabs, who, however, amused us by singing songs. Arabs, Kurds, and Persians, seem to think the chief excellence of music is loudness: these fellows sang, or rather roared, with wonderful vigour, but not unpleasingly: their favourite song was that of their tribe, in praise of the nobility, courage, generosity, and hospitality

* Gold River.

of their Sheïkh. Each line seemed to consist of three or four words, and then a chorus of the word Sheïkhá: the music was very monotonous, but without any of the disgusting thrilling used in Persian singing, which I believe is in imitation of the nightingale. Three miles before reaching Kerkúk we passed several naphtha pits, which diffused a disagreeable odour to a considerable distance. In Kerkúk, naphtha is used for lights and fire. Kerkúk is a large open town in a plain, and, like all the towns in this part of the world, is in great part in ruins: plague, famine, and, I believe, cholera have almost destroyed it. Near to it is a fort built on a mound, not very high but steep. It is said to have no manufactures except a coarse calico, but there is a considerable trade in gall-nuts, which are brought from the Kurdistán mountains. The river of Kerkúk, called the Kháṣeh Cháï, was now dry: here we saw, for the first time, date trees, which would have reminded us, if it were necessary, that we were now in a very hot climate. The inhabitants are Arabs and 'Osmánlís, with some Christians and Jews, but no Kurds. The women wear immense turbans, which has a very strange effect to a person not accustomed to see females in the East with that head-dress.

August 19.—We left Kerkúk after sun-set in the evening of the nineteenth, and travelling in the general direction of east, arrived before daylight at an O'bá or summer encampment of Kúrds, whose village was behind one of the neighbouring hills. The distance was about twenty miles, of which half was among hills, with a good deal of ascent.

August 20.—We left the O'bá on the evening of the 20th of August, and, travelling still in the same direction, arrived in the morning at a large village, which belongs to Suleïmániyeh, distant about twenty-eight miles. The road was exceedingly bad, winding among defiles, with a great deal of ascent and descent: about the twentieth mile we crossed a wide torrent called the Waï Ṣú, which probably afterwards becomes the Diyálah, flowing to the south-west. All the people in this village were encamped in huts at a short distance from the village: we were lodged in a most comfortable hut, close to the tomb of a holy man, a descendant of the famous 'Abdúl-kádir Gílání, who is interred at Baghdád. Almost all our party had been for some days unwell, but here there was a great increase of the sickness. The extreme heat was sufficient to account for this. We mounted at midnight, and travelled over an extremely bad road, intersected by ravines and hollows: at the fifth mile we entered a long and exceedingly strong defile with high mountains on both sides, and a road through it, not exceeded in difficulty by any we had travelled over. At the twentieth mile, after having crossed a very fatiguing ascent, the extreme heat obliged us to halt under a few trees, with very imperfect shade. In the evening we marched eight miles to Suleïmániyeh, by a good road, through a tolerably level, well-cultivated country. Suleïmániyeh is a small town, of about 1000 houses, the capital of a district of the same name, in the centre of which it is placed, and which extends forty miles in every direction. It is possessed by the Kurd tribe of Bebah, who are esteemed excellent cavalry, and have many horses. I saw a mare for which the owner wanted 500 tómáns; and I have no doubt, that if even so large a price were offered him, he would be very reluctant to take it. The town is situated at the end of a plain under some hills: it contains few good houses, many of which are in ruins; and has a large and well supplied bázár of fruit, meat, and vegetables.

From Suleïmániyeh I travelled in a N.N.E. direction about 200 miles, by a well-known road, to Sardasht, Láhiján, Ṣó-úḳ Búláḳ, and by Marághah to Tabríz.

Ṭehrán, Feb. 12, 1837.

V.—*Memoranda to accompany a Sketch of part of Mázanderán, &c., in April,* 1836. By E. D'Arcy Todd, Major, serving in Persia. Communicated by John Backhouse, Esq.

This sketch of Mázanderán on the scale of $\frac{1}{380160}$, or of *six* British miles to an inch, was made in the month of April, 1836, from observations taken with a Schmalcalder's compass, the distances being calculated from the pace of a horse walking on an average three and three-quarters statute miles an hour.

The lines of road here followed were corrected by frequent *magnetic* bearings* of the peak of Demávend, which is visible from Ṭehrán, and from most of the principal points in the routes from that city, through Mázanderán, to the southern shore of the Caspian.

Wheeled carriages are not used in any part of the road here laid down.

On quitting Ṭehrán in an E.N.E. direction, at fifteen miles the Jáj-rúd river is crossed by a ford; but from the month of April to the middle of June, when the mountain snows are melting, it is often so much swollen that laden mules make a circuit of several miles to cross the river by a bridge, said to be about three miles above the caravanserai at the ford.

From Jáj-rúd to Demávend, a distance of twenty-five miles, the road is crossed by several small streams, upon each of which

* The variation of the compass at Ṭehrán in 1837 was 2° westerly.—Ed.

is situated a village surrounded by cultivation. A range of hills sprinkled with snow bounds the view to the south, at the distance of about ten miles. Demávend is a large village pleasantly situated in a high, rich, and well-watered valley, the streams flowing to the southward.* From Demávend to Serbendán† fifteen miles: the road passes through several well-supplied villages; but from that place to Fírúz-kúh‡ no supplies are to be had, except in small quantities at the wretched huts which are called the caravanserais of Arú, Dalí-chăï,§ and Amín-ābád.|| At Arú the road gradually inclines towards the range of hills to the south. The Dalí-chăï river is a rapid mountain-stream flowing from the north-west, in a deep ravine, through a mountainous, broken country. To the east of the ford is a steep and rocky ascent over hills sprinkled with juniper. From Amín-ābád to Gházán-chăï, also called the Arjamand (from a village of that name up the stream, the residence of the Governor of Fírúz-kúh), the road is good, except at the descent to the river, which is abrupt and rocky. The Gházán-chăï flows from the north; it is a considerable stream, but fordable at all seasons. The range of mountains, along the southern skirt of which the road runs from Demávend, here terminates; and a few miles beyond the river the peak of Demávend is seen bearing N. 66° W.

The village of Fírúz-kúh is situated at the base of a rock, along its northern and eastern sides. The rock of Fírúz-kúh is about 750 feet in height above the plain, being the termination of a range of hills running from the south along the western bank of the Ḥablah-rúd¶ river. The rock was formerly crowned by a fortress, the remains of which are yet visible. Tradition reports it to have been considered to be impregnable in former days, but it is commanded from several points. There are some fine pasture-lands in the vicinity of Fírúz-kúh, and the valley of the Ḥablah-rúd is richly cultivated.

The road from Fírúz-kúh into Mázanderán leads over a pass, about ten miles to the N.E. of that place. Near the summit of the pass is a large and well-built, though now ruined caravanserai. On the 9th April the snow was deep on the summit of the ridge for about half a mile, and the road had only recently been opened. Immediately below the snow on the northern side is a forest of stunted oak, and the path runs in the bed of a small stream, the source of the Tálár river. This bed descending becomes a narrow ravine, and in the course of a few miles the oak gives place to

* Demávend village is found by late observations to be 6,000 feet above the sea. —Ed.

† Head of the dams or embankments.—F. S.

‡ Mount Victorious, or Blue Mountain.—F. S.

§ Mad-stream.—F. S.

|| Ruler's Town, or Peaceful Abode.—F. S.

¶ Rope-river.—F. S.

the juniper, elm, chesnut, beech, box, and rhododendron, springing from every crevice of the rocks. At thirty-two miles below the remains of a hill fortress, called by the people of the country the Castle of the Dív Sefíd,* the ravine contracts to a narrow gorge which was formerly defended by a stone wall, the remains of which and of a gateway are still visible. In the vicinity of this place are a few patches of barley. One mile beyond Surkh-rabát† (a wooden hovel said to have been erected for the accommodation of the late Sháh) rice is cultivated. The road here runs through a dense thicket, and before we reached the Puli Sefíd,‡ which is a well-built stone bridge over the Tálár, we came upon the commencement of this branch of Sháh 'Abbás's causeway. At Puli Sefíd the road turns N.N.W., and continues along the eastern bank of the river, and as far as Shírgáh§ is execrable. The causeway of Sháh 'Abbás has been a magnificent work, formed by a trench twenty feet deep and fourteen wide, cut in the side of the mountain, and then filled with large stones. In consequence of the incessant rains which fall in this part of the country, and of the want of all repair, this causeway has been nearly destroyed. In many parts the stones have been washed away, and their place having been supplied by logs and branches of trees, it is difficult even for mules to pass. Zír-áb|| and Shírgáh are two groups of wretched, wattled huts, which are only inhabited during the season of rice-planting. At other times one or two villagers only remain to watch the fields, and sell provisions to passing muleteers. The bulk of the inhabitants reside higher up in the mountains.

Four miles beyond Shír-gáh the road leaves the hills, and enters upon a fine level tract of country, covered with cultivation, pasturage, and wood. The road here diverges to the N.N.E. from the Tálár river. Between 'Alí-ābád¶ and Sárí, fifteen miles in a N.E. direction, the causeway for a mile or two is in good repair, and leads through a natural avenue of magnificent trees. All traces of it are then lost until within three miles of Sárí. Travellers are obliged to pick out a pathway for themselves through swamps, brushwood, and rice-fields, at some distance from the original line of road. Here and throughout Mázanderán, as in Gílán, we find the houses scattered in little groups, concealed among the woods, so that it is almost impossible to form any correct idea of the amount of the population.

Sárí, the capital of Mázanderán, is surrounded by a ditch and a mud wall, flanked by pentagonal brick towers. The gateways

---

* The White Demon. Ouseley, iii. 231.—F. S.
† Red Resting-place. Ouseley, iii. 232.—F. S.
‡ White Bridge.—F. S. § Lion-place.—F. S. || Under river.
¶ 'Alí's Abode.—F. S.

have fallen down, and roads have been broken through the wall in every direction. The state of the defences proves that the Sárians have for many years enjoyed security from without. The appearance of the town differs essentially from that of any other in Persia south of Elburz. The houses are built of burnt brick, and neatly tiled; some of the streets are well paved, and although marks of ruin and neglect are everywhere visible, Sárí has something the appearance of an English village, or a small market-town. The place was nearly depopulated about four years ago by the plague.* The peak of Demávend bears S. 50° W. from Sárí. The river Tejin† flowing from the S.E. passes about a mile to the eastward of Sárí, and falls into the Caspian at Faraḥ-ābád,‡ about twenty miles north of that town. The Tejin probably takes its rise in the same mountains as the Tálár. From Sárí to Bálfurúsh the road returns S.W. to 'Alí-ābád, and thence strikes to the N.W. and N.N.W. The Tálár river is forded at the third mile from 'Alí-ābád. The causeway has there fallen to decay, and is described as being impassable. Travellers are obliged to make a circuit through a forest of magnificent oak, beech, and elm, interspersed with villages and rice-fields. The country becomes again open within two miles of the town.

Bálfurúsh, or more correctly Bárfurúsh (the mart of burdens), is an open town of considerable extent, built in the midst of a forest; the houses are scattered over a vast surface of ground, surrounded by gardens and cultivation. The population of Bálfurúsh is described as being considerably greater than that of Sárí, but from the straggling condition of the town it is difficult to form an estimate of the number of houses.§ Bálfurúsh is slowly recovering from the devastations of plague and cholera, by which it has been of late nearly depopulated. It has a considerable trade, its bázár is good, and well supplied; its port, Meshhedi Ser,‖ on the Caspian, being the place at which all Russian goods destined for Mázanderán are landed.

To the south of the town, at the distance of 500 yards, is the Baḥru-l-Irem (or sea of Paradise),¶ a palace built by the late Governor of Mázanderán, Mohammed Ḳulí Mírzá,** on an island of about half a mile in circumference, formed by a stream brought from the river Bábul,†† by means of ḳanáts, or subterranean canals.

* Sárí contained from thirty to forty thousand inhabitants in 1822.—See Fraser's Travels on the Shores of the Caspian, p. 14.—Ed.

† Or Tejín. Ouseley, iii. 269. ‡ Abode of Pleasure, built by Sháh 'Abbás.

§ Mr. B. Fraser, in 1822, states it at 200,000, and considers it rather underrated. Travels, &c. p. 84.—Ed.

‖ Place of Martyrdom of Ser, or evidence-place of the head.—F. S.

¶ The Garden of Irem, a sort of Paradise.—F. S.

** Prince Mohammed's Slave.—F. S. †† Bahbul, Ouseley, iii. 291.

The island was formed by Sháh 'Abbás; but of the palace which he built upon it there are no remains, and the present building, although a modern one, is now a ruin. The surrounding water, at the broadest part about 300 yards, is now a stagnant pool, covered with reeds and rushes, the abode of numberless wild fowl. A wooden bridge, the perilous planks of which have parted company, connects the island with the main-land, and a square building, the gateway of the place, forms a kind of *tête de pont* to the position. A hard, level, excellent road, passable for wheeled carriages at all seasons, leads, in a north-by-west direction fifteen miles, from Bálfurúsh to Mesh-hedi Ser, on the Caspian. About four miles from the town the road comes upon the right bank of the river Bahbul, which it thence follows to the village of Mesh-hedi Ser, situated at the mouth of the river.

The Bahbul is about fifty yards broad in the vicinity of Bálfurúsh, and flows at this season with a current of not more than two miles an hour. It is said to be navigable for boats to within three or four miles of the town. Several small villages are scattered along each of its banks. Flax and cotton are here cultivated in large quantities. Goods are not conveyed by the river, which may be accounted for either by the excellence of the road, or by the ignorance of the people in the arts of boat-building, or by the trouble and delay of re-embarking and again disembarking the goods after they have once passed the custom-house. Iron and naphtha are the chief imports from Russia. Mesh-hedi Ser is a small village, at which the customs are collected; its houses are scattered on both sides of the river. Vessels of about 200 tons lie in an open roadstead, about three miles from the shore.

Retracing our steps to Bálfurúsh, the road, which is good and passable at all seasons, leads thence over a level, well-cultivated country, to A'mol, the general direction being south-west. Richly-wooded hills, backed by a lofty range of snowy mountains to the south, gradually close in to the distance of five to six miles towards the road. At about a mile from Bálfurúsh, the Bahbul is crossed by a bridge of masonry, of eight arches, in excellent repair, built by Moḥammed Ḥasan Khán, grandfather of Fat-ḥ-'Alí Sháh. The banks of the river at this spot are precipitous, the level of the country being thirty or forty feet above the stream.

The town of A'mol is without walls or defences of any kind. The river *Herhaz,** a mountain stream, which at some seasons is said to be of considerable volume, runs through it, flowing from an opening in the hills immediately to the south of the town. It is crossed by a narrow but substantial bridge of masonry. A'mol is of less extent than Bálfurúsh, being considered, in point of

* Ouseley, iii. 295.

size and importance, the third town* in Mázanderán. It is easily approached from the sea.

On quitting A'mol for Párus the road leads in a south direction for eighteen miles along the western bank of the Herhaz, and at the sixth mile enters a valley 400 yards broad, through which the stream flows to the plain. A low range of wooded hills slopes to the water's edge on either side of the river. The road then runs along the bed of the stream, and at the eleventh mile, the valley closing in on both sides, the road ascends the left bank of the river by a narrow causeway, in some places not more than three or four feet broad, cut in steps on the side of the hill, and formed of layers of wood and stone placed on deep clay, the natural soil of the hill. This causeway, although lately repaired at considerable expense by one Ḥájí Ṣáleḥ, a merchant of A'mol, is almost impassable to horsemen and laden mules, and is fast falling into utter decay, numberless mountain streams and the incessant rains of the country having washed away in many places the materials of which it is made. The remains of an ancient and more substantial road, built up against the solid rock on the other side of the stream, are visible, and are probably the work of Sháh 'Abbás; but earthquakes and torrents have nearly destroyed it, and travellers prefer the modern causeway, although the distance by it is greater. At the thirteenth mile the road again descends to the bed of the stream, but occasionally runs over causeways similar to the one above described, but shorter in extent. The hills cease to be wooded within about two miles of Párus, a ruined and deserted building, which may once have been a caravanserai. Supplies, in small quantities, are procurable here, but with great difficulty.

Beyond Párus, still pursuing a southerly course, the road improves from the hardness of the rock over which it passes and from the dryness of the climate. It is, however, in some places only a narrow pathway, built upon or scooped out of the face of a perpendicular rock overhanging the torrent. Frequent accidents are said to occur from land-slips and fragments loosened by sudden thaws in the mountains. Between Párus and Karú † the stream is crossed in two places by wooden bridges, near the remains of stone ones, which have been swept away by the torrent. At Karú some caves, cut in the side of the hill, afford shelter to the traveller, but supplies are not to be procured.

Four miles beyond the caves of Karú the mountains close in on both sides of the Herhaz, which here runs in a deep and narrow channel between walls of perpendicular rock. The pathway, in

* In 1822 said to contain from thirty-five to forty thousand persons.—Fraser's Travels, p. 104.

† Kharoe, in J. Arrowsmith's map.—F. S.

some places not more than three feet broad, is scooped out of the face of the rock about 200 feet above the torrent. This strong natural defile, about a mile in length, is said to be the only entrance on the northern side into the district of Lárìján. Beyond the defile the road improves, and after the twelfth mile, at Waneh, it turns S.W., and passes through several fine villages at the immediate south-eastern foot of Mount Demávend, forming the district of Amírí.* Before it reaches Ask,† eight miles farther, the stream is crossed in six places by wooden bridges. The village or town of Ask is said to contain a thousand or fifteen hundred houses, and is the principal of seventy-two villages, which form the district of Láriján. Its position is remarkable. The Herhaz river runs below the town through a deep and narrow channel of rock, which is crossed by a wooden bridge, the stone one having been swept away. From the water's edge the town is built on natural steps, rising to the height of several hundred feet, upon the south-eastern side of the mountain of Demávend, which forms one of the natural impassable barriers of the place. A chain of lofty mountains shuts in the valley on every side; the only ingress and egress to which is at the points where the river enters, and leaves the narrow basin, on the southern side of which is situated the town of Ask.‡

At a short distance from Ask the road leaves the river, and ascends the southern shoulder of Demávend. The ascent is steep and rocky. The point at which the road turns the shoulder of the mountain is about 1300 feet above the stream. This road is impassable in winter, when horsemen cannot approach Láriján in this direction, but foot passengers contrive to scramble over the rocks immediately above the bed of the river. The descent is comparatively easy, and much shorter than the ascent; but the road is execrable, being almost blocked up with masses of rock and half-frozen snow. Near the foot of the mountain the river is joined by a rapid stream called the Lár, flowing from the north-west. This is crossed by a stone bridge. The road thence ascends the bed of the Herhaz, which is here a mountain torrent; and for the last two or three miles before we arrive at Imám-Zádeh Háshim,§ which marks the summit of the pass,‖ the steep and rocky pathway is scarcely practicable to a laden mule. The snow

---

* Commander's District.—F. S.

† Asek (J. Arrowsmith). Ouseley, iii. 329.—F. S.

‡ Ask is situated about 5900 feet above the level of the sea.—Ed.

§ The Imám's son Háshim.

‖ This is the line of separation of waters flowing north to the Caspian, and south towards the plain of Persia, and may probably be estimated at 7000 feet above the sea, or 3000 nearly above Ţehrán.—Ed.

was deep on the northern face of these mountains on the 21st of April.

From Imám-Zádeh Háshim the descent is gradual, the road takes a westerly direction, and after four miles turns to the south into a rich valley, with a fine stream running from the north, on which is situated the village of A'lí, also called Ṭáḳi-zumurrud,* from a garden-palace, now in ruins, erected near the spot by Fat-ḥ-'Alí Sháh. Near the village of Rúdehán, about six miles beyond 'Alí, the road joins that which leads from Jáj-rúd to the village of Demávend.

From the above description of the two most practicable routes, north and south, through the province of Mázanderán, the natural strength of the country may be estimated. The lofty chain of Elburz † is at present an impassable barrier on the south. Roads might, of course, be constructed with enormous labour and expense over the passes; but even the great causeway of Sháh 'Abbás, in its best days, could have been hardly practicable for heavy-wheeled carriages. The climate and soil of the wooded belt of hills which fringes the northern skirt of Elburz render the construction of roads difficult, and their permanence, without constant and expensive repair, almost impossible.

The route between Fírúzkúh and Keláteh,‡ in a general N.E. direction about one hundred miles, was followed in August, 1836. The road ascends the stream of Ḥablah-rúd for about eight miles in an easterly direction; the river then enters a defile and turns to the south. Quitting the Ḥablah-rúd the road passes over some fine pastures. Fourteen miles beyond Gúri Sefíd § a road leads by a more northerly direction to Fúlád-Maḥallah.|| From this point the general direction is eastward, over a steep pass into the valley of Sháh-mírzád,¶ a large village with abundance of water. The road then turns to the N.E., and passes among low hills with pastures, and occasionally strips of cultivated ground, in the valleys. The villages of Fúlád-Maḥallah, Surkh-deh,** Chesh-meh 'Alí,†† and Keláteh, contain but twenty or thirty houses each. The road is in most places good, and practicable for wheeled carriages. Keláteh is about thirty miles west of Bastám,‡‡ and the same distance S. by W. of Astarábád.

* The Emerald; Portico or Cupola.—F. S.
† Properly El-burj, the watch-tower.—F. S.
‡ Hill-top. *Kuláteh* signifies a hamlet.—F. S.
§ White Ass (Onager).—F. S.
|| Steel District.
¶ Prince Royal.
** Red Village.
†† 'Alí's Spring.
‡‡ Ouseley, iii. 226.

VI.—*An Account of the Ascent of Mount Demávend, near Tehrán, in Sept.,* 1837. By W. Taylor Thomson, Esq., serving in Persia; with Notes by W. Ainsworth, Esq., M.R.G.S., &c. Read, Jan. 8, 1838.

Sept. 4.—Having received the sanction of His Excellency the Ambassador, I started from Gúlúhek, about seven miles north of Tehrán, with the intention of attempting the ascent of the peak of Demávend. My object in doing so, was to endeavour to fix by astronomical observations, the position of the mountain, which from its superior height and peaked summit, forms an excellent point for correcting the bearings taken while surveying the adjacent country. I also intended taking a correct series of bearings and angular distances, to assist in laying down a plan of the country of Mázanderán, and by barometric observations to determine the altitude of the Peak; and lastly, to make any general observations as to the geological structure and mineral productions of the mountain, which might occur to one who does not possess more than an ordinary knowledge of that subject.

In any European country, the ascent of a mountain of equal altitude with this, would be attended with no greater difficulty than a little exertion of physical force, but here it is very different; the great obstacle to be overcome is the suspicion and jealousy of the natives, who suppose that a European can have no other object in making such an exertion of strength, than by his superior knowledge to discover some hidden stores of metals and wealth. On this account no European had hitherto succeeded in making the ascent; but through the foresight and prudent arrangements of the ambassador, in my case that difficulty was overcome.

On my arrival at Ask, a village about 42 geographical miles east-north-east of Tehrán, situated on the left bank of the river Herház, and at the foot of the mountain, I delivered the letter and presents with which the ambassador had provided me, to the mother of 'Abbás Kulí Khán, the chief of Láriján (who was absent), and arrangements were immediately made to furnish me with guides and whatever was necessary for the ascent. From Ask I proceeded to Germah about an hour distant, and the highest village on the south side of the mountain, where I was provided with four guides, only one of whom, I afterwards found, had previously made the ascent. On the morning of the 8th, I ascended two hours beyond Germah; but the weather, which had all along been lowering and sultry, broke into heavy rain accompanied with thunder, which forced us to take a partial shelter under a ledge of rock, where we remained during the rest of the day and succeeding night. In the morning, the snow line which had been the night before a long way above us, now reached to within a few feet of the place where we had bivouacked. During the night the storm had expended itself, and notwithstanding the cold and wetting we had got, we started with daylight in high spirits and with strong expectations of reaching the summit about noon.

I had been informed that there was a cave at the top of the mountain, where, if pressed for time, I might remain during the night. I therefore provided myself with a change of clothes, and four days' provision of bread for the whole party, and determined on remaining on the top for that time, in case the weather should prove cloudy. I had not ascended however above an hour, when two of the men refused to go any farther, and in consequence, the provisions and additional articles of dress had to be left with them. With the remaining two I continued the ascent, and although one of them made great complaints of pain in his head, and palpitation of the heart, he was too necessary to allow him to return. By dint of promises of reward, and threats of representing his conduct to the Khán, I succeeded in getting him to the top. It was evening, however, before we reached it, and as the cold was excessive, as soon as we had done so, we repaired to the sulphur cave, which is on the east side, and within a few feet of the summit. The cone, for about 100 feet from the top, is entirely composed of a soft rock, from which the pure sulphur is dug, with the assistance of a piece of stick, and afterwards carried down in bags on the shoulders of the men who gather it. Long before our arrival at the top, the whole mountain was wrapped in clouds, which rendered it impossible to make any observation, except barometrical, even had I been willing to risk the danger of exposing the body, while in a heated state, to the piercing cold wind which blew from the Caspian sea, distant 50 miles north. The cave is of small dimensions, having two divisions; the inner one, which is the largest, will not contain more than five or six men, and in it the temperature is very high, so much so, that in one corner I could not bear to expose my hand to the hot current of air which flowed from the rock. The whole of the bottom of this part of the cave is highly heated. On entering, we scraped together a quantity of the dust, and stripping off our goats'-hide shoes, and goats'-hair stockings, which were wet with passing through the snow, we covered our feet with the ashes, which kept them sufficiently comfortable during the night. In the outer division, where I made observations of the altitude with the barometer, the temperature was at 56° of Fahrenheit, but this was at least 20° below that of the place where we slept.

The mouth of the cave is so low, as to oblige one to enter on all-fours, but as it opened to the east, and the wind blew the fine particles of snow into the interior of the cave, they were immediately melted, and fell upon us like a heavy dew. In the morning, my caoutchouc cloak and clothes, with the exception of the side which came in contact with the bottom of the cave, were completely drenched—had we ventured to close the entrance, we must have soon been suffocated from breathing a heated atmosphere highly impregnated with sulphureous particles—even as it was, we all awoke with severe headache and sickness.

The sun in the morning shone clear into the mouth of the cave, and when I supposed that it had reached a sufficient altitude to make the cold endurable, I went out with the intention of proceeding with the observations, but before I had gone a few yards from the mouth of the cave, my cloak and clothes being full of moisture, were frozen stiff, and the wind felt so piercingly cold, that a few minutes' rest would have benumbed us; so that, however much I regretted it, and however anxious I was not to leave the place without finishing what I intended, yet I was obliged to run at full speed down the side of the mountain to prevent being frozen. Had the four guides who started with us from Germah accompanied us to the top, the additional clothing which they carried might have enabled us to stand the cold at the summit; but I would recommend to any one making the same trip, to do it two months earlier in the season. By leaving the cave at the foot of the mountain, about twelve o'clock on a moonlight night, he would be enabled to pass nearly the whole day on the summit, without exposing himself to the noxious and heated vapour of the cave, on leaving which, at any season, there is so sudden a transition from a high to a low temperature, as to endanger very much the health of the person doing so.

As to the volcanic nature of the mountain * there can be no doubt. The sulphureous cone with its crater-shaped summit, the heated air and steam issuing from its crevices, the hot springs at its base, the scoriæ and pumice found upon its side—all show, that it not only was, in former times, the mouth of an extensive volcanic district, but that its fires are not yet extinguished—again they may be lit up, and its action, whether by earthquake or eruption, may cause most extensive changes on the surface of this part of the country. The volcanic action through the whole of this part of the country has at one time been very great; strata of rocks are contorted and twisted about in all directions. In the neighbourhood of Germah are the hot springs of Germsír, the

* We learn from Mr. Baillie Fraser that on approaching the mountain from the west, after leaving the limestone mountains that surround it, he found a large inclined plain below the cone, covered with pumice-stone and scoriæ, and having dark basaltic rock starting up through them to the surface; the cone seemed to consist of pumice and scoriæ, with debris of basalt, and the ridges and peaks of the latter stared through the surface all the way to the summit.—Ed.

hottest of which shows a temperature of 148° of Fahrenheit. The springs, however, are not confined to this place alone; at Ask they are very numerous, both in the village and its neighbourhood, although they are of a temperature much inferior to those of Germsír.

On my return from the top of the mountain to Gúlúhek, I passed through Rayrah, Ask, village of Demávend, Abú-mehán, and Jájrúd, making a sketch survey of the country as I passed along. I annex a table of barometric observations made at different stations, and a section of the rocks from the base at Germah to the top of Demávend [see Map].

	Barom. Eng. inches.	Therm. Fahr.
Lavassán . .	23. 1	75°
Ask . . .	24.11	71
Germah . .	23. 2	68
Demávend Peak (sulphur cave)	15.05	56
Demávend village . .	23.45	66
Jájrúd . .	25.03	81
Ṭehrán . .	25.97	80

The geological formation of Demávend immediately about Germah would seem to be a deposit of calcareous tufa—above this occur beds of sandstone of the coal formation, with one seam of coal of inferior quality, for about 1000 feet—then limestone for a thickness of nearly 1200 feet; above this again, greenstone coloured with iron, to within 100 feet of the summit, which is a deposit of pure sulphur.

At the request of Mr. Thomson, Mr. Ainsworth adds the following remarks:

The mean height of the barometer at the level of the Persian Gulf, in the month of September is, from my own observations, as much as 30.600 inches, which compared with the observations made by Mr. Thompson, would give a rough amount of 4,300 feet elevation to Ṭehrán, and 15,000 to the cave near the summit of the Demávend; but there can be no doubt that this mean is much too high for the latitudes of the Persian Taurus, and probably the best approximate mean would be obtained by taking the difference between the mean of the level of the sea in our own latitudes, or 29.600, and that of the Persian Gulf, which would give 30.100; and hence the elevation of Ṭehrán would be 4,000 feet; of Jájrúd, 4,900; of Lavassán, 6,800; of Ask, 5,800; of Germah, 6,700; of Demávend village, 6,000; and of the cave near the summit of the peak, 14,700.

It is to be observed here, that there are no corrections except for temperature, and hence, if these elevations approximate to

Pubd. for the Journal of the Royal Geographical Society by J. Murray, Albemarle Str. London 1838.

within 100 feet of the truth, it is the most that can be expected of them. They serve to give a good idea of the height of what may probably be considered as the culminating point of the Persian Taurus; and it is to be hoped that the Society will, before long, be in possession of more accurate observations.

The geological observations made by Mr. Thomson possess great interest by establishing the existence of a pseudo-volcano in these central districts of Western Asia, and ally themselves to the observations which the Baron von Humboldt has made upon the evidences of volcanic action, whether in ignivomous mouths, (extinct or active,) or canals of communication between the surface and the interior of the earth, and of the same action as evidenced in lines or circles of elevation, and which he has traced throughout the great continent of Asia. The geological relations of the Caspian sea are connected with the same phenomena, belonging, as they undoubtedly do, to the latest changes which have taken place on the surface of the earth.

It is a remarkable fact, that throughout those districts of Taurus, Amanus, Kurdistán, and the Persian Apennines, (Bakhtiyárí-Luristán mountains,) which I have travelled, I have never as yet met with rocks of the secondary series. The absence of every member between the chalk and the primary formations is one of the most remarkable features in the geology of Western Asia, and from this phenomenon obtaining so very generally over such extensive districts, it is most probable that the sandstones which occur at the base of Demávend belong to the supracretaceous or tertiary series, and are either sandstones with lignite coal belonging to the plastic clay, (the ostracite sandstone of Kupffer,) or what is equally probable, belong to the *Terrain marno-charboneux* of Brogniart. As there are no fossils transmitted, it is impossible to determine the age of the superimposed limestones. The sulphur deposits of Músúl in Mesopotamia, and of Sú Khúmatú in Kurdistán, are both of them in the Cerithia limestone, corresponding to the London clay; but the sulphur formation of the Demávend appears to be of even a more recent date.

Mr. Thomson found thermal springs in the neighbourhood of Germah, in one of which the thermometer of Fahrenheit indicated a temperature of 148°. I observed a curious phenomenon in some of the thermal springs in Syria, not far from Antioch, which, according to a well-authenticated tradition, had made their appearance at different times, and with different shocks of earthquakes, so frequent in those countries, viz., that the temperatures of the most recent springs exceeded by several degrees those of the most ancient ones. I shall call the attention of my friend, Mr. Thomson, to this fact, in order that he may observe if any law of a decreasing temperature in thermal springs can be established by observations made in different sources, in different countries, and during a prolonged period of time.

With regard to the rocks designated as green-stone in the above account, their position is very various in Western Asia. I have, however, never observed rocks of augite and feldspar, or other rocks which are more modern than the hornblend series, intercalated in, or tilting up very recent sedimentary formations, as is the case with the diallagic formations; but it is very evident that the connexions are no longer the same where these rocks are associated with pumice-stone and scoriæ, as in the Demávend, just as the same rocks are, near 'Osmánjik in Taurus from my own observations, associated with argillaceous rocks containing mica and vitreous feldspar.

VII.—*Abstract of the Historical Evidence for the Discovery of America by the Scandinavians in the Tenth Century.* Extracted from the "ANTIQUITATES AMERICANÆ."

[THE labours of the Royal Society of Northern Antiquaries at Copenhagen have long since obtained for them an honourable station in the republic of letters, but the recent publication of the work named "ANTIQUITATES AMERICANÆ" entitles them to the gratitude of the whole civilized world. Rumours of the fact that the Scandinavian Northmen were the original discoverers of America have long been prevalent, and several Icelandic authors of great intelligence and respectability have investigated the subject of western discovery by their adventurous countrymen; but it has been well observed by Baron Humboldt, that the information which the public as yet possesses of that remarkable epoch in the middle ages is extremely scanty, and he has expressed a wish that the northern literati would collect and publish all the accounts relating to the subject. To that wish the Royal Society of Northern Antiquaries has fully responded, and the result is this collection of documents compiled from the numerous and valuable MSS. now extant, and accompanied by a Danish and also a complete Latin translation—by prefatory remarks and archæological and geographical disquisitions—and further, by an abstract of the historical evidence in English.

As such a subject must deeply interest all those who occupy themselves with ancient discoveries in geography and history; and as from the nature of the work its circulation will be extremely limited, it has been considered that an abstract of the historical evidence would appropriately find its place in the London Geographical Journal, with the two-fold object of more widely diffusing the knowledge of the facts, and of inducing the reader to refer to the original work for the complete evidence that has been brought to light on the subject.—Ed.]

Biarne Heriulfson's Voyage in the Year 986.

Eric the Red, in the spring of 986, emigrated from Iceland to Greenland, formed a settlement there, and fixed his residence at Brattalid in Ericsfiord. Among others who accompanied him was Heriulf Bardson, who established himself at Heriulfsnes. Biarne, the son of the latter, was at that time absent on a trading voyage to Norway; but in the course of the summer returning to Eyrar, in Iceland, and finding that his father had taken his departure, this bold navigator resolved "still to spend the following winter, like all the preceding ones, with his father," although neither he nor any of his people had ever navigated the Greenland sea. They set sail, but met with northerly winds and fogs, and after many days' sailing they knew not whither they had been carried. When the weather again cleared up, they at last saw a land which was without mountains, overgrown with wood, and having many gentle elevations. As this land did not correspond to the descriptions of Greenland, they left it to larboard, and continued sailing two days, when they saw another land which was flat, and overgrown with wood. From thence they stood out to sea, and sailed three days with a S.W. wind, when they saw a third land, which was high and mountainous and covered with icebergs (glaciers); they coasted along the shore, and saw that it was an island. They did not go on shore, as Biarne did not find the country to be inviting. Bearing away from this island, they stood out to sea with the same wind, and after four days' sailing with fresh gales, they reached Heriulfsnes in Greenland.

Discoveries of Leif Ericson, and First Settlement of Vineland.

Some time after this, probably in the year 994, Biarne paid a visit to Eric Earl of Norway, and told him of his voyage, and of the unknown lands he had discovered. He was blamed by many for not having examined these countries more accurately. On his return to Greenland there was much talk about undertaking a voyage of discovery. Leif, a son of Eric the Red, bought Biarne's ship, and equipped it with a crew of thirty-five men, among whom was a German of the name of Tyrker, who had long resided with his father, and who had been very fond of Leif in his childhood. In the year 1000 they commenced the projected voyage, and came first to the land which Biarne had seen last. They cast anchor, and went on shore. No grass was seen; but everywhere in this country were vast ice mountains (glaciers), and the intermediate space between these and the shore was, as it were, one uniform plain of slate (*hella*): the country appearing to them destitute of good qualities, they called it Helluland. They put out to sea, and came to another land, where they also went on shore. The country was level (*slètt*), and covered with woods, and wheresoever they went, there were cliffs of white sand (*sandar hvítir*), and a low coast (*ósæbratt*); they called the country Markland (Woodland). From thence they again stood out to sea, with a N.E. wind, and continued sailing for two days, before they made land again. They came to an island which lay to the eastward of the mainland, and entered a channel between this island and a promontory projecting in an easterly (and northerly) direction from the mainland. They sailed westward. There was much ground left dry at ebb tide. Afterwards they went on shore at a place where a river, issuing from a lake, fell into the sea. They brought their ship into the river, and from thence into the lake, where they cast anchor. Here they constructed some temporary log huts, but afterwards, when they had made up their mind to winter there, they built large houses, that were afterwards called Leifsbúdir (*Leifsbooths*). After they had finished the building of their houses, Leif divided his people into two companies, who were to be employed by turns in keeping watch at the houses, and in making small excursions for the purpose of exploring the country in the vicinity: his instructions to them were that they should not go to a greater distance than that they might return in the course of the same evening, and that they should not separate from one another. Leif took his turn also, joining the exploring party on the one day, and remaining at the houses on the other. It so happened that one day the above-named German Tyrker was missing. Leif accordingly went out with twelve men in search of him, but they had not gone far from their houses when they met him coming towards them. When Leif inquired the reason why he had been so long absent, he at first answered in German, but they did not understand what he said. He then said to them in the Norse tongue, "I did not go much farther, still I have a discovery to acquaint you with; I have found vines and grapes." He added by way of confirmation that he had been born in a country where there was plenty of vines. They had now two occupations to employ themselves, viz., to hew timber for loading the ship, and to collect grapes; with these last they filled the ship's long-boat. Leif gave a name to the country, and called it Vínland (*Vineland*). In the spring they sailed again from thence, and returned to Greenland.

Thorwald Ericson's Expedition to more Southern Regions.

Leif's Vineland voyage was now a subject of frequent conversation in Greenland, and his brother Thorwald was of opinion that the country had been far too little explored. He therefore borrowed Leif's ship, and aided by his brother's counsel and directions, he commenced the voyage in the year 1002. They arrived in Vineland at Leifsbooths, where they spent the winter and employed themselves in fishing. In the spring of 1003 Thorwald sent a party in the ship's long-boat on a voyage of discovery southwards. They found the country beautiful and well wooded, there being but little space between the woods and the sea, and extensive ranges of white sand: there were many islands and shallows. They found no traces of men having been there before them, excepting on an island lying to the westward, where they found a wooden shed. They did not return to Leifsbooths until the fall. In the following summer, 1004, Thorwald sailed eastward with the large ship, and then northward past a remarkable headland enclosing a bay, and which was opposite to another headland. They called it Kialarnes (Keel-

Cape). From thence they sailed along the eastern coast of the land, into the nearest firths, to a promontory which there projected, and which was everywhere overgrown with wood. There Thorwald went ashore with all his companions. He was so pleased with this place that he exclaimed, "Here it is beautiful, and here I should like well to fix my dwelling." Afterwards, when they were preparing to go on board, they observed on the sandy beach within the promontory three hillocks. They repaired thither, and found three canoes, and under each three Skrellings (Esquimaux); they came to blows with them, and killed eight of them, but the ninth escaped with his canoe. Afterwards, a countless number came out from the interior of the bay against them. They endeavoured to protect themselves by raising battle screens on the ship's side. The Skrellings continued shooting at them for a while, and then retired. Thorwald had been wounded by an arrow under the arm: when he found that the wound was mortal, he said, "I now advise you to prepare for your departure as soon as possible, but me ye shall bring to the promontory, where I thought it good to dwell; it may be that it was a prophetic word that fell from my mouth about my abiding there for a season; there shall ye bury me, and plant a cross at my head and also at my feet, and call the place Krossanes (Crossness) in all time coming." He died, and they did as he had ordered. Afterwards they returned to their companions at Leifsbooths, and spent the winter there; but in the spring of 1005 they sailed again to Greenland, having important intelligence to communicate to Leif.

Unsuccessful Attempt of Thorstein Ericson.

Thorstein, Eric's third son, now resolved to proceed to Vineland to fetch his brother's body: he fitted out the same ship, and selected twenty-five strong and able-bodied men for its crew; his wife Gudrida also went along with him. They were tossed about the ocean during the whole summer, and knew not whither they were driven: at the close of the first week of winter they landed at Lysufiord, in the Western Settlement of Greenland. There Thorstein died during the winter, and in the spring Gudrida returned again to Ericsfiord.

Settlement effected in Vineland by Thorfinn.

In the following summer, 1006, there arrived in Greenland two ships from Iceland: the one was commanded by Thorfinn, having the very significant surname of Karlsefne (*i. e.* who promises or is destined to be an able or great man), a wealthy and powerful man, of illustrious lineage, and sprung from Danish, Norwegian, Swedish, Irish, and Scottish ancestors, some of whom were kings or of royal descent. He was accompanied by Snorre Thorbrandson, who was also a man of distinguished lineage. The other ship was commanded by Biarne Grimolfson of Breidefiord, and Thorhall Gamlason of Austfiord. They kept the festival of Yule at Brattalid. Thorfinn became enamoured of Gudrida, and obtained the consent of her brother-in-law Leif; and their marriage was celebrated during the winter. On this, as on former occasions, the voyage to Vineland formed a favorite theme of conversation, and Thorfinn was urged both by his wife and others to undertake such a voyage.

It was accordingly resolved on. In the spring of 1007 Karlsefne and Snorre fitted out their ship, and Bjarne and Thorhall likewise fitted out theirs. A third ship (being that in which Gudrida's father Thorbiörn had formerly come to Greenland) was commanded by Thorward who was married to Freydisa, a natural daughter of Eric the Red; and on board this ship was also a man of the name of Thorhall, who had long served Eric as huntsman in summer and as house-steward in winter, and who had much acquaintance with the uncolonized parts of Greenland. They had in all one hundred and sixty men. They took with them all kinds of live stock, it being their intention to establish a colony, if possible. They sailed first to the Westerbygd, and afterwards to Biarney (Disco). From thence they sailed in a southerly direction to Helluland, where they found many foxes. From thence they sailed again two days in a southerly direction to Markland, a country overgrown with wood, and plentifully stocked with animals. Leaving this, they continued sailing in a S.W. direction for a long time, having the land to starboard, until they at length came to Kialarnes, where there were trackless deserts and long beaches and sands, called by them Furdustrandir. When they had past these, the land began to be indented by inlets. They had two Scots with them, Hake and Hekia, whom Leif had formerly received from the Norwegian King Olaf Tryggvason, and who were very swift of foot. They put them on shore, recommending them to proceed in a S.W. direction, and explore the country. After the lapse of three days they returned bringing with them some grapes and some ears of wheat, which grew wild in that region. They continued their course until they came to a place where a firth penetrated far into the country. Off the mouth of it was an island past which there ran strong currents, which was also the case farther up the firth. On the island there were an immense number of eyderducks, so that it was scarcely possible to walk without treading on their eggs. They called the island Straumey (Stream-Isle), and the firth Straumfiördr (Stream-Firth). They landed on the shore of this firth, and made preparations for their winter residence. The country was extremely beautiful. They confined their operations to exploring the country. Thorhall afterwards wished to proceed in a N. direction in quest of Vineland. Karlsefne chose rather to go to the S.W. Thorhall, and along with him eight men, quitted them, and sailed past Furdustrandir and Kialarnes, but they were driven by westerly gales to the coast of Ireland, where, according to the accounts of some traders, they were beaten and made slaves. Karlsefne, together with Snorre and Biarne, and the rest of the ships' companies, in all 151 (CXXXI.) men, sailed southwards, and arrived at the place, where a river falls into the sea from a lake. Opposite to the mouth of the river were large islands. They steered into the lake, and called the place Hóp (*í Hópe*). On the low grounds they found fields of wheat growing wild, and on the rising grounds vines. While looking about one morning they observed a great number of canoes. On exhibiting friendly signals the canoes approached nearer to them, and the natives in them looked with astonishment at those they met there. These people were sallow-coloured or ill-looking, had ugly heads of hair, large eyes, and broad cheeks. After they had gazed at them for a while, they rowed away again to the S.W. past the cape. Karlsefne and his company had erected their dwelling-houses a little above the bay; and there they spent the winter. No snow fell, and the cattle found their food in the open field. One morning early, in the beginning of 1008, they descried a number of canoes coming from the S.W. past the cape. Karlsefne having held up a white shield as a friendly signal, they drew nigh and immediately commenced bartering. These people chose in preference red cloth, and gave furs and squirrel skins in exchange. They would fain also have bought swords and spears, but these Karlsefne and Snorre prohibited their people from selling them. In exchange for a skin entirely gray the Skrellings took a piece of cloth of a span in breadth, and bound it round their heads. Their barter was carried on this way for some time. The Northmen then found that their cloth was beginning to grow scarce, whereupon they cut it up in smaller pieces, not broader than a finger's breadth; yet the Skrellings gave as much for these smaller pieces as they had formerly given for the larger ones, or even more. Karlsefne also caused the women to bear out milk soup, and the Skrellings relishing the taste of it, they desired to buy it in preference to everything else, so they wound up their traffic by carrying away their bargains in their bellies. Whilst this traffic was going on, it happened that a bull, which Karlsefne had brought along with him, came out of the wood and bellowed loudly. At this the Skrellings got terrified and rushed to their canoes, and rowed away southwards. About this time Gudrida, Karlsefne's wife, gave birth to a son, who received the name of Snorre. In the beginning of the following winter the Skrellings came again in much greater numbers; they showed symptoms of hostility, setting up loud yells. Karlsefne caused the red shield to be borne against them, whereupon they advanced against each other, and a battle commenced. There was a galling discharge of missiles. The Skrellings had a sort of war slings. They elevated on a pole a tremendously large ball, almost the size of a sheep's stomach, and of a bluish colour; this they swung from the pole upon land over Karlsefne's people, and it descended with a fearful crash. This struck terror into the Northmen, and they fled along the river. Freydisa came out and saw them flying; she thereupon exclaimed, "How can stout men like you fly from these miserable caitifs, whom I thought you could knock down like cattle? If I had only a weapon, I ween I could fight better than any of you." They heeded not her words. She tried to keep pace with them, but the advanced state of her pregnancy retarded her. She however followed them into the wood. There she encountered a dead body. It was Thorbrand Snorrason; a flat stone was sticking fast in his head. His naked sword laid by his side. This she took up, and prepared to defend herself. She uncovered her breasts, and dashed them against the naked sword. At this sight the Skrellings became terrified, and ran off to their canoes. Karlsefne and the rest now came up to her and praised her courage. Karlsefne and his people were now become aware that, although the country held out many advantages, still the life that they would have to lead here would be one of constant alarm from the hostile attacks of the natives. They therefore made preparations for departure, with the resolution of returning to their own country. They sailed eastward, and came to Streamfirth. Karlsefne then took one of the ships, and sailed in quest of Thorhall, while the rest remained behind. They proceeded northwards round Kialarnes, and after that were carried to the north-west. The land lay to larboard of them. There were thick forests in all directions, as far as they could see, with scarcely any open space. They considered the hills at Hope and those which they now saw as forming part of one continuous range. They spent the third winter at Streamfirth. Karlsefne's son Snorre was now three years of age. When they sailed from Vineland they had a southerly wind, and came to Markland, where they met with five Skrellings. They caught two of them (two boys), whom they carried away along with them, and taught them the Norse language, and baptised them; these children said that their mother was called Vethilldi and their father Uvæge; they said that the Skrellings were ruled by chieftains (kings), one of whom was called Avalldamon, and the other Valdidida; that there were no houses in the country, but that the people dwelled in holes and caverns. Biarne Grimolfson was driven into the Irish Ocean, and came into waters that were so infested by worms, that their ship was in consequence reduced to a sinking state. Some of the crew, however, were saved in the boat, as it had been smeared with seal-oil tar, which is a preventive against the attack of worms. Karlsefne continued his voyage to Greenland, and arrived at Ericsfiord.

Voyage of Freydisa, Helge, and Finnboge: Thorfinn settles in Iceland.

During the same summer, 1011, there arrived in Greenland a ship from Norway commanded by two brothers from Austfiord in Iceland, Helge and Finnboge, who passed the following winter in Greenland. Freydisa went to them, and proposed a voyage to Vineland, on the condition that they should share equally with her in all the profits which the voyage might yield. This they assented to. Freydisa and these brothers entered into a mutual agreement that each party should have thirty able-bodied men on board their ship, besides women; but Freydisa immediately deviated from the agreement, and took with her five additional men whom she concealed. In 1012 they arrived at Leifsbooths, where they spent the following winter. The conduct of Freydisa occasioned a coolness aud distance between the parties; and by her subtle arts she ultimately prevailed on her husband to massacre the brothers and their followers. After the perpetration of this base deed, they in the spring of 1013 returned to Greenland, where Thorfinn lay ready to sail for Norway, and was waiting for a fair wind: the ship he commanded was so richly laden, that it was generally admitted, that a more valuable cargo had never left Greenland. As soon as the wind became favorable he sailed to Norway, where he spent the following winter, and sold his goods. Next year, when he was ready to sail for Iceland, there came a German from Bremen, who wanted to buy a piece of wood from him. He gave for it half a marc of gold; it was the wood of the Mazertree from Vineland. Karlsefne went to Iceland, and in the following year, 1015, he bought the Glaumbœ estate in

Skagefiord, in the Northland quarter, where he resided during the remainder of his life, as did also Snorre, his American-born son, after him. On the marriage of the latter, his mother made a pilgrimage to Rome, and afterwards returned to her son's house at Glaumbœ, where he had in the mean time caused a church to be built. The mother lived long as a religious recluse. A numerous and illustrious race descended from Karlsefne, among whom may be mentioned the learned Bishop Thorlak Runolfson, born in 1085, of Snorre's daughter Halfrida, to whom we are principally indebted for the oldest ecclesiastical Code of Iceland published in the year 1123; it is also probable that the accounts of the voyages were originally compiled by him.

A Survey of the Preceding Evidence.

1. *Geography and Hydrography.*—It is a fortunate circumstance that these ancient accounts have preserved not only *geographical* but also *nautical* and *astronomical facts*, that may serve in fixing the position of the lands and places named. The *nautical facts* are of special importance, although hitherto they have not been sufficiently attended to, viz., statements of the course steered and the distance sailed in a day. From data in the Landnama and several other ancient Icelandic geographical works, we may gather that the distance of a day's sailing was estimated at from twenty-seven to thirty geographical miles (German or Danish, of which fifteen are equal to a degree; each of these accordingly equal to *four* English sea miles). From the island of Helluland, afterwards called Little Helluland, Biarne sailed to Heriulfsnes (Ikigeit) in Greenland, with strong south-westerly gales, in four days. The distance between that cape and *Newfoundland* is about 150 miles, which will correspond, when we take into consideration the strong gales. In modern descriptions it is stated that this land partly consists of naked rocky flats, where no tree, not even a shrub, can grow, and which are therefore usually called *Barrens;* thus answering completely to the *hellur* of the ancient Northmen, from which they named the country.

Markland was situate to the south-west of Helluland, distant about three days' sail, or about from eighty to ninety miles. It is therefore *Nova Scotia*, of which the descriptions given by later writers answer to that given by the ancient Northmen of Markland: "the land is low in general;" "the coast to the seaward being level and low, and the shores marked with white rocks;" "the land is low with white sandy cliffs, particularly visible at sea;" "on the shore are some cliffs of exceedingly white sand." Here "*level*" corresponds completely to the Icelandic "*slètt*," "*low to the seaward*" to the short expression "*ósæbratt*," and "white sandy cliffs" to the "*hvítir sandar*" of the Northmen. Nova Scotia, as also New Brunswick and Lower Canada situate more inland, which probably may be considered as all belonging to the Markland of the Northmen, are almost everywhere covered with immense forests.

Vínland was situate at the distance of two days' sail, consequently about from fifty-four to sixty miles, in a south-westerly direction from Markland. The distance from Cape Sable to Cape Cod is stated in nautical works as being W. by S. about seventy leagues, that is, about fifty-two miles. Biarne's description of the coasts is very accurate, and in the island situate to the eastward (between which and the promontory that stretches to eastward and northward Leif sailed) we recognise Nantucket. The ancient Northmen found there many shallows (*grunnsæfui mikit*); modern navigators make mention at the same place "of numerous riffs and other shoals," and say "that the whole presents an aspect of drowned land."

Kialarnes (from *kiölr*, a keel, and *nes*, a cape, most likely so named on account of its striking resemblance to the keel of a ship, particularly of one of the long ships of the ancient Northmen) must consequently be *Cape Cod*, the Nauset of the Indians, which modern geographers have sometimes likened to a horn, and sometimes to a sickle or scythe. The ancient Northmen found here trackless deserts (*öræfi*), and long narrow beaches and sandhills or sands (*strandir lángar ok sandar*) of a very peculiar appearance, on which account they called them Furðustrandir (Wonder-Strands, from *furða*, res miranda, and *strönd*, strand, beach). Compare the description given of this Cape by a modern author, Hitchcock: "The *Dunes* or *sandhills*, which are often nearly or quite barren of vegetation, and of snowy whiteness, *forcibly attract the attention on account of their peculiarity*. As we approach the extremity of the Cape, the sand and barrenness increase; and in not a few places it would need only a party of Bedouin Arabs to cross the traveller's path, to make him feel that he was in the depths of *an Arabian or Libyan desert*." A remarkable, natural phenomenon, which is observed there, has also most probably had a share in giving rise to that peculiar name. It is thus described by the same author: "In crossing the *sands* of the Cape, I noticed *a singular mirage* or deception. In Orleans, for instance, we seemed to be ascending at an angle of three or four degrees; nor was I convinced that such was not the case, until turning about I perceived that a similar ascent appeared in the road just passed over. I shall not attempt to explain this optical deception; but merely remark, that it is probably of the same kind as that observed by Humboldt on the Pampas of Venezuela: 'all around us,' says he, 'the plains seemed to ascend towards the sky.'" Thus we observe that the appellation given by the ancient Northmen to the three strands or tracts of coast, *Nauset Beach*, *Chatham Beach*, and *Monomoy Beach*, is remarkably appropriate.

The great *Gulf Stream*, as it is called, which issues from the Gulf of Mexico, and runs between Florida, Cuba, and the Bahama Isles, and so northwards in a direction parallel to the eastern coast of North America, and of which the channel in ancient times is said to have approached still nearer to the coast, occasions great currents precisely at this place, inasmuch as the peninsula of Barnstable offers opposition to the stream, as it comes from the southward. The Straumfiörðr of the ancient Northmen is supposed to be *Buzzard's Bay;* and Straumey, *Martha's Vineyard;* although the account of the many eggs found there would seem more precisely to correspond to the island which lies off the entrance of Vineyard Sound, and which at this day is for the same reason called *Egg Island*.

Krossanes is probably *Gurnet Point*. It must have been somewhat to the northward of this that Karlsefne landed, when he saw the mountain range (*the Blue Hills*) which he considered as forming part of the same range that extends to the region where we recognise the place named Hóp (i Hópe).

The word Hóp, in Icelandic, may either denote a small recess or bay formed by a river from the interior, falling into an inlet from the sea, or the land bordering on such a bay. To this Mount Hope's Bay, or Mont Haup's Bay, as the Indians term it, corresponds, through which the Taunton River flows, and, by means of the very narrow yet navigable Pocasset River, meets the approaching water of the ocean at its exit at Seaconnet. It was at this Hópe that Leifsbooths were situate; it was above it, and therefore most probably on the beautifully situate elevation called afterwards by the Indians Mont Haup, that Thorfinn Karlsefne erected his dwelling-houses.

2. *Climate and Soil.*—Concerning the climate of the country and the quality of the soil, and also concerning some of its productions, the ancient writings contain sundry illustrative remarks. The climate was so mild that it appeared that cattle did not require winter fodder, for there came no snow, and the grass was but slightly withered. Warden uses similar expressions respecting this region: "La température est si douce que *la végétation souffre rarement du froid ou de la sécheresse.* On l'appelle *le paradis de l'Amérique* parcequ'elle l'emporte sur les autres lieux par sa situation, son sol et son climat." "An excursion from Taunton to Newport, Rhode Island, down Taunton River and Mount Hope Bay, conducts the traveller among scenery of great beauty and loveliness," says Hitchcock; and when he adds that the beautiful appearance of the country "and the interesting historical associations connected with that region, conspire to keep the attention alive, and to gratify the taste," he will find that this last remark is applicable to times much more remote than he thought of, when he gave expression to the above sentiment.

A country of such a nature might well deserve the appellation of "THE GOOD," which was the epithet the ancient Northmen bestowed on it; especially as it yielded productions whereon they set a high value, and of which their colder native land was for the most part destitute.

3. *Produce and Natural History.*—*Vines* grew there spontaneously, a circumstance which Adam of Bremen, a foreign writer of the same (that is, of the eleventh) century, mentions that he had learned, not from conjecture, but from authentic accounts furnished by Danes. As his authority on this occasion he cites the Danish King Sveyn Estrithson, a nephew of Canute the Great. It is well known that vines still grow in that region in great abundance.

Spontaneously growing wheat (*sjálfsánir hveitiakrar*).—On the subsequent arrival of the Europeans, maize was found growing here, or Indian corn as it is called, which the natives reaped without having sowed, and preserved in holes in the earth, and which constituted one of their most valuable articles of food Honeydew was found on the island which lies off it, as is also still the case.

Mazer (*mausur*), a species of wood of remarkable beauty, probably a species of the ***Acer rubrum*** or ***Acer saccharinum***, which grows here, and which is called "bird's eye," or "curled maple." Wood for building was also obtained here.

A great number of forest animals of all kinds.—It is understood that the Indians chose this region in preference for their abode, chiefly on account of the excellent hunting. Now-a-days the forests are for the most part cut down, and the animals have withdrawn to the interior and woodland regions. From the natives the Northmen bought squirrel-skins, sables, and all kinds of peltries, which are still to be found in abundance in this district.

Eyderducks and other *birds* were found in great numbers on the adjacent islands, as is also at present the case, on which account some of them have the name of Egg Islands.

Every river was full of *fish*, among which are mentioned excellent *salmon*. On the coast was also caught a great quantity of fish. The Northmen dug ditches along the shore, within the high-water mark, and when the tide receded, they found *halibuts* in the ditches. On the coast they also caught *whales*, and among these the *reiðr* (*Balæna physalus*). In the modern descriptions of this region it is stated, that "all the rivers are full of fish," and of the waters in that neighbourhood it is said, "il y a une grande abondance de poissons de presque toutes les espèces." Salmon may be mentioned as one of these. Not long ago, the whale fishery was, in that very region, an important branch of industry, especially for the inhabitants of the adjacent islands. Very possibly the adjacent Whale Rock has its name from the same circumstance.

4. *Astronomical Evidence.*—Besides the nautical and geographical statements, one of the ancient writings has preserved an *astronomical* notice, where it is said, that the days there were of more equal length than in Iceland or Greenland: that on the shortest day the sun rose at half-past seven o'clock and set at half-past four, which makes the shortest day nine hours. This astronomical observation gives for the place latitude 41° 24′ 10″. We thus see that this statement corresponds exactly with the other data, and indicates precisely the same region.

Discoveries of more Southern Regions.

The party sent by Thorwald Ericson in the year 1003 from Leifsbooths to explore the southern coasts, employed from four to five months in the expedition; they therefore most likely examined the coasts of Connecticut and New York, probably also those of New Jersey, Delaware, and Maryland. The description of this range of coast is accurate.

Are Marson's Sojourn in Great Ireland.

In those times the Esquimaux inhabited more southerly regions than they do now-a-days. This is both evident from the ancient accounts, and seems besides to gain corroboration from the skeletons of ancient times which have been dug up in regions even more southerly than those in question; a circumstance which however merits a more accurate examination. In the neighbourhood of Vineland, opposite the country inhabited by the Esquimaux, there dwelled, according to their reports, people who wore white dresses, and had poles borne before them on which were fastened lappets, and who shouted with a loud voice. This country was supposed to be Hvítramannaland as it was called (the Land of the Whitemen), otherwise called I'rland It Mikla (Great Ireland), being probably that part of the coast of North America which extends southwards from Chesapeak Bay, including North and South Carolina, Georgia, and Florida. Among the Shawanese Indians, who some years ago emigrated from Florida, and are now settled in Ohio, there is preserved a tradition which seems of importance here, viz., that Florida was once inhabited by white people who were in possession of iron implements. Judging from the ancient accounts, this must have been an Irish Christian people, who previous to the year 1000 were settled in this region. The powerful chieftain Are Marson, of Reykianes in Iceland, was in the year 983 driven thither by storms, and was there baptised. The first author of this account was his contemporary Rafn, surnamed the Limerick-trader, he having long resided at Limerick in Ireland. The illustrious Icelandic sage, Are Frode, the first compiler of the Landnama, who was himself a descendant in the fourth degree from Are Marson, states on this subject, that his uncle, Thorkell Gellerson (whose testimony he on another occasion declares to be worthy of all credit) had been informed by Icelanders, who had their information from Thorfinn Sigurdson Earl of Orkney, that Are had been recognised in Hvítramannaland, and could not get away from thence, but was there held in high respect. This statement therefore shows, that in those times there was an occasional intercourse between the Western European countries (the Orkneys and Ireland) and this part of America.

Voyages of Biörn Asbrandson and Gudleif Gudlaugson.

It must have been in this same country that Biörn Asbrandson, surnamed Breidvíkingakappi, spent the latter part of his life. He had been adopted into the celebrated band of Jomsburgh warriors under Palnatoke, and took part with them in the battle of Fyrisval in Sweden. His illicit amatory connexion with Thurida of Frodo in Iceland, a sister of the powerful Snorre Gode, drew upon him the enmity and persecution of the latter, in consequence of which he found himself obliged to quit the country for ever; and in the year 999 he set sail from Hraunhöfn in Sniofelsnes with a N.E. wind. Gudleif Gudlaugson, brother of Thorfinn, the ancestor of the celebrated historian Snorre Sturluson, had made a trading voyage to Dublin; but when he left that place again, with the intention of sailing round Ireland and returning to Iceland, he met with long continuing north-easterly winds, which drove him far to the southwest in the ocean, and at an advanced period of the summer he and his company arrived at last at an extensive country, but they knew not what country it was. On their landing, a crowd of the natives, several hundreds in number, came against them, and laid hands on them, and bound them. They did not know anybody in the crowd, but it seemed to them that their language resembled Irish. The natives now took counsel whether they should kill the strangers, or make slaves of them. While they were deliberating, a large company approached, displaying a banner, close to which rode a man of distinguished appearance, who was far advanced in years, and had gray hair. The matter under deliberation was referred to his decision. He was the aforesaid Biörn Asbrandson. He caused Gudleif to be brought before him, and addressing him in the Norse language, he asked him whence he came. On his replying that he was an Icelander, Biörn made many inquiries about his acquaintance in Iceland, particularly about his beloved Thurida of Frodo, and about her son Kiartan, supposed to be his own son, and who at that time was the proprietor of the estate of Frodo. In the mean time, the natives becoming impatient and demanding a decision, Biörn selected twelve of his company as counsellors, and took them aside with him, and some time afterward he went towards Gudleif and his companions, and told them that the natives had left the matter to his decision. He thereupon gave them their liberty, and advised them, although the summer was already far advanced, to depart immediately, because the natives were not to be depended on, and were difficult to deal with, and moreover conceived that an infringement on their laws had been committed to their disadvantage. He gave them a gold ring for Thurida, and a sword for Kiartan, and told them to charge his friends and relations not to come over to him, as he was now become old, and might daily expect that old age would get the better of him; that the country was large, having but few harbours, and that strangers must everywhere expect a hostile reception. They accordingly set sail again, and found their way back to Dublin, where they spent the winter; but next summer they repaired to Iceland and delivered the presents; and everybody was convinced that it was really Biörn Asbrandson whom they had met with in that country.

Bishop Eric's Voyage to Vineland.

It may be considered as certain that the intercourse between Vineland and Greenland was maintained for a considerable period after this, although the scanty notices about Greenland contained in the ancient MSS. do not furnish us with any satisfactory information on this head. It is, however, recorded that the Greenland Bishop Eric, impelled probably by a christian zeal either of converting the colonists or of animating them to perseverance in the faith, went over to Vineland in the year 1121. As we have no information of the result of his voyage, but can merely gather from the above expression that he reached his destination, we must presume that he fixed his permanent residence in Vineland. His voyage, however, goes to corroborate the supposition of a lengthened intercourse having been kept up between the countries.

Discoveries in the Arctic Regions of America.

The next event in chronological order, of which accounts have been preserved in ancient records, is a voyage of discovery in the Arctic regions of America, performed during the year 1266 under the auspices of some clergymen of the bishopric of Gardar in Greenland. The account of it is taken from a letter, addressed by a clergyman of the name of Halldor, to another clergyman named Arnold, formerly established in Greenland, but who had then become chaplain to the Norwegian King Magnus Lagabæter. At that time all men of any consequence in Greenland possessed large vessels built for the purpose of being despatched northwards in the prosecution of hunting and fishing expeditions. The northern regions which they visited were called Nordrsetur; the chief stations were Greipar and Króksfiardarheidi. The first of these stations is supposed to have been situate immediately to the southward of Disco; but that the ancient Northmen went much farther north on this coast may be inferred from a very remarkable runic stone, found in the year 1824 on the island of Kingiktórsoak, lying in the latitude of 72° 55′ N. The latter mentioned station was to the north of the former. The object of the voyage is stated to have been, to explore regions lying more to the northward than those they had hitherto been accustomed to visit, consequently lying farther north than Króksfiardarheidi, where they had their summer quarters (*setur*), and which they were therefore regularly accustomed to visit. Relating to this voyage of discovery the following particulars are mentioned:—

They sailed *out of* Króksfiardarheidi, and after that encountered southerly winds, accompanied by thick weather, which obliged them to let the ship go before the wind. On the weather clearing up, they saw many islands and all kinds of prey, both seals and whales and a great many bears. They penetrated into the innermost part of the gulf, and had icebergs (glaciers) lying also to the southward as far as the eye could reach. They observed some vestiges indicating that the Skrellings had in former times inhabited these regions, but they could not land on account of the bears. They then put about and sailed back during three days, and now again they found traces of the Skrellings having been on some islands lying to the southward of a mountain by them called Sniofell. After this—on St. James's day—they proceeded southwards, a great day's rowing. It froze during the night in those regions, but the sun was above the horizon both night and day, and when on the meridian, in the south, it was not higher than that when a man lay down across a six-oared boat, stretched out towards the gunwale, the shadow formed by the side of the boat nearest the sun reached his face: but at midnight the sun was as high as when it was (highest) in the north-west in the Greenland colony. Afterwards they sailed back again to their home at Gardar. Kröksfiardarheidi, as we have observed above, had been for some time previous regularly visited by the Greenlanders. The name shows that the firth was surrounded by barren highlands (*heidi*), and the description of the voyage shows that it was a firth of considerable extent, in and through which there was room for several days' sail. It is stated, for instance, that they sailed *out of* this firth or sound into another sea, and into the innermost part of a gulf, and that their returning voyage occupied several days. As to the two observations mentioned as having been taken on St. James's day, the first of them leads to no certain result, as we have no sure means of ascertaining the depth of the boat, or rather the relative-depth of the man's position, as he lay across the boat, in reference to the height of the side of the same, so as to enable us to deduce the angle formed by the upper edge of the boat's side and the man's face, which is the angle measuring the sun's altitude at noon on St. James's day, or 25th July. If we assume, as we may do with probability, that it was somewhat less than 33°, and yet very near that measure, the place must have been situate near N. L. 75°. There seems no probability that it was a larger angle, and consequently that the place lay more to the southward. The result obtained from the other observation is however more satisfactory. In the thirteenth century, on the 25th July, the sun's declination was + 17° 54′; inclination of the ecliptic + 23° 32′. If we now assume that the colony, and particularly the episcopal seat of Gardar, was situate on the north side of Igaliko Firth, where the ruins of a large church and of many other buildings indicate the site of a principal settlement of the ancient colony, consequently, in 60° 55′ N. L., then at the summer solstice the height of the sun there, when in the north-west, was = 3° 40′, equivalent to the midnight altitude of the sun on St. James's day in the parallel of 75° 46′, which falls a little to the north of Barrow's Strait, being in the latitude of Wellington's Channel, or close to the northward of the same. The voyage of discovery undertaken by the Greenland clergymen was therefore carried to regions, which in our days have been more accurately explored, and their geographical position determined by Sir Edward Parry, Sir John Ross, and Captain James Clark Ross, and other British navigators, in the no less daring and dangerous expeditions conducted by them.

Newfoundland re-discovered from Iceland.

The discovery next recorded was made by the Iceland clergymen Adalbrand and Thorwald Helgason, well known in the history of Iceland as having been involved in the disputes at that time prevailing between the Norwegian King Eric Priesthater and the clergy, and which in Iceland were chiefly headed by the governor, Rafn Oddson, and Arne Thorlakson, Bishop of Skalholt. Accounts drawn up by contemporaries contain merely the brief notice that in the year 1285 the above mentioned clergymen discovered a new land to the westward of Iceland (*fundu ny'ja land*). This land, to which by command of King Eric Priesthater, a voyage was some years afterwards projected by Landa-Rolf, is supposed to have been *Newfoundland.*

A Voyage to Markland in the Year 1347.

The last piece of information respecting America, which our ancient MSS. have preserved, refers to a voyage in the year 1347, from Greenland to Markland, performed in a vessel having a crew of seventeen men, being probably undertaken for the purpose of bringing home building timber and other supplies from that country. On the voyage homeward from Markland the ship was driven out of her course by storms, and arrived with loss of anchors at Straumfiord in the west of Iceland. From the accounts, scanty as they are, of this voyage, written by a contemporary nine years after the event, it would appear that the intercourse between Greenland and America Proper had been kept up to so late a date as the year above mentioned; for it is expressly said that the ship went to Markland, which is thus named as a country that in those days was still known and visited.

* * * * *

After having perused the authentic documents themselves, which are now accessible to all, every one will acknowledge the truth of the historical fact, that during the tenth and eleventh centuries, the ancient Northmen discovered and visited a great extent of the eastern coasts of North America; and will besides be led to the conviction, that during the centuries immediately following, the intercourse was never entirely discontinued. The main fact is certain and indisputable. On the other hand there are, in these as in all other ancient writings, certain portions of the narrative which are obscure, and which subsequent disquisitions and new interpretations may serve to clear up. On this account it seems of importance that the original sources of information should be published in the ancient language, so that every one may have it in his power to consult them, and to form his own judgment as to the accuracy of the interpretations given.

With regard to such traces of the residence and settlement of the ancient Northmen as, it is presumed, are still to be met with in Massachusetts and Rhode Island—the countries which formed the destination of their earliest American expeditions—we will content ourselves for the present with referring to the hints, which are contained in the work itself. This matter will continue to form a subject for the accurate investigation of the Committee of the Royal Society of Northern Antiquaries on the Ante-Columbian History of America; and the result of this investigation, together with such additional elucidations of the ancient MSS. as we may have it in our power to furnish, shall be communicated in the Annals and Memoirs of the Society.

VIII.—*Sketch of the Island and Gulf of Symi, on the South-western Coast of Anatolia, in February,* 1837. By James Brooke, Esq.

[The island of Symi* lies about 18 miles N.W. of the northern point of Rhodes, at the mouth of a gulf of the same name—the ancient *Sinus Doridis*. The island is mentioned, in modern times, by Hasselquist in 1750; by Van Egmont and Heyman in 1700; and by Dr. Clarke in 1801. But of the Gulf of Symi nothing accurate is known; its outline, as well as that of the Gulf of Cos, from which it is separated only by the peninsula formerly called *Triopium*, is most incorrectly marked on our charts:—this part of the coast lies between the western termination of Captain Beaufort's survey of Karamania, which only came as far as the Gulf of Makri, and the survey of the coast of Asia Minor, now executing by Lieutenant Graves, R.N.—but as it is probable that officer may not be enabled to reach these gulfs before the close of the year, we are glad to profit by the information afforded by Mr. Brooke, who visited Symi in his yacht during the spring of the past year.]

On the authority of Pliny, V. 28, the Sinus Doridis was included between the Triopium Promontory and that of Cynos-sema, and this space, with which the ancients were well acquainted, is scarcely known to modern geography. Even the Island of Symi, lying directly in the track of vessels bound along the coast, is so imperfectly described, that the following brief account may possess some degree of interest as relating to the least-visited portion of the shores of Asia Minor, until a more complete description be obtained.

After visiting various places between Smyrna and Rhodes, we reached the harbour of Panerimiotis,† at the south-western point of Symi, early in the morning of February 2, 1837.

* Syma or Syme, 'Ε Σύμη, Meletius, Geograph. Ant. et Nov. Ven. ii. 222.
† Perhaps ἡ πανερημιώτισσα, "the most solitary."

Symi is a high rocky and barren island of grey limestone, rising about 1000 feet above the sea, its general outline irregular, and its coasts bold and steep. Deep indentations mark both the northern and southern extremities of the island. The soil is extremely unproductive, but the inhabitants cultivate every available patch of ground between the rocks with great assiduity. Commerce and enterprise have rendered this unpromising spot a thriving mart. It is a depôt for sponge and wood. The former article is found in the neighbouring seas, and employs the male population during the summer to procure it. France is the great outlet for the coarser, and England for the finer kind of this article, but some of it is carried direct to its destination. Greek boats come from Rhodes, Greece, Smyrna, &c., to convey the sponge previous to its shipment in European vessels. This intermediate carrying trade is useful and lucrative to the boatmen of Symi in particular, and generally to those of the whole coast and islands. The second export, wood, is collected in the Gulf of Symi, and along the coast to the eastward towards Makri and Marmorice,* and exported from the island even as far as Alexandria, whence ships come for their cargoes. The possession of three harbours affords Symi these resources, and it is pleasing to observe the active use they make of these natural advantages. The island is included in the Páshálik of Rhodes, and their annual tax amounts to 27,000 piastres, about 270*l.*,—a very light burden divided among a population of not less than 7,000 persons.

At the southern extremity of Symi is the island of Kiskilies, a picturesque and fertile spot, about a mile or more in extent. Off the Kiskilies (so called in the charts) on the S. E. side is a detached cluster of rocks and an islet near them. To the westward is likewise an islet, with a narrow but deep passage between it and the larger island. A passage of a mile or more divides Kiskilies from Symi; it is deep and clear from danger, but the wind is not to be trusted, on account of the high land. At the S. W. angle of the island of Symi is the harbour of Panerimiotis:—a deep bay incloses the mouth of the port, which is narrow and formed by natural piers. It faces N. W. In coming from the southward, after rounding Kiskilies,† the outer bay will easily be discerned from its depth, and the right shore may be kept on board until the harbour's mouth opens. The best leading marks into the port, however, are four islands which run out from the westernmost point of Symi. From the outer one of these islands the compass course E. by S. will conduct to the harbour's mouth. The headland on the left hand going in, which forms the western

* Mermeríchah, from the Slavonian and Greek Marmaritza, a diminutive of Marmara.

† Probably from the Latin Quisquilliæ, "chips."

extremity of the bay, is bold and projecting, but a rock or shoal is reported to lie off it; and it will, therefore, be advisable to give it a berth on coming from the northward and westward. The harbour's mouth is about two hundred yards wide and deep, within a short distance of either shore. The soundings just within the entrance are six fathoms, then five, four, and three. The harbour has three, and a quarter less three fathoms all over it—is oval in shape, and secure from all winds, but the holding ground is not very good. Its breadth is about half a mile, and it forms a convenient shelter for small vessels. At the bottom of the bay is the monastery of Panerimiotis, inhabited by a few Greek monks: they were very civil and obliging, and furnished us with a bullock, some poultry, and vegetables, but, generally speaking, wood, water, and provisions must be brought from the capital in boats, as the Caloyers are the only inhabitants, and the road to the town is over steep and rugged mountains.

Symi, the capital and only town on the island, is a thriving place, situated on an acclivity above the Scala* or landing-place in the bottom of the bay at the north-eastern angle of the island, and facing the Gulf of Symi, which from the centre of the bay bears E.N.E. The sailing directions for the Archipelago place this harbour at the wrong end and wrong side of the island, evidently confounding the Scala of Symi with the port of Panerimiotis.

The harbour is long and narrow, with deep water, and vessels lie close in with their warps made fast to the shore. To make the port from the northward, round the headland with a small islet off it, and steer S.W. ½ S. (by compass) for the anchorage. Approaching from the southward, round the Windmill Hill, which forms the south-eastern point of the bay, and keeping the southern shore on board, steer directly for the Scala. On coming from the southward, however, the town of Symi will be seen on the hill at the head of a deep bight, before rounding Windmill Hill, but this bight must not be mistaken for the harbour.

Here is a castle with remains of Cyclopean walls, and on the Windmill Hill is a circular Hellenic building, said to be the trophy erected by the Lacedemonians on their defeating the Athenian squadron. These, with fragments of other walls and terraces, are sufficient perhaps to identify the present with the site of the ancient town.

The harbour called Nimborio by the natives, *alias* Emporio, is to the northward of the Scala, and divided from it by a rocky

---

* Scala, a staircase or ladder, has since the time of the Genoese and Venetian predominance in the Archipelago, been the common term for a harbour throughout the Levant; the Turks have converted it into Iskeleh, and the French into Echelle. —F. S.

ridge; it is situated at the bottom of a bay, but as we did not visit it, I can say nothing beyond pointing out its situation.

My impression of the island of Symi is altogether favourable. Nothing can exceed the barrenness and sterility of its limestone mountains, but its harbours afford a source of wealth which supplies the defects of its soil. Every thing bespeaks commercial vigour and prosperity. Provisions of all descriptions can be obtained—we even procured *rum and potatoes,* articles unknown at Rhodes.

The population of Symi may be about 1000 persons. The women are diminutive and plain; their costume is a long, loose quilted jacket, and a scanty red petticoat; a huge turban composed of manifold handkerchiefs surmounts the head, and their legs are adorned with large boots. From the neck to the waist they wear a line of round silver ornaments, and appear fond of decking themselves with trinkets. Of the men it is difficult to judge, since natives and foreigners are confounded; as agents and merchants from various sea-ports live here to purchase sponge, and together with Greek skippers and pilots, form a miscellaneous race.

After a stay of a few days in the island our party started on a boat excursion for the Gulf of Symi. Passing between Symi and the Kiskilies we stretched across to the main-land, where, the wind dying away, we took to our oars and pulled gently along. If Symi be little known, what can I say of the main-land? The charts are an absolute dead letter, not giving the faintest resemblance to the coast. There are *two* headlands where the charts lay down Cape Volpo;* one stretches to the southward, and the other approaches the island of Symi. From the latter of these points (probably the Cape Volpo of the charts) the coast forms a capacious bay, *before* reaching the larger gulf. Two other headlands, nearly on the same meridian as the former, lie between the above-mentioned bay and the Gulf (so called) of Symi. Off the southernmost of these is a dangerous patch of rocks, which was barely above water when we saw it. The bearing of the N. point of Symi from the shoal is W. by N. The nearest headland towards the Gulf of Symi N.E. The east point of Kiskilies S.S.W. Distance from the shore from half to three quarters of a mile.

Rounding the two headlands already mentioned, the Gulf is open. Five islands lie on the right or southern shore; the three first small, the two last much larger. The larger islands are so shut in with the main as apparently to form a separate bay. Within these islands the main-land is bold and indented. Passing

* Cynos-sema (Dog's Tomb) of the ancients, called Alepú (Fox) by the modern Greeks. Volpo is the vulgar word for Volpe.—F. S.

a deep bight opposite the fourth island, we pulled along under a bold and precipitous cliff. The scenery was grand and imposing, and the various openings and curves of the bay before us kept our attention on the stretch. There are patches of cultivation on the larger islands; an islet lies between the fifth island and main. At the termination of the cliff a large bay opens, affording a most attractive view. On the further shore of this bay are the ruins of a city and castle crowning the summit of a hill, from 800 to 1000 feet high. Considerable remains of Cyclopean walls exist amid the later works of the Byzantine period, and traces of tombs and foundations are numerous in the valley below. The situation is steep and difficult of access. The castle, from its extent, must have been a place of importance, and the ancient city by no means insignificant. Leaving this bay we proceeded into the bight, where, about one mile and a half from the beach, the village of Bidelos, containing about 200 persons, is situated, and through which flows a small stream. A short distance further the shore recedes and forms another deep and varied bay; on one side is a rocky island. There are ancient remains all along the shore of this bay, and on the neighbouring hills. They consist entirely of terraces and portions of wall chiefly Cyclopean, but too numerous and too much scattered to admit the belief that they all belonged to one city. From the nature of the materials, there is little hope of finding inscriptions: one, however, I discovered in the ruins of a Greek church. The scenery here is of the finest and most picturesque nature: the valley lies enclosed by an amphitheatre of pine-clad and craggy mountains, through which a stream winds its course, and discharges itself into an inner basin formed by a neck of sand. There is another village, of about forty houses, situated rather more than a mile up the stream. In this bay we passed our second night, in a situation as cool and as picturesque as the former—our huge fire burned clearly against a few blocks of an ancient wall, and the remains of a more recent building partly protected us from the night air. At the further extremity of the Gulf is another castle, and probably an ancient city: we did not land, for our time was limited. Glancing therefore at the extremity of the Gulf, we crossed to the northern shore and sailed along it. It is barren, red, and not so deeply indented, forming from its colour a strong contrast with the opposite coast.

The narrowest part of the Gulf is nearly abreast of the fifth island, perhaps about two miles, and its depth may be roughly estimated from twelve to fourteen miles.

We did not follow the shore further than a spot which we called Gothic Isle; but could perceive many bays and curves towards the entrance of the Gulf. From Gothic Isle the land trends away towards the isthmus which forms the Triopium promontory. This neck of land between the Gulfs of Cos and Symi is narrow throughout its whole extent; but though we had no opportunity of landing to ascertain the exact point where the Cnidians proposed dividing their promontory from the main, I feel no hesitation in saying that the spot is, where the land sinks into a bay, bearing N. by W. from Symi. This bay is deeper than generally represented; the coast comparatively low, and the water of the Gulf of Cos visible from Symi and other places. It appears very narrow, but whether exactly five stadia across, I am unable to say, and it would, perhaps, be best to leave the naming of the various features to those who are better acquainted with the ancient geography of the Gulf. I will, however, venture to glance at the ancient account of the Sinus Doridis. The limits of the Gulf may be fixed between the Triopium Promontorium * and the headland called Cynos-sema. Within the Gulf were the three subordinate bays of Bybassus,† Schænus, and Thymnias.‡ The Triopium Peninsula met the Bubassian or Bybassian Peninsula, and at the junction was the proposed cut of the Cnidians. Nothing can agree better with our observations. A deep bay is formed at the junction of the two Peninsulas, which it is just to suppose is the Bubassius Sinus. The city of Acanthus was in this bay,§ and a Greek of Symi informed us of some ruins situated in the bight on the Triopium Peninsula. I regret we did not examine this more minutely, but the deep bay and the narrow isthmus joining the two Peninsulas is beyond doubt. The bay we did examine (called the Gulf of Symi) may be either Thymnias or Schænus; on Cramer's authority I call it the latter. Several towns were situated within it. It may be observed that Colonel Leake places Asseréna at the extremity of the bay near the shore; Cramer omits this town,|| but remains do exist in the place marked by Colonel Leake. Hyda was likewise within this bay, and the ruins of the city on the hill being considerable may be referred to that place. There are likewise other remains which might justly have been looked for.

If this slight sketch of the topography of this Gulf be correct, and it appears probable, the Aphrodisian Promontory will be found distinctly marked between the two bays.¶ The Sinus Thymnias (entirely omitted in modern charts) will likewise be evident, and Cape Volpo will be the Cynos-sema Promontorium.

I will here conclude these remarks of a hasty visit to a most interesting locality, which added, and still adds, the charm of novelty to the higher interests of antiquity and natural beauty.

* Now Cape Krio.

† Βύβασσος.

‡ Cramer, Asia Minor, ii. 189.

§ In the Peninsula of Cnidus (Steph. Byz.).

|| Because it is modern, not ancient.—F. S.

¶ "Three bays in succession," says Mele (i. 16), "Thymnias, Schænus, Bubassius. Aphrodisium is the Promontory of Thymnias; Schænus surrounds Hyde, Bubassius (*i. e.* Bubassian bay), Acanthus." He is travelling from S.E. to N.W. Dr. Cramer has overlooked the order of these bays.—F. S.

IX.—*On the difference of level between the Black Sea and the Caspian.* Communicated by the Baron ALEXANDER VON HUMBOLDT.

THE kindness of M. Kupffer, Corresponding Member of the Society at St. Petersburg, enabled us to give, at p. 426, Vol. vi. of the Geographical Journal, some account of the Russian expedition about to be sent forth, in 1836, to determine the long-contested point of the difference of level between the Black and the Caspian Seas. That expedition has fulfilled its object.

The trigonometric levelling along the line of country between Novo-Tcherkask, by Stavropol, to Kisliar, is now completed, and although the detail of the survey has not yet reached London, yet the fact of the determination of so remarkable a point in physical geography is too important not to find a place in the Geographical Journal, especially as the Society is indebted for the communication of it to one of its most distinguished Foreign Members.

"*Berlin, Jan.* 10, 1838.

"The beautiful trigonometric levelling between the Black and the Caspian Seas is at length finished. There is a depression, but a much less depression than M. Parrot had announced after his first barometric levelling by stations; this always appeared to me probable (see my letter to M. Parrot, inserted in his 'Voyage à l'Ararat,' vol. ii. p. 192), on account of the elevation of Kasan (level of the junction of the Kasanza and the Wolga), and on account of some corresponding observations which I obtained during my journey to the Caspian. The levelling of MM. Fuss, Sabler, and Sawitch, shews that the level of the Caspian is 101.2 Russian feet, equal to 94.9 Paris feet (old measure), lower than the level of the Black Sea.

"The height of Kasan has been much discussed, and all travellers have assumed different heights; at page 639 of the volume published by M. Gustave Rose, I have given the latest results of the labours of the astronomer Simonoff, professor at Kasan. The cistern of the barometer in the building of the University (Universitäts Gebäude) at Kasan, has an elevation of 181 Paris feet; the junction of the Kasanza and the Wolga is 53 feet (8.9 toises) above the level of the ocean. The fall of the Wolga from Kasan to the Caspian would be then 148 feet, or 24.7 toises. In a direct line it is about 157 geographical leagues (of 15 to the degree) from Kasan to the Caspian. This is a greater inclination than that of the Amazon or the Nile, and almost as great as that of the Oder. The most striking fact is the little elevation of part of the interior of the East of Europe above the level of the ocean, since from Kasan to the Icy Sea it is 135 geographical leagues. The city of Berlin, so near the Baltic, is 100 feet above the level of the sea, according to an excellent trigonometric levelling, which the chief of the staff corps has, at my request, caused to be executed during last year, by the able observer M. Bäyer.

"I cannot sufficiently congratulate the Geographical Society on having found so excellent a traveller as M. Schomburgk: his latest labours, the ascent of the Rivers Corentyn and Berbice, in Guayana, place him very high in my opinion; and the zone of hieroglyphic figures, sculptured in the rocks from Encamarada, in 66° 50′ W. longitude, as far as the eastern limit of British Guayana, a distance of nearly 600 miles, is an ethnographical phenomenon which daily increases in interest.

"The astronomical geography of the North of Asia will shortly be set right by the publication of the important labours of M. Federow, *élève* of M. Struve, who has recently returned, after a five years' absence. Should I still publish the detail of my astronomical observations in Siberia, it will only be in order to fix more accurately the points where I have made observations on terrestrial magnetism.

"I learn with the greatest satisfaction that my letter to the Duke of Sussex, on the subject of Magnetic Observatories, has produced some useful results. As we make observations here both with the needle of Gambey, furnished with microscopes and with the new apparatus of Gauss, an apparatus furnished with a mirror, we have an opportunity of convincing ourselves more and more of the great perfection of this mirror apparatus, which doubtless requires greater skill and more instruction on the part of the observers; the tracing of the curves of horary variation, based upon observations made across all Europe for every five minutes of time (see Gauss und Weber, *Resultate aus den Beobachtungen des Magnetischen Vereins im Yahr* 1836; Göttingen, bei Dietrich, 1837), will prove what advantage is derived from making use of the apparatus of Gauss, which doubtless will soon be employed in all our great observatories.

"As I think that this subject is not without importance to Seamen, I beg you to invite the leading Members of the Geographical Society to be good enough to propagate Gauss' manner of observing in all new stations, where intelligent persons can be found. Points near the Magnetic Equator, and those which are in high latitudes in the southern hemisphere, as the Cape of Good Hope, Australia, Van Diemen's Land, would be the most desirable, if they would observe at the same epochs indicated by M. Gauss, and followed throughout the North of Asia, in Germany, in Sweden, and at Milan."

X.—*Notes of a Journey in Asia Minor, in* 1837. By W. J. Hamilton, Esq. Read 26th March, 1838.

A short account of my journey in Asia Minor, in 1836, extracted from private letters, having been read before the Royal Geographical Society, I beg leave to submit for the same purpose the following sketch of that which I made through the more southern portion of that country in 1837.

I left Constantinople on the 24th of May, and again crossing the Propontis, landed at Modániyah; from thence I proceeded in a S.S.W. direction to Abulliont, situated on a small island near the north-eastern extremity of the lake of the same name, and connected with a narrow peninsula by a long wooden bridge. I copied a few inscriptions in the ruined walls of the town, ascertained the position of the theatre, and observed the foundations of several small buildings outside the town, which were probably the tombs. Although no inscriptions mention its name, there can be no doubt that the modern town of Abulliont stands upon the site of Apollonia ad Rhyndacum. On one of the small islands near it are the remains of considerable massive walls of Hellenic construction. From hence I proceeded four hours along the northern shore of the lake to U'lúbád, where are the ruins of a large Byzantine fortress, and where the Rhyndacus, which I crossed by a long wooden bridge, flows out of the lake. Two hours more, nearly W. by N., over low marshy ground, brought me to Míkhálich,* a large and straggling town of 1500 houses picturesquely situated on some low hills near the junction of the Rhyndacus and the Macestus.

May 26th.—Míkhálich to Aïdinjik,† ten hours in a westerly direction, over an uninteresting undulating country. Three hours before we reached Aïdinjik I came in sight of the lake of Miletopolis, now called Mániyás Gól,‡ situated in a low flat grassy country to the south: near Aïdinjik the scenery improves, and the ground is highly cultivated. The town is full of ancient fragments brought from the ruins of Cyzicus, called Balkís § by the Turks, and the

* Commonly pronounced Múhalích.—A. For all the notes not marked A. the Foreign Secretary is responsible. The orthography of the names of places is according to the standard adopted by the Society.—See vol. vii., p. 245.

† Little Moonshine. Jihán-numá, p. 669.

‡ Mániyás gól (pronounced ghieul, according to the French orthography), not mentioned in the Jihán-numá (Speculum Mundi), p. 656.

§ Balkís-seráï. It is known, says Kátib Chelebí Hájí Khalífeh (Jihán-numá, p. 669) by the name of Tamáshálik (Showplace, because Suleïmán Páshá there *shewed* the shores of Europe to O'rkhán, and urged him to cross over and conquer them) and Balkís Seráyí, *i. e.* "The Palace of Balkís," the queen of Sheba.

Colonel Leake observes that *Bal* is often used in the names of places to signify *old*, being, perhaps, a corruption from *πάλαιος*. Kis is evidently the first syllable of *Cyzicus*, pronounced as in Greek with a *K;* Balkís would therefore mean Old Cyzicus.—A.

Aghá showed me several sepulchral monuments or votive tablets with inscriptions, all from the same place, which he was anxious I should purchase, but they were not sufficiently interesting to tempt me.

May 27th.—Visited Artákí* and Cyzicus. At the former place the plague was raging violently, so that I did not stop there long. I visited the harbour, where are the remains of an ancient mole, and returned to the ruins of Cyzicus, visiting Cape Mélanos in the way. The ruins on the Cape do not appear to be of very great antiquity, and are certainly neither Hellenic nor Cyclopean. At Balḳís I pitched my tent under the walls of the town, near a beautiful fountain under some fine plane trees, and near what was probably the western harbour, but which, being choked up with sand, is now a mere swamp. The whole site of Cyzicus is so covered with gardens, or overgrown with underwood and a luxuriant vegetation, that it is extremely difficult to make out anything distinctly. I discovered the remains of a very large theatre, nearly overgrown with ilex and other shrubs, besides the amphitheatre or naumachia mentioned by former travellers, and several other large buildings, but from their ruined state it is impossible even to guess at their original destination. On the whole, considering the great renown of the architects of Cyzicus, I was much disappointed at the poor and meagre style of all its buildings, which do not at all bear the solid and imposing character of Hellenic masonry. No doubt there is much buried under the soil, the constant accumulation of which is greater here than usual, in consequence of the nature of the rock on which the town was built, which is a soft and rapidly decomposing granite, which, from the quantity of feldspar contained in it, and the smallness of its grains of quartz, produces a fine rich soil, and with it a luxuriant vegetation. To the eastward of the town, and partly on the narrow isthmus, the large and closed harbour may be traced, as well as a narrow passage or canal of solid masonry, by which it communicated with the sea, and which was probably the part which, as we learn from Strabo, could be closed. The whole country between Balḳís and Artákí is covered with flourishing vineyards, and the wine of Artákí is much esteemed, even at Constantinople.

May 28th.—After again exploring the ruins, and waiting until I could obtain a meridional observation, which places Cyzicus in latitude 40° 23′ N., I returned to Aïdinjiḳ, and on the following day started for the purpose of tracing, if possible, the course of

* Ἔχει πηγὴν, says Meletius (Geogr. iii. p. 130) ἡ ὁποία τοπάλαι ἐκαλεῖτο τοῦ Ἔρωτος, καὶ ἑτέρα πηγὴ περὶ αὐτὴν εἶναι, Ἀρτακία λεγομένη. "It has a spring which was anciently called the 'Fountain of Love;' and there is another spring near it, called 'Artakía,' named perhaps from Artáki (Ἀρτάκη, p. 131), formerly a city, now a country town (κωμόπολις), near to which runs the river Priapus." This is the ὄρος εὔδενδρον Ἀρτάκη of Strabo, xii. p. 576.

the Ṣúṣigherlí-sú* or Macestus, which, as I learnt at Aïdinjiḳ, takes its rise in a large lake near Símáwul.†

Three hours from Aïdinjiḳ, while travelling round the western end of the lake of Miletopolis, I passed through an interesting settlement of Cossacks,‡ who have been there ever since the capture of Ismá'íl, in 1770,§ and preferring Turkish to Russian rule, have been protected and encouraged by the Turks. They live principally on fish, which they catch in great quantities in the lake, and the neighbouring hills and low plains afford excellent pasture for their flocks: they pay no taxes to the Turkish government, choose their own chief, and have a small church. Their fair Teutonic looking features contrasted strongly with the dark complexion of the Turks, and the long embroidered white smock-frock of the peasants first called my attention to the peculiarity of their appearance. The plague had lately broken out among them, and as we left the village, our road led us through an enclosed space, in which the sick were placed in small huts made of rushes; their relations brought them provisions, leaving them on the ground near the hut, and carefully avoiding all contact. In the evening we reached, after crossing several streams flowing into the lake from the wooded hills to the south, the small village of Meulvíkőï,|| eight hours from Aïdinjiḳ.

May 30th.—An hour's ride E.S.E. brought me to the village of Mániyás,¶ which lies a little way off the high-road towards the south. Here I found abundant traces of an ancient site, and several inscriptions, some of which are built into the walls of a Byzantine fortress, which now occupies the hill, once the Acropolis of the ancient town. Part of this wall is entirely composed of ancient blocks and pedestals, altars, &c. Whole courses in some of the towers consist of columns laid transversely across the wall, whilst others are formed of pedestals, some of which probably have inscriptions. I am inclined to think that these ruins mark the site not of Miletopolis, but of Pœmanenus,** which was

---

* Properly Ṣú-ṣighir-lí ṣúï, *i. e.* Buffalo-water. Ṣú-ṣighir or ṣú-ṣighír, water-ox is the Turkish word for buffalo.

† Símál or Símán is mentioned in the Jihán-numá (p. 632) as one of the districts of Kermiyán (Phrygia Epictetus). In the text it is spelt Símáú; in the map Símán or Símál; w, n, and l, when carelessly formed, nearly resemble each other in the Arabic character. Símáú might be a corruption of Synnao, as the modern Greeks would pronounce Synnaus.

‡ Ḳázáḳlí; nine miles from Aïdinjiḳ.

§ Hammer, Gesch. des Osmanischen Reichs viii. 363.

|| Mevleví-kọï, five hours from the Ḳazáḳs, twenty-four miles from Aïdinjiḳ. Kőï is pronounced as *kieui* in the French orthography.

¶ Mániyás, J. n. p. 656, three miles from Mevleví-kőï, twenty-seven from Aïdinjiḳ.

** Pœmanēnum is the termination given by Stephanus Byzantinus; the vowels are ascertained by the coin (Sestini Lett. Num. iv. 76). Phemenio is the ablative of Pœmenion, a corrupt contraction of Pœmanēnum introduced in the times of the Lower Empire. (Cramer's Asia Minor, i. 56-58.)

afterwards called Phemenio. This name, leaving out the first syllable, greatly resembles the modern name, which, combined with the remarkable position of the Acropolis, so completely answering to that of *πολίχνιον ἐρυμνότατον*, as described by Anna Comnena (p. 440), leaves in my opinion no doubt upon the subject. From thence I proceeded four hours S.S.E. to Ṣú-ṣighir-lí, descending into the valley through which the Macestus flows, by some fine wooded hills. Soon after leaving Ṣú-ṣighir-lí, we reached the banks of the river, and proceeded several miles up it, through very beautiful woodland scenery, to a place called Ḳayá Ḳapú, or Ṭásh-kapú, Rock-gate,* or pass of the rock, where the road winds along a narrow path cut on the steep side of the rock overhanging the deep torrent below. A ruined castle on the height above defends the pass, which, from the remains of a massive bridge over the river, appears to have been once of considerable importance. The rich meadows and valleys which we crossed, leaving the river on the left, were covered in many places with the white Iris in full flower. About three hours from Ṣú-ṣighir-lí I reached the small village of Ildiz,† where I halted for the night.

June 1.—Ildiz to Kefsút‡ or Kespít, four hours S. by W. Leaving the Macestus winding through a broken hilly country to the left, our road led us across a range of high and partly wooded hills by a steep and winding path. Before reaching Kebsúd, I crossed the Ṣú-ṣighir-lí Ṣú again, which flows from W. by S. to E. by N., and is joined immediately above the town by another river from the S.E., the sources of which are said to be at a place called Bólát,§ twelve hours off, nearly S.E. At Kebsúd, which is in latitude 39° 39′ 0″ N., I found a great number of inscriptions and other remains of antiquity scattered about the town; but none of the inscriptions throw any light upon the ancient name of the place, nor could I learn from the natives whence they had been brought; but its situation, in a rich plain at the confluence of two streams, renders it not unlikely that it is the site of some ancient town. Here I was again told that the main branch of the river rose near Símáwul, and was better known by the name of Símáwul Ṣú. From Kefsút [Kebsúd] I ascended the valley of the Macestus through a hilly country, for seven hours, nearly S., with the river on my right as far as Bigaditza,‖ situated

* I believe this to be the Demir Kapou in Wheler's Narrative.

† Or Yildiz, *i. e.* Star, from Ṣú-ṣighir-lí, nine miles.

‡ The Turks are so vague in their pronunciation, that I have sometimes heard the same place pronounced quite differently by different persons.—A.

Kebsúd, (Jih. numá, p. 660,) pronounced *Kepsút*.

§ Bólád. The Turks always sound the final medials as tenues, or, in other words, convert final *sonants* into *surds*.

‖ Bíghádíj (Jehán-numá, p. 661), though spelt thus, is pronounced Bighadich, with a stròng emphasis on the *ich*. *Μπιγαδίτζα* in Greek.

in a fine plain, and with some insignificant remains of a castle of the middle ages near it. From Bigaditza to Singerlí, four hours, nearly S. Here I again crossed the river, flowing from the E., and continued up its bed, eighteen hours, almost due E., to Símáwul. Some of the scenery up this valley is very beautiful; the plains and banks of the river, and particularly such parts as are subject to frequent inundations, being covered with magnificent plane trees.

Three hours E. from Singerlí, I visited some hot springs, by the Turks called I'líjah,* a little way out of the road, which rise in several places out of an igneous rock, a grey porphyrytic trachyte. Their heat is nearly that of boiling water, and they are heard distinctly bubbling up under the rocky and treacherous crust. The water deposits remarkably white stalactitic and stalagmitic concretions, and is accompanied by a strong sulphureous smell. The united springs form a stream large enough to turn several mills, and near the road, about a mile from the sources, the water is still hot enough to form a hot bath, much used by the natives.

From Símáwul I made an excursion to the lake which forms the sources of the Símáwul Ṣú. It is about five miles N.W. from the town, and the river flows at once from the western extremity of the lake a considerable stream. This lake is supplied by subaqueous springs, for no river of any consequence flows into it. The many small streams which I had crossed in my way to Símáwul flowing down from the high range of Demirjí, on the S., are all absorbed by irrigation before they reach the lake. Near the western end of this lake is a small insulated hill, round which are considerable remains of wall, but too much ruined to ascertain their age; and near it, as well as in the neighbouring village of Kilíseh kői,† are many large marble blocks and fragments of broken columns, friezes, &c., and one or two imperfect inscriptions. I have no doubt that this hill was the Acropolis of the Phrygian Ancyra. In a Greek inscription which I found in the wall of the mosque at Símáwul the word Synaus may be read. That some ancient town existed in this neighbourhood is proved by the number of marble blocks and other fragments, which we saw in every village in the plain.

From Símáwul I crossed a high range of mountains, the eastern prolongation of the range of Demirjí, between four and five thousand feet above the sea, which has formed in former ages the northern boundary of the vast lacustrine deposit of the tertiary

* I'líjah, "hot-springs" in Turkish.

† Kelíseh kọ̈ï, church-village; Kilisyá, as it is most correctly spelt, is from the Greek "ecclesia."

period which covers a large portion of Asia Minor; and in twelve hours, nearly S., we reached Selentí.* The river on which this town stands is not the Hermus, as it is called in all the maps, but the Aïneh-chấï,† which does not join the Hermus until twenty or twenty-five miles lower down.

June 8*th*.—Selentí to Ḳúlah,‡ eight hours, nearly S.W. After leaving the valley of the Aïneh-chấï, and crossing the mountains between it and the Hermus, which are chiefly horizontal beds of lacustrine and cretaceous marl, I entered the district of the Katakẹkaumené, and remained some time at Ḳúlah, examining the remarkable volcanic phenomena which this district presents, and making excursions to ascertain the extent of the various streams of lava which have flowed from the three modern cones or craters, and which correspond with the three pits described by Strabo.§

In one of my excursions I discovered the ruins of Saittæ, a celebrated town of Lydia, and of which a few well preserved coins are in existence. These remains are situated about nine hours N.N.W. of Ḳúlah. They consist of a stadium, one end of which is nearly perfect, between two low hills; but the northern end, which extends into the plain, is quite destroyed: numerous tombs and sepulchres in the hills around, and many massive remains of temples and other public buildings. Large broken columns lie about the fields and plain in all directions, and prove the magnificence of its ancient decorations. The modern name of these ruins is Sidàs ka'léh,|| which is probably a corruption of Sitas, the accusative of Sitæ, the name by which this town is mentioned in the Synecdemus of Hierocles. In a neighbouring village I found a few inscriptions, but none which contained the name of the ancient city, nor could I procure any coins there, although at Ḳúlah I had obtained several of Saittæ.

The height of Ḳúlah above the sea is considerable, perhaps 2250 feet, as the average height of the barometer whilst I was there was 27.680 inches; attached therm. 76°, and detached therm. 78° Fahr. The height of the volcanic cone is 530 feet above the town. It is situated in latitude 38° 31′ N.

June 17.—I left Ḳúlah for Afiyún-Ḳará-ḥiṣár,¶ with the intention of finding a middle road between the two I had already travelled, and to the north of the Mæander. It proved, however, a more difficult task than I had expected; for the country

* Selentí or Selendí, J. n. p. 632, 633.
† Mirror-stream, or rather Aïneh River, *i. e.* the river from Aïneh.
‡ Ḳúlah, J. n. p. 633.
§ Strabo, xiii. p. 628. Δείκνυνται δὲ καὶ βόθροι τρεῖς οὓς φύσας καλοῦσιν, ὅσον τετταράκοντα ἀλλήλων διεστῶτες σταδίους. "Three pits are shown, which they call 'bellows,' distant from each other about forty stadia (five miles)."—A.
|| Siṭás ḳal'eh, Sitás Castle. This T is pronounced D by the Turks.
¶ Opium-black-castle. Ḳarah Ḥiṣár-Ṣáḥib. J. n. 631.

between the Mæander and the hills of Taḳmák, which is a continuation of the extensive undulating plain between Góbek and Sejikler, is cut up by many deep and precipitous ravines, some of which are from 400 to 500 feet deep. Through two of these ravines, rivers flow into the Mæander, of which the Bánás-cháï* is the most considerable. It rises at the S.W. foot of the high mountains near Kútáhiyah, called Morád Ṭágh; the other is much smaller, and its course is more to the westward.

June 18.—After travelling all day over this plain, we reached a large village called Medereh Kṓï, situated at the S.E. extremity of the plain, and distant by road eighteen hours from Kúlah nearly E.S.E. Very near this place, the Mæander, after winding through a chain of hills of mica-schist and saccharine highly crystalline limestone, which separates this large plain, which I think may possibly be the Cyri Campus,† from the Chál‡ district, enters the southern part of the plain, flowing through a rocky gorge, 600 or 700 feet deep.

The Chál district, through which I proceeded to Demirjí Kṓï,§ the residence of the governor, is a rich and well-cultivated oval plain, watered by the Mæander: it measures about fourteen miles from N. to S., and four or five from E. to W. Its principal produce is the vine. The plain is entirely surrounded by high and steep hills, to which the ground of the valley rises gently all round, and it bears undoubted evidence of having been a large lake. The Mæander both enters and leaves this plain through very narrow ravines.

June 20.—Crossing the high hills which form the eastern boundary of this plain, I descended into another larger and more level plain, equally watered by the Mæander, at a higher level than the former; this plain, which extends the whole way to Isheklí,|| distant nine hours from Demirjí Kṓï, my course being nearly N.E., is not improbably the Peltenus Campus. At its N.E. extremity, immediately behind Isheḳlí, rise the fine springs mentioned by Pococke, probably the ancient Glaucus, which at once form a considerable river, and flowing S.E., soon join the Mæander, flowing E.N.E. from Dineir. These two branches alone constitute the Mæander, or Menderez-ṣú, with the exception of a few small springs which rise at the foot of the high limestone-mountains, which extend the whole way between Isheḳlí and Dineir. The Obrimas, as marked in Cramer's map, does not exist.

June 22.—Crossing another high range of mountains, and

* River of Bánás.

† Cyri campus.—Cramer, ii. 22.

‡ Chál.; J. n. p. 633.

§ Demirjí kṓï, Blacksmith's village.

|| Isheḳlí kói, cleft-town, or sheïkhlṹ, *i. e.* Sheïkh's town, J. n. p. 633.

partly ascending by the dry bed of a winter torrent, I reached the plain of Ṣandúḳlí;* the elevation of which above that of Ishiḳlí is very considerable. At Ṣandúḳlí itself are no remains of any importance, but in the centre of the plain the road passes through the undoubted ruins of an ancient town, situated eight miles south of Ṣandúḳlí. The remains consist of straight lines of the foundations of houses marking the lines of streets, and built of large blocks of stone; also part of a wall round the Acropolis, which stands on a low hill, in the rocky sides of which a few tombs are excavated, one of which had a Greek inscription. I also saw a few more sepulchral monuments in the burial ground of the village, which is built up among the ruins, but none of the inscriptions throw any light on the ancient name of the town.

From Ṣandúḳlí I crossed over a succession of high mountain-ridges, and passed through several perfectly flat alluvial plains, surrounded by the mountains, till I reached Afiyún Ḳará-ḥiṣár, situated in latitude 38° 44′. One of the most remarkable physical features of this part of the country is this constant succession of level plains, completely surrounded by high mountain-ridges, and bearing such strong evidence of having formed large lakes or inland seas at some very distant period.

June 24.—Started from Afiyún Ḳará-hiṣár for Ḳóniyeh.† This road has been described by so many travellers, that I need not enter into many particulars; but there are some points of which the geography still requires much correction.

Between Afiyún Ḳará-ḥiṣár and Bóláwádín,‡ which latter place I did not visit, are remains of several ruined towns on the hill-side to the north of the plain; but although they contain fragments of marble columns, broken sarcophagi, and mutilated fragments of inscriptions, it is difficult to say whether they can safely be assigned to an earlier period than the Turkish conquests.

At I'lghún§ the lake, supposed to be that of the forty martyrs, and which, in Colonel Leake's map, is called Trogitis, and is placed E.N.E. of the town, is, in truth, three or four miles to the N. by W. Aḳ-shehr|| is Philomelium; and I much doubt whether I'lghún be on the site of any ancient town.

The Turkish burial-grounds of Ḳhánum khánah,¶ and walls of the khán, are full of numerous inscriptions, chiefly sepulchral, and either themselves mark the site of an ancient town, or are derived from a site which I visited in the hills to the south, about six miles off, where I found the ruins of an ancient castle,

* Ṣandúḳlí, Trunk-town, J. n. p. 641. † Ḳóniyeh (Iconium), J. n. p. 615.
‡ Búlávádín, J. n. p. 631, 671. § I'lghún, J. n. p. 619. I'lghín, Ibid. p. 671.
|| Aḳ-shehr, White-town, J. n. p. 619. ¶ Khánum khánah, Lady's House.

crowning the summit of a lofty hill, which may have been an Acropolis, and among the stones of which were several inscriptions, which resemble, both in character and appearance, those at Ḳhánum khánah.

At Ládík,* the ancient Laodicea Combusta, are also numerous inscriptions, but chiefly sepulchral. It has been supposed that this town received its appellation of Combusta from the volcanic nature of the surrounding country, in the same way as the Katakecaumené: this is, however, a mistake, as all the mountains, which almost surround the town, consist of a blue semi-crystalline limestone, and a few associated beds of shale and schist.

On the 29th of June I reached Ḳóniyeh, the ancient Iconium, in latitude 37° 56′ 30″. The mean of the barometer during several days was 26·400 in.; the weather oppressive, with constant thunder storms. From thence I intended to strike due north for the salt lake† of Ḳóch-ḥiṣár.‡ I had in vain attempted to leave the high road for that place at Afiyún Ḳará-ḥiṣar, Aḳshehr, and I'lghún, having everywhere found it impossible at this season of the year, when, I was told, the springs across the intervening country were dried up, and the nomade tribes had left the plains for their summer residences in the mountains to the north and east. At Ḳóniyeh I found the same difficulties, and was obliged to go from thence to Ḳará-buńár,§ on the road to Ereglí; thence strike north to Aḳ-serái, and from thence north-west to Ḳóch-ḥiṣár.

I may here mention, that the high-peaked mountain which Colonel Leake saw from near Ḳóniyeh, and believed to be Mount Argæus, cannot be that mountain, but is the summit of Ḥasan Ṭágh, situated eighteen miles S.S.E. of Aḳ-serái, and presenting an appearance very similar to that of Argæus.

At Ḳará-buńár I left the great Ḥajj, or Pilgrim road, which leads from Iconium to Ereglí and Tarsus, and proceeded nearly N. by E. sixteen hours to Aḳ-serái. I may observe, that when I crossed it, the great plain to the eastward of Ḳóniyeh was perfectly dry; but in winter it is flooded and impassable. The route, therefore, is then along the foot of the hills to the north. I was much struck with the remarkable appearance of the *mirage* in crossing this flat plain. Several times we thought ourselves on the very borders of a lake, which appeared not half a mile off; and on the surface of which I saw the reflection of the distant

---

* Lázikiyehi ḳaramán, also called Yúrugán Ládík (J. n. p. 616), *i. e.* Ládhiḳíyeh (Laodicea) of the country of Ḳaramán, or the Gipsies' Ládíḳ.

† Memliḥah, or Túzlah.

‡ Kóch-hisár, or Champion-Ram castle, J. n. p. 620, 626, 646.

§ Or Ḳarah-bińár, Black Spring, J. n. p. 616.

mountains, but the whole appearance vanished as we approached. I always, however, found that some slight irregularity of ground was necessary to produce the appearance, and also that the hills which bounded the horizon must be very low. Ḳará-búṅár is undoubtedly upon the site of Baraté, of the Itineraries, which I am inclined to look upon as a corruption of Barathra—a name derived from the deep volcanic craters or pits which surround the town. The town of Ḳará-buṅár was completely abandoned, except by the Menziljí* and his people; the inhabitants had all gone off to their summer quarters in the mountains, for the sake of pasture for their flocks and herds, and to avoid the plague of gnats, which made their town almost uninhabitable.

Aḳ-serấï,† in lat. 38° 20′ N., is situated in an open and well-cultivated valley, through which a small stream called the Beyáẓ ṣú‡ flows into the salt lake of Ḳóch-ḥiṣar. Madder is grown in great quantities in its neighbourhood; and I also saw a few fields of rye (chávdár).

July 8.—I made an excursion to some interesting ruins about eighteen miles S.E. of Aḳ-serấï, at the foot of Ḥasan Ṭágh, and on the road from Ak-serấï to Bór,§ or Tyana. If Aḳ-serấï is Archelais, it is probable that these ruins are the remains of Nazianzus. The place is called Vírán-shehr,‖ or Yúrán-shehr (ruined town), and is situated on a rocky platform immediately above some fine copious springs, which form a long and deep lake, out of which flows a small river, which joins the Beyáẓ ṣú of Aḳ-serấï, and also falls into the Salt lake.

The streets and houses of a great part of the town are still standing, the walls in some places being twenty or thirty feet high, entirely composed of rough cyclopean blocks, without any cement or mortar. The walls of the Acropolis may be distinctly traced, besides some curious vaulted buildings of more regular masonry. The tombs, which are very numerous, are generally better built, and more in the Hellenic style. Those which I entered had usually two rows of stone benches round them for receiving the dead bodies, and in one I found several human bones. Besides these buildings, I saw the ruins of three Byzantine churches of very ancient and rude construction, but evidently of a much more recent age than the rest of the town, being built with a good deal of mortar and small rubbly stones.

July 11.—I started from Aḳ-serấï for Kóch-ḥiṣar, along the plain, following the course of the Beyáẓ ṣú, about sixteen miles the first day, nearly W.N.W. A short way farther on, the river

* Innkeeper.
† White Palace, J. n. p. 620.
‡ White water.
§ Bór, J. n. p. 617.
‖ Yúrán is a corruption of the Persian word Vírán or Uïrán. Bel-vírán, J. n. p. 617.

becomes salt, and on the following day we could find no fresh water until we reached Ḳóch-ḥisár, about thirty-two miles N.W. by N., and situated in latitude 38° 54′ N.

The Salt lake* begins about five miles to the west of Ḳóch-ḥiṣár, and is said to be thirty leagues in circumference. The remains of the causeway which was built across an arm of it by Sultán Selím, are nearly concealed under an incrustation of salt; and at the spot where I reached the shore, the bed of the lake consists of a thick crust of solid salt. The salt itself is a government monopoly, and is farmed by Ḥájí 'Alí Páshá of Ḳóniyeh, for twenty-six purses a year (135*l.*), who sells it again for about thirty-four purses (177*l.*). The salt is collected at only four places round the lake, at each of which a receiver is stationed. The Turks who come for it, pay forty párahs,† equal to 1 piastre, or not quite 2½*d.*, for an 'arabah or cart-load drawn by two oxen; 10 párahs for a horse, mule, or camel-load; and 6 párahs for an ass-load.

The water of this lake is so extremely salt that no fish or other animals can live in it: birds dare not even touch the water,—their wings become instantly stiff with a thick crust of salt; and anything thrown into the lake is soon covered with that substance.

July 14.—Started for Cæsarea. The country, more hilly and better watered for some distance, is only inhabited by Turkománs, who, although like Yúruks‡ and Kurds, they live in tents during the summer, have generally fixed villages for their winter quarters. They are by far the most numerous and most civilised of the nomade tribes of Asia Minor. The Yúruks live in tents all the year round, but dwell almost exclusively in the mountains; and when in the neighbourhood of large towns, generally act as charcoal-burners, and supply the inhabitants of the town with that article: they, as well as the Turkománs, sometimes cultivate a little ground. The Kurds differ much in manners and in language§ from the two last tribes. They are a more wealthy, independent people, and live entirely in tents. They dwell on the eastern flanks of Mount Argæus and in the great plain of the Háïmáneh,|| towards Angora.

July 16.—Reached Tátlar, a remarkably-situated village in a deep ravine of white, sandy, volcanic tuff, the steep sides of which are, in places, covered with large fragments of black basalt, with which the flat summit of the hill is capped. Many curious caves, and even modern dwellings, are excavated in this soft sand, in

---

* Túzlah, *i. e.* Saltern, in Turkish; Memlihah, in Arabic.

† Though spelt párah, *i. e.* "piece," this word is pronounced *pará*.

‡ Wanderers, from yúrumek, "to march about."

§ The Kurds speak an ancient Persian dialect; the Yúruk and Turkománs, Turkish.

|| Háyimáneh, J. n. 620, 644.

which many of the inhabitants now dwell. Some of them appear to have been ancient tombs, others are evidently chapels of the earliest Greeks, and may perhaps have served as places of refuge during the persecutions in the early ages of Christianity. In one of them I found a curious old Greek manuscript, much worn and damaged, respecting which the Turks had told me many strange and marvellous tales.

July 18.—I left Nembshehr,* and passed through U'ch-ḥiṣár†, and Urgúb,‡ near both which places, I was much struck with the remarkable conical and pointed hills, varying from 1 to 300 feet in height, which may be said almost to fill the valleys, they are so close and numerously wedged together.§ Most of them are excavated either as ancient Greek tombs or chapels of the early Byzantine period, or serve as modern houses and dove-cotes. Many entire villages are built among them. The only tree which grows upon this pumiceous soil is the apricot, which appears to be indigenous, and grows in great quantities. Some of the varieties of the fruit, although small, are very high-flavoured.

From Urgúb I ascended several miles, nearly due east, up a deep and narrow ravine watered by a small stream, on each side of which, were rich and well-cultivated gardens and orchards, extending as high up the hills on each side as they could carry a supply of water, which everywhere marks the limit of cultivation in this part of Asia Minor.

July 19.—At 5 A.M. I left Baḳtásh, which is called half-way between Nemb-shehr and Cæsarea, or Ḳaïṣaríyeh, the whole distance being computed at eighteen hours or fifty-four miles. Leaving the valley, which I had followed up from Urgúb, I crossed a high ridge of volcanic sand-hills capped with basalt, and descended upon Injú-sú,|| from whence, winding round the N.W. foot of Argæus, I reached Cæsarea soon after 3 P.M. Six miles to the westward of Cæsarea, and in the plain at the foot of Mount Argæus, is a small lake abounding in fish and water-fowl, from which all the maps, except Major Rennell's, make a river flow into the Euphrates, passing either to the north or south of Cæsarea. This is quite wrong. A considerable river falls into the north end of this lake, flowing through a rich and well cultivated plain from the north-east, and the river which issues from it flows through a deep and narrow valley to the W.N.W., and falls into the Halys. It is called the Ḳará-sú, or Black-water, and is undoubtedly the Melas of Strabo, which

* Nemb or Neb-shehr, a corruption of Nev, Neû, or Nau-shehr (New-town).

† U'ch-ḥiṣár, Three castles.

‡ U'rgúb, J. n. 617.

§ The description of these conical hills given by Paul Lucas is not in the least exaggerated.—A.

|| Injú ṣú, Pearl-water. Injú is often pronounced Injí.

was dammed up by King Ariarathes, and which, on the bursting of the dyke, flooded the lands of the Galatians by causing the Halys to overflow its banks, and for which damage the King was condemned by the Romans to indemnify the Galatians. Strabo, or some one of his copyists, has evidently written Euphrates instead of Halys, by mistake, for how could the rising of the Euphrates have flooded the lands of the Galatians?

The heat at Cæsarea was very great while I was there, from the 19th to the 26th July, the thermometer in-doors generally averaging 86°—88° Fahr., notwithstanding its elevation above the sea, which is about 4200 feet. The barometer varied during my stay there from 25.568 in. to 25.328 in., and at Smyrna during the same time from 29.986 in. to 29.820 in., without corrections. The boiling point of water was 204°. 45. Cæsarea is situated in lat. 38° 42′.

July 27.—After many inquiries and contradictory reports as to the easiest side for ascending Mount Argæus, and having satisfied myself that this could only be done from the southern side, and that Everek-kői, a large village at the S.E. foot of Argæus, was the only place where I had any chance of procuring guides, I started from Cæsarea, and halted for the night at the Greek convent, where I obtained much curious geographical information from the monks and the bishop, or despotes,* and reports of many ruins existing to the east and south-east of Cæsarea, but in a wild, inhospitable, and mountainous country, nominally under Ibráhím Páshá; all the inhabitants of which are reported to be robbers, and who often make incursions into the Turkish territory, and carry off the cattle of the more peaceable inhabitants of the plains. This convent is ten miles E.S.E. from Cæsarea, and about 1000 feet above the plain.

July 28.—My journey this day was round the eastern side of Mount Argæus. On this side, the mountain does not rise at once from the plain, as on the other sides, but is connected by rugged hills, with several lower ranges of mountains, which stretch away far to the eastward. I reached Everek-kői in the evening, after eight hours' journey, and with the assistance of the governor (who lodged me in his own house, and gave me every facility in his power for effecting my purpose of ascending the mountain), soon found some Armenians who said they knew the way up the ridge, which forms, as I had seen on this day's journey, the lip of a large and extinct crater, but added that they had never been to the summit.

July 29.—With my Armenian guides and some Turkish

* Dhespótis, Δεσπότης.

guards, whom the governor insisted on my taking, and my interpreter, I started for the mountain early this morning. It rises up almost to a single peak from a broad and extended base, consisting entirely of volcanic rocks and scoriaceous cinders of different kinds. Its sloping sides are studded all round with numerous cones and craters, the effects of volcanic action at different periods. Near the foot of the mountain is a little cultivation, but a few solitary wild pear-trees or stunted oak-coppice are the only trees upon it. Its appearance is, therefore, peculiarly barren and rugged, which, added to the black and cindery nature of its rocks, gives it a wild and inhospitable look. I was unable to reach the summit this evening, and indeed there is considerable danger in ascending the steep part of the cone after the sun has thawed the surface of the sloping sides, when large masses of rock are detached, and roll down the ravines or over the snow. We halted for the night under a projecting rock, where I found a hollow space just big enough for me to lie down in, while my followers accommodated themselves beneath a similar one. This was at the foot of the snow; indeed we had left several patches of it in the ravine below us, and a cold place it was during the night. The thermometer at 6 P.M. stood at 47°; the barometer at 20.198 inches; indicating a height of about 10.300 feet as the lowest limit of the snow line, in the parallel of 38½°, N.

July 30.—5 A. M. Bar. 20. 246.; ther. 35°: det. ther. 33° Fahr. Such was the dilatoriness of my Turks and Armenians, that the sun was well above the horizon before we started, notwithstanding the absolute necessity which they said existed for our being up before the ground thawed. After half an hour's walking up a steep, inclined plain, at an angle of 15°—20°, we reached the snow, which continued unbroken to the summit. On this southern side several sharp ridges rise through the snow, over which we contrived to climb, occasionally crossing the intervening ravines of snow, where, as the warmth of the sun was felt and thawed the surface, many masses of rock, detached from the soil by the melting of the ice, came bounding past us at a most rapid rate. This is the only danger attending the ascent of the mountain. Two hours' more very steep walking and climbing brought us to the summit. This consists of a narrow ridge, the highest point of which is nearly the point of junction of two large and contiguous craters, both of which are broken down on the north side. The snow in them is very deep and unbroken, and descends much lower than on the southern flank, forming extensive glaciers, resembling those of Switzerland; but such is the porous nature of the rocks and soil, that, however fast the snow melts, no streams of water flow down the sides of the mountain, but are all instantly absorbed. The bar. on the summit was just below 18 inches, and the ther. 40° Fahr. This gives a height of 13,300 feet, which I believe to be nearly correct, as it agrees very closely with the result of two angles of elevation taken from different spots below the mountain; one of which, taken from the Greek convent (the height of which, calculated by the barom., is 5200 feet above the sea), makes the height of the mountain 13,242; and the other, taken from Ḳará-ḥiṣár (the height of which, measured by the barom., is 4300 feet above the sea), makes the whole height of the mountain 12,809 feet.

Taking a mean of these three observations, we shall not be far wrong in estimating the height of this celebrated mountain at 13,100 feet above the sea.

At the spot where we spent the night, scarcely any vegetation was to be seen: a few small stunted Alpine plants grew among the stones, and I found one small plant, with a flower resembling the thistle, but growing low and spreading, and the leaves smelling strongly of musk, which, the Turks declared, was found nowhere else. They ascribed some rare medicinal virtues to it.

I was disappointed in my expectation of a distant view. The day was hazy, and a sea of clouds floated far below us, obscuring the surrounding country from our sight, except when high ranges of hills appeared, like islands above the waving clouds; consequently I was unable to take many angles or bearings of distant places.

Descending from the mountain I found, near the edge of the plain, the ruins of a town, which, at some period of the Byzantine empire, must have been of considerable importance, to judge from the remains of several old Greek churches, columns, and tombs which appear on the hill-side near the modern village of Gerameh. Returning from these ruins to Everek-kőï, we had a narrow escape from a large band of well-mounted Kurds, returning from a predatory excursion, whom we saw descending from the mountain-side at full gallop, and driving before them across the plain large herds of beasts and cattle which they had just been plundering.

July 31.—From Everek-kőï to Ḳará-ḥiṣár, eight hours, across a flat plain, wet and marshy in places, and which is almost entirely under water in winter. At Ḳará-ḥiṣár I found no antiquities; but on the following day, August 1, I made an excursion about eight miles to the westward to some ruins, called Sowánlí Dereh, where I found a very remarkable valley, the steep and almost perpendicular sides of which were, for nearly two miles, excavated into an immense number of chambers, grottoes, dwelling-houses, tombs, and chapels of the Byzantine age. From its position and similarity of name, I have no doubt it occupies the site of the ancient Soandus, which is mentioned by Strabo (xiv. p. 663) as on the great road from Phrygia through Lycaonia, to the capital of Cappadocia.

August 2.—Starting from Ḳará-ḥiṣár, I proceeded along the usual road to Nígdeh, Bór, Kilísá Ḥiṣár, and Ereglí, and reached Mislí* in five hours, at nearly eight the same evening.

August 3.—I left Mislí early for Nígdeh, five hours, S.S.W. Mislí is a small village of Greeks, subject to the Bishop of Nígdeh, and quite independent of the Turkish authorities. They pay no taxes to the government, in lieu of which they formerly worked the lead-mines of Ma'den Ṭágh, about six hours off, E. by S. Now, however, they pay their contributions to these mines, instead of working them, and the miners are procured from Gúmish-khánah, between Trebizond and Erz-rúm, the place which supplies most of the miners of Asia Minor. These Greeks of Mislí never leave their village, and neither man nor woman is allowed to marry a stranger, or any one not belonging to the village.

Seven miles before reaching Nígdeh, some very copious springs rise in the low flat valley, and form a stream which flows through Nígdeh and Bór to the S.W. This stream serves to irrigate some rich meadows and gardens; and about five miles above Nígdeh to the N.N.E. is a place called Eskí Andavál, or Old Andavál. There now only remains a small Byzantine church, dedicated to St. Constantine; but the surrounding country at once strikes the traveller as being the spot alluded to by the author of the "Jerusalem Itinerary," in the words following Mansio Andavalis, sixteen miles from Sasima, and on the road to Tyana, from which last place the "Antonine Itinerary" gives it a distance of sixteen miles.† These words are—"*Ibi est villa Pampali unde veniunt equi curules.*" The valley is most remarkable for its freshness and verdure. There is no other spot in the country more fitted to the breeding of horses; and here was probably the stud of Pampalus, or (as they are supposed by the critics to have been one and the same person) of the Palmatius, mentioned in the Glossæ Nomicæ of the Lower Empire, whose breed of horses, extensive landed property, and magnificent palace at Cæsarea nearly equalled the splendour of the Emperor Valerian. There is also a modern village of Andavál, about two miles off to the east of Nígdeh, amongst the mountains.

August 4.—Twelve miles S.S.W. of Nígdeh, I passed through the town of Bór, on the banks of the same river; and two miles

* Brazen or Brass-place.

† See Wesseling's Note to the "Jerusalem Itinerary," p. 577.—A.

farther south I reached the small village of Kiz-hisár, or Kilísá-hisár. The remains of antiquity in the village and its neighbourhood have long since made it probable that it is the site of Tyana; but no inscriptions have been found there which throw any light upon its ancient name. Nevertheless, one or two material features, which I observed in its vicinity, are so remarkable as to leave no doubt upon the subject.

One of these is the circumstance that the town is built upon a small rising mound or hill in the middle of the plain, which exactly corresponds with the description of Strabo, who says* that it was built upon the mound of Semiramis. It was in order to convey water to the summit of this hill that the aqueduct built apparently by the Romans, and which extends for several miles across the plain, from a rich and copious spring to the eastward, was originally intended. Numerous remains and foundations of ancient buildings exist amongst the walls of the small cottages upon this hill; and amongst others is part of the floor of an ancient Doric temple, with one slightly fluted column still standing *in sitû*.

The other circumstance is the existence of a very singular lake in the plain about two miles south of the town, which answers to the description of the fountain of Asbamæus, recorded by Ammianus Marcellinus,† and by Philostratus in the life of Apollonius Tyaneus.‡

Ammianus says that there is a fountain which rises in a marshy plain near Tyana, which swells with the quantity of water, and again disappearing, never overflows its banks.

Philostratus says that near Tyana is the fountain of Asbamæus, sacred to Jupiter, which rises very cold, but it bubbles up exactly like a boiling cauldron.

The apparent discrepancy of these two accounts vanishes on seeing the real phenomenon, which perfectly bears out both descriptions. About two miles to the south of the village several small springs of brackish water rise, and a little way on is a small lake or pool, about thirty or forty feet in diameter, of turbid brackish water, which appears to be boiling up all over, but particularly in the centre, where a violent jet of water rises to a height of nearly a foot, and about a foot and a half in diameter, with considerable noise. Notwithstanding this quantity of water which is constantly boiling up, the lake never rises or overflows its banks, nor does any stream of water escape from it, although the ground around is perfectly flat. There is a slight smell of sulphuretted hydrogen gas around it; and I think it probable that the jet in the centre of the pool is partly caused by the escape of a large quantity of gas, and not solely by water.

* Τὰ δὲ Τύανα ἐπίκειται χώματι Σεμιράμιδος τετειχισμένῳ καλῶς. Lib. xii., p. 538.—A.

† A. Mar. lib. xxiii. c. vi.—A.

‡ Phil. de Vit. Apol. lib. i. c. 6.—A.

Another evidence of this being the site of Tyana may be found in the name of a low chain of hills to the north of the town, in which are several caves and tombs excavated, which hill is called Ifti'yánkas, or Ifti'yán-keler—the last syllable being perhaps a corruption of the Latin *castrum*, or the Arabic *kal'éh*.

August 6.—Between Ereglí or Ereklí and the Bín-bir-kilísá,* on Ḳará-ṭágh (Black Mountain), there is a large swampy lake in the great plain, which appears to be a continuation of that of Ḳóniyeh. This plain is bounded on the south by a low chain of hills of secondary limestone, which stretches down from Mount Taurus; and in passing between these mountains and the lake, I was surprised to find a stream flowing to the south out of the lake, and escaping through a chasm or Katabothron at the foot of the cliffs, thus affording an outlet to the superabundant waters of this plain, which was not suspected to exist.

The stream was not very considerable when I passed it; but it was evident, from the wide bed of the torrent, and the unusual circumstance of the Turks having built a substantial bridge over it, that in winter and spring a very large quantity of water must escape by this subterranean channel.

I halted this night at an encampment of Turkománs, near the lake which is called Ak-Gól.† The winter residence of these Turkománs is at a village eight hours off to the south, called Devlí, which is probably the ancient Derbe, which we learn from Hierocles was called Delbia, from which the change to Devli is most simple—Delbia, Delbe, Delve, Devle. I was assured that no water is found within eight hours' journey of it at this time of year, which prevented my visiting it.

August 7.—On my way to Ḳárá-dagh due west, I passed through the ruins of two ancient towns, one of which is marked by the many ancient tombs excavated in the rocks, on the side of one of which I found the remains of a Greek inscription; and the other by the numerous ancient blocks and broken columns of marble built into the walls of every cottage.

August 8.—Visited the ruins of the Bín-bir-kilísá on Ḳará-dagh. They are very extensive and interesting, covering a large space of ground; but, with the exception of the many large sarcophagi and tombs, resembling those of Hierapolis, appear to belong entirely to the early ages of Christianity. The ruins consist chiefly of the remains of Byzantine churches, evidently of great antiquity, and some of very considerable size. They are without exception built of the red and grey porphyritic trachyte of the neighbouring hills. I am inclined to attribute these ruins to Lystra, which we know was an Episcopal See under the Byzantine Emperors, which well accords with the existence of so many ruined churches;

* Bín-bir kilísá, *i.e.* 1001 Churches. † Aḳ gọl, White Lake.

whereas Derbe, which has hitherto been supposed to be at this place, is not even mentioned by the Ecclesiastical Notices.

August 11.—I left Ḳaramán* for Beg-shehr,† and at Kaṣabah,‡ four hours from Ḳaramán, leaving the road to Ḳóniyeh, turned off due west, with the intention of passing through the ancient province of Isauria, and with the hopes of finding some remains of the old city of Isaura, which I had understood from M. Texier was at Beg-shehr. I halted this night at a small village, called El-másún, four hours west from Ḳaṣabah, and learnt, during the evening, that eight hours farther west, and almost on the line of road to Beg-shehr, were some extensive ruins on a lofty hill, near the villages of Ulú-buńár § and Ḥájílar. ||

The most remarkable feature on our journey this day was the number of ruined Turkish villages and towns, the date of whose prosperity was probably during the time of the Sulṭáns of Iconium and Ḳaramán-ílí; and their destruction appears owing to the numerous Nomad tribes, who now inhabit these plains during the winter.

El-másún is situated almost at the foot of the high hills, which may be supposed to form the eastern boundary of the mountainous district of Isauria, which commences immediately to the westward. Low undulating hills extend to the N.E., sloping gradually down to the plain of Ḳóniyeh.

August 12.—El-másún to Ḥájílar, eight hours. The greater part of the road through a wild and thickly-wooded country, with rocky hills and deep ravines, agreeing with the historical description of the fastnesses of Isauria. On reaching Ḥájílar I was disappointed at the contradictory accounts which I heard of the ruins in its neighbourhood; but, thinking that it would be best to judge for myself, I procured a guide and started for them at once. Fortunately I was quite rewarded by discovering them to be the undoubted remains of Isauria, the new town built by Amyntas, and surrounded by a massive wall, with lofty hexagonal towers, beautifully constructed, and of a very peculiar style of architecture. The buildings within the walls are all in the same style, consisting of alternate courses of very thick and very thin blocks of marble. Amongst them is a handsome triumphal arch, with a Greek inscription, stating that it was erected in honour of Adrian by the senate and people of the Isaurians. The town is built on the highest point of a high range of hills, extending from N.W. to S.E., and commanding a most extensive view, even as far as the plains of Ḳóniyeh and the lake of Beg-shehr. Outside the walls are many tombs, from which, and in the neighbourhood of the forum, I copied several inscriptions. Some of these

* Karamán, or Ḳarahman, *i. e.* Black House.—Otter, Voy. i. 59.

† Pronounced Bey-shehr, *i. e.* Lord's Town. ‡ Kaṣbah, *i. e.* Market Town.

§ Ulú-bunár, Great Spring. || Ḥájílar (Kọï). Pilgrimville.

tombs are excavated in the rock, others are large massive buildings erected with considerable magnificence on the side of the hill towards the south.

It would appear that the town was inhabited even in Christian times, as I observed on each side of the ancient road which leads up to the eastern gate of the town many rude sepulchral stones, with large crosses carved upon them.

I remained here the whole of the following day, and in the evening went about six miles W.S.W. to Tirís ma'den. The chief occupation of the villagers is smelting lead, the ore of which is brought in its rough state from the mines in Mount Taurus, about ten hours off to the south. A considerable stream flows through this valley to the N.N.E., but is absorbed or exhausted by cultivation before it reaches the plains of Ḳóniyeh. In winter only does it flow through into that plain.

August 14.—Tirís ma'den to Ḳaráorán* seven hours. Ḳaráorán is situated near the N.E. end of a large lake, to the S.E. of that of Beg-shehr, and into which a large stream from the lake of Beg-shehr † empties itself. It is sometimes called the lake of Seïdí-shehr,‡ and sometimes of Sóghlah. It is certainly the Trogitis of Strabo, as that of Beg-shehr is the Caralitis, the modern town of Kerálí being on its shore. This smaller lake of Sóghlah § is said to be sometimes dried up, the water escaping by a subterranean chasm near the foot of Mount Taurus, a range of which forms its southern boundary.

August 15.—Ḳaráyurán to Seïdí-shehr four hours, almost the whole way along the borders of the lake. Seïdí-shehr contains from 400 to 500 houses. I here learnt that the plague was raging with great violence all the way to Smyrna, which, as I advanced, I found to be true.

August 16.—Seïdí-shehr to Beg-shehr six hours. A high range of hills intervenes between the two lakes, and the river makes a great detour to the north in flowing from that of Beg-shehr into that of Soghlah. The water of the lake of Beg-shehr is perfectly fresh. Paul Lucas has evidently confounded this lake with some other, when he says that he saw the salt collected into heaps round its borders. There are several islands upon it, particularly at the northern end; but, as there was no boat, I could not visit them. The existence of these islands agrees with the account of the lake of Pusgusa alluded to by Dr. Cramer,|| and it is probably the same. At Beg-shehr the plague was bad; at Ḳérálí, which I reached the following day, still worse; and at Ḳará-aghách ¶ three fourths of the population had died

---

* Pronounced Eurán according to the French orthography.

† Beg-shehrí. J. n. p. 618. ‡ Seïdí-shehrí. *Ibid.* § Soghlah. J. n. p. 619.

|| As, Minor, ii. 77. ¶ Black Tree.

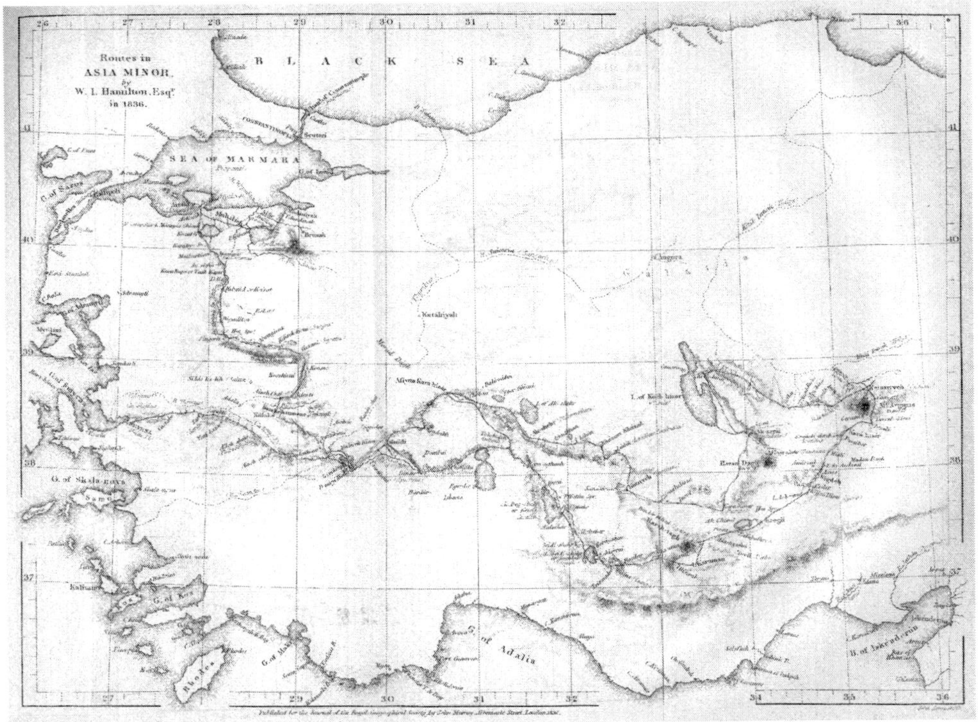
Routes in
ASIA MINOR
by
W. I. Hamilton, Esqr.
in 1836.
BLACK SEA
SEA OF MARMARA
CONSTANTINOPLE
Scutari
G. of Saros
G. of Skala-nuova
Samos
Rhodes
G. of Kos
Kalymno
G. of Adalia
B. of Iskanderun
Kutahiyah
Angora
Kaisariyeh
Karaman
Hasan Dagh
L. of Kodj hissar
Published for the Journal of the Royal Geographical Society by John Murray, Albemarle Street, London 1836.

within the last three months, and the corn for many miles round the town remained uncut or uncarried. A more striking instance of the destroying character of this dreadful malady cannot be imagined, than this vast extent of uncut corn rotting on the ground, when you are told that not only there exists no one to claim it, but no one even to carry it away without a claim. The Governor might seize it as his own, but he could not find people to cut it, or carry it, or thrash it out. The very cattle have perished when tied up in the stables, because, when the owners were dead, there was no one either to feed them or to release them.

Under these circumstances, and hearing that the malady was raging with equal violence at Aṭáliyah, on the sea-coast, and throughout the intervening country, I determined to give up that part of my plan, and to return immediately to Smyrna by Ulúbúrlú (the site of the ancient Apollonia), Díneir, Ishekli, and Philadelphia; and, after a hurried journey, reached Smyrna on the 25th August.

XI.—*Considerations on the Political Geography and Geographical Nomenclature of Australia.* By Captain Vetch, Royal Engineers, F.R.S.

Political Geography.—The extent and boundaries of empires, states, and provinces, and the still lower political divisions of the earth's surface, have but too generally arisen out of accidental circumstances, mere caprice, or have been determined by impure and interested motives; so that the want of wise design and systematic arrangement in this department of politics is as manifest as it is unfortunate: for if we consider these divisions and subdivisions of States to be the organs and channels of government and links of social institutions, their defects cannot but prove highly detrimental to the true and complete administration of the laws, the peace and prosperity of the people.

It is indeed true that circumstances in many cases exist to prevent the establishment of political divisions on principle and system, but it must also be admitted that, in other cases, where neither difficulties nor obstacles presented themselves, the introduction of order has been equally neglected, and the omission not unfrequently only begins to be lamented when the measure is no longer practicable: much labour, expense, and trouble have then to be bestowed to remedy or palliate defects which a timely organization would have entirely obviated.

In no division of the globe could a system of political geography be introduced with so much ease, or with so many prospective advantages, as in that of Australia; and it is under the above considerations that the writer would humbly submit to her Majesty's government the advantage of an immediate adoption of a systematic arrangement for that country, by establishing at once the whole of the grand divisions into which the continent of Australia may be most appropriately apportioned, and of laying down at the same time the plan upon which the minor divisions and subdivisions ought to be formed.

In approaching this subject, we have, in the first place, to consider that Australia is a continent 2500 statute miles in length, with an averge breadth of half that quantity; that it contains an area of 3,000,000 square statute miles, and that it is only one-sixth part less than the whole of Europe; and, if we reckon the population of Europe at 186,000,000, Australia may at a future day, on the same scale of density, possess a population of 153,000,000.

It is very probable that the native population of Australia has been stationary for a great lapse of years, and that there is small chance of its increasing. But if we look to the activity of British colonization, and the progress which the colonists are making in multiplying themselves, we may, with great safety, believe that the increase of souls will follow an equal ratio to that which has occurred in the United States of America, or even greater; because in Australia the ground requires much less clearing, and the obstruction from the natives is much less formidable; there is, therefore, nothing unreasonable in the expectation, that the population and power of the Australians may, in two centuries, equal those of the United States of America of the present day.

From the above general view it will be apparent that such a vast territory with a proportionate population can only be ultimately well governed by the division of the whole into a number of distinct States, and that, whether these rival communities become entirely independent of each other, or whether they be held together by a confederation, it will in either case be important for their general peace and individual prosperity, that they be pretty nearly balanced in power and natural advantages, and that in forming the grand divisions the following points should be attended to:—

1st. That each should possess areas nearly equal.

2nd. That each should be as compact as circumstances will permit.

3rd. That each should possess a tract of sea-coast.

With the above objects in view, and with the knowledge premised that Australia is of an oblong figure, twice the length of its breadth, it will be apparent that there is only one mode of division that will attain the required conditions. 1st. By dividing the continent longitudinally into two equal parts, and transversely into four equal parts, from which will result eight equal grand divisions; at least such would be the case if this continent were somewhat more regular in its form. It happens, however, that its breadth towards the eastern shore is so considerably expanded as to offer the facility of dividing the eastern sea-board into three portions as conveniently as into two, should such become, on other considerations, a more desirable arrangement; or, in other words, the form of the continent presents a facility of division into either eight or nine provinces, and of affording to each the conditions and advantages already proposed to be secured to them. But it also happens that to one province of Australia boundaries have been already assigned, by charter and by Act of Parliament, which could not be materially altered without much difficulty, and therefore rendering it necessary to endeavour to combine what is proposed to be done with what has already been performed.

If the limits and extent of the province named South Australia, as assigned by Act of Parliament, be assumed as fixed conditions in a system of grand divisions of the continent, it will not require much study of the map to show that a distribution of the land into nine rather than eight parts will most harmonise with the step which has been already taken.

Should it be proposed to divide the continent into a greater number of parts than eight or nine, it could only be accomplished by depriving some of the divisions of a sea-coast position, or otherwise, by constructing them of a long and narrow form, and departing from the compact shape so desirable for the easy and cheap administration of government. On the other hand, should it be proposed to make the number of divisions less than eight or nine, it must not be overlooked that, with a division into eight parts, each individual State would possess an average area of 375,000 square statute miles, or be of an extent one-fourth greater than Spain and Portugal combined, and that each may possess a population of 19,000,000 of souls, and consequently that a greater extent of territory and people would not be likely to ensure so good a government, or so much happiness to the people.

If from the foregoing reasons it be admitted that, in laying the foundations of a number of contiguous empires, careful design and systematic arrangement would tend to promote the future welfare of the inhabitants, and that an essential part of such design and system must consist in forming the grand divisions upon the principle above proposed, then it must also follow that the plan should be adopted and reduced to practice ere conditions or circumstances may arise to render the same either impracticable or difficult.

The accompanying sketch of a study for the grand divisions of Australia will serve to illustrate the author's views and aid his explanations. On this subject he has at present only further to propose, that, should his views be deemed correct, not only should the adoption of some such similar division take place, but also that a colony or garrison, however small, should be at once established on some convenient part of the coast of each division.

By means of small garrisons, commanded in some instances by a person of no greater rank than serjeant, did Russia maintain some possessions or settlements on the west coast of America, although her original right to them might be somewhat questionable.

The utility of establishing these small garrisons of even a company of soldiers, or less, would be as follows:—

1st. The taking and maintaining possession, and consequently preventing disputes or even war with foreign powers in respect to right.

2nd. Serving as points of refuge and refit for vessels coasting either for trade or discovery, and these must shortly become numerous.

3rd. The acquiring useful information concerning the natives, and the capabilities of the country for future settlements.

4th. Constituting an organ of government, ready at any time to give effect to what measures may be deemed necessary in the progress of future settlements; and,

Lastly. They would be useful for advancing the geographical knowledge of large territories at present too little known.

In the accompanying sketch of the grand divisions of Australia, the boundaries are formed chiefly by means of meridian lines and parallels of latitude, and these, though not the most eligible as a frontier between hostile States, have several compensating advantages; they are easily described and precisely defined, and furnish the means of giving the required size and shape better than natural boundaries; and, lastly, they facilitate the system of allotment of the land, which has been followed with so much advantage in the United States of America.

Nevertheless, should it be found, in the progress of geographical discovery, that natural boundaries do exist somewhat conterminous with those of the imaginary lines, then it would, no doubt, be advisable to substitute the former for the latter. The only good natural boundary with which we are at present acquainted is perhaps that of the River Murray towards its mouth, where it might be advantageously substituted for the meridian line boundary at least, from the point of intersection of the two, down to the sea.

GEOGRAPHICAL NOMENCLATURE.—This is a branch of geography generally left to chance or caprice, and it will not be easy to find any department so left which has been more abused.

It might not be worth while to take much pains for the mere purpose of introducing a complete and faultless nomenclature. But good taste, and even common sense, is concerned in rescuing Australia from some barbarous and nonsensical names, which nothing but a positive necessity should tolerate.

Wherever native names exist, and where these names may have existed for a number of ages, it appears something like sacrilege to disturb or change them; such names, besides the sacredness of antiquity, are often significant, and contain in themselves useful information as to the migrations of the human race, and the former connexion which existed between tribes now far separated. Thus rivers in Australia may be found with names identical with others in America; native names possess this advantage also, that they are seldom vulgar or ridiculous, and they furnish a copious fund of distinctive terms to obviate the confusion which arises to geographical nomenclature in the repetition for the hundredth time of the rivers Thames, Trent, Tyne, &c., and it fortunately happens that in no country, however barbarous or thinly peopled, are the great features of nature, as rivers and mountains, without names; and the name of a river or mountain may be appropriately applied also to the district in which it occurs.

That Australia is not defective in native names, and of good sounding names, may be shown by the following random specimens, the freedom having been taken of curtailing the double letters, with which the English are so apt to encumber their orthography of unlettered tongues, and apparently to so little purpose:—

Monam	-	Downs.	Anglicised	Brisbane Downs.
Colaiè -	-	a Morass.	,,	Dalrymple's Marsh.
Morumbigi	-	a River.		
Molonglo	-	ditto.		
Paramata	-	a Town.		
Waragamba	-	a River.		
Kernok	-	a River.		
Gelong	-	a River.		
Tramo -	-	a Down.	,,	Oxley's Plains.
Pannara	-	a Mount.	,,	Maclachlam Mount.
Belubula	-	a River.		
Waragong	-	a Range of Mountains.		

We may now contrast the above with the following new names:—

The New Year's Range.	Tunbridge Hill.
Cockburn River.	Darling River.
McQuarrie River.	Reid's Mistaken Hill.
Encounter Bay.	Sir Joseph Banks' Island.
Peaked Hill.	Mount Disappointment.
Back Stairs Passage.	

It is presumed the above will suffice to show the inexpediency and impropriety of explorers placing their own names, or those of their patrons, in lieu of *existing names,* in every way preferable; and it would only be necessary for the ruling powers to express a wish that the native names should be ascertained as far as possible, and adopted, to establish that line of future proceeding.

Where towns and villages are to be founded, there can be no remedy against bad taste; but the founders may then indulge their fancy without interfering with names already sanctioned by age.

The Spaniards rarely attempted to change the Indian names in America, but contented themselves with providing a patron saint, and prefixing his name to the towns, which prefixes are now becoming obsolete; and if we contrast the Indian names of Peru and Mexico with the jargon of new and old names introduced in the United States of America, we shall have good reason to admire the better taste of the Spaniards.

That good native names were to be found in the United States as well as in Australia, we have only to recur to some that have been preserved—as the Ohio, Oswega, Ticonderaga, &c.—and to regret that a greater number should not have been handed down in place of the Brandy-Wine, and such like.

It is not, however, to be expected that we shall find any native name used to designate the entire of Australia, or even of any large portions of it, as the circumscribed knowledge and power of the present native races cannot be supposed to reach, or to have any motive for reaching, to so great a grouping of land.

The names of the whole continent and of its grand divisions must, therefore, necessarily be of modern invention or application.

The whole continent long went by the name of New Holland, until Pinkerton pointed out the absurdity of it, and Flinders suggested that of Australia, which has happily come into general use.

There can be no impropriety in calling a new town New Carthage, or New Orleans, &c., as colonised or founded from parent cities of those names; but the term "New" cannot with the same propriety be applied to the face of a country, or its features. The New Thames, or the New Tiber—the New Alps, or the New Grampians—would not be tolerated; and, upon the same principle, New Holland, New England, and New South Wales have only ceased to excite the smile (from use) which New Turkey or New Egypt would readily create.

The name Holland is descriptive of a low country, and the term is not applicable to Australia; but the further absurdity was embodied of naming a great continent from a small province—an objection which applies also to the term *New South Wales,* an example of a name consisting of a sentence, and implying the necessity of designating the inhabitants by the names of New South Welsh men and New South Welsh women.

There are two names which have arisen lately to denominate two provinces of Australia which are hardly less objectionable than that of New South Wales. These are South Australia and West Australia. These names consist of two words in two different languages, and if put entirely into English would imply in one case a repetition, and in the other a contradiction, viz., South Southland and West Southland. Surely one language and one word might supply a better appellation than a confusion both of tongues and terms.

If we adopt the term Australia, as distinguishing it in its southern position from other continents, then, according to the usual system of classification, some other distinguishing character ought to be used for the genera than that for the classes. If, however, South Southland is a term meant to describe and distinguish a province, it will not effect its object, since the same name will be equally descriptive of three other provinces of Australia, and West Australia will be equally so of two of them.

The name in present use for the whole of the continent (Australia) is admitted to be good, nor would it appear a matter of much difficulty to supply simple and distinctive terms for the provinces, and if any such proceeded from a quarter of authority, no one would feel disposed to subject them to criticism, unless involving some manifest impropriety. The case, however, is much altered when a private individual ventures to cater for the public taste; he must be both prepared for criticism, censure, or sarcasm. The writer, having no predilection of his own in the matter, would have left the task to other hands, could he have seen any hope of its being so performed. If he, therefore, now ventures to undertake it himself, it arises from the consideration that we should otherwise *still* remain without the means of classifying the various parts of Australia; whereas by means of a system of division, and corresponding terms, we can at once refer any spot of that continent to its admitted place in the system, and which, if established for no other purpose than that of facilitating geographical description, cannot but prove of obvious utility.

The principle followed in selecting names for the great divisions has been that of doing honour to the most original discoverers of the respective portions, but preserving, at the same time, names which have been long attached to various parts; and if this principle has been departed from in two instances, the inducements for so doing will, it is hoped, be duly appreciated. Taking, therefore, the divisions as numbered in the accompanying map, and proceeding on the principle alluded to, the following terms are very humbly submitted for adoption:—

I. DAMPIERIA. In honour of the intrepid and celebrated navigator who, in 1688 and 1699, explored some of the coast in this quarter, and whose name is still attached to an Archipelago, and a tract of the shore comprised in this division.

II.—VICTORIA. In honour of our gracious Queen, under whose auspices it is to be expected, the great Australian empire will receive a form and development corresponding to its magnitude and prospective importance.*

III.—TASMANIA. In honour of the celebrated Dutch navigator, *Abel Janz Tasman*, who, in his second voyage in 1644, discovered the whole of the N.W. coast from lat. 11° to lat. 18° south.

IV.—NUYTSLAND. In honour of *Pieter Nuyts*, who, in the ship, Gulde Zeepaard, in 1627, discovered, and ran along 1000 miles of the south coast, and to which his name has long been most justly affixed.

V.—CARPENTARIA. So named from the great Northern Gulf, said to have been so called from the Governor-General *Carpenter*, who, in 1625, was Governor over the Dutch East India possessions.

VI.—FLINDERSLAND. In honour of *Flinders* the eminent navigator, to whose enterprise and surveys we are so much indebted for our knowledge of the shores of Australia, and whose name as a discoverer is especially connected with this division.

VII.—TORRESIA. In honour of the Spanish navigator *Torres*, who discovered this portion of Australia in the year 1606, and whose name is deservedly attached to the straits situated at the north cape of this division.

VIII.—COOKSLAND. In honour of our celebrated circumnavigator, who, in 1770, discovered and examined a great extent of the eastern coast, and from whom, therefore, this division may with much propriety derive its name.

IX.—GUELPHIA. In honour of King George the Third, the zealous and munificent patron of geographical discovery, and under whose auspices the first settlement of the British in Australia took place, in the year 1788.†

Before quitting the subject of Australian geography it may not be misplaced to express regret and even surprise that so little

---

* The names of the navigators Edels, Vlaming, Vancouver, and D'Entrecasteaux, claim to be recorded in the subdivision of this province, but as it would be difficult to apply any one of them to the whole division without committing an act of injustice to the others, a fit opportunity presents itself of doing honour to our august Sovereign and Patron of the Geographical Society.

† The name of New South Wales, long associated with this quarter of Australia, has been shown to be in many ways objectionable, and a fit occasion therefore occurs of honouring the memory of our Third George, in whose reign possession of the country was taken, and a new continent added to the British empire.

progress has been made in the inland discovery of a country where the general climate is so mild, and where the native hostility is so little formidable.

It is now half a century since the first settlement of the British in Australia was effected, and we may still say that we know little more than the sea-coasts. The face and features of the central regions are still matters for conjecture, and the geographer is still anxious to learn whether the great hiatus has to be filled up with lofty mountains or with lakes and marshes, whether dismal swamps or arid wastes may forbid the settlements of man, or fertile plains and rich valleys invite his approach.

British enterprise and perseverance have explored the frozen shores of North America amongst all the dangers, privations, and difficulties which the extremity of cold can inflict, and the same qualities have carried our daring travellers into the heart of the burning and pestilential climate of Africa, in the midst of a dense, savage, and bigoted population, while the genial climate and thinly-peopled tracts of Australia have failed to excite the enterprise of discovery, when to all appearance it would be attended with so many more prospects of success and utility.

To those who have been accustomed to traverse the vast spaces of continental America, the undertaking of crossing Australia in various directions would appear a matter of no great magnitude or difficulty if furnished with the hardy horses and mules and the well-trained muleteers of Spanish America.

But travelling with success and safety over great spaces in wild and savage countries is an art which can only be acquired by considerable practice; and the British travellers in Spanish America must often have remarked the insufficiency or uselessness of European servants in their journeys, while the natives, trained to travel, seemed prepared for every contingency.

In contemplating the causes which may have prevented or damped the spirit of discovery in Australia, the most important one appears to be the want of points of *appui* for rest and refitment, or for succour when the journey is accomplished. The traveller must have some haven in view to which to direct his steps, and where, at the end of his toils, he can expect safety and repose; but if this haven be the one he started from, then his resources and toils will but procure half of the discovery which might have been obtained by proceeding to a different terminus.

Settlements and points of succour are now becoming numerous on the coast, but in a continent like Australia, without inlets of the sea and great navigable rivers, it must be manifest that no considerable progress can be made in a general knowledge of the interior, until some permanent posts are there established to which the traveller can direct his steps, either by design or in case of need: the number of such posts need only be three, at or near the points marked *a*, *b*, and *c*, on the accompanying map, and as a measure of facility, the post marked *a* might be furnished by the governor of the Swan River settlement, the post *b* by the governor of South Australia, and the post *c* by the governor of New South Wales.

These three posts established, it will be seen, from the imaginary routes traced on the map between them and some nine points on the sea-coast, that the face of this continent may be very generally ascertained and traversed in many directions from sea to sea without subjecting travellers to any greater journey than one of about 500 miles without coming to a place of rest and succour; nor must it be considered that the establishment of such inland posts with a small garrison would be so much expense incurred for pure motives of curiosity, since much advantage must result to the settlements on the coast from a knowledge of the natural resources of the interior, from knowing and bridling the numbers and power of the natives. Such posts and garrisons would also prove useful in checking the movements of roving Europeans disposed to lead a life of plunder and aggression, as has so unfortunately happened in Diemen's Land.

The establishment of three posts in the central portion of Australia may appear at first sight a measure attended with difficulties; but it is to be considered that the country is generally free from thickets and dense forests, that on the contrary it is essentially open and transitable, not only for foot and horsemen, but also for carriages, so that the transport of a small body of armed men in a fine climate, for a distance of 500 miles, with a supply of provisions, could not, in any degree of probability, be an enterprise either of difficulty or danger. The servants of the North West Company, in a rigorous climate, would not estimate such an undertaking highly, neither would it be so deemed in many other parts of America.

Thirty soldiers, with a portion of settlers, might readily maintain themselves in provisions by means of an original stock of cattle, sheep, and swine, in a country so peculiarly pastoral, and might, moreover, in a short time have a superabundance for barter and profit; and where horses are so soon likely to abound, the making communications with head-quarters could be effected (when required) in a space of ten days or so, in favourable seasons of the year. The only precaution necessary to be taken would be the selection of a good site for the post, and route to it, by means of a previous exploring party.

Attention of late has been much turned to preserve the indigenous population from injuries by the colonists, and to draw them gently into social and civilised habits; and if the person in charge of such a post was a man of judgment, he might, with the assistance of a missionary, be mainly instrumental in forwarding these objects. On the whole, I can view the establishment of such inland posts in no other light than a duty to all interests,—to the government, the colonists, the natives, and the acquisition of geographical knowledge. Nor do I see how any considerable progress is ever to be made in inland discovery without such means. Are we to wait till population gradually extends from the shore to the centre? or till some rival nation, establishing a colony on the shore, shall push on discoveries in the interior? All difficulty and delay would in that case speedily vanish, and we would yield to pride, if to no other inducement, in the performance of a duty which science, no less than policy, demands of that nation to whose guidance the destinies of Australia are committed.

In relation to the early history and migrations, not only of the human, but also of the brute races, the geography of Australia exhibits some strange anomalies, which are the more interesting and deserving of study in proportion to the difficulty of their explanation.

Of all continental land, Australia (as far as it is known) presents man the most scantily disseminated, in the lowest state of civilization and knowledge, and the most abject in the means of bodily comfort or enjoyment, the most deficient of all history or tradition, and with the fewest monuments of past times; circumstances directly contrary to all *à priori* deductions from the geographical position and physical condition of the country, which—enjoying a most favoured climate, both for the rapid increase of the human race and the development of its best powers, and immediately connected by numerous islands with China on the one hand, and India on the other, the most anciently populous and civilised of nations—would well have warranted the conclusion of an ancient numerous and civilised people, in possession of all the knowledge and arts which India and China could supply; and if, through the accident of time, the present race had sunk into barbarism, that we should still find the monuments of former prosperity and past ages, marking the ruin of empires and revolutions of nations. But as far as the most recent researches permit us to speak, there is no indication during the space of four thousand years, or since the origin of the human race, that colonies had been established in Australia by any civilised nation until the present epoch; whereas Java, at little greater distance from Australia than England is from Spain, appears to have been early populated and civilised, as so emphatically shown by the following statements of Sir Stamford Raffles respecting that island:—"The numerous and interesting remains of former art and grandeur which exist in the ruins of temples and other edifices, the abundant treasures of sculpture and statuary with which some parts of the island are covered, give evidence of a former state of religious belief and national improvement, which are represented in images, devices, and inscriptions." And again—" The grandeur of their ancestors sounds like a fable in the mouth of the present degenerate Javan; and it is only when it can be traced in monuments which cannot be falsified, that we are led to give credit to their traditions concerning it."

Of such easy access to the maritime nations of South Asia, and in the vicinity of a country so anciently peopled and civilised as Java appears to have been, it can hardly be ascribed to accident that Australia should have remained during the great flood of time unvisited and uncolonised; that, amidst the pursuit of conquest and discoveries by some nations, and flight from oppression by others, Australia should have continued unreclaimed from the state of nature, a sealed and secluded land to the migrations of man, so much nearer to the presumed cradle of his race, while the far distant countries of Peru and Mexico offer proofs of ancient and civilised nations.

The inference would seem to be, that in former times some physical difficulties or conditions existed, which do not now exist, which placed a barrier to the colonization of this continent, or else that some convulsions of nature destroyed in this division of the globe both the race and the works of man. These are, however, inferences which we cannot venture to adopt until the progress of inland discovery be farther advanced, for we are still allowed to expect in the interior traces and proofs of the ancient dominion of civilised nations.

But, on the other hand, the absence of most of the larger quadrupeds of Asia, and the possession of a few peculiar to itself, rather confirms the presumption of the long seclusion of Australia from the rest of the world.

Some have supposed that Australia is a more recently formed land than the other continents, but if even such a notion was admitted for the sake of argument, it could only hold good in respect to mineral composition, while in respect to the epoch of man, there are sufficient geological proofs of an equal antiquity with the other divisions of the globe.

The curious problem of the long-secluded state of Australia from the rest of the world, while apparently so easy of access, must remain for the present matter of conjecture, though we may justly entertain the hope that the progress of inland discovery will throw considerable light on the subject. We shall learn the present physical state of the country, and also a probable knowledge of what that state has been during the last four thousand years, We shall learn in what state the indigenes of the interior remain—their languages, customs, and manners—what connexion may be traced between them and the tribes of other

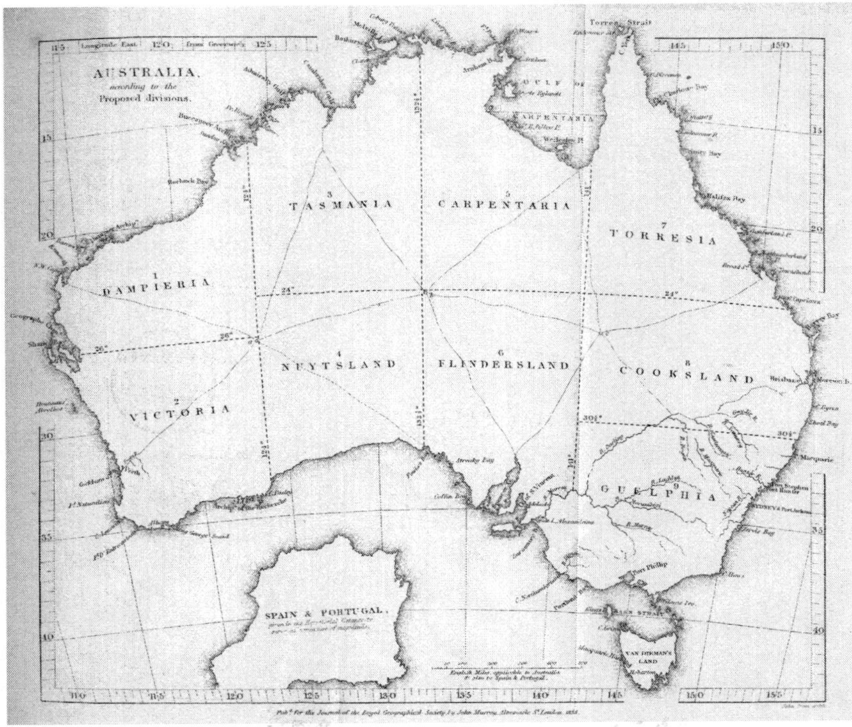
AUSTRALIA,
according to the
Proposed divisions.
1 DAMPIERIA
2 VICTORIA
3 TASMANIA
4 NEYTSLAND
5 CARPENTARIA
6 FLINDERSLAND
7 TORRESIA
8 COOKSLAND
9 GUELPHIA
GULF OF CARPENTARIA
Torres Strait
VAN DIEMAN'S LAND
SPAIN & PORTUGAL
Port Phillip
Brisbane
Streaky Bay
Halifax Bay
Longitude East 120 from Greenwich 125
English Miles applicable to Australia & also to Spain & Portugal.

countries—and whether there exist any remains of a more ancient and more enlightened people.

But while we remain in want of more decided facts on which to ground deductions, too great care cannot be taken to preserve the vocabularies of the various tribes, their manners, customs, and rites, and the names of places, which, in other countries, are sometimes significant in languages no longer spoken on the spot.

Much credit seems due to Mr. Scott Nind for his attention to the above subjects, as exemplified in a paper communicated to the Geographical Society,* in which he describes a law of the natives near King George's Sound, of so extraordinary a character, and yet so similar to one adopted by some tribes of British Guayana, as may well justify the temptation of placing the two accounts together in this place for the purpose of showing the value of such researches.

"The whole body of the natives," says Mr. Nind, "are divided into two classes, *Erniung* and *Tem*, or *Taaman;* and the chief regulation is, that these classes must intermarry, that is, an *Erniung* with a *Taaman:* those who infringe this rule are called *Turedangers*, and are subject to severe punishment. The children always follow the denomination of the mother—thus, the man who is *Erniung*, will have all his children *Taaman*, his sister's children will be *Erniung*. This practice is common to all tribes in the neighbourhood, with the exception of the *Murram*."

"The Arrawack [Arahuac] tribe of Indians, in Guayana," says Mr. Hillhouse,† "consists of twenty-seven families. The cast of blood is derived from the mother, and the family genealogy is preserved with the greatest care, as a preservation from incestuous intercourse, one family (clan) not being allowed to intermarry within itself. The children of a *Maratakayu* father cannot therefore be *Maratakayu;* but if the mother be *Queyurunto*, the children are also *Queyurunto*, and can marry into the father's family, but not into the mother's."

It does not appear probable that a law so extraordinary, but so well devised for the regulation of marriages, could have been adopted accidentally by savage nations, so far apart, and in opposition to the general habits, the want of restraint, and hasty passions of man, in a barbarous state of existence, and we might therefore infer that this institution has been derived by each of these tribes from some people in a more advanced stage of knowledge than themselves; and if at any time we shall be able to trace this custom to its source, and observe its greater extension, these circumstances may tend much to elucidate the migrations and connexions of the early families of man.

* Journal, vol. i. p. 37. † Ibid. vol. ii. p. 228.

XII.—*Descriptive Sketch of the Islands and Coast situated at the Entrance of the Persian Gulf.* By Lieutenant WHITELOCK, Indian Navy. Communicated by the Bombay Geographical Society.

[As the chart of the Persian Gulf, resulting from a trigonometrical survey by the Officers of the Indian Navy, between the years 1821 and 1829, has now been published for some years, and has been made use of in all good modern maps of that region, it has not been thought necessary to reprint the portion of it here referred to.

This descriptive sketch will be found to be a useful supplement to a Memoir by Lieut. Kempthorne, of the Indian Navy, in vol. v. p. 263 of the Geographical Journal; and here, perhaps, it may be permitted to regret that—although this important survey of the Persian Gulf has been finished more than eight years—no complete account of it, nor any description of the coasts, has yet been published.—ED.]

PASSING two rocky islets on the Arabian shore, which are called the Quoins, you enter the Gulf of Persia, and there are few parts within it which present a higher claim to attention than this, for the whole region on every side abounds in historical and classic interest.

On the right hand, beneath a lofty mountain, called by the Arabs Jebal Shamál,* which is seen towering far above the other hills on the Persian shore, with its summit clad with snow, even in the spring-season, lies the far-famed island of Hormuz; on the other hand Lárek; † and only a few miles further on, the the town of Gamrún, which in opulence and magnificence was only inferior to Hormuz. Kishm,‡ also, the ancient Oaracta, and Mínáú, near which took place the meeting of Alexander and Nearchus, are situated in this vicinity.

The former renown of the island of Hormuz has often occupied the descriptions of earlier travellers. My object in this sketch is to describe the island as it is at present; and this, when considered with reference to its former opulence and splendour, may not be deemed wholly uninteresting. Hormuz is twelve miles in circumference. Its form is nearly circular, and its appearance from seaward is broken and rugged. The surface, entirely denuded of soil, exhibits the various tints of its singular stratification—which, with the conical shape and isolated position of the numerous small hills composing the island, gives the former a highly volcanic aspect, and would induce us to attribute the origin of the island itself to the same agency.

* Mount North. The orthography of the names of places has been reduced to the standard adopted by the Society—(see vol. vii. p. 245).

† Or Lárij.

‡ Sometimes Kishm and Jism—(Niebuhr, Besch. von Arabien, p. 328). It seems to have been anciently called Ḳeïs or Ḳeïsh—(see Jihán-numá, pp. 275, 532). It is also called Jezírah ṭawílah, or Jezírah Diráz, *i. e.* Long Island.—(Niebuhr, l. c.)

With a pilot, Hormuz may be approached from either hand without apprehension. The harbour, situated on the N.E. side, is both secure and convenient. To this, and to its insular and otherwise advantageous position, must be attributed its former importance.

The fort, in lat. 27° 6′ N., long. 56° 29′ E., is situated about 300 yards from the beach, on a projecting point of land, which is separated from the body of the island by a moat. The position is remarkably well chosen; and the whole, with the exception of the ordnance, which has been destroyed by time and rust, is still in good condition.

A few hundred yards from this, now tottering in ruins, stands the lighthouse, which must formerly have been a fine building: its spiral staircase still exists, but it would be dangerous to ascend it. A level plain extends for some distance to the N.E. of this building, having its surface scattered over with mounds and ruins of former habitations. Several tanks and wells have also been sunk here: the former, though now out of repair, are covered over with an arched roof; they are about fifteen yards in length, and seven or eight in breadth. As there are no fresh-water springs on the island, the inhabitants are wholly dependent on the supplies which are collected in these reservoirs during the rainy season.

Across this plain, towards the rugged hills which line the eastern shore of the island, a singular phenomenon presents itself, which strikingly resembles the "Mer de Glace." The hills, for a considerable distance from their bases, are covered with an incrustation of salt, which in some places has the transparency of ice; in others its surface is partially covered with a thin layer of a dusky red-coloured earth, receiving its tinge from oxide of iron, with which the whole surface of the island is deeply impregnated.

As we ascended the ridge, our progress was continually impeded by deep pits, on the sides of which the saline crystalizations have assumed a stalactitic form. From the summit you obtain a noble view of the whole of the lower parts of the Gulf: the Quoins, Cape Musandam (Rás Maskaṭ*), Lárek, and the island of Kishm, are all distinctly seen.

The Imám of Maskat has possession of Hormuz at present: he farms it from the King of Persia, and retains in the fort a garrison of 100 men, commanded by an officer, who is styled Shaïkh. A small sum is collected on account of the salt, which is exported in large quantities, and conveyed to different parts in the Gulf.

* Also spelt Maskat-Jihán-numá, p. 496.

When the island was surveyed in 1827, the number of inhabitants, who had no other employment than that of collecting this salt and fishing, was estimated at 300.

A few fowls and some sheep, brought from the main, may be obtained here, but no other supplies; nor is its port at any time visited by vessels for other purposes than to obtain salt, or for shelter during the prevalence of the westerly winds.

Such are the few remains that are left to denote the former opulence of Hormuz. The wretched habitations of its present occupants, and the dreary and barren aspect of the surrounding hills destitute of vegetation, would not lead us to recognise this spot as a fitting site for a city which contained 4000 houses and 40,000 inhabitants, whither merchants from every quarter of the globe resorted, outvying each other in the display of wealth and luxury.

The kingdom of Hormuz, or Hormuzeïn,* situated on the adjacent main, gave its name to this island, which, according to some authors, was previously called Jerún.† It is impossible to ascertain at what period this island was first occupied, but there are various authorities to prove that it has often served the inhabitants from the main as a retreat, when suffering either from civil commotions or foreign invasion. The advantages of its harbour, joined to its insular position, converted it from a barren rock, to which Nature has denied even water, into the emporium of the East.

Nearly opposite Hormuz, on the Persian shore, the river of Mínáú‡ enters the sea, in lat. 27° 7′ 48″ N., long. 56° 49′ E. Following the course of this stream, which is very tortuous, we reached the town of Sháh-Bander,§ which stands on the bank, at a distance of fourteen miles from the sea by the winding of the stream, but only eight in a direct line.

To this point, which forms the extreme limit which the tide reaches, the river is navigable at high water for vessels of twenty tons; its average width being 100 yards, and its general depth about six or seven feet. At low tide, its bed is laid almost entirely bare, and it then has the appearance of a foul, muddy creek.

There is a custom-house, besides a few other houses, at Sháh-Bander, as boats either receive or land their cargoes there, which are conveyed by land carriage to and from Mínáú. Leaving Sháh-Bander, and after proceeding for about two hours over a fertile plain, we reached a small town named Ḥájíábád.‖ Here we put up in a small house, which had been prepared for our reception by the Shaïkh; but, although situated in the most

* The two Hormuzes (Hormisdases). † Sir W. Ouseley's Travels, i. 156.
‡ Or Mínáv. § King's Port. ‖ Pilgrim's Abode.

respectable part of the town, it had more the appearance of a store-room for grain than a human habitation, and we found the heat very oppressive, notwithstanding it was at a cool period of the year.

From the appearance of the houses, and the state of the bázárs, I do not conceive that this town either possesses wealth, or is of any commercial importance. The number of its inhabitants may be estimated at 600 or 700, and they are principally engaged in agricultural pursuits. Bullocks, sheep, and goats, are very numerous; and when a dearth occurs on the island of Ḳishm a great number are sent over there.

The fort of Mínáú, distant about a mile from the town, is situated on elevated ground on the southern bank of the river, which winds round its base. It is of a quadrangular form, flanked by round towers at the corners, in which there are a few old guns, bearing inscriptions in Portuguese and Dutch.

A draw-bridge, thrown across a moat, leads to a gate thickly studded with iron knobs and spikes on the south-western side.

The walls are strong, and the fort is generally in good condition. The garrison consists of about 100 men, well appointed, who are obliged to be constantly on the alert, in consequence of the numerous marauding bands who rob and plunder the country. The fort, however, is commanded by a hill on the N.E. side; but, in a country where the use of artillery is nearly unknown, this is of little consequence.

The river at this point is little more than a mountain stream: its width is about 130 yards, and the water is clear and deep. It takes its rise from the mountain called Jebal Shamál, distant about thirty miles. Notwithstanding its present insignificance, when the snow melts on the hills, or heavy rain falls, it swells into a large and rapid stream.

In some parts, where the river is fordable, I observed its bed to be composed of coarse gravel, with small pebbles of primitive rocks, which have been brought during the floods from the surrounding mountains. The steep banks near the sea exhibit a succession of alluvial deposits. The district comprehended between the fort and the sea bears the general name of Mínáú; and to the river it owes a great portion of its fertility. Numerous artificial rills conduct the water over the face of the country, and afford near the banks a constant supply; but in tracts more remote the grounds are irrigated from wells. The water is drawn up by bullocks, either by the mote, as practised in India, or, when the wells are sufficiently shallow, it is raised by the lever, as on the banks of the Nile and Euphrates.

The soil is of a rich alluvial nature, and yields, with little labour to the husbandman, a plentiful crop. From its loose nature it requires but little ploughing, and the instrument used is rude and simple. From Sháh-Bander to Ḥájíábád, the whole of the country is cultivated, yielding large crops of wheat, fruit, and vegetables. Melons are common, and onions are raised in large quantities; plums, cherries, frequently fine apples, and dried fruits, are brought from the interior. The Indigo plant is also cultivated here to a considerable extent.

Although the site of the town is low and badly chosen, yet it does not, excepting at the close of the date season, appear to be considered unhealthy; but near the fort the air is said to be very salubrious. During the hot months many of the better classes from Bander-'Abbás* and Ḳishm resort hither, when, in addition to its superior climate, they enjoy the luxury which its light and pure water affords, which can only be duly appreciated in such a country.

In the better parts of the town of Mínáú the houses are constructed of rough stone, cemented together with mud. In the windows talc is substituted for glass. A small open space, serving for their cattle and for various domestic purposes, is sometimes enclosed by a wall, but more generally a fence, constructed with branches of the date-palm. With the same material, the lower classes construct their huts, which are afterwards covered over with a layer of mud.

We found the inhabitants civil and obliging, but very anxious to cheat us on every opportunity. I believe this feeling, which is common in other places along the Gulf, arises from an idea that we are ignorant of prices, or indifferent to the value of money. Some individual usually attaches himself to a stranger, reserving to himself the exclusive right of taking advantage of him. They are shrewd and intelligent, and this compensates in some degree for their extortion, as they possess considerable information, which they are very willing to impart, and are very useful to a visitor when walking abroad, by keeping off the crowd, or as messengers.

Gamrún, or, as it is now styled, Bander-'Abbás, appears to have been a town of little importance until 1622, when Sháh 'Abbás, assisted by the English, drove the Portuguese from the island of Hormuz, and transferred its commerce to this port. Here, instead of being carried in ships to Baṣrah and the northern ports of the Gulf, a very considerable portion of the imports from India and Africa were landed, and transported by means of caravans to the interior parts of Persia and the adjacent countries; so that Gamrún became for a time the emporium of Persia.

The English, Dutch, and French had factories here. Merchants from all parts resorted to it, and it seemed destined to

---

* 'Abbás's Port—more correctly, Bander-'Abbásí.

attain the former opulence and splendour of Hormuz; but its commercial career was far more brief. Towards the close of the seventeenth century, the internal commotions and distracted state of Persia frequently interrupted the route for very long periods, and the current of the trade became diverted to the northern ports. It should still, however, be remembered, that this route conducts by one of the natural passes into the heart of Persia; for when Bushire,* a few years ago, remained for some time in a disturbed state, commerce found its way again into this channel; and, if Bushire had not been speedily restored to peace, Bander-'Abbás would very soon have recovered a considerable portion of its former importance.

Even at present the trade is not inconsiderable, and it is said to be still increasing. In 1827 the Imám of Maskat, to whom the port at present belongs, collected a revenue of from eight to ten thousand dollars.

Persian carpets, tobacco, and dried fruits, form its exports; and piece goods, Indian cloths, and China-ware, constitute its principal imports. The annual importation of these articles at the same time was estimated at nearly three lakhs of rupees.

The town is situated on a slope, which approaches close to the sea; the houses are few, and wretchedly constructed, and the people are mostly lodged in huts. They are a mixed population, composed of Persians, Arabs, Kurds, a few Armenians, and Bedowins. Their number, though constantly fluctuating, may be estimated at from four to five thousand. Some portions of the English factory-house are still standing; but that erected by the Dutch is in better repair, and still serves his highness the Imám of Maskat as a residence during his visits to the port.

The tombs of the former European inhabitants are just without the town. In their vicinity there are some tanks, which were excavated with extraordinary labour by the Portuguese; the length of the most extensive cannot be less than half a mile. These are intersected at right angles towards the extremity by two others, so that they assume the shape of a cross.

Between Gamrún and Linjah† there is little on the sea-coast of Persia to attract attention. The range of mountains extending from Jebal Shamál, which is distant thirty miles from the sea, gradually approaches the shore towards the latter port, where they are not more than three miles distant. The maritime plain throughout the whole distance is low and barren, though not without occasional spots of cultivated ground. Abreast of Laft, the coast is fronted by swamps, thickly covered with mangrove thickets, and within this, close to the margin of the sea, stands the small village of Khamír. In this vicinity there are mines of sulphur, which

* Properly Abú-Shehr (Father Town).

† Or Lunjeh and Lung. Nieb. p. 333.

are extensively worked, and the produce imported in large quantities to Maskat. Between Khamír and Linjah there are two small towns, one called Bandar Hallam, containing about three hundred inhabitants, who trade in salt; the other Kung, where the Portuguese had formerly a small factory established, principally with a view to command the copper-mines in its vicinity, which were worked by them.

Abreast of Básidóh* the height of the coast range was ascertained, by trigonometrical measurement, to be 3498 feet above the level of the sea. Although the ascent is very laborious and difficult, it has been accomplished by several of our officers. Crossing over the maritime plain, which is here not more than three miles and a half in breadth, they found at the foot of the hills a mineral spring, the waters of which are highly beneficial for cutaneous eruptions, as well as rheumatism, scurvy, &c.

Of the islands which, besides Hormuz, form the group situated in this part of the Gulf of Persia, that of Kishm is the largest, and indeed surpasses in size all the islands of this inland sea. Kishm stretches along the Persian shore, from which it is separated by a channel thirteen miles in (maximum) width, but contracting in the middle of its length to three. The channel is studded with islets, and bears in the new charts the designation of Clarence's Straits.

In its form, Kishm bears a striking resemblance to a fish, the town of the same name being situated at its head, which faces the eastward; Laft and the island of Anjár† to the northward and southward of either fin; and Básidóh to the westward, at the extremity of its tail. Its length is fifty-four miles, and its width, at the broadest part, twenty miles. On the southern side, a ridge of hills extends from one extremity to the other, while the remaining space to the northward is occupied by arid plains and deep ravines. The greater part of the surface of the island is sterile, and in some places incrusted with a saline efflorescence; but the most striking features in its structure are some singular-shaped table-hills, which occupy insulated positions in the plains. These are of a circular form, principally composed of sandstone, and are broader at the upper part than at the base. Their average height is from two hundred to four hundred feet: their surface and sides, worn into hollows by the weather, give them the appearance of having been subjected to the action of a powerful stream, an illusion still further increased by observing the plains and the sides of the hills, which, in the form of banks, bound what seem to be the beds of deserted water-courses. In a coun-

* Or Básidó, as Niebuhr spells it (p. 329), is probably a corruption of the Portuguese Baxador for Embaxador.

† Hinjám. Niebuhr, p. 328. "The English," he says, "call it Angar." No doubt the Persians say Angár, and the Arabs Anjár.

try where earthquakes are frequent, we might infer, from the general appearance of the whole, that these isolated masses denote the original level of the island, and that the plains have sunk in every direction around them.

At Básidóh, in March, 1829, for six hours during the night, successive shocks were felt. The inhabitants were in great alarm, and even the cattle evinced symptoms of fear; nothing serious, however, occurred.

The northern part of the island is the most fertile, and on this account the most populous. The soil consists of a black loam, and on it is reared wheat, barley, vegetables, melons, grapes, &c.: dates are produced in large quantities: cattle and poultry are also reared; but, unless their crops fail them, the inhabitants are indifferent about disposing of the former. The whole number of inhabitants on the island may amount to about five thousand. They employ themselves in fishing, in cultivating the soil, and in making cloth. They reside in villages and hamlets scattered along the sea-coast.

The only towns on the island are Ḳishm (the largest), Laft (next in importance), and Básidóh.

Ḳishm. The town of this name is situated near the sea at the eastern point of the island, its site being remarkably well chosen. A wall flanked by turrets surrounds it, and affords the inhabitants security from robbers or pirates. Some of the houses are large, and, for this country, neatly fitted up; the roofs are flat, and the apertures for light are partially filled with curious devices formed of a fine cement.

Ḳishm has the appearance of having been formerly of greater commercial importance than it is at present. Even when I first visited it, in 1821, the bázár was abundantly supplied with vegetables of various kinds, fruits, apples, and pomegranates from the interior of Persia. Very good wine and every description of dried fruit could be then obtained, as well as silk and cotton cloths; together with very fine carpets soft as silk, and of the richest pattern and dye. These latter we purchased at the rate of twenty dollars each; they were six or seven feet long, by three feet broad. At this time, the British force was encamped near the town, and the demand was in consequence considerably increased; yet the supply was in general fully equal to it.

Ḳishm is frequently visited by native vessels, which touch here for wood and water, or to engage pilots for the Ḳishm channel, and the town has, in consequence, a bustling appearance. A few bagalahs * are constructed here with timber brought from the Malabar coast.

* Coasting-vessels.

Captain Brucks computes the number of inhabitants at two thousand. In the plains to the westward of the town there are several patches of cultivated ground, interspersed with clumps of date-trees. Our force encamped about one mile from the town, in a strong position, on an elevated tabular ridge, which presents a steep face on either side. The situation was found to be so hot and unhealthy, that, after losing several men from fevers, they were obliged to quit it.

Laft, when in the possession of the Juwáśimí* pirates, was a place of considerable strength, to which they resorted, and the tortuous nature of the channel and numerous shoals (then unknown) in Clarence's Straits rendered it very difficult to follow them. During the expedition under Colonel Smith and Captain Wainwright in 1809, these Arabs beat back, with considerable loss, a storming party, but surrendered when the vessels came close in, and had battered their walls. The town is at present in a miserable state, built on the slope of a hill on the northern side of the island and surrounded by a wall.

Básidóh, in latitude 26° 39′ N. and longitude 55° 22′ E., at the western extremity of the island, was formerly in the possession of the Portuguese, and the ruins of the town and fort which they erected may be still traced. This station has been happily selected; for the British, after various attempts to fix on other spots, were compelled to abandon all, and finally to settle here; and, after similar attempts to establish a rendezvous at other ports, the naval squadron became also finally stationed here.

On account of its salubrity, and the local advantages it enjoys, this is the most elegible spot which could have been chosen in the lower part of the Gulf. An hospital and storehouse, a guard-room, cooperage, &c., have been erected at the public expense. Five or six private houses, a billiard-room and fives-court, erected by subscription by officers of the Indian navy, were soon afterwards raised, and a very respectable bázár was subsequently established.

The few vessels now stationed in the Gulf, from being constantly employed in various parts of it, visit Básidóh less frequently than formerly, and the place is in consequence going to decay.

The bázár affords some scanty supplies from Mínáú and Linjah; a few merchants, who emigrated hither from Bushire when the place was more flourishing, still remain. They contrive to increase their income by hiring out jaded horses and asses to our seamen. Some Indian washermen derive a more

* Jawáthimí, or Jawáśimí, i. e., belonging to the Jawáśim or Jaśúms, the aboriginal natives of the country.

certain profit; for within the Gulf, with the exception of Bushire, this is the only place where clothes can be washed well. Some Jews also reside here; they are principally goldsmiths, and occupy themselves in making rings and bangles for the females.

Básidóh is scantily supplied with water. The wells dry up in April or May; and the few tanks, which are similar in construction to those of Hormuz, and have, it is supposed, the same origin, although kept in repair by the English, do not afford more than is sufficient for the station, and the ships fill up either on the opposite coast, or further up the channel.

Although nothing can exceed in barrenness the appearance of the country in the vicinity of Básidóh, yet there are several places, only a few miles distant from it, which often exhibit all the verdure of more fertile regions; such are the plains contiguous to Górì and those near Dastagán. The former cover a space of eight miles in length, and three in width, and contain groves of the date-palm, verdant plots of cultivated ground, and, after the rains, a luxuriant crop of high grass.

The few productions of this island do not differ from those to be found on the main: a few grapes are grown in wells, or the vines are permitted to climb around the branches of the banian (*ficus indica*); a few mangoe-trees are also found at Dastagán, but in no other part of the island. Salt is found on the southern side, rising up into hills, or formed into caves. In the centre of one of these caverns, about fifty yards in length, and twelve in height, flows a stream of water; and from the roof and sides hang stalactites of salt, which are sometimes eighteen or twenty inches in length. The surrounding plains are covered with a saline crust, which the natives collect and convey to Dastagán. Towards the centre of the island there is an insulated rock about three hundred feet in height, which is steep on every side, and seems to have formerly served the purpose of a retreat to some bands of pirates or robbers. The summit can only be gained by climbing up through a narrow aperture resembling a chimney. Some of our officers, who ascended by this way, found at the top the ruins of several houses and two tanks.

The natives have a tradition, that this singular spot was formerly taken possession of by the crew of a Portuguese ship wrecked on the island, who for a long time resisted the attempts of the inhabitants to destroy them.

Sandstone appears the predominant rock on the island. On many of the arid plains in the centre of Kishm are found fragments of mica, varying in size from three or four inches to even a foot square.

Good hunting is obtained in several parts, and a small and very beautiful description of antelope is found during the day on the plains. At night they retreat to the hills. They are very shy, and, in order to give the greyhounds any chance of success, a party must be stationed to turn them from the hills, for which they invariably make upon the slightest alarm. They are sometimes taken after a run of two or three miles, completely tired, and unable to proceed further; but they more generally escape. I have known them in the former case to be taken alive and unhurt, the dog standing over the poor animal, but unable from fatigue to harm it. Their flesh in the cold weather is much esteemed, but in the hot season it is lean and tasteless. Hares and small rabits are also found on this island. Jackalls and foxes afford occasionally a good chase, but, as they are favoured by the country, they more frequently take to earth. Camels and asses are employed as beasts of burthen: in the rutting season the former are very savage.

The principal birds are vultures, cranes, grey partridges, hawks, pigeons, the kingfisher, houppoe,* and jay. There are several others remarkable for the beauty of their plumage. Several varieties of fish are caught on Básidóh bank: prawns, lobsters, and large crabs are also abundant. There are several varieties of snakes; some of the most venomous kind. I have seen death follow in two hours after the person had been bitten.

From the irregular outline of the island, and the existence of numerous banks and islets, the direction of the channel which separates Ḳishm from the main is varied and tortuous.

Commencing from the westward, about midchannel between Básidóh and the main, there is a sandbank with about ten feet water on it. Across this, towards the Persian coast, you carry a depth of two or three fathoms; but towards the Ḳishm side the channel varies both in its nature and depth. In some places you have soft mud over hard rocks, in others a mixture of clay and mud very tenacious, and in other parts a clear bottom of sand. Proceeding up the channel towards Gorún† the deepest water is near the island, and its depth is indicated by the appearance of the shore; if the cliffs rise up boldly from the beach, the water is deep close to the shore: on the other hand, where the plain slopes down to the sea, extensive mud flats run off it to a considerable distance. Beyond Gorún, approaching towards Laft, two channels branch forth; one near the Persian shore used by ships, and another, although more narrow and winding, preferred by boats, on account of its being free from rocks or banks: the space included between these two channels is nearly blocked up with mud flats dry at low, and but partially covered at high water. Narrow streams intersect these flats, and form them into groups

* Upupa.

† Garún for Jarún (?).

of islets. These islets are covered with a dense thicket of mangrove-trees; and the lively green of their foliage, in a country so destitute of vegetation, presents a refreshing and pleasing effect.

During our stay amidst these islets we were apprehensive, from the close nature of the woods, and the thick fogs which we observed hovering over them, particularly after sunset, that severe sickness would have prevailed amongst us; but, notwithstanding that we were three weeks engaged in surveying this part of the channel, suffering much, though in the winter season, from exposure and fatigue, we had not a single case of fever.

Beyond Laft the forest disappears, but for about sixteen miles the channel continues equally intricate: from this point it runs along the Ḳishm shore, and eventually opens out into the Gulf of Hormuz, where all is clear.

There is a point of some interest connected with the set and direction of the tide in this channel; the flood enters at both extremities of the channel, and meets at Laft, where the rise and fall is about fourteen feet. This affords great facility in navigating the straits; for a vessel quitting the town of Ḳishm with the first of the flood may reach, and start from Laft at high water, and have the whole of the ebb tide to carry her to Básidóh.

The Island of Anjár is situated on the south side of Ḳishm, opposite to the town of Laft, which stands on the northern shore. This island was formerly inhabited, but since the destruction of the town by the pirates, it has been deserted. Vessels occasionally seek shelter here from north-westers. Water also can be procured from wells and reservoirs situated near the anchorage.

A ruined mosque, which stands near the site of the former town, is still conspicuous. We found its geographical position to be, latitude 26° 41′ N., longitude 55° 56′ E.

This island is formed of bare rocks, and has the volcanic appearance which is commonly observed in the other islands of this Gulf.

About twenty-four miles to the south of Básidóh there are two uninhabited islands, called the Great and Little Tomb.* The former is well stocked with antelopes, and much resorted to by the officers stationed at Básidóh, for the purpose of hunting. In the winter months the island is well covered with grass, and the water is very good.

Lárek is the last island to be mentioned. It lies in latitude 26° 53′ N., longitude 56° 23′ E. It is of a volcanic character, and in size, as well as in the colouring of its strata, is very similar to Dálmah,† one of the islands near the Arabian coast, called by us Maud's Group.

* Ṭunb and Ṭunb Námiyúh. Niebuhr, p. 328. Námiyúh or Námiyóh is a corruption of some European or Hindí word.

† Ḍálimah or Ẑálimah, 'dark.'

A rocky ledge, extending to the average width of half a mile, surrounds Lárek; beyond that the water suddenly deepens over a bottom of rocks and sand to eighteen or twenty fathoms. The island has neither harbour nor any secure anchorage near it, so that, from whatever quarter the wind may blow, the sea rises and breaks with much fury over the rocks which gird its shore. It is therefore highly dangerous to land in unsettled weather, and for these reasons Lárek is rarely visited.

We found it inhabited by a few fishermen, who, to the number of about 100, reside in wretched huts, within the walls of an extensive fort. They live together as one family, and are a poor and insulated race, bearing some resemblance to the tribe (to be described hereafter) who reside in the vicinity of Rás Musandam, with whom, and in this they are singular, they maintain a friendly intercourse. They have a great aversion to mixing with their neighbours, and rarely ever visit the town of Ḳishm, though only six miles distant.

They subsist on fish and dates. No part of the island is cultivated, and the few cattle they rear, for the sake of their milk, partake in general of the same food as their masters.

I shall now conclude this sketch with a brief description of the land about Rás Musandam, and of the inhabitants who are found residing on the shores of the deep inlets and coves in its vicinity.

Rás Musandam lies in latitude 26° 23′ N., and longitude 56° 35′ E. It forms the outer point of an island bearing the same appellation, but the true promontory of the coast is called Rás Gabr Hindí,* or Rás el Jebal.† From this the island is separated by a deep and narrow channel.

The Cape is about 200 feet in height, and rises abruptly from the sea. It is composed principally of basalt, which gives it a black and gloomy aspect.

On both sides of this promontory the coast line is indented in a most singular manner into deep coves and inlets, extending as far as Rás Sheïkh Mas'úd on the western side, and to Rás Ḥaffár to the southward. The two most remarkable of these inlets are named in the new charts after Mr. Elphinstone and Sir J. Malcolm: the former inlet lies on the western side, and runs in a most tortuous course for nearly eight miles. At the bottom it is separated from Malcolm's inlet, which lies on the opposite side of the promontory, by a mountain ridge which is 500 feet high, and difficult to ascend, but only 100 yards broad at the summit.

The depth of water in the coves varies from thirty to forty

* If rightly spelt, this name signifies Cape Hindú Pársí or Gabr.
† Cape Mount.

fathoms in the centre, shelving towards the rocks on either side over a bed of branching coral and fine sand. The water is exceedingly clear, and the various kinds of rock-fish may be seen sporting amongst the coral in eight and ten fathoms' depth close to the shore.

At the entrance, and inside the coves, there are several curious rocky islets; some of them with deep water close to their base.

The hills, in general, rise perpendicularly from the sea, and average in height from 200 to 800 feet; they are extremely rugged and barren, and in some places deep caverns have been formed at their base by the action of the waves. They are principally composed of basalt and granite, in a state of decomposition, which renders it dangerous to ascend in many places, as, by the slightest pressure, large masses of rock are detached. Quartz is met with very commonly, and likewise slate-stone. Some stunted shrubs and grass grow on the side of the hills, and likewise the senna plant.

Khasáb* bay (the fort in latitude 26° 13′ N., longitude 56° 20′ E.) is the only one of these inlets in the vicinity which differs sufficiently from the others to merit a separate notice. Nature appears less forbidding in this spot, for at the bottom of the bay there is a plain of considerable extent, which is covered with a rich soil, and yields a tolerable crop of wheat, barley, and onions, besides dates; and the verdure, which is everywhere surrounded by naked rocks, produces a pleasing effect.

The fort is large, and strongly built of the usual form, with turrets at the corner; but it is considerably out of repair, and will soon fall into ruin. Good water is plentiful, and easily procured; and we obtained some cattle and other supplies from the natives.

Fish is procured in great abundance in all the coves: mullet, sír-fish, and the different kinds of rock-fish, are the most common; oysters are found attached to the cliffs, and are very good. The natives procure a beautiful large conch-shell from a great depth, and eat its fish when cooked.

The isolated condition of the inhabitants of these hills and coves has rendered them remarkable for their primitive state of ignorance and poverty, which is, however, compensated in a great measure by their love of home, and general contentment. They are principally found residing in the little sandy bays situated at the extreme end of the inlets, living in small stone huts, and surrounded by a few palm-trees: they subsist on fish, barley, cakes, goat's milk, and dates. They are badly clothed, but their dress is not otherwise remarkable.

They profess the Mohammedan religion, and practise its laws

* Kaṣṣáb, *i. e.* Butcher (?). Casaab in the charts of the Persian Gulf.

as far as they understand them. They speak Arabic, a corrupt jargon certainly, and difficult to be understood even by Arabs; but I do not believe that they have a distinct language, for, when Mr. Wellsted put the question to the Imám of Maskat, he decidedly said they had not; that he had seen a few of them at Maskat, but he believed it was very seldom that they ever left their native hills, and they were a singular, but a poor and inoffensive race.

It is impossible to say what their number may be, as they shift about at different seasons, and sometimes quit their valleys and live on the summit of the hills. At a place called Limah we found them residing in natural excavations on the side of a steep hill, the front part only being partially built up with loose stones. It had a most singular appearance. The caverns were in ranges one above the other; the children were usually seen tied with cords, to prevent them tumbling down the precipice.

They are too ignorant to be even inquisitive; and when some of them were induced to come on board, idiotic surprise for a moment, and indifference immediately afterwards, formed the principal characteristic of these poor people. Watches, pictures, and looking-glasses were shown to them, which they had evidently never seen before; but the chain-cable and the pigs were the only objects that fixed their attention. Their interest in the cable arose from the following circumstance. We anchored in the first deep cove, about ten o'clock at night, in forty fathoms, and the chain in running out of the hawse certainly made a noise, which reverberated amongst the hills to such a degree that the inhabitants fled in terror with their wives and families, and could not be induced to return again, until the cause of their alarm was explained to them.

The natives are very indolent and slovenly, and never work more than is necessary for their maintenance; fishing and making nets are their only occupations. The women do the house work, and milk the goats, which ramble about the hills. We conceived at first that their goats were wild, indeed they were so to us, and afforded some good sport; however, it was explained by the natives that they were individual property, and we paid liberally for our mistake. We found the people exceedingly civil and good-natured, and they seldom allowed us to leave a village without inviting us to feed on dates and milk.

The men possess the faculty of pitching the voice to a remarkable, shrill note, which can be heard over the hills and valleys to a distance which would be considered incredible.

XIII.—*Notes upon the Comparative Geography of the Cilician and Syrian Gates.* By WILLIAM AINSWORTH, Esq. Read April 9, 1838.

THE Gulf of Iskenderún is nearly surrounded by mountains. To the south it is bounded by Rhosus,* which attains an average elevation of 5000 feet, and terminates rather abruptly in the Jebel Kheserík and Rás el-Khanzír,† on the sea, scarcely leaving room for the passage of an army; but, if this difficulty could have been overcome, another presents itself in the Jebel Músa, a more southerly chain which advances precipitously into the sea, its south western base bearing the ruins of Seleucia in Pieria, by the galleries of which it is cut through.

At the northern foot of Rhosus, an aqueduct of considerable length and many ruins attest the site of an ancient city, called Arsús‡ by the natives: it is probably the Rhosus of Strabo.

The chain of Rhosus extends east a little north to the parallel of Iskenderún, where it is separated by a pass from the range of Amanus, which runs from south-west to north-east.

This pass is called the Beïlán Pass. It is the only pass commonly practicable from Cilicia into Syria; and near its foot are the ruins of Pagræ,§ which, according to Strabo (lib. xvi. p. 751), overlooked the plain of Antioch, now called El 'Umk; and on the verge of which is Khán Karamúrt or Palamút Khán, a large khan occupied by Ibráhím Páshá's troops, when he carried this pass in his advance into Asia Minor. There is also in the same neighbourhood the ruined castle of a chieftain who levied tribute in the mountains. The castle is called Bedri Káfir, where, before an arch-enclosed tomb, said to contain the body of the bandit, were offerings of miniature bows and arrows. Near the summit level of the pass, but a little to the north-west of the crest, is the modern town of Beïlán,‖ much esteemed for its fine air and water. Its khán was built by Sultán Suleïmán,

---

* Rhosus. See Strabo, xvi., Casaub. 751; sometimes Rhossus, but the best MSS. and coins have only one *s*.

† Cape Hog.

‡ Or Ka'bah (the Cube).

§ Bakrás Kal'ah-sí (the Castle of Bakrás in M. Rousseau's map). "Baghrás" (Abúl-fedá, Tabull. Syriæ, p. 119), "with a lofty castle, springs, and a river; twelve miles from Antioch and as many from Iskenderún, overlooking the valley (úmk) of Herem, which lies to the east of it."

‖ The town of Beïlán was found, by barometrical measurement, to be 1584 feet above the Mediterranean. The Christian church (zone of quercus ægilops and Valonía oak) 2696 feet; temperature of an abundant spring in limestone 56° Fahr. (12.6 cent.); air 62; Kurlú (Wolves') Pass, 4068; summit of Beïlán Mount (limestone), 5337. Jebil Kheserík, according to Capt. Beaufort, is 5550 feet above the sea; the first cone west of it, by barometer, was found to be 5216 feet, and the second cone west 5091 feet high. The lower limit of the pine forests was 2975 feet; temperature of springs 51° Fahr. (10.4 cent.); springs from diallage rocks at 4353 feet 48° Fahr. (8° 8 cent.).

and its mosque or jámi' by Sulṭán Selím. Upwards of 500 feet above the town is a longitudinal valley, communicating with Báyás, by a road which passes by the side of the Mount, called Beïlán. In this valley are the ruins of a Christian church. Between the north-western foot of the Beïlán-pass and the sea is a rocky site with abundant springs and caves. This is supposed by Rennell to be the site of Myriandrus.

Beyond is the almost depopulated Alexandria, Alexandretta, Iskenderún, or Scanderoon; and between it and Myriandrus, the ruins of Godfrey of Bouillon's castle and some stone fortifications, which, in the 'Mecca Itinerary,' are said to have been built by Ibn-Abí Dáwud in the time of the Khaliph Wáthiḳ. Scymnus of Chios and Strabo attribute the foundation of Alexandria to the conqueror of Darius. A traveller of the thirteenth century, Willebrand of Oldenburg (Itin. Terr. Sanct. p. 135), says that, according to the tradition of the natives, the town was built in one day by Alexander for his horse Bucephalus, and that he gave it its name.

The bay at Iskenderún extends still further east than the town of the same name to the foot of the mountains, and the traveller who proceeds by land to Báyás must pursue a circular direction till he reaches a ruined marble gateway, where the mountain acclivity descends with a gentle slope, covered with brushwood, into the sea. Over this narrow pass the road is carried with care, and, although steep, is paved throughout. This ruined gateway, presenting at sea the appearance of two columns, has been called by sailors "Jonas' Pillars." *

The road, which was formerly carried through the marble gateway, has, in its southern continuation, been swept away by the sea; and the present one is carried higher up on the hill-side. Both are constructed in a coarse limestone conglomerate. This pass, in the 'Mecca Itinerary,' is called Ṣaḳál Ṭútán.†

* The ruins consist simply of two crumbling walls 13 feet 6 inches in length, and 3 feet in thickness, the length of the plinth being 17 feet. The width of the gate walls was 20 feet. The walls are built of blocks of white and black marble, from 2 to 3 feet in length, 1 foot 10 inches in depth, and 18 inches in width. This ruin has been noticed by many travellers, as Pococke, Niebuhr, Kinneir, Drummond, and, indeed, it must be observed, by all who travel along thi sroad.

Also, in a slight outline of M. Callier's travels (p. 225, Vol. V., of the *Bulletin de la Soc. de Géog.*), it is said, "to the north of the defiles, the acclivities of Amanus recede from the shore, the plain develops itself, and is soon a league in width: here M. Callier thought he recognised the field of Issus." This is very general, and may apply to the plain of Báyás, as well as to that of the Pinarus. It is much to be regretted that so long a space of time, now nearly four years, has elapsed since M. Callier's journey in the East, before any account of it has been published.

In No. 24 of the Views in Syria and Palestine, published by Fisher and Co., there is a well-executed sketch of part of this coast; it is taken from immediately south of the Sakal Tútán, and embraces the approach of Amanus to the shore, the plain of Báyás, and in the distance the rise of the country towards the Pinarus.

† "Beard-catcher." (Turk.)

Immediately beyond the marble gates the plain begins to widen, and there is a modern Turkish castle called Merkez,* upon the top of a hill about 300 feet high, which commands the pass,—but the fort is now dismantled.

Half a mile beyond, is a stone wall, which crosses the plain from the mountains to the sea, where it terminates in a tower; and a short distance beyond this wall, is a river about forty feet in width. Following the river up to the point where it issues from the mountains, a distance of nearly two miles, more ruins are met with, and also traces of a double wall, between which the river flowed. Above is a steep and remarkable pass (Baghrás Belí).† There is also at this point a pretty village, which is called Merkez, and a little beyond the village, abundant springs issue and give birth to another river, which loses itself in the sea about three miles to the north of the former. The inhabitants called the rivulet which comes from the mountains Merkez-ṣúï. It is the ancient Kersus. Between it and the more northerly river there is a small pile of ruins, which appear to be a mass of Roman brickwork.

Between the Kersus and Báyás, or Páyás, two headlands advance into the sea, known as Rás Báyás and Eskí Rás Báyás. Beyond the most northerly is a little gulf with a few feet of water only, in which are the remains of a pier and tower: close by, is a small village, and, in the plain and commanding the harbour, a modern castellated building.

The public buildings belonging to the town of Báyás lie due south of the castle, half a mile across the plain. They are abandoned, but in perfect repair, and consist of a mosque or mesjid, castle, khán, bázár, and baths of superior construction and considerable beauty and solidity. The houses are all destroyed. A mile to the north-east there is a village with a governor and a body of soldiers placed in a secluded position near the entrances of a transverse valley, which has an opening into the chain of Amanus, and affords a summer pass over the mountains.

The public buildings of Báyás were erected by Sakulí-Mohammed Páshá, known by the name of Ibráhím Khán-Zádeh, one of the viziers of Sultán Suleïmán. There can be no doubt that Báyás is the Baiæ of the Romans, a favourite bathing-place. Pococke supposed it to be Issus and its river the Pinarus, and Mr. Williams identifies it with Myriandrus. The shore is high and dry, the vegetation truly beautiful, and the scenery magnificent.

There are two villages between Báyás and the Issus,—the

* "Centre." (Arab.)

† "Baghrás, its fork." (Turk.)

most southerly Yúzler,* the northern Kői Cháï or "village-stream."†

The Pinarus or Issus flows between the latter and the village of Urzin or Urtsulí: it is called the Delí Cháï or Mad River, and, at the time we visited it (January, 1836), was about forty-five feet wide, on a stony bed: it comes from the Amanus, flowing across the plain in a direction a little south of west.

The plain to the south of Delí Cháï is about seven miles wide; and around the villages and near the sea it is much cultivated and diversified by orange and lemon groves. It is crossed by the stony beds of several mountain torrents, and a considerable brook runs through Kői Cháï.

North of Delí Cháï the country begins to rise. The plain is uncultivated, but everywhere covered with green sward, and generally with shrubs: toward the higher part it is intersected by ravines. The valley of the river penetrates to some depth into the hills, and to this point the Amanus preserves its direction of S.W. and N.E., gradually receding more and more from the shore; but here the base of the mountains advances farther to the westward into the plain.

At this point the gulf bends round to the west, leaving at its north-eastern point a low plain which is bordered to the north by nearly insulated hills of no great height, and which themselves are separated from the Amanus by a valley which is prolonged to the north-east to some distance, until closed up by hills, which stretch along the foot of the mountain-chain.

On the western side of this valley, at the foot of a hill, and about seven miles from the sea, are the ruins of a considerable town, in which many public buildings may still be traced, and where an acropolis and aqueduct still exist in some perfection. This is probably the town of Nicopolis, which, according to Stephanus Byzantinus, was first called Issus by the Macedonians, in honour of the victory gained there; but two much better authorities, Strabo and Ptolemy, speak of Issus and Nicopolis as two distinct places.

To the west the plain begins to narrow: near to the sea, south of Issus, is a *tell* or mound, called Ḳará-Ḳayá,‡ composed of black lava-pebbles, and having ruins of lava-walls upon the summit. About four miles to the west of this *tell* a river traverses the plain. It is called the Búrnáz Ṣú, and is crossed by a bridge. The plain is here covered by a sand-flood, and is not above two miles in

* Probably Gőzler. "Eyes or springs."

† Near Yúzler are some extensive indications of ruins which might have been supposed to belong to Issus, if Arrian had not described Darius as advancing, after marching upon Issus, next day to Pinarus. If Issus and Nicopolis were two distinct places, one of them remains to be discovered.

‡ "Black-rock." (Turk.)

width, being bounded to the north by a range of low sandstone hills. Ruined arches are seen peeping in two or three places out of the sands. They are constructed of tile-bricks.

The hilly country is soon united with the sea by loftier mounds of Plutonic rocks, and the direction of the shore changes to the south-west.

In this plain, thus enclosed between hills and the sea, are many ruins of former times: a little brook runs through its centre, and passes by the foot of a round *tell*, in part artificial, having the remains of fortifications on its top; remnants of forts and arches occur in the plain around. These ruins, belonging probably to the Castabalum of the Romans, are designated in the 'Mecca Itinerary' as 'Uzeïr (Esdras), otherwise called Matakh.

To the north, a pass through the sandstone range is guarded by a gateway and tower of tile-brick; ruins of a peculiar character, consisting of two masses of an imperfect obelisk-like form, having four sides, with re-entering angles at each corner, and slightly tapering towards the summit. A building in the same style occupied the other side of the pass; and the two appear to have been connected by a limestone wall. The pass here was upwards of 500 feet in width. To the north-west, the great Constantinopolitan road follows the direction of the rivulet through a valley where the sandstone chain bends round to meet the higher hills of Plutonic formation. Half-way up this pass, about 300 feet above the level of the sea, and where the pass is scarcely 500 feet in width, is an arch of elaborate workmanship; polygonal stones fitting with great nicety, arranged in courses and of the same height, and rather noble dimensions, built of limestone and flanked by walls of angular masses of lava, closely fitted, and of the third era of Cyclopian architecture, according to the divisions made by Mr. Hamilton,* and as in Argolis and other Phocian and Bœotian cities. The remains of a causeway are also still in existence. These gates are called Ḳará Ḳapú (or black gate), Tímúr Ḳapú, or Demir Ḳapú (or the iron gate), in the 'Mecca Itinerary.' The country was examined on two different occasions from these gates to the south-west to 'Ayás, and to the north-west by Ḳúrd Ḳúláḳ† (Tardequeia) to Miṣṣíṣah (Mopsuestia), where other ruins occur to throw light upon the character of those observed surrounding the gulf of Issus from point Rás el Khanzír to the mouth of the Pyramus.

In discussing the questions of historical geography connected with the country we have just described, one of the most immediate causes of error has been a passage of Strabo (xiv. Casaub. p. 676), in which he says, "after Mallus (Mopsuestia),

* Archæologia, vol. xvi.

† "Wolf's ear." (Turk.)

comes Ægæ ('Ayás), a small town with a road-stead; then the Amanian gates with an anchoring station." The Amanian gates may either apply to the Beïlán-pass, to the gates of Kersus, or to the marble gateway of Ṣakál Ṭútán, both near Iskenderún; but there can be no doubt, from another passage (lib. xvi. p. 751), when he says "Pagræ is situated on the road which, traversing Amanus, leads from the Amanian gates into Syria," that one of the last two is meant. "After *Ægæ*, Issus," continues Strabo, (xiv. p. 676) "a small town with an anchoring station, and the Pinarus,* where the battle was fought between Alexander and Darius, and the gulf called Issic. On it (the gulf) are the cities Rhosus and Myriandrus, and Alexandria, and Nicopolis, and Mopsuestia, and the place called 'the Gates,' the boundary between the Syrians and the Cilicians." The gates referred to in this passage appear to be the Ḳára Ḳápú, or black gate.

Ptolemy's description of the sea-coast is,—" In Cilicia, the Cydnus (Társús River), the Sarus (Saiḥán), the Pyramus (Jaïḥán), Mallus (Miṣíṣ), the village Serrepolis (unknown), Ægæ (A'yás), and Issus. Inland the Armenian gates" (Ḳulleh Búgház pass in Taurus from Adanah to Kóniyeh). "In Syria after Issus and the Cilician gates (Ṣakál Ṭútán), Alexandria near Issus (Iskenderún), Myriandrus, and Rhosus. Inland the Syrian gates (Pass of Beïlán). Few difficulties present themselves, where there is an accurate knowledge of the position of places, in assigning the localities of what, in historical geography, has often been confusedly described and variously named.

It is well known that Cyrus, in the expedition of which so admirable an account has been transmitted to us by Xenophon, led his army by these passes. According to the narrative of that general and historian,† Cyrus marched from the Pyramus (Jaïḥán) in two days' march, fifteen parasangs, and arrived at Issus, the last town of Cilicia, situated near the sea, a large city, rich and well situated, where he staid three days. "Hence Cyrus made, in one march, five parasangs to the gates of Cilicia and Syria. There were two walls of which the inner, next to Cilicia, was occupied by Syennesis with a guard of Cilicians, and the outer, next to Syria, was said to be defended by the king's troops. Between these two walls runs a river called Kersus, 100 feet in breadth. The interval between them was three stadia, or 625½ yards, through which it was not possible to force a way, —the pass being narrow, the walls reaching down to the sea,

* Pindus in the MS. of Strabo and the text of Casaubon. This variation from the form in Arrian has been pointed out to the author by Mr. Long, to whom he is anxious to express his obligations for several valuable hints and corrections. Pinarus, in Fischercke's edition of Strabo, is a correction on the authority of Plutarch (Vita Alexandri, i. 20) as well as Arrian.

† Xenophon, Anab., I. iv. 4.

and inaccessible rocks above. In both these walls stood the gates. Hence Cyrus proceeded through Syria, five parasangs in one march, to Myriandrus, a city near the sea, inhabited by Phœnicians, at which, being a trading town, where many ships lay at anchor, they continued seven days."

The river Kersus corresponds with the Merkez of the present day; it traverses ruined walls at its entrance into the plain, but now falls into the sea at a short distance from where the walls terminate. Rennell supposed a wide space of alluvia to have been deposited at or near this pass, but such is not the case, for the walls terminate at the present day in the sea. As the two walls were distinguished as outer and inner, and in 'Cilicia' and 'Syria,' the description has reference to ruins observed at about half a mile to the north, and the walls and ruins which stretch from the sea up to the rocks south of the Kersus.

It may be observed, in connexion with this, that, according to his historian,* "In order to gain this pass, Cyrus sent for his ships, that, by landing his heavy armed men both within and without the gates, they might force their passage through the Syrian gates if defended by the enemy."

The next most important texts are those of the historians of Alexander, who also invaded the East by the same road, and there met and conquered the Persian king.

Q. Curtius (iii. 7), who is supposed to have derived his history from the memoirs of Aristobulus and Ptolemy, relates that "Alexander, having moved, and thrown a bridge across the Pyramus, arrived at the city of Mallus:" in two days more he reached Castabalum. From this it would appear that Mallus, afterwards Mopsuestia, was the present Missiṣah. Castabalum appears to have been at or beyond the Ḳárá Ḳapú. There he met Parmenio, who had been sent forward to examine the road through the defile (Ḳárá Ḳapú) which lay between them and Issus. This general, after having made himself master of the passes, left a sufficient guard there, and then captured Issus, whence the barbarians had fled. He then advanced from Issus, dislodged the enemy who occupied the interior heights (Amanus?), placed there strong bodies of troops, and having hurried back, announced his own success to the king. From Castabalum, Alexander advanced to Issus. According to Arrian (ii. 6), "before Alexander had quitted Mallus, he was informed that Darius, with all his forces, was encamped at Sochi. This place is in the Assyrian territory, and distant about two days' march from the Assyrian gates." The gates here alluded to are the pass in Amanus, which is to the north-east of Issus. The road is still used in going from Báyás to Mar'ash, but is not practicable in the winter season, and in the

* Anab., I. iv. 5.

summer could only be so to the light (Parthian) horse, which did not alone constitute the army of Darius.

"Next day," says Arrian, "Alexander advanced (from Mallus) to meet Darius and his Persians; and after surmounting the pass, encamped on the second day at Myriandrus." The omission which occurs here of the march to Issus, renders it doubtful whether the pass alluded to means that between Mallus and Issus (Ḳará Ḳapú), or that between Issus and Myriandrus (Ṣaḳál-ṭútán), but most probably the latter.

"By chance," to continue the narrative of Curtius, "on the very same night Alexander arrived at the pass by which Cilicia is entered"—(This alludes evidently to where Cilicia is entered from Syria (Saḳal-ṭútan), which is corroborated by the continuation of the narrative by the same historian)—"and Darius at the spot called the Amanian gates. Nor did the Persians doubt that the Macedonians had fled, as Issus, captured by them, had been left unguarded." No great importance can be attached to the distance of an army's out-post given to Alexander, under circumstances of considerable anxiety, if not alarm; but still the distance, if we suppose the Olympic stadium to have been that used by the Macedonians, as I have uniformly found to be the case in Babylonia, will amount to about ten miles, which is not very far from what might *à priori* be supposed to be the position of the out-posts of an army occupying the southern bank of the Issus. "Alexander could scarcely believe them, and sent scouts, who ascertained the truth; he then ordered his men to prepare for battle, and marched back at twelve o'clock at night. At break of day they arrived at the narrow pass which they had determined to occupy."

Arrian (ii. 7) gives the following account of Alexander's march at this interval:—"Darius crossed the mountain by the pass called the Amanian Gates, marched upon Issus, and thus placed himself in the rear of Alexander, who was ignorant of his movements.—Next day he advanced to the Pinarus. When Alexander heard that Darius was in the rear, as he did not think the account credible, he embarked some of the accompanying troops on board a thirty-oared galley, with orders to examine into the truth of the report. These sailed up in the galley, and as the sea here forms a curve or bay, they more easily discovered the Persians encamped, and made their report that Darius was at hand.

"Alexander ordered his troops to refresh themselves, sent a few of the cavalry and archers in the direction of 'the gates,' in order to reconnoitre the road, and placing himself, as soon as it was night, at the head of his army, set out in order to occupy 'the gates' a second time. About midnight he again made himself master of the pass, and after carefully stationing sentinels upon the rocks, allowed his army to repose for the remainder of the night."

Throughout this detail Arrian keeps in excellent consistency with his first statement, that Alexander went to Myriandrus, and that "the gates" which he occupied a second time were those which occurred between Myriandrus and Issus. The Ṣákál-ṭútán was the midnight halt. Had Myriandrus been the present Báyás, as advanced by Mr. Williams,* there would have been no necessity to send a reconnoitring galley, since the army on the ridge above Pinarus would have been easily recognisable. There are no gates between Báyás and Issus.

"With the dawn he descended from the gates along the road, and as long as the pass was narrow he led his army in columns; but as the defile expanded" (which it does considerably about four miles beyond) "he regularly formed his column into line, by bringing up his heavy-armed troops, successively to occupy the vacant space between the main column and the mountain on the right, and the sea on the left."

There is also a subsequent description of the arrangements which Alexander made to prevent his right wing from being flanked by the left wing of Darius's army, which corresponds with the condition of the country, the mountains descending so rapidly to the south-west from the valley of the Pinarus, that the army which occupied that valley would, in the deploying of the line, be enabled to command the extreme wings of the enemy on the south.

Mr. Williams has objected to these distances as too great; although they are the same as those Alexander marched when not expecting Darius from Tarsus to Mopsuestia, on account of the mountain to be crossed between the Pinarus and Issus, and the gates to be surmounted between Issus and Myriandrus. There is some hilly country at Ḳúrd Ḳúlak to be crossed, and a low mountain-chain, the pass of which is not however 500 feet in height, and the slope is gentle to the Jebel El-núr, 'Mountain of Light,' between Missisah and Castabalum; but the Ḳará Ḳapú is not 200 feet, and the Ṣakál scarcely 100 feet above the Mediterranean. The points in historical geography of the greatest importance which Mr. Williams advocates are, that Kará Kapú was the midnight halt of Alexander. Now Issus is to the east of Ḳará Kapú, and would in that case be before Alexander. How could Darius by crossing Amanus, then, have placed himself in the rear of the Macedonians? How could he, by descending to Issus, have been in the rear, for by marching to the Pinarus, he would have placed himself still farther in advance of his enemy, instead of approaching them from the rear, as Arrian relates?

According to Xenophon, the Greeks marched in one day from

* On the Geography of Ancient Asia. London, 1829.

the pass (Kersus) to Myriandrus; and according to Arrian, Alexander and his army quitted Myriandrus as soon as it was night, and arrived at the pass (Ṣáḳál-ṭútán, which was not quite so far) at midnight. If, therefore, says Mr. Williams, the pass was at Kará Kapú, Myriandrus must be represented by the modern Báyás. If this were the case, the armies of Cyrus and of Alexander must have marched a distance of thirty miles, that of Alexander between sunset and midnight; and where is Issus, which, according to Mr. Williams himself, is to the south-east of Ḳará Ḳapú?

The opening which I have described as occurring in the Amanus above, or to the east of Báyás, has been noticed by Captain Corry, and also by most modern map-makers* as the pass by which Darius came down to Issus; but if this were the case, it certainly is not that by which he effected his retreat after the battle, and which yet is stated to be the same as that by which he approached from Sochi to Issus; for he would have had to force his way through Alexander's victorious army, which occupied the plain on the Pinarus, between Báyás and Issus. Pococke calls it the middle of the three passes into Cilicia. But the rivers which flow into the lake at Antioch are no more avoided by passing from Báyás to Aleppo, than they are by passing from Beïlán; and the silence of Xenophon upon this subject remains in the same mystery.

It has been supposed that Abrocomas, in his retreat before Cyrus, retired along the sea-coast by Rhosus and the site of Seleucia; and Mr. Williams has also asserted, that it was by the same road that the Macedonian Amyntas, and his Greek mercenaries, reached the Phœnician Tripolis after the battle of Issus; but it is to be objected to this line of retreat, that the Jébél Músa terminates in abrupt and nearly perpendicular cliffs over the sea, not far to the north of Seleucia; and though there is a horse-road over Rhosus, it is tedious and difficult.

Cicero, in his letter to Cato,† says, "There are two passes from Syria into Cilicia, both of which, from their narrowness, can easily be defended by a few troops; nor can anything be better protected than Cilicia on the Syrian side." These passes are evidently the upper Amanian pass (the pass of Darius) and the lower Amanian pass (the pass of Beïlán). Cicero, also, in the same Epistle, when he gives the reasons, why he led the army which he commanded, as Proconsul, into Cappadocia, rather than into Cilicia, says—"Duo enim sunt aditus in Ciliciam ex Syriâ."

The Antonine Itinerary‡ furnishes us with a line of road between Nicopolis and Zeugma, by the upper Amanian pass.

---

* See Modern Traveller. Syria and Asia Minor. † Cic. Ep. ad Fam. xv. 4.
‡ Cilicia, Ptolemæi Geographia, Ed. Bertii, p. 15.

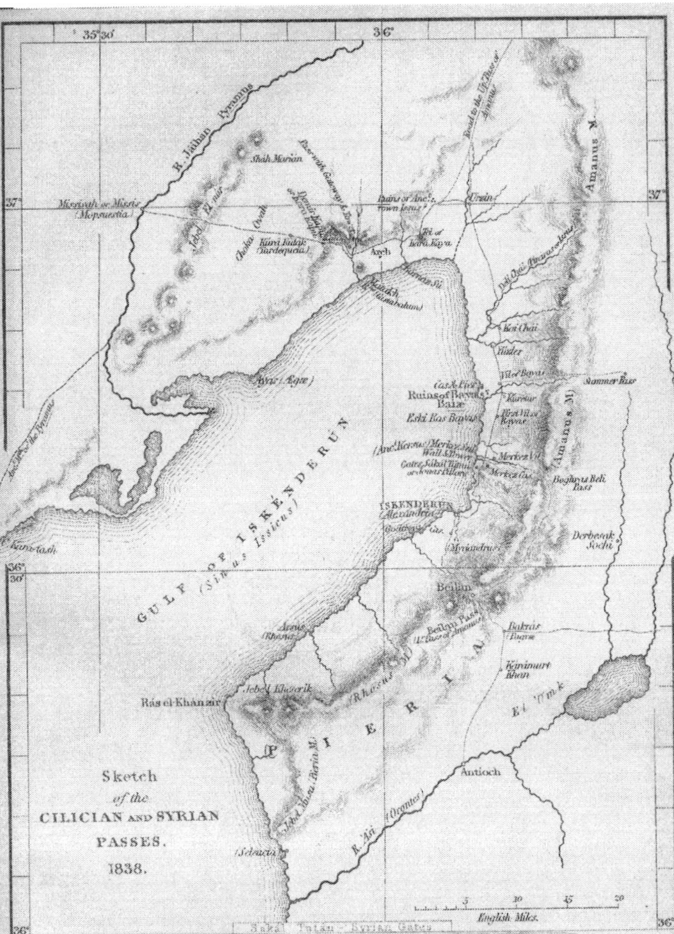

Saka Tutan – Syrian Gates

Publᵈ. for the Journal of the Royal Geographical Society, by John Murray, Albemarle Sᵗ. London, 1838. J. Arrowsmith.

Nicopolis . . .			
Aliaria	13	. . .	Unknown.
Gerbedissus . .	15	. . .	Unknown.
Doliche . . .	20	. . .	Doliche, north of 'Aïntáb.
Zeugma . . .	24	. . .	Rúm-ḳál'ah.

Major Rennell, it may be remarked, in his 'Illustrations of the History of the Expedition of Cyrus,' p. 38, *et seq.*, distinguishes, with his usual critical sagacity, four passes. The first is the Kuli Bóghâz, which he calls the Cilician pass; the second is that formed by the near approach of Mount Amanus to the Sinus Issicus, and which he calls the Maritime pass, but very properly observes that the descriptions of Xenophon and Arrian refer to distinct sites, although very near to each other. He considers Strabo as having been ignorant of the existence of this pass. The third pass is that of Beïlán, which he calls the lower or Southern pass of Amanus, and the fourth is the Upper or Northern pass of Amanus. The only omission here is the pass of Ḳará Ḳapú and the hills beyond it, so essential to the true understanding of the movements of Alexander and his general Parmenio. In the Appendix to the same work these gates are, however, alluded to, without feeling their importance; but he judiciously advances that the causeway met near them is the pavement of a street of Castabalum. Indeed the Ḳará Ḳapú appears to have been one of the gates of that city; the piers of a gateway in the valley to the east, to have marked another; and the arch in the sand hills to have formed a third or eastern gateway.

Colonel Leake (*Journal of a Tour, &c.*, p. 209) being acquainted with the Ḳará Ḳapú, and viewing Strabo and Ptolemy as naming the same two Pylæ or passes, identifies these with the Ḳará Ḳapú, which he calls the Northern or Amanic pass, and with the Maritime pass, at or near which Pococke noticed the Pillars of Jonas.

It will be seen, then, that many questions of high interest in comparative geography here present themselves within a very small extent of territory, and in which one celebrated traveller states that he saw neither defiles nor passes. There are, nevertheless, in that circumscribed spot, gates, walls, rivers, and ruins, which have almost every one some association of ancient times connected with them; and to unravel the importance to be attached to each of these, has been the humble endeavour of the author of this essay.

It remains only to acknowledge the obligations due to Colonel Chesney, for the use of the geographical points obtained by Lieutenant Murphy, R.E., in the neighbourhood of Iskenderún, and which have been used in the construction of the accompanying sketch map.

XIV.—*On the Bore, or Rushing Tide, in the Gulf of Cambay, and at the entrances of the Máhí and Sábarmatí Rivers.* By Lieut. R. Ethersey, Indian Navy. Communicated by the Bombay Geographical Society.

The Gulf of Cambay lies between 21° 5′ and 22° 17′ north latitude, and 72° 19′ and 72° 51′ east longitude; it is seventy-two miles long, and varies considerably in breadth. At the entrance between Vaux's Tomb and Gópnát'h Point it measures thirty-two miles across, which in a distance of ten leagues narrows, between B'aróch Bar and the island of Perím, to eight miles; it then opens out again to nineteen miles, between the entrance of the Dhádar River and Bhaunagar on the Káthwár coast; this space, with the exception of three channels, is occupied by extensive shoals. The Gulf contracts again to ten miles between Góngwá and the western coast, on the same parallel with a remarkable spit of land which has formed within the last twenty years. Sand-banks extend from the shores on either side, which leave a channel between them towards the centre of the Gulf from three and a half to four and a half miles wide.

From the head of the Malacca banks* in lat. 21° 10′ N., to the parallel of Lóhárá, a point on the northern side of the entrance of the Narbadá River, in 21° 38′ N., the Gulf is, with the exception of the Bhagwá Sands,† clear of shoals, with irregular soundings from eight to thirty fathoms; but above this parallel it is filled with extensive shoals and sand-banks, having several deep channels between them, all of which are liable to shift, particularly during the rains. These channels all take a northerly direction, and the two principal ones unite in 22° 7′ N. and 72° 36′ E.; a little below *G* on the plan, forming a channel three and a half miles wide, the greatest depth of water in this channel being six fathoms; it takes a north-easterly direction, runs close past Cambay Creek, from which it crosses to the south-eastward, and enters the Máhí River.

Another channel extends in a winding direction from the Sábarmatí River close past the Amlí Creek, from whence it keeps pretty close to the shore, and joins the main channel in about 22° 9½′ N. below the Bore rocks; in both of these channels the flood-tide makes with a bore or wave, caused, wherever it is observed in this gulf, by a rapidly flowing tide forced through a narrow obstructed passage.

* Lieutenant Ethersey, in addition to his survey of the Gulf of Cambay from Diu Island, along the Kát'híwár Coast, around the head of the Gulf, and down its eastern shore as far as Surat, has lately examined, in detail, these extensive shoals, and thereby rendered an important service to hydrography.—Ed.

† So called from Bhagwá, a village on the north bank of a small stream called the Sená River, which takes its rise at Segwá and Sehwán.

The rivers which empty themselves into this gulf are the Narbadá, Dhádar, and Máhí,* from the eastward; the Sábarmatí from the north; and the Bhádar, or Gómá, from the westward: the three latter only can have any effect on the Bore, which, in the fine season, is trifling, as the discharge from them is then very inconsiderable; for the channel opposite Dehwán, on the north bank of the Máhí, is only 300 yards wide, a few inches deep, and the stream scarcely perceptible. Again, at the small pagoda marked *H* in the chart, in 22° 24′ N., on the east bank of the Sábarmatí River, the channel at low water is only 150 yards wide, with a mean depth of two feet, and the stream is too weak to be ascertained by the log.

The tides throughout the Gulf are extremely rapid, and their rise and fall very large; the whole coast is low, overflowed for some distance inland at high spring-tides, and intersected by numerous small creeks and inlets.

Its situation being such as to receive the full force of the tide-wave coming from the southward and from Díú Head, the south point of the peninsula of Gujerát, lying between the Gulfs of Cambay and Cutch, along the Kát'híwár coast, together with its peculiar shape, will, I think, sufficiently account for the strong tides which are experienced here; for at Perím, twenty-six miles from the entrance of the gulf, the stream is forced through a space four times less than it occupied between Vaux's Tomb and Góp-nát'h Point; and again, at *G*, below the Bore rocks, thirty-two miles further north, it flows into a channel only one-ninth of its original width, being not quite half of its breadth at Perím; from which circumstances the velocity of the tide is not only considerably increased as it flows towards the northern parts of the gulf, but the water is also forced up to a higher level.

The eastern or principal bore rises five miles to the west-south-westward of Cambay Creek, and is not perceptible in the neap-tides without the previous spring-tides have been very high, when it may be observed slightly through the quarter. It generally commences when the springs begin to lift; the wave increasing daily in height, as the tides gain strength; and it is at its greatest height about two days after new and full moon; it also varies with the night and day tide, because the higher the tide the greater is its velocity; and as the two tides differ from six to eight feet, and the flood of both runs the same length of time, the highest tide must have the greatest velocity, and hence the wave of the bore will be highest with the greatest tide; the night tide both of new and full moon is the highest.

By a reference to the chart it will be seen that the channel

* Vulgó, M'háï.

between the Bore rocks and *A* on the eastern sand-bank is quite clear, and free from shoals. It is four miles and a quarter wide, and the greatest depth six fathoms, where I found the velocity of the tide to be as follows, the direction being N.E. and S.W.:—

		K.	F.	Rise and Fall.
High Springs	Flood	6	0	33 feet
	Ebb	7	2	
Ordinary Springs	Flood	4	4	25 „
	Ebb	6	2	
Neaps	Flood	3	6	18 „
	Ebb	5	0	

From this point (*A*) the channel begins to narrow with a decreasing depth of water, until at *B*, seven miles higher up, it is only 550 yards wide, the greatest depth of water being seven feet. It is at this point that the tide first rises in a wave; when the flood makes, this may be seen running along both sides of the sand-banks, and it soon spreads across the channel, rapidly increasing in height; for by the time it passes *C*, a distance of not quite one mile and a half, it is nearly as high and has as great a velocity as it attains in any part of its course. From *C* it runs close to the high cliffs as far as *D*, spreading entirely across the channel, and rushing along with a loud roar. The small sand-bank which commences at *D* turns it to the southward of east; and when abreast of Cambay Creek it is divided into two parts by another sand-bank, the most considerable of the two, taking a south-easterly direction towards the south bank of the Máhí River, and continues on this side as far as *E*, where it crosses to the north bank, near Dahwárá, close to which it pursues its course to the village of Dehwán, where it is entirely broken and interrupted by a number of sand-banks, but proceeds several miles further up, although with greatly diminished height and velocity.

The other part of the divided stream runs to the eastward, and is soon exhausted, there being no free channel for it, and the banks are greatly elevated with gradual slopes, over which it flows very slowly.

The following data show the result of my observations on the ordinary spring-tide on December 24, 1836, between the point *C* and Cambay Creek; the highest part of the wave being three feet and a half, its velocity nine knots, which was the utmost strength of the tide after the bore had passed:—

H.	M.		K.	F.	H.	M.		K.	F.
0	5	after	3	6	0	30	after	6	0
0	10	„	5	7	0	40	„	7	0
0	20	„	4	4	0	50	„	6	2

Seven knots was the strongest.

The flood ran three hours: the rise and fall of the night tide was twenty-three feet; that of the day tide sixteen feet six inches; giving a difference of six feet six inches.

During the first hour the rise of tide was fifteen feet; in the second, six feet; and in the third, two feet.

In the first hour the water rose six feet during the first ten *minutes*, which will give some idea of the rapidity with which the tide rises.

January 7, 1837.—Very high spring-tide, nearly up to the mark of the tide in the rains; between the point *C* and Cambay Creek the highest part of the wave rose six feet, and its velocity was ten knots and a quarter, the strength being as follows:—

H.	M.		K.	F.	H.	M.		K.	F.
0	20	after	4	6	0	50	after	8	0
0	30	„	4	6	1	00	„	7	0
0	40	„	7	6	1	20	„	6	0

The flood tide ran three hours and two minutes. The rise and fall of the night tide was thirty feet eleven inches, the day tide twenty-three feet; the difference being seven feet eleven inches.

			FT.	IN.
Rise of tide during the first hour		. . .	18	4
Ditto	„	second hour . .	8	4
Ditto	„	third hour . .	3	6

The ebb tides run steadily, but do not acquire their greatest strength until more than half-tide has passed, when the high banks are uncovered, and the stream is confined to its proper channel.

Jan. 10, 1837.—Near the village of Dehwán, on the north bank of the Máhí, the highest part of the wave was seven feet, and its velocity ten knots. The rise and fall of the night tide was twenty-two feet; but by the mark of the tide on the shore, this was one foot lower than it had been a day or two before, which is too trifling to have diminished the effect of the Bore. I was not able to measure the rise of the tide here, having no convenient place to erect a scale; it ran only two hours fifty minutes; the greatest velocity of the flood was seven knots, and the ebb six knots two furlongs.

The Western Bore is so nearly similar to that just described, in almost every particular, except its direction and effect, that a very few additional remarks will be sufficient to point out the small difference between them; and even that would have been superfluous had I not found a very great alteration in the banks and channel subsequently to my former report.

I found that from 700 to 1300 yards of the western shore had been washed away through the whole space from *I* to the east point, which is 1300 yards farther south than its former position; the bank all along, being a sandy cliff, continues still to fall with such frequency, from the action of the tide undermining it, that, in a short time, the coast will be quite straight from the Amlí Creek to below that which is now the east point. Tons of this sandy bank are detached from the shore at one time, and these masses continue to fall into the water at different parts in such quick succession, that the report sounds at a distance like heavy artillery; the top of these cliffs, at two-thirds ebb, being from eighteen to twenty feet above the level of the water.

Again, the high sand-banks to the eastward of the Amlí Creek, which were only partially covered by the neap tides, have been swept away; nothing but low sand-banks remaining in their place, and the channel leading to the Sábarmatí River, which before took a N.N.E. direction from Amlí Point direct for that river, now turns to the N.N.W. at Amlí Point, and runs along the shore close past Amlí Creek, where it turns to the N.E., for the Sábarmatí River.

The Bore on this side is now greatly diminished, the cause of which will, I think, be evident, when it is considered that formerly, instead of the whole force of the tide setting, as at present, to the N.E. past the Bore rocks, it took a N.N.W. direction, below the above rocks, right into the Western Channel, which is only now fed by a portion of the tide from the main stream.

At present the wave of the Bore on this side, as on the other, is not perceptible as formerly at the neap tide, but is confined to the springs, at which time it takes its rise at *F*, where the channel is only 150 yards wide, with three feet of water in it; its course is close to the shore as far as *B*, where it turns, and has to encounter a stream running at the rate of two miles an hour. I found its height here increase to four feet. After passing this, the channel widens a little, having low sand-banks, which the water spreads over, the wave decreasing to one and a half and two feet, and continuing at this height past the Amlí Point, where it again runs close to the steep bank, and increases to three feet; it thus rushes along until it comes near to the Amlí Creek, where it has to encounter a stream of two knots and a half, and, in consequence of the channel's turning, its whole force is directed to that part of the high sand-bank marked *b*, where the wave was five feet; after passing this, it gradually diminishes to a few inches, for the banks are very low and the water so shallow that it can scarcely be called a channel, so that the tide is not confined, but flows over the banks in all directions. It forms again, however, at the entrance of the Sábarmatí, where the highest part was two feet: it continues its course (from this to one foot and a half) nearly to *H*, one mile past which it is lessened to a few inches.

Dec. 9, 1836.—Between *b* at Amlí Creek and Amlí Point, the highest part of the wave was five feet, its velocity eight knots and a half. The velocity of the flood tide after the Bore had passed was as follows:—

H.	M.		K.	F.	H.	M.		K.	F.
0	10	after	3	2	0	50	after	5	6
0	20	„	4	7	1	00	„	5	4
0	30	„	4	4	1	20	„	5	0
0	40	„	5	0					

Rise and fall of the night tide, twenty-eight feet; day tide, twenty-one feet two inches; the difference being six feet ten inches. The flood ran three hours and five minutes.

The velocity of the wave was ascertained by measuring a distance from two to three and a half miles in those parts where the waves ran close to the shore, and noting the time it took to go from one station to another by a good watch. The height of the wave off Cambay, Dehwán, and the Amlí Creek, was ascertained by a pole, having feet marked on it, and in other places by estimation. The velocity of the flood was measured by the common log-line and glass—the patent log being of no use, from the irregularity of the stream, as it would only have given a mean rate.

I have given the greatest heights of the wave during each spring tide; but this does not continue to be the same throughout its course, being affected by several circumstances, which cause it to vary at different distances; for instance, at those points marked in the chart *a a a*, in the Máhí, and b *b*, on the western side, where the bank is steep, forming a concave, and the stream is strong, while the direction of the Bore is towards the steep bank, the wave at these points will be greatly increased, in consequence of the force of the tide being directed to one point, as well as by the resistance of the stream. It is at these places that the wave frequently flows quite perpendicularly, having the appearance of a wall, when it curls and breaks with a thundering roar.

In those parts of its course, where the sand-banks are low on both sides, with a strong stream running in the centre (which is the case in several parts on both sides), the middle part of the wave will be retarded, and increased in height by the resistance of the ebb; while towards each side, which is out of its influence, the wave proceeds with undiminished velocity, the whole forming in a crescent shape across the channel.

By the foregoing observations it will be seen that the tide does not attain it full strength until forty or fifty minutes after it has made: it is until that time very irregular, coming with a sudden burst, with great velocity, and then decreasing again slightly. The tide never attains the same velocity as the wave of the Bore; why this is the case I am not able to determine.

In *February*, 1835, in order to try the effect of the Bore on a large-sized bander boat, and at the same time to ascertain the strength of the stream after the wave had passed, at spring tide I anchored the boat half a mile to the northward of what was then the last cape on the western side of the Gulf, and it proved a more dangerous experiment than I had anticipated; although I anchored in five fathoms, the boat grounded at low-water, and was left high and dry. A few hours afterwards, the noise of the Bore was heard, when every precaution was immediately taken for the safety of the boat. The night was still and calm, and its roar, as it approached, echoing among the neighbouring cliffs, was truly awful. It struck the boat, lifted her, and threw her violently round on her bilge; in which position she was forced before it, broadside on, for the space of five minutes, the grapnel being of no use, for it was carried faster than the boat. I fully expected she would go to pieces, so violently was she shaken. However, no accident happened; for, on getting to a hollow in the sand-bank, which was quickly filled, she righted, much to my satisfaction. About twenty minutes after this I hove the log, and found the stream running seven knots and a half, at which time the boat, I think, could not have been driving less than two knots and a half per hour, which would make the velocity of the stream ten knots. This I did not think too great; for, although calm, the water frequently came over the boat's bows.

Additional Observations on the Remarkable Tides in the Gulf of Cambay. By Captain Thomas Best Jervis, E. I. C. Engineers, F.R.S.

The situation of this gulf at the extreme north-western limit or angle of the Malabar coast, and consequent accumulation of the head wave at flood tide, propelled, as it is, through a comparatively contracted channel, are the immediate causes of the singular influx of the tide described in Lieutenant Ethersey's paper. The Orinoko, the Indus, and the Ganges, present similar phenomena, though the height or head with which the bore rushes into these rivers is considerably modified by their respective situations, and the directions of their principal channels. The gulf of Cambay is, in point of fact, the great or common outlet of the whole of the waters of Gujarát, which are discharged by seven principal streams; the Sábarmatí, the Máhí, the D'hádar, the Nurbhuddá,* and Taptí, and on the Káthiwár coast, the Bhádar or Góma river, and the Gailá, which are better known as the Dhólárá and Bháónuggur rivers, these

* The orthography of the names of places is *not* according to the system usually adopted in this Journal.

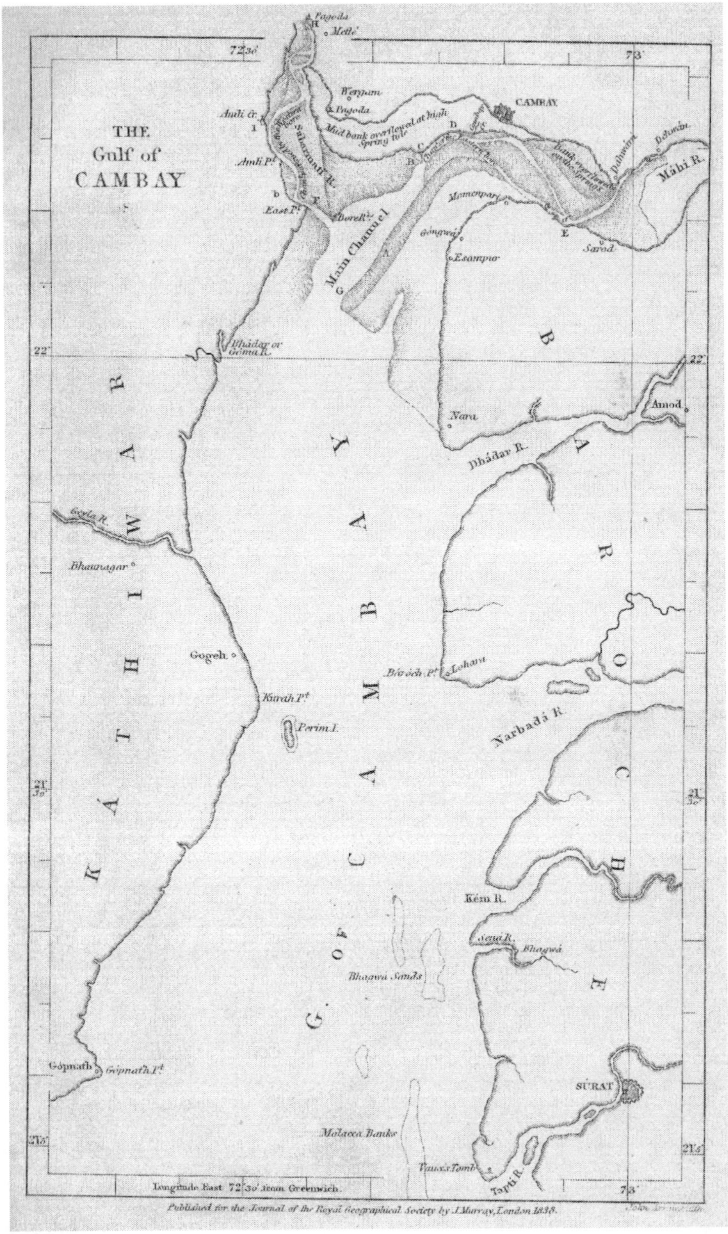
THE Gulf of CAMBAY
CAMBAY
Mahi R.
Main Channel
Bhádar or Goma R.
Bhaunagar
Gogeh
Kurah Pt.
Perim I.
Broach Pt.
Lohara
Narbadá R.
Dhádar R.
Amod
Nara
Esampur
Sarod
Kém R.
Bhagwa Sands
Malacca Banks
SURAT
Tapti R.
Vaux's Tomb
Gópnath
Gópnath Pt.
Sabarmati R.
East Pt.
Amli Pt.
K A T H I W A R
G. OF C A M B A Y
B A R O C H E
Longitude East 72° 30' from Greenwich
Published for the Journal of the Royal Geographical Society by J. Murray, London 1838.

being the great ports or towns on their banks where the cotton vessels usually take in cargo. The province of Gujarát, as defined in the geographical distribution of the Moghul empire, under the emperor Akbar, comprised the whole of Káthiwár; and the Hindús, at an earlier date, considered it variously, either as constituting part of Sauráshṭrá (Súrat), or (Gurjara) Gujarát. The province, at present, however, is understood to comprehend those districts only, on either side of the gulf, which are dependent on the civil jurisdiction of Ahme'dábád, Khéṛá, Baródá, Broach, and Súrat, the remainder being distinguished as Káthiwár, and under separate authority. In this restricted sense the entire territory of Gujarát is, generally speaking, a completely dead level; insomuch, that the mountain-torrents which accumulate in the provinces of Rájpútánah, Málwah, and Khándésh, immediately on its frontiers, have at different periods cut for themselves channels, of various breadth, depth, and direction, according to the force or quantity of the water which has fallen in particular seasons during the great periodical rains. Sometimes these have taken the direction of the main and ordinary channel, or, on occasions perhaps, of any unusual and continued fall of rain, some new course, according as artificial or local changes may have been favourable to such alteration. Such, in short, is the simple character of this natural drainage of a tract, composed, in most parts of a loose rich alluvial, in all others, of a deep, sandy, or mixed soil—resting on a remarkable calcareous formation, similar in all respects to that in the plains of Bengal and Bahár, and which is termed by the natives *Kankar*. The Bhádar and Gaïlá rise in the Girnár mountains; the Sábarmatí, with its tributary streams, the Mhéswá, Wátrak, Sírí, &c., in the A'rawallí hills; the Máhí river, on the eastern face of the Udeypúr range, which separates Hindústhán from Gujarát; the Dhádar, Nurbhuddá, and Taptí, disemboguing into the gulf north of the Malacca banks, effect the entire drainage of the territories stretching west of the Rájpiplá hills. Of all these channels, the Nurbhuddá and Taptí can alone be considered as rivers, in other words, channels supplied by springs throughout a considerable extent of country, and flowing uninterruptedly during the year with greater or less depth and rapidity. The Sábarmatí and Máhí, which have a considerable breadth and depth from bank to bank, with respect to the country they traverse, may rather be termed great watercourses or drains, as there are so few springs in their whole course, and, indeed, in the entire surface which their tributary channels intersect, that they are almost completely dried up at intervals within three or four months after the periodical rains have ceased.

The obvious consequence of such drainage is to detach an

enormous quantity of alluvium and sand year after year, which, being propelled farther down the channel, would be carried out directly to sea were there no predisposing cause for its deposition. The progressive formation of the Malacca and Bhágwá banks is incidental to the deposition of silt brought down by those streams, and the opposing influx of the tide, conjointly with strong south-westerly gales. By such slow and gradual process the opposite shores of the gulf, as well as the intermediate channels of these deposits have been gradually contracted; to which cause the singular influx of the tide is principally to be attributed. About twenty-five years ago Colonel Monier Williams made a survey of the mouth of the Bhádar river and the neighbouring coast of Cambay (Kambáyet), since which a long spit of mud which projects several miles into the gulf has gradually been formed, and the main channel of the Sábarmatí has also been considerably contracted and altered. There is little reason to doubt that this continual accession of mud will still further contract the channel; though artificial aid and much science would be essentially necessary to the recovery of any such land for the purposes of agriculture. A careful examination of the revenue surveys of the Broach and Súrat Collectorates by Colonel Williams, and the vast accession of the richest lands observable on either bank of the Nurbhuddá, below Unklésur, and on the sea-coast of Súrat within the last two centuries, the period of the latest Moghul assessment, would lead us to conclude that a like process had been going on, time immemorial, in that curious tract called the Run, containing many hundred thousand square miles. Although volcanic agency is supposed to have been the immediate cause of this sinking in of the land, or, with equal probability, the uplifting of the original bed of a great backwater or inlet of the sea, the progressive deposition of silt, brought down by the streams which empty themselves into its bed, must have materially assisted such a change, and favoured further volcanic agency by additional superincumbent matter.

In March and April, 1816, I crossed the gulf from Cambay to Jambúsír, and thence to Gógeh or Gógó, and back to Cambay, and had an opportunity of witnessing the rushing influx of the bore, and observing its direct progress up the channel for several days with great advantage. I had a small schooner belonging to the Collector of Khéra, Colonel Archibald Robertson, of about thirty tons burthen, which was manned by sailors from Gógeh, a class of people who are remarkably courageous and expert at sea. The coolness and dexterity with which they secured and righted the vessel on the rush of the first wave is still fresh in my memory, and I remarked that in casting anchor every day, as the tide went out, when the vessel grounded in the mud, that the Tindál, or master of the vessel, invariably took the precaution of selecting some spot in the direct line of its progress, that is, in the main channel, where, he said, there was less danger to be apprehended than in the neighbourhood of the shores, in consequence of the recoil or curl of the tide along shore, and the falling in of the loose banks. The bore appeared to set in like a straight wall of water with a head of five or six feet, each succeeding wave decreasing more and more, till the whole gulf was reduced to the same level with the sea without. We heard it approach several minutes before it came upon us, when we were fairly lifted up, and afloat in an instant. The shifting of the sand-banks, the gradual filling in of the rivers, and the formation of extensive bars at their entrances, render new surveys of the whole western coast continually necessary for the safety of the navigation to foreign vessels—but the most important service would be rendered to science by an exact registration of the tides, and observations of the gradual accession of land on the one, or detrition of land on the other coast of India; a service which the East India Company alone can originate and execute satisfactorily.

XV.—*Remarks on the supposed Situation of Minoa and Nisæa.* By Mr. T. Spratt, of H.M.S. Beacon. Communicated by Captain Beaufort, Royal Navy, F.R.S., Corr. Inst. France.

The fortress Minōa of the Megareans, which Thucydides* speaks of as an island, and Strabo † as a promontory or peninsula, appears to agree with a rocky hill surmounted by a ruined fortress, and standing on the margin of the sea south of Megăra, at the distance of little more than a geographic mile; thus agreeing with the distance given by Thucydides, who calls it about eight stadia.‡

That this hill was once a peninsula, appears evident from the dry beds of two rivers which pass close to its base; one on each side. The eastern bed winds round the back of the hill, leaving only a narrow neck of elevated ground between it and that on the west side: and it is therefore clear, that when these two rivers had communication with the sea, the intermediate neck of land, with this hill, would have been a peninsula, or promontory.

These two river-beds were once the only outlets of the moun-

---

* Thucyd. iii. 51.

† Beyond [*i. e.* east of] the rocks called Skirōnides, stretches out the point Minōa forming the post at Nisæa; but Nisæa is the harbour (or dock) of the Megarenses, eighteen [read "eight"] stadia distant from the city [of Megara], connected with it by "legs" [skelé, *i. e.* walls] on both sides. This port is also called Minōa. (Strabo, ix. pp. 391-2.)

‡ Thucyd. iv. 66.

tain streams which issue from the valleys on the north side of Mount Geraneia; for the ancient course of the eastern bed, although now ploughed over and cultivated, can be traced through the plain to the northward as far as its junction with that river, whose torrent at present flows in an easterly direction towards the shallow bay of Tikho, crossing the site of the long walls which connected Megara with Nisæa and Minoa, and losing themselves in the swamps bordering that bay. Although vestiges of the walls are not found in the bed of the river, yet, on examining the ground near it, the evidence is convincing that its present course does cross their site, as at a short distance from it, on the Megarean side, their foundations may be traced in a direction transverse to the course of the river, and towards the castellated hill before mentioned.

The dry water course on the western side of this isolated hill can be traced to within two or three hundred yards of the eastern one, and having no communication with any other mountain stream, it may not be unreasonable to suppose that formerly the river split there into two branches or mouths. This hill would then have been an island, as Thucydides calls Minoa; and if it was so in his time, it is not more than probable that the subsequent deposit of earth brought down by this stream, during the four succeeding centuries, which elapsed between his time and that of Strabo, should have joined it to the main by a neck of land or isthmus. To accomplish this, no very great deposit was requisite, as the historian says* "the bridge of communication between Minoa and Nisæa was over a swamp, or morass." But to identify this hill still more clearly with the site of Minoa, it is necessary to discover near it the ruins of the town of Nisæa, which, it appears from Thucydides, was situated at the harbour formed by the island Minoa.

That any considerable remains of the many important edifices mentioned by Thucydides should now be found, cannot be expected, where there is such extensive evidence of the changes undergone in the neighbourhood from the action of the rivulets when swollen by the mountain floods in the rainy season. Indeed, from the level form and small elevation above the sea of the whole plain, it is very probable that the greater part of it is an alluvial deposit. Yet that a town, besides the fortress on the hill, was here, is evident from the many vestiges that are seen on its

* The island was *very* near the continent, for the expression "to the island not far distant from the continent" (Thucyd. iii. 51, near the end) certainly appears from the context to signify *less than a mile.*

The word translated marsh (τέναγος) signifies, according to the Greek lexicographers, "a wet muddy place; mire with a little water on the surface, and plants appearing on the water."

eastern side, between the sea and a low rock which stands in the plain a short distance to the northward. Among these remains are four small heaps of ruins with massive foundations, in one of which there are three broken shafts of small columns erect, and wanting apparently only the fourth to complete the original number. Probably they were monuments, or *temples*;* and two Greek churches, which are now in ruins, but standing on two ancient foundations, will not be unfavourable to the supposition. Another church, Agios Nicolaos, which is perfect, also occupies the site of an ancient building, but it stands nearer to the sea.

The wall built by Nicias,† the Athenian general, in order to prevent ingress or egress to or from Minoa by the bridge, was hastily thrown up, and it is therefore natural to suppose that it was not a work in which durability was studied; and, consequently, its entire disappearance does not weaken the conjecture that the above-mentioned ruins were those of Nisæa. On the other hand, the bridge of communication may well be supposed to have been of greater strength, and therefore likely to have some part still remaining. Now, between the base of the hill on its north side, and the opposite bank of the dry bed of a former river, there are three platforms of heavy buildings, one of which lies immediately at the foot of the hill, another on the edge of the opposite bank, and the third nearly central; and as the course of that former river-bed clearly and indisputably passes between them, it is more than probable that the bridge of communication may be recognised in these ruins; the general appearance of which in nowise militates against this supposition.

The extent and number of the ancient remains in this plain are, however, too great to be considered as all belonging to Minoa; and the fortressed hill is equally inconsistent with the idea of the whole being Nisæa, for no mention is made of its having an acropolis, though such a conspicuous and important a strong-hold as this hill could not have remained unnoticed. A division, therefore, of these ruins, appears the only means of determining the site of either Minoa or Nisæa; and, in the first place, it may be assumed as almost conclusive evidence that the fortified hill was Minoa, and that the ruins in the plain are those of Nisæa, when we state that distinct remains of an ancient mole are to be seen extending from the south-eastern end of this hill, and curving to the eastward, so as to have formed a harbour between the hill and those ruins; for Strabo expressly says that the port of Nisæa was formed by the island or promontory of Minoa.‡

* Posidonium, Thucyd. iv. 118.

† Thucyd. iii.

‡ Strabo, ix. 391.

The fortress on the hill, which I shall now call Minoa, was originally built of courses of quadrangular blocks of limestone quarried from the hill on which it stands; but the greater part of the present ruins are of a more recent date, exhibiting no regularity in either the size or shape of the materials, and intermixed with travertine, and with a soft shelly sandstone often found in the Morea and in Attica. A high tower of the same style of building projects from the north wall, and being similar to those commonly erected by the Venetians, it may help to indicate that the repairs of the fortress were made at the time the country was in their possession.

Although there exists so much evidence in favour of the sites of Minoa and Nisæa being here, a few remarks on the adjacent coast may be satisfactory. At the distance of nearly half a mile to the eastward of these ruins, there is a small rocky peninsula, and further off two islands, the inner one of which affords shelter to a few of the small class of coasters. It has therefore been supposed, as this is a port of the Megareans of the present day, and formed by an island, that this island was Minoa.

That any encroachment either of the land or sea could have taken place, so as to alter the general features of the island or peninsula, is disproved by their rocky shores; for they both, as well as the ridge of hills from which the latter extends, are entirely of a limestone formation, which in its nature is extremely hard. No conjectured uplifting of the land, or fretting away of the coast, can therefore plausibly account for any difference in the locality of this place from the descriptions of Thucydides or Strabo. Now the inner island is distant from the promontory about two hundred yards, with seven fathoms water between them; they could not therefore have been connected by a bridge: nor was the peninsula ever an island, as its isthmus is of equal height with the extremity. Consequently, as both differ so materially from the descriptions of Minoa, in two points so necessary to determine its position, and as both are destitute of any remains of either a fortress or habitations, it must be sought for elsewhere.

The long range of hills that terminate in the promontory of Tikho,* and from the western end of which the little peninsula juts out, from its proximity to the ruins (supposing the whole of them to be Nisæa), might be assumed to be Minoa, because it also may have been at one time an island. But Pausanias,† who

* *Τεῖχος*? a fort.

† The passage in Pausanias (Attic. xliv.) is as follows:—

"And as they go down [from Megara] to the harbour, still called in our time Nisæa, they come to the Temple of Ceres, the sheep-bearer (Ceres ovifera, Demeter malophera)......Any one might perceive that the roof [of the temple] has fallen

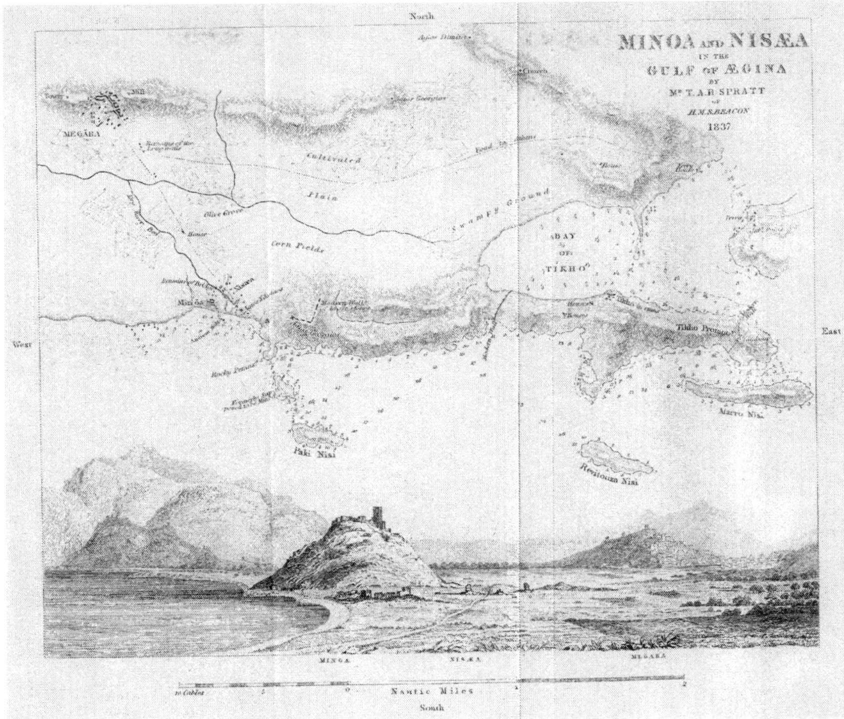
North
MINOA AND NISÆA
IN THE
GULF OF ÆGINA
BY
Mr T.A.B. SPRATT
OF
H.M.S. BEACON
1837
MEGĀRA
Cultivated
Plain
Swampy Ground
Olive Grove
Corn Fields
BAY
OF
TIKHO
Tikho Promy.
Macro Nisi
Paki Nisi
Revitoura Nisi
Rocky Point
West
East
MINOA
NISÆA
MEGARA
10 Cables
Nautic Miles
South

alone speaks of the relative magnitude of Minoa, calls it a *small* island, whereas this range of hills is three miles in length, and could never have been less as an island, for its lowest part, near the head of the bay of Tikho, is twenty-five feet above the level of the sea; the size of this ridge, therefore, as an island, is a strong objection to the supposition.

As the places described both before and after Minoa and Nisæa are found in succession in Strabo's description, it may perhaps be some guide towards fixing their situations to consider them in that order. First, Crommyon is described; then the Skironides rocks; next Minoa; afterwards Nisæa; the five islands follow; and Salamis after them. The promontory of Minoa in that case should be found between the Skironides rocks and Nisæa. This, then, will also be a good argument against the supposition of the long range of hills having been Minoa, for Nisæa would then have been a few stadia nearer the Skironides than to it; and had Nisæa been so situated, it appears likely that it would have been mentioned before Minoa. Besides, as no ruins remain on any part of the range to denote its ever possessing a fortress, the supposition has little to support it. There are, indeed, on each extremity, some portions of a modern wall of loose stones, perhaps for defensive works; and also a very perfect wall across the hill, from the bay of Tikho to the coast on the opposite side, but this was built by the Megareans during the late war. In short, there are no ruins of antiquity.

There is yet some difficulty in accounting for the five islands which Strabo says are passed in sailing from Nisæa towards Attica, before arriving at Salamis. The probability is that he included the islet which lies close off the south-west extremity of Salamis; but if that should not be admissible, and that we are thereby driven to suppose the above-mentioned long range of hills to have been an island when he wrote, then there is at once an end of the hypothesis that those hills were Minoa; for Minoa, *he says*, was a promontory.

---

in through lapse of time. There is also an *acropolis* there, itself in like manner called *Nisæa*......Near to Nisæa there is an island, not large, where Minos, in his war against Nisus, stationed the Cretan fleet."

The most accurate version of Thucydides is that by Hobbes; Dr. Bloomfield's, though his notes are useful to scholars, does not appear to be equal to Hobbes' in fidelity and closeness to the original.

XVI.—*On the Ground Ice or Frozen Soil of Siberia.* By Professor BAER. Communicated by Admiral KRUSENSTERN. Read February, 1838.

IT was remarked long ago that there is a large extent of country in Siberia where the soil never thaws. In summer, when the temperature is in some districts considerable, the ground becomes warm to a greater or less depth, and thawed, but lower down the ice is again met with. Gmelin, the elder, in his travels in Siberia, states that at Yakutsk, shortly after the foundation of that town, towards the end of the seventeenth century, the soil was found frozen at a depth of thirteen sashes, or ninety-one feet, so that the inhabitants were obliged to give up the idea of sinking a well. Many similar cases were brought forward by persons sent out by the Academy of Sciences of St. Petersburgh about the middle of last century, but these do not appear to have been generally believed. In 1825 Leopold von Buch, a philosopher of high authority upon all questions relating to the physical condition of the earth, rejected them as erroneous; they have, however, been corroborated in our time by Erman and Humboldt.

In his paper published in the transactions of the Academy of Berlin, Von Buch says, " I am fully convinced that the accounts of the soil being frozen in summer to the depth of many feet, in districts capable of maintaining the growth of shrubs and bushes, are not to be relied on, and that Gmelin's statement that the soil was frozen in a well at Yakutsk at the depth of 100 feet, ought no longer to be quoted in elementary works upon natural philosophy. The accounts given by the Cossacks, who in all probability did not like the labour of sinking the well at Yakutsk, and who were dead before Gmelin extracted the information from reports at Yakutsk, cannot be sufficient proof of a fact so extraordinary and so much at variance with physical science. In Hudson's Bay, where the average degree of temperature is much below the freezing point, there are springs which run under a crust of snow and ice all through the winter." (*Captain James,* 1631; *Memoirs of the Royal Academy of Berlin,* 1825.)

The objections of this philosopher are founded partly on his opinion that cold cannot penetrate so far under the surface, and partly on the story of the springs at Hudson's Bay flowing the whole year round; but if this statement is correct, may not these springs, flowing from under the ice, be thermal springs? Other persons have questioned the existence of perpetual ice in Siberia, as being in direct opposition to the modern experiments, which show that the heat of the earth increases directly as the distance from the surface. It was not known till very lately how deep the frozen bed extended: for the first information on this subject we are indebted to a merchant of Yakutsk, of the name of Schargin, who carried on an experiment for several years for this purpose. His object at the beginning was to sink a well, but when there was no longer any chance of getting water, and he was about to discontinue his work, Admiral von Wrangel requested him to go on until he reached the bottom of the ice: he did this, and kept a journal of the progress of the work, which he sent to the admiral, and a communication to the Academy of Sciences respecting it was made by M. von Helmerson.

The well was sunk to a depth of 54 sashes, or 382 feet, when the soil became so loose that it was impossible to proceed without timbering, which had been unnecessary nearer the surface. At this depth M. Schargin ascertained the temperature of the earth to be $-\frac{1}{2}^{\circ}$ Réaumur, but nearer the surface it had been much lower:—

-6° Réaumur at some feet below the surface.

$-5\frac{1}{2}^{\circ}$ R. at 77 feet.	$-1\frac{1}{2}^{\circ}$ R. at 305 feet.
-4° „ 119 „	$-\frac{1}{2}^{\circ}$ „ 350 „
-2° „ 217 „	$-\frac{1}{2}^{\circ}$ „ 382 „

It is probable on looking at this table that in the last observation a greater degree of cold is assigned to the ground than it really has. Besides, as the work was carried on in winter as well as summer, and the mouth of the pit was eight feet square, so large a volume of cold air must have penetrated as to chill the sides—I am therefore inclined to believe that the freezing point must have been reached. The immense thickness of the layer of ground ice (which at Yakutsk is not less than 382 feet) proves that Siberia must have been in the same physical condition for a long period of years as it is at present.

It is impossible to determine accurately, in the present state of our information on this subject, what is the boundary of this layer of ground ice; we only know enough to say that it extends over an immense tract of country. Humboldt found the soil frozen at a depth of six feet at Bogoslovsk, in 59° 44′ N. lat., near the Uralian chain. Near Beresov, Erman found the temperature of the soil, at a depth of twenty-three English feet, to be $+ 1^{\circ}$; but in 1821 a dead body was found which had been buried upwards of ninety-two years in a bed of ground ice, and which showed no signs of decomposition. It has long been known that the soil at Obdorsk, a town situated some degrees further north, is always frozen. There is no ice at Tobolsk, but the further we proceed to the east the more the direction of the ice is to the south. Georgi found it on the banks of Lake Baïkal, which, however, is in a mountainous country, and it is said to be found east at Nertchinsk, on the banks of the river Argún, but here also in an elevated tract of country. No ice is found at Okhotsk, and the soil is in general warmer on the shores of the Pacific.

The ice, of course, melts more or less every summer, in proportion to the greater or less degree of heat, but old authors assert that the warmth has only the effect of thawing the surface: this is a mistake, which arises from the measurements having been taken at too early a period of the summer, and having then been looked on as the mean of the effect of the heat during the summer months, and as the greatest depth to which the heat penetrated. I have seen ice at a depth of six inches under a soil covered with moss and lichens in the month of July at Novaïa Zemlïa, whilst in a contiguous situation, where the soil is without any signs of vegetable life, it occurs at a depth of two feet. But when winter recommences, which is generally about the beginning of September, the soil in the southern division of Novaïa Zemlïa is thawed to the depth of $2\frac{1}{4}$ to $2\frac{3}{4}$ feet, and it may be true that the soil on the coasts of the Icy Sea, at the mouths of the rivers Obi and Yeniseï, is sometimes free from ice to more than a foot in depth, and the surface is covered with vegetation.

I have no doubt that the Academy of Sciences at St. Petersburgh will repeat the measurements of temperature at the different depths in a better manner and on better principles than M. Schargin could adopt, and endeavour to ascertain the depth to which the heat penetrates at Yakutsk as well as in other places, and also the extent of the ground ice.

It would be very desirable, also, that the Geographical Society of London should collect information respecting the extent of the layer of ground ice in North America, the thickness it attains, and how much of it disappears, particularly in those countries over which the factories of the Hudson's Bay Company extend.

As connected with the same subject, the following extract from a letter of Prof. ADOLPH ERMAN, dated Berlin, March 5, 1838, is subjoined:—

"I see by a report recently published of one of your meetings that some members doubt the reality of the fact that the soil in some parts of Siberia does not thaw till a depth of 400 feet from the surface is reached. Permit me to draw your attention to the observations I have made on this subject, recorded in the Second Volume of my Journey round the World, p. 248, *et seq.* The well at Yakutsk, a notice of which Admiral Krusenstern has sent you, existed when I was in that town; it had then a depth of 50 feet, and in plunging my thermometers into the clods of earth which were dug up before me, and guarding them carefully from the influence of atmospheric temperature, they constantly marked —6° of Réaumur.

"The latitude of the place, however, is only 62° $1\frac{1}{2}'$ N., according to the result of all my observations; a mean temperature, therefore, even lower than that which Mr. Scoresby assigns to the north of Spitzbergen, might well surprise me until I had seen it perfectly established by observations on the temperature of the air which were made during several consecutive years, and with thermometers compared with my own.

"I enclose the observations taken three times a day for the year 1827, whence it results that the mean temperature of the atmosphere at Yakutsk is — 5°.9 Réaumur, which agrees very well with the temperature which I had found near the surface of the ground.

"I may remark that I have selected a *temperate winter*, for in 1828 the cold in the month of January was much more severe, as the mean of the observation then gave

	At 6 A.M.	2 P.M.	9 P.M.
January, 1838	— 38°.3	— 35°.7	— 37°

and the mercury did not thaw for *three months together*: in ordinary years it is only solid for two months.

"Now the mean temperature of Yakutsk being — 6°, it follows necessarily that if we dig deeper into the earth we must not expect to find the ground thawed till the increment of heat due to the approach towards the centre should amount to 6° of Réaumur.

"The *data* which we hitherto possessed on the increase of the internal heat of the globe, and which have been collected together by Mr. Delabeche in his excellent treatise on Geognosy, indicated from 90 to 100 French feet for an increase of 1° of Réaumur; I did not therefore expect to find the ground thawed at Yakutsk until at a depth of from 500 to 600 French feet (see p. 251 of vol. ii.), and if the actual fact of a thaw at the depth of 400 feet has surprised me, it is only because it has occurred *too soon;* and that it thereby indicates for the strata that compose the ground at Yakutsk a greater faculty for conducting heat than is possessed by the strata hitherto examined in Europe."

XVII.—*An Account of the Recent Arctic Discoveries by* Messrs. DEASE and T. SIMPSON. Communicated by J. H. PELLY, Esq., Governor of the Hudson's Bay Company.

[Just half a century has elapsed since Alexander Mackenzie, in 1789, first descended the great river which so justly bears his name, and reached the waters of the Polar Sea. Thirty-seven years later, in 1826, Franklin and Back followed Mackenzie's course to the mouth of the same river, and coasted 370 miles to the westward, tracing the northern shore of America till within 160 miles of Point Barrow, which was reached from the westward by Mr. Elson, Master of H.M.S. Blossom, only four days after Franklin was obliged to return. The intermediate portion has remained a blank on our maps till the last few days have brought us the gratifying intelligence of an expedition headed by Messrs. Dease and Simpson, two enterprising officers in the Hudson's Bay Company's service, having successfully traced the unexplored country between Point Barrow and Franklin's farthest; and thus a continuous line of 60 degrees in extent, of the northern coast of America, from Point

Turnagain on the east to Behring's Strait on the west, has been explored by British hardihood and perseverance. The only account of it yet received is contained in the simple narrative of the gentlemen who conducted this expedition, which was with much kindness immediately communicated to the Geographical Society by the Governor of the Hudson's Bay Company.]

Fort Norman, 5th September, 1837.

Honourable Sirs,—We have now the honour to report the complete success of the expedition this summer to the westward of Mackenzie's River.

Our arrangements up to the departure of the expedition from Fort Chipewyan were stated in full to Governor Simpson and the Northern Council, who, no doubt, communicated them to your Honours; we shall, therefore, present a brief detail of our subsequent operations.

On the 1st June we quitted Fort Chipewyan, with two small sea-boats, accompanied by a luggage-boat and a party of hunters, for Great Bear Lake; visited the salt plains, and arrived at Great Slave Lake on the 10th, where we were detained by ice until the 21st. The same cause prolonged our passage across that inland sea, and, having been for two days stopped by a strong contrary wind at the head of Mackenzie's River, it was the 1st July when we reached Fort Norman. Our Indians cast up on the following day, and the crews and cargoes were finally divided and arranged. Our boat-builder, John Ritch, received his instructions to proceed immediately, with a fisherman, two other labourers, and the hunters, to Great Bear Lake, and at its north-eastern extremity to erect our winter-quarters, and lay in a stock of provisions against our return from the coast. We then took our departure, and on the 4th reached Fort Good Hope. There we found an assemblage of Hare Indians and Loucheux. The latter informed us that three of their tribe had been killed and a fourth severely wounded by the Esquimaux in the preceding month, which at once put an end to our intention of procuring an interpreter from among them, although several volunteered to accompany us in that capacity. They at the same time earnestly cautioned us to beware of the treacherous arts of their enemies. On the 9th July we reached the ocean by the most westerly mouth of the Mackenzie, which Sir John Franklin sought for in vain. It is situated in lat. 68° 49½′ N., long. 136° 37′ W., and perfectly answers the description which the Esquimaux messengers gave of it to that officer when they came to apprise him of the intended attack by the Mountain Indians.

We had proceeded but a short distance to seaward, when a party of nineteen men came off to us from Tent Island. We gave each of them a small present, a practice which we continued throughout the voyage, and employed our vocabularies to the best of our ability, and to their great surprise, to explain the friendly feeling of the whites towards their tribe. Being a lively and communicative people, we in the course of the season acquired some facility in our intercourse with them, and when words failed, signs supplied their place, so that we seldom experienced much difficulty in making ourselves understood, or in comprehending their meaning. When indulged, however, they always became daring and excessively troublesome, and they were ever on the look-out for plunder. On this first meeting they made several unsuccessful attempts in that way, and it was no easy matter to induce them to return to their camp, after we had finished our business with them. They said they wished to accompany us to our encampment, where they would have soon been joined by fresh parties, and we had a shoal and dangerous navigation before us that night. We therefore peremptorily ordered them back; but it was not until we fired a ball over their heads that they put about and paddled off. A storm soon after arose, but we made the land in safety the following morning at Shingle Point, in lat. 69°, where we were detained until the 11th. The thermometer had already fallen to 48° (Fahr.), being 30° lower than on the evening we left the Mackenzie River, and, instead of the bright and beautiful weather we enjoyed in our descent of that noble stream, we were now doomed to travel in cold dense fogs, which enveloped us during nearly the whole of our progress along the coast. But although they perplexed and retarded us, we never allowed them to arrest our course, nor did we ever throughout the voyage encamp but when compelled to do so by ice or contrary winds, to which line of conduct may, under Providence, be ascribed the early and successful accomplishment of our undertaking. In the afternoon of the 11th July we reached Point Key, where we were detained by a compact body of ice occupying Phillip's Bay until the 14th. There we were visited by another party of Esquimaux, whose tents were pitched at no great distance from us. They live in the country bordering on Babbage River, and informed us that, except when flooded by the melting of the mountain snows, it is an insignificant stream, not fifty yards in breadth: of this we had ocular proof in a clear day on our return. A handsome flora was collected in this neighbourhood. Having found a passage through the ice in Phillip's Bay, we reached Herschel Island the same evening (14th July), and had intercourse with other parties of the natives, who were pretty numerous along this part of the coast. We found on the island the skull of a whale eight feet in breadth; and whalebone is everywhere an article in extensive use among the natives, especially for the making of their nets and the fastenings of their sledges. We continued our route before an easterly wind, along and through the ice, with very little interruption till two A.M. of the 17th, when an unbroken pack, extending to seaward, made us seek the shore in Camden Bay, near a considerable camp of Esquimaux. As soon as the fears of the latter were removed, an amicable meeting took place, and, having made them the usual presents, we purchased a good many of their mouth ornaments, weapons, and other articles, which will be forwarded in due course to the Hudson Bay House. Three of the men were remarkable for their good looks, upright figures, and a stature of from five feet ten to six feet. They gave us a specimen of their dances, and one of them afterwards won the palm from all our people at leaping. They informed us that they have two sources of trade; the first and most regular with their countrymen, who come annually from the westward; the other with the Mountain Indians, who use fire-arms, and travel a great way across land from the direction of the Russian settlements. They showed us the knives, iron kettles, beads, and other things thus procured, which we have no doubt are of Russian manufacture. Their means of repayment appeared to us very limited, consisting in seal-skins, whalebone, ivory, and a few inferior furs, viz., wolvereens, foxes, and musk-rats. A pair of indifferent beaver gloves was purchased from them, which they had probably procured from the Mountain Indians, for we saw no other symptom of the existence of that valuable animal near the coast, though it doubtless abounds at some distance up the large wooded rivers which we subsequently discovered. In the afternoon there appeared a narrow lane of water stretching outwards, and we immediately embarked. We had advanced about three miles from the land, when the ice suddenly closed upon us, before a strong north-east wind; one of the boats got squeezed, and it was only by throwing out the cargo upon the floating masses that she was saved from destruction. By means of portages made from one piece to another—the oars serving as bridges—the cargo was all recovered, and both boats finally hauled up on a large floe, where we passed an inclement and anxious night. Next morning the gale abated, the ice relaxed a little around us, and by a long circuit we regained the shore, about a league to the eastward of our former position. There we were detained till midnight of the 19th, when a favourable wind enabled us to round the body of ice at a distance of four miles from the land, and, continuing, carried us on the 20th into Foggy Island Bay. There we were stopped by the ice and a violent north-east wind until the 23rd, having on the preceding day made an ineffectual attempt to weather Point Anxiety, in which we narrowly escaped with a thorough drenching. The latitude ashore was 70° 10'. From this situation we had the satisfaction of discovering, during a clear afternoon, a range of the rocky mountains to the westward of the Romanzoff chain, and not seen by Sir John Franklin, but, being within the limit of his survey, we called it the "Franklin Range," as a just tribute to his character and merits. On the 23rd we again set sail, rounded the pack of ice, which extended six miles to seaward from Yarborough Inlet, then, abruptly turning in, we supped near Return Reef, and the survey commenced.

Return Reef is one of a chain of reefs and islets which runs for twenty miles parallel to the coast, at the distance of about half a league, affording water enough within for such light craft as ours. The mainland is very low. From "Point Berens" to "Cape Halkett" (named after two members of your Honourable Board) it forms a great bay, fifty miles broad by a third of that depth, which in honour of the Deputy-Governor was named "Harrison's Bay." At the bottom of this bay another picturesque branch of the Rocky Mountain range—the last seen by us—rears its lofty peaks above these flat shores: we called them "Pelly's Mountains," in honour of the Governor of the Company. At their base flows a large river, two miles broad at its mouth, which we named after Andrew Colville, Esq. This river freshens the water for many miles, and its alluvial deposits have rendered Harrison's Bay so shallow, that it was not till after a run of twenty-five hours, during which we had repeatedly to stand well out to seaward, that we could effect a landing on a grounded ice-berg, nine miles to the south-west of Cape Halkett. A north-east gale kept us there the whole of the following day. The country extending to the foot of the mountains appeared to consist of plains covered with short grass and moss, the favourite pasture of the reindeer, of which we saw numerous herds. Observations were obtained, determining our position to be in lat. 70° 43′ N., long. 152° 14′ W.; variation of the compass 43° E. Next morning (26th July) the tide rose nearly two feet at six A.M., and enabled us safely to cross the shoals. At no great distance from our encampment we passed the mouth of another large river, one mile broad, whose banks were thickly lined with drift timber. We named it the "Garry," in honour of Nicholas Garry, Esq. Cape Halkett forms the extreme point of a small island, separated from the main shores by a narrow channel too shallow for boats. Its situation was likewise found by observation to be in lat. 70° 48′ N., long. 151° 55′ W. It appears to be a place of resort to the Esquimaux, for we found a spot where they had been building their baidars last spring. We suppose them to have been part of a very large camp, which we saw in the bay of Staines River, as we sailed past the east end of Flaxman Island on the 20th July; that this camp consisted of the western traders of that tribe on their annual journey to meet their eastern brethren at Barter Island; and that we missed them on our return, from the circumstance of their being then dispersed along the rivers, lakes, and in the skirts of the mountains, hunting the reindeer.

From thence the coast turned suddenly off to the W.N.W. It presented nothing to the eye but a succession of low banks of frozen mud. The ice was heavy all along, but there were narrow channels close to the shore; the soundings on these averaged one fathom on sandy bottom. In the evening we passed the mouth of a considerable river, which was named after William Smith, Esq. From thence, for about six miles, the coast-line is formed of gravel reefs, near the extremity of which, at "Point Pitt," (called after another member of your Honourable Board,) the land trends more to the westward. The ice lay much closer here; numerous masses adhered to the bottom under the water, which obliged us to search for a passage out from the shore. The night was dark and stormy, and we were in considerable danger; one of the rudders gave way, but we at length effected a landing on a place near an immense reindeer pound. This was ingeniously formed by the Esquimaux with double rows of turf, set up on a ridge of ground enclosing a hollow four miles by two, the end farthest from the beach terminating in a lake, into which the unsuspecting animals are driven and there despatched with spears. The vegetable soil in this vicinity was barely four inches in depth, beneath which the clay was frozen as hard as rock, so that our tent pickets could not be driven home. The men had to go a good mile to find a log or two of drift-wood for fuel, the scarcity of which essential article is doubtless the chief cause of the want of inhabitants along so great a portion of the coast. We were detained at this place till the following afternoon (27th), when the ice opening a little enabled us to resume our route. It blew a cutting blast from the north-east, and the salt water froze upon the oars and rigging. "Point Drew," called after Richard Drew, Esq., seven miles distant from our last encampment, is the commencement of a bay of considerable size, but extremely shallow, and much encumbered with ice, in pushing through which the boats received several blows; and we had on this, as on many other occasions, to admire their excellent workmanship. To seaward the ice was still smooth and solid as in the depth of a sunless winter. At midnight we reached a narrow projecting point, across which the peaks of some high icebergs appeared, and were from a distance mistaken for lodges of the natives. This point we named "Cape George Simpson," as a mark of respect for our worthy governor. It was destined to be the limit of our boat navigation, for during the four following days we were only able to advance as many miles. The weather was foggy and dismally cold, the wild fowl passed in long flights to the westward, and there seemed little prospect of our being able to reach Point Barrow by water. "Boat Extreme" is situated in lat. 71° 3′ N., long. 154° 26′ W.; variation of the compass, 42° E.

Under the above circumstances Mr. Thomas Simpson undertook to complete the journey on foot, and accordingly started on the 1st of August with a party of five men. They carried with them their arms, some ammunition, pemmican, a small oiled canvas canoe for the crossing of rivers, the necessary astronomical instruments, and a few trinkets for the natives. It was one of the worst days of the whole season, and the fog was so dense that the pedestrians were under the necessity of rigidly following the tortuous outline of the coast, which for twenty miles formed a sort of irregular inland bay, (being guarded without by a series of gravel reefs,) the shore of which was almost on a level with the water, and intersected by innumerable salt creeks, through which they waded, besides three considerable rivers or inlets, which they traversed in their portable canoe. They found at one place a great many large wooden sledges, joined with whalebone and strongly shod with horn. Mr. Simpson conjectures that these vehicles were left there by the western Esquimaux, already spoken of, on their eastward journey, to be resumed again on their return when winter sets in. The tracks of reindeer were everywhere numerous. Next day the weather improved, and at noon the latitude 71° 10′ was observed. The land now inclined to the south-west, and continued very low, muddy, and, as on the preceding day, abounding in salt creeks whose waters were at the freezing temperature. The party had proceeded about ten miles when to their dismay the coast suddenly turned off to the southward, forming an inlet extending as far as the eye could reach; at the same moment they descried, at no great distance, a small camp of the western Esquimaux, to which they immediately directed their steps. The men were absent hunting, and the women and children took to their boats in the greatest alarm, leaving behind them an infirm man, who was in an agony of fear. A few words of friendship removed his apprehensions and brought back the fugitives, who were equally surprised and delighted to behold white men. They set before the party fresh reindeer meat and seal-oil, and besought them for tobacco (tawăccah), of which men, women, and even children are inordinately fond. Mr. Simpson now determined to adopt a more expeditious mode of travelling, and applied for the loan of one of their "oomiaks" or family canoes, to convey the party to Point Barrow, with which, from a chart drawn by one of the most intelligent of the women, it appeared that these people were acquainted. The request was immediately complied with; four oars were fitted with lashings to this strange craft, and the ladies declared that our party were true Esquimaux and not "Kabloonan." Before starting the hunters arrived, and were likewise gratified with tobacco, awls, buttons, and other trifles. "Dease's Inlet" is five miles broad at this place, yet so low is the land that the one shore is just visible from the other in the clearest weather. It now again blew strongly from the north-east, bringing back the cold dense fog, but the traverse was effected by aid of the compass. The waves ran high, and the skin boat surmounted them with a buoyancy which far surpassed that of our boasted north canoes. The party encamped on the west side of the inlet. The banks there were of frozen mud ten or twelve feet high; the country within was perfectly flat, abounded in small lakes, and produced a very short grass, but nowhere had the thaw penetrated more than two inches beneath the surface, while under water along the shore the bottom was still impenetrably frozen. Not a log of wood was to be found in this land of desolation, but our party followed the example of the natives, and made their fire of the roots of the dwarf-willow in a little chimney of turf. Next morning (August 3rd) the fog cleared for a while, but it was still bitterly cold, and the swell beat high on the outside of a heavy line of ice which lay packed upon the shore. The latter, after extending five miles to the northward, turned off to the north-west, beyond which the latitude 71° 13′ was observed. From this point the coast trended more westerly for ten miles, until the party came to what appeared a large bay, where they stopped for two or three hours to await the dispersion of the fog, not knowing which way to steer. In the evening their wish was gratified, and from that time the weather became comparatively fine. The bay was now ascertained to be only four miles in width: the depth half way across was one fathom and a half on a bottom of sand; that of Dease's Inlet was afterwards found to be two fathoms, muddy bottom, being the greatest depth between Return Reef and Point Barrow, except at ten miles south-east from Cape Halkett, where three fathoms on sand were sounded on our return. After crossing "Mackenzie's Bay" the coast again trended for eight or nine miles to the W.N.W. A compact body of ice extended all along and beyond the reach of vision to seaward; but the party carried their light vessel within that formidable barrier and made their way through the narrow channels close to the shore. At midnight they passed the mouth of a fine deep river, a quarter of a mile wide, to which Mr. Simpson gave the name of "The Belle Vue," and in less than an hour afterwards the rising sun gratified him with the view of Point Barrow stretching out to the N.N.W.

They soon crossed Elson Bay, (which in the perfect calm had acquired a coating of young ice,) but had no small difficulty in making their way through a broad and heavy pack that rested upon the shore. On reaching it, and seeing the ocean extending away to the southward on the opposite side of the Point, they hoisted their flag, and with three cheers took possession of their discoveries in his Majesty's name.

Point Barrow is a long low spit, composed of gravel and coarse sand, which the pressure of the ice has forced up into numerous mounds, that, viewed from a distance, assume the appearance of huge boulder rocks. At the spot where the party landed it is only a quarter of a mile across, but is broader towards its termination. The first object that presented itself on looking round the landing-place was an immense cemetery. The bodies lay exposed in the most horrible and disgusting manner, and many of them appeared so fresh that some of the men became alarmed that the cholera or some other dreadful disease was raging among the natives. Two considerable camps of the latter stood at no great distance on the Point, but none of the inmates ventured to approach till our party first visited them, and with the customary expressions of friendship dissipated their apprehensions. A brisk traffic then commenced, after which the women formed a circle and danced to a variety of airs, some of which were pleasing to the ear. The whole conduct of these people was friendly in the extreme; they seemed to be well acquainted with the character if not the persons of white men, were passionately fond of tobacco, and when any of the younger people were too forward, the seniors restrained them, using the French phrase "c'est assez," which, like "tawăccah," they must have learned from the Russian traders. They designate the latter "Noonatagmun," and a respectable looking old man readily took charge of a letter addressed by Mr. Simpson to them or to any other whites on the western coast, containing a brief notice of the proceedings of the expedition. To the northward enormous icebergs covered the ocean, but on the western side there was a fine open channel, which the Esquimaux assured the party extended all along to the southward; and so inviting was the prospect in that direction that, had such been his object, Mr. Simpson would not have hesitated a moment to prosecute the voyage to Cook's Inlet in his skin canoe. The natives informed him that whales were numerous to the northward of the Point, and seals were everywhere sporting among the ice. These Esquimaux were well clothed in seal and reindeer skins; the men all used mouth ornaments, and the "tonsure" on the crown of the head was universal among both men and boys; the women had their chins tattooed, but did not wear the lofty top-knots of hair which are fashionable to the eastward. They were very inquisitive about the names of our party, and equally communicative of their own. A number of their words were taken down, some of which are different from the corresponding terms given by Sir Edward Parry, but the greater part are either the same or dissimilar only in their terminations. They lay their dead on the ground, with their heads all turned to the north. There was nothing else either in their manners or habits remarked as differing from the well-known characteristics of the tribe, except an ingenious and novel contrivance for capturing wild fowl. It consists of six small perforated ivory balls attached separately to cords of sinew three feet long, the ends of which being tied together, an expanding sling is thus formed, which, dexterously thrown up at the birds as they fly past, entangles and brings them to the ground.

Mr. Simpson could not learn that there had been any unusual mortality among this part of the tribe, and is of opinion that the concourse of natives who frequent Point Barrow at different periods of the year, coupled with the coldness of the climate, sufficiently account for the numerous remains already noticed. It was high water between one and two o'clock A.M. and P.M.; the rise of the tide was fourteen inches; and the flood came from the westward. Observations were obtained which determined the position of the landing-place to be in lat. 71° 23½′ N., long. 156° 20′ W., agreeing closely with the observations of Mr. Elson. Then, bidding adieu to their good-humoured and admiring entertainers, the party set out on their return. They were arrested that evening by the ice, but next morning (August 5) it opened and allowed them to proceed. At a late hour they reached the camp of the Esquimaux in Dease's Inlet, and, after liberally recompensing them for the loan of their canoe, directed some of the men to follow to Boat Extreme, where it would be left for them. Then continuing their route all night, at five A.M. on the 6th the party rejoined the main body of the expedition.

We commenced our return the same afternoon, and, being favoured by a light wind and an almost open sea, we sailed all night, and next day (Aug. 7) at noon reached Cape Halkett. We then steered across Harrison's Bay: the wind increasing to a gale we shipped much water, but persevering, under treble-reefed sails, at three P.M. of the 8th we landed safely at "Fawn River," within view of the point where our survey commenced. The position of this encampment was ascertained by good observations to be in lat. 70° 25′ N., long. 148° 25′ W. The gale having moderated we re-embarked the following afternoon, and, running without intermission before a fresh breeze, we reached Demarcation Point to breakfast on the 11th. Several showers of snow fell during this run, and it was piercingly cold. The Romanzoff and British mountains had assumed the early livery of another winter. The ice in Camden Bay was still very heavy, but it protected us from the dangerous swell to which we became exposed after passing Barter Island. Soon after leaving Demarcation Point the ice became so closely wedged that we could no longer pursue our way through it. The following day (August 12) it opened a little, and the weather becoming fine we put out and advanced for a few hours, when the mountainous swell and heavy ice obliged us to seek the shore, which we reached with some difficulty between Backhouse River and Mount Conybeare. There we were detained until the 15th. The icebergs which begirt the coast were of great size, and of every imaginable shape, but from the summit of a hill, six miles in-land, we had an unbounded prospect of the blue ocean stretching to the north. The pasture in the deep valleys among the mountains was luxuriant; herds of reindeer were browsing there, and we procured some venison. In the night of the 14th the stars and aurora borealis were first visible. The following morning we resumed our route, and the weather continuing nearly calm we reached the western mouth of the Mackenzie on the 17th, and there encamped. The first Eskimaux seen during our return from Boat Extreme were at Beaufort Bay, but from thence to the vicinity of the Mackenzie we were continually falling in with small parties, many of whom we had seen on the outward voyage. We maintained a friendly intercourse with them all, and they were very anxious to know whether we were soon to visit them again. The habitations on Tent Island were abandoned in consequence, we understood, of an alarm that the Loucheux meditated a descent to revenge the murder of their friends.

We have but few general remarks to add to the foregoing narrative. The tides all along the coast were semi-diurnal; the flood coming from the westward. The rise, however, was strongly affected by the winds and ice; and our opportunities of observing were but few. At Boat Extreme the average rise was fifteen inches; high water from one and to two o'clock, A.M. and P.M. The rise generally decreased to the eastward, and at Point Kay it was only eight or nine inches. That the main sea is clear and navigable by ships during the summer months the long rolling swell we encountered on our return, and the view obtained from the mountains, furnish tolerable proof. We likewise saw some whales on our return. The prevalence of east and north-east winds during the early part of the summer is a remarkable fact. We were indeed favoured by a westerly wind for five days on our return, but this was almost the only exception. At a more advanced period of the season, however, the winds blew more from the west and north-west. It is now certain that from Kotzebue's Sound to Cape Parry there is not a harbour into which a ship can safely enter, but it must be a very unpropitious season that would not admit of achieving that portion of the Arctic navigation; and another year ought certainly to suffice for the remainder, whether the voyage were commenced from Barrow's or from Behring's Strait. On this subject, however, we shall be better able to offer an opinion if successful in our next summer's operations.

The natural history of the coast from Return Reef to Point Barrow is meagre in the extreme. In the botanical kingdom scarcely a flower or moss was obtained in addition to the collection made on other parts of the coast. In zoology, reindeer, arctic foxes, one or two limmings, seals, white owls, snow buntings, grouse (Lagopus salicite et rupestres), and various well-known species of water-fowl, were the only objects met with; while in the mineralogical department there was not a rock *in sitû* or boulder-stone found along an extent of more than two hundred miles of coast. The variation of the compass was found to have increased from one to three degrees at the corresponding points where observations had been made by Sir John Franklin. At Boat Extreme, on the other hand, it was only twenty-one minutes greater than that stated by Mr. Elson at Point Barrow, where, by continuing the proportion, the quantities would coincide. The moon was never once visible during the whole outward and homeward voyage, till our return to the western mouth of Mackenzie, where a set of distances was obtained, and the longitudes of the other points reduced back from thence by means of a very valuable watch generously lent to the expedition by Chief Factor Smith.

The map of our discoveries will be prepared and transmitted to your Honours in the spring.

Our ascent of the Mackenzie has not been characterised by any circumstance of particular interest. The weather continued calm and beautiful; and the journey was performed entirely by towing, in which manner we advanced at the rate of from thirty to forty miles a-day. The river has fallen very low, and the fisheries have, in consequence, been unproductive, causing a scarcity of provisions both at Fort Good Hope and among the natives. We saw a good many of the Loucheux, but the Hare Indians were all dispersed in the interior, searching for subsistence. From the coast up to Point Separation moose-deer were numerous—being quite undisturbed, but from our first falling in with the

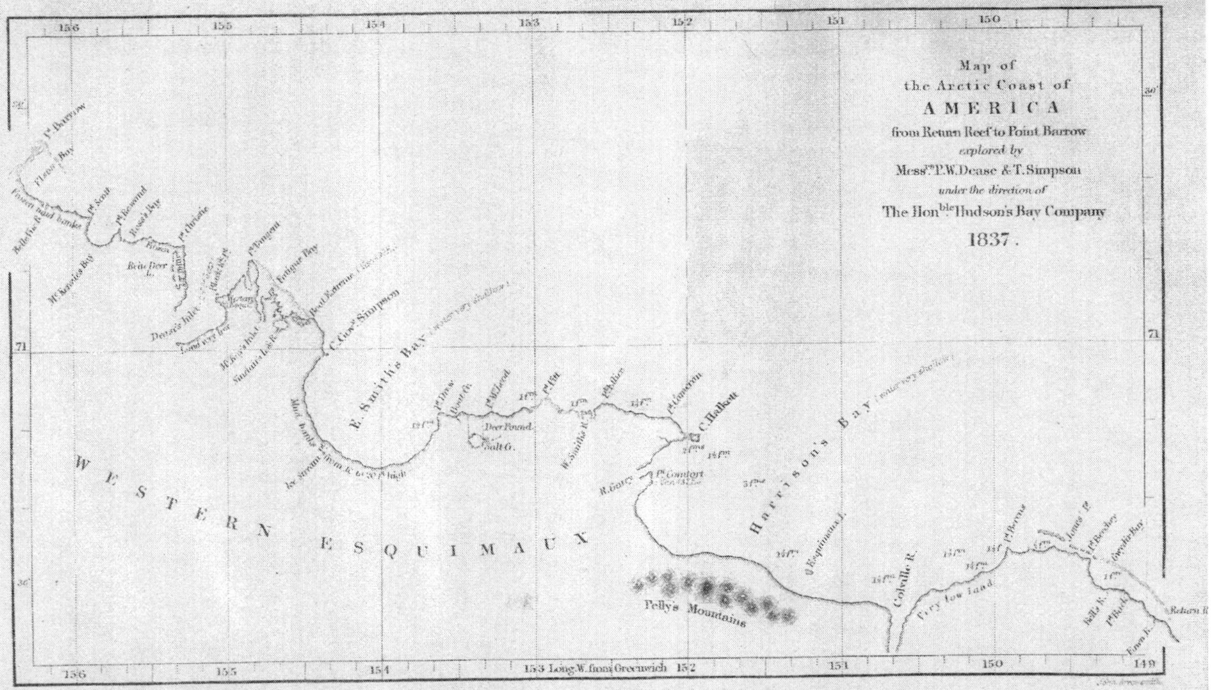

Published for the Journal of the Royal Geographical Society by J. Murray, London 1838.

Loucheux no vestiges of either moose or reindeer have been seen. We reached this place yesterday with half of our summer's stock of provisions still forthcoming, and are now awaiting with impatience the arrival of our outfit and despatches.

Some Indians from Great Bear Lake have brought us intelligence of the party sent to establish our winter-quarters. They were stopped in Bear Lake river by ice during the whole month of July, lost one of their canoes, and it was not till about the 6th ultimo that they passed Fort Franklin, after which they had the prospect of an unimpeded passage across the lake. The continued easterly winds were the cause of this vexatious detention, during which the Dogribs kindly supported our people with the produce of their nets.

Sept. 8.—We have this morning received Governor Simpson's letter, dated London, 11th November last, and have to offer our lively acknowledgments to your Honours for the interest you have been pleased to express in the success of the expedition, and in the welfare of ourselves and party.

We have duly received the journals of Captain Back's last expedition, and are glad to find that his new undertaking is in no way to interfere with our original instructions.

Your Honours may rest assured that our efforts in the cause of discovery and science next summer, to the eastward of the Coppermine river, will not be less zealous than they have this season been in another field, and we are sanguine in the hope that they will be crowned with equal success.

Our supplies for next season have come to hand. They were delivered in very indifferent order at Portage la Loche, and there was a considerable deficiency in the weights of the pemmican; but with the quantity saved of this year's stock there will be provision enough for next summer's operations, and we have no further demands to make upon the Dépôt for goods.

We send two men express to Great Slave Lake with this despatch, and to meet the spring packet, the non-arrival of which causes us some anxiety; and to-morrow we take our departure for winter-quarters.

We are, &c.

PETER W. DEASE
THOMAS SIMPSON

The Governor, Deputy Governor,
and Committee of the
Honourable the Hudson's Bay Company, London.

XVIII.—*Heights obtained during the Campaign to Ḳosṭanṭinah, in September,* 1837. By Monsieur De Falbe, Captain in the Danish Royal Navy, late Consul General at Tunis and at Athens. Communicated by Sir Grenville T. Temple, Bart.

As a valuable supplement to the account of his journey from Bónah to Ḳosṭanṭinah, inserted at p. 39 of this volume, Sir G. Temple has transmitted the following table of elevations obtained by M. de Falbe, who has kindly sanctioned their being published in the Journal of the Society: most of the positions are marked in the map which illustrates Sir G. Temple's route.

Results of Barometric Observations.

	English feet above the sea.
The Camp at Ed-dere'án	280
——— Neshmáyah	820
Summit of Jebel Fejúj	1021
Redoubt at Ḥammám el Berda'ah . . .	980
Fortress at Ḳalemah	924
Camp at Mejáz 'Amár	863
Level of the river Sebús near Mejáz 'Amár . .	795
Jebel Ḥasaníyah (northern spur of Rás el 'Aḳabah) .	1923
Pass of Rás el 'Aḳabah (by the new road) . . .	2718
Bed of Wád-ez-Zenátí (at ford of Sídí Tamtám) . .	2086
Ḳobbeh of Sídí Tamtám	2208
Summit of hill near Rás-ez-Zenátí, on right bank of river	2842
Level of Wád el Boḳrah at the point where the road crosses it	2700
Line of separation of waters flowing N.E. to the Zenátí and Sebús, and westerly to the Abú Merzúḳ and the Rum .	2867
Summit of the Mountain Umm es-Seṭáḥ el Sharḳíyah .	3575
Ḳosṭanṭinah at the north point of the Ḳaṣbah . .	2300
Sídí Rashíd at south point of the city . . .	1968
The monument at Es-ṣúm'ah	2666
Summit of Jebel Sídí Meshíd (1600 yards N.N.E. of centre of city)	2758
——— Seṭáḥ el Manṣúrah (1200 yards S.E. of centre of city)	2486
——— Jebel Emṣálí (1640 yards S.S.W.) . .	2344
Marábút of Sídí Mabrúk (2300 yards E.S.E. of centre of city)	2340
Summit of Kudyaṭ 'A'tí (840 yards S.W.) . .	2338

	English feet above the sea.
The causeway opposite Báb el Wád, connecting the plateau of the city with Kudyaṯ 'A'tí (440 yards S.W. of centre of city)	2230
El Ḳanṭarah over Wád er-rummel (550 yards E. by N. of centre)	2082
Confluence of Wád er-rummel and the Abú Merzúk (one mile S. by E. of centre)	1867
Surface of Wád er-rummel under the great bridge (550 yards E. by N. of centre)	1741
———————— above the cascades . .	1705
———————— below the cascades and the mills (730 yards N.N.W.)	1548
———————— at its junction with Wád el Melaḥ (2200 yards W.N.W.)	1413

The mean of several observations gives the latitude of Ḳosṭanṭinah at

The Ḳaṣbah, near the north end of the city	36° 21′ 3″ North.
The Breach at the south-west angle of the city	36° 20′ 41″ ,,
The fountain at Sídí Mabrúk . .	36° 20′ 29″ ,,

The longitude of the Ḳaṣbah obtained by the total eclipse of the moon on the 13th of October, 1837 = 6° 0′ 30″ East of Greenwich.

[N.B. As this longitude differs from 25 to 30 miles from all former determinations of Ḳosṭanṭinah, it is recommended that it should be received with caution; Shaw, who was there in 1726, assumed the longitude of Ḳosṭanṭinah at about 6° 30′ East; M. D'Avezac,* in 1836, was disposed to adopt 6° 24′ East.

It may be worthy of remark that Shaw, with his limited means, fixed the latitude of Ḳosṭanṭinah within one mile of the recent determination by M. de Falbe.]

Besides the above notes, and the account by Sir G. Temple given at p. 39 of this volume, Colonel Harding, of the Royal Engineers, has had the kindness to communicate to the Society the narrative of his Journey from Bónah to Ḳosṭanṭinah in 1826, which contains much valuable information relative to the country at that time, and would certainly have been published in the Geographical Journal had not the later route by Sir G. Temple reached the Society at the same moment.

Colonel Harding's MS. is preserved in the library for reference, and may be consulted with advantage.

* In his excellent work *E'tudes de Géographie Critique*, p. 58.

XIX. — *On the Bay of Antioch, and the Ruins of Seleucia Pieria.* By Lieutenant-Colonel Chesney, Royal Artillery, F.R.S. Read May 14, 1838.

The Bay of Antioch is spacious, free from rocks, and well sheltered on every side, with the exception of the south-west, where, in the distant horizon, is seen the lofty island of Cyprus; the anchorage, however, is good, and the water deep, almost to the very beach.* This was the spot selected, in order to avoid the Beïlán mountains, for the disembarkation of the party destined to proceed on the expedition to the Euphrates. On the 3rd of April, 1835, H.M.S. Columbine, followed by the George Canning, under all sail, led the way from the offing towards the anchorage. To the south, as we proceeded, was the lofty Jebel el 'Akrab, rising 5318 feet above the sea, with its abutments extending to Antioch. To the north, the Beïlán range (5337 feet), well stocked with fine forest-trees, chiefly oak, walnut, and fir; and in front the broad expanse of the bay backed by the hills of Antioch, Mount St. Symeon, or Bín-kilíseh, covered with myrtle, bay, and arbutus, altogether forming a striking and magnificent Panorama. We all felt on many accounts a deep interest, as we approached the bottom of the bay. The pilot of the Columbine (a Greek), although selected as knowing this part of the coast, admitted just then, that he had never entered the bay before; and there was but one person in the expedition who had visited this part of the country, and that only by land. The vessels continued their course for a considerable time, finding no bottom with twenty-five fathoms of line; and as there was no appearance of the river Orontes, great uneasiness was felt lest we might have mistaken the intended bay; until, after some suspense, and a good deal of anxiety, the extensive line of excavations along the sides of the rocky hills behind the ruins of Seleucia were discovered to the left, but still at some distance. In less than an hour the Columbine began to shorten sail; being then in ten and a half fathoms water, and sufficiently near to the land, she hauled her wind and stood towards the south side of the bay, until the inner part of the Orontes was seen, from the topmast-head, winding towards its æstuary; which, owing to its slanting direction, was still hidden from view: both vessels then tacked, and stood towards the ruins of Seleucia, near which they anchored soon after sunset.

The southern horn of the bay of Antioch trends inwards east by north about seven miles to the beach. Near its outer extre-

* At one and a half cable's length from the beach there are three fathoms water, and it gradually deepens to seven and a half. Opposite the bar, at three quarters of a mile distant, and close to the land on the south-west side, are the best places for anchoring.

mity is the little bay or fissure called Kaṣab, and three miles nearer to the main-land, that of Kará Mayor; which is rather larger, and has a good anchorage off it close in to the shore;* the rest of the distance along the foot of Mount Cassius being precipitous, and for the most part inaccessible, as far as the beach, beyond which, the range of Jebel el 'Akrab runs towards Antioch in the previous direction east by north, with the rich picturesque valley of the Orontes at the foot, and the celebrated fountain of Daphne on its slope. The bight of the bay forms an almost imperceptible curve. At four miles north by west, half west, from the foot of Cassius, is the present æstuary of the Orontes; having what is called the Custom House about one mile from the bar in a strait line east by south, and the ancient port of Antioch about a mile further.

According to the reports of the natives, as well as the appearance of the marshy ground, the river entered the sea formerly by a second branch close to the mountain on the south side, which probably formed the small island we find marked on some older maps. Eight miles and a half north by west, half west, is the other horn of the bay, which is formed by Jebel Músa; on the base of which opening north-west are the ruins of the well-known city built by Seleucus Nicator to celebrate his victory over Antigonus; but it has a much deeper interest to the Christian, from being the spot where Paul and Barnabas embarked for Cyprus.

On the eastern side of Jebel Músa is the large and scattered village of Sweïdíyah, situated in a rich picturesque valley filled with fruit gardens, chiefly mulberry,† and producing a variety of good fruits, as well as the finest silk in Syria. Southward from the village is the low mountain called Bín-kilíseh,‡ the sides of which are thickly covered with myrtles, interspersed with arbutus. On the crest of the hill are the ruins of the convent of St. Symeon Stylites. Here is part of a column of very large dimensions cut in the solid rock, and close by are the scattered fragments of the extensive convent dedicated to St. Symeon, which was destroyed in the sixth century, now almost covered with shrubs. Close at the foot of this low mountain, on the south-east side, the Orontes breaks through a deep valley, and makes a very winding course both above and below. The summit is about five miles from the sea, and commands a beautiful view westward, over a very rich plain extending to the sea, closed in by Mount Cassius to the southward, and the Jebel Músa range to the northward; whilst to the east is the valley of the Orontes, terminated by the castellated hills of Antioch; the general view being closed to the north-east

* An Egyptian sloop of war lay here at one time, and also several small merchantmen at intervals during our stay.

† The bay-tree and myrtle are particularly large; and the oranges, peaches, nectarines, and figs are particularly good.

‡ 1000 churches.

by the Beïlán mountains. It was hoped that the stores might have been carried altogether by water to Antioch, but, as the river proved rapid, usually without even the assistance of a path along its bank, and there was besides little inclination to accommodate us with a temporary passage through the fish-weirs, this part of the plan was abandoned after tracking up the keelsons of both vessels only. The windings give a distance of about forty-one miles, whilst the journey by land is only sixteen miles and a half. Ibráhím Páshá talked of making the river navigable, which might be done by blasting some rocks in its bed, and by removing the wooden fish-weirs which traverse the river in several places near Antioch; it would also be necessary to cut a towing-path for horses through the woods along its banks. Lieutenant Cleaveland and the other officers were of opinion that a short tug-steamer of sufficient power would certainly go up the river to Antioch, which was, in fact, done by the Columbine's boat for the greater part of the way; and if a row of piles were to be driven into the sea, in the line of the river, extending beyond the bar, so as to enable the current of the river to carry the sand and mud farther out into deep water, the Orontes would then admit vessels of 200 tons, instead of being obstructed by a bar, over which there is a depth of water of from three and a half to nine feet, in winter. At any rate, it might be made navigable for boats, as the average fall of the river between Antioch and the sea scarcely exceeds five feet and a half per mile; and boats would then go twenty-seven miles above the town to Murád Páshá, and different parts of the lake of Antioch.

'Alí Páshá, the present governor of Bagdad (once governor of Aleppo), had, however, a different project, when he turned his thoughts to the means of increasing the commercial prosperity of this part of Turkey. The foundation of his plan was to be the restoration of the once magnificent port of Seleucia, the masonry of which is still in so good a state, that it merely requires trifling repairs in some places, and to be cleared out, which might have been done for about 31,000*l.*, and partially for 10,000*l.** On the south side of the entrance there is a very substantial jetty, formed of large blocks of stone secured by iron cramps. It runs north-west for seventy yards to the sea, and it may still be traced curving more to the north under water, and overlapping the northern jetty, which is in a more ruinous state, but appears to have taken the direction of W.S.W., forming a kind of basin with a narrow entrance tolerably well protected, and altogether suited for the Roman galleys. The ancient flood-gates are about fifty yards east of the south pier. The passage for the galleys, &c., is cut through the solid rock, on which are the remains of a defensive tower on each side; apartments below, with the remains of staircases to the

* According to the estimate of Mr. Vincent Germain.

top of each, are sufficiently distinct, as well as the places where the gates had been suspended between the towers.

Immediately on passing the gateway the passage widens to about 100 yards; it takes the direction of S.E. by E. between two solid walls of masonry for 350 yards, to the entrance of the great basin, which is now closed by a garden-wall. The port or basin is an irregular oval of about 450 yards long by 350 in width at the southern extremity, and rather more than 200 at the northern. The surrounding wall is formed of large cut stones solidly put together, and now rising only about seven feet above the mud, which during the lapse of ages has gradually accumulated so as to cover probably about eight feet above the original level. The exterior side of the basin is about one-third of a mile from the sea; the interior is close to the foot of the hill. The walls of the suburb touch the south-western extremity of the basin, and extend S. by E. from thence parallel to the sea for three quarters of a mile, when they turn eastward for the same distance, flanked at short intervals by square towers. These walls form a triangle, touching the basin at one end, and the walls of the principal city at the other, so as to inclose what is described by Polybius, and subsequently by Pococke, as the market-place and suburbs.* The walls of the interior part of the city appear to have had, as usual in Roman fortresses, a double line of defence, sweeping round to the north, where they rest against the hill, which seems to have had a castellated citadel on its summit. On the S.E. side of the walls is the Gate of Antioch, adorned with pilasters and defended by towers: this entrance must have been very handsome: near it, and parallel to the walls, are the remains of a double row of marble columns. The space within the walls of the town and suburbs, which have a circumference altogether of about four miles, is filled with the ruins of houses. A short distance from the town, on the east side, are the remains of a large amphitheatre tolerably distinct. About fourteen rows of seats may be traced in a semicircular form, filling up the whole of the valley in which the amphitheatre is placed, with its opening to the west commanding a fine view of the bay. To the S.E. and behind the hill (on which is the amphitheatre) are the remains of two temples; the fragments of pilasters, shafts, &c., are numerous; one seems to have been of the Corinthian order, in good taste, but I could not make out the plan of either of the buildings. The range of hills behind the ruins extends almost two miles, and contains, along its sides as well as in the valleys, numerous excavations, which are almost continuous throughout this distance. Generally speaking, they form only a single row and of small size, but occasionally

* Pococke, vol. ii. p. 184.

there is a second line of them, above or below the others, for part of the distance. These grottoes (evidently sepulchral) are generally of two kinds; the larger about twelve feet long by seven wide, having the front supported by pilasters left in excavating the solid rock, and within are three niches for bodies, viz., one on each side and one at the back of the same dimensions, viz., two feet and a half high and the same width, with a raised place left in the niche, of solid stone about four inches high, like a pillow for the head to rest upon: these niches are sometimes arched, but generally flat above. The smaller grottoes have a niche at each side with a narrow space between them. One set of grottoes is called the Tomb of the Kings; it consists of a façade-entrance and several apartments one within the other, with columns and a staircase leading to another range of rooms above. In addition to these, which are the most striking, there is another single grotto of large dimensions in one of the valleys along the side of the hill: this excavation is 100 paces by 60 wide and 25 high in the centre, the rock being excavated so as to form an arch springing from the ground on each side, that is, without side-walls. In addition to these sepulchral grottoes, of which some hundreds cover the face of the hills and all their valleys, there are many sarcophagi scattered about in every direction, always of good workmanship and tolerably perfect; although they have been opened in almost every instance probably in search of money.

But the most striking part of the interesting remains at Seleucia is a very extensive excavation cut through the solid rock from the north-eastern extremity of the town, almost to the sea;—part of which is a deep hollow way, and the remainder regular tunnels—executed with great skill and considerable labour.

In following the line from its eastern to its western extremity, a distance of 1088 yards, we have, in the first place, a hollow way of 600 feet long by 22 wide, through compact tertiary lime-stone, and of different heights at the sides, as it enters the hill. Next to this we have a regular tunnel, 293 feet by 22 wide, and 24 high. Then follows a second hollow way, 204 feet by 22, having on its south side an aqueduct cut in the rock 18 inches wide, which takes its departure from the bottom of the passage; and, by continuing onwards almost level, whilst the excavation itself descends by a more rapid slope, it reaches the surface eventually, and is carried southward to supply water to the town. In this portion of the cut which is 110 feet high, and narrowing towards the top, there is a narrow stone staircase from the surface to about 14 feet from the bottom; which probably was the usual water line. The opening above is well shaded by trees overhanging the aperture, which give it a very picturesque effect. The hollow way just mentioned is succeeded by a second tunnel 102 feet long by 25 feet 6 inches wide, and nearly the same height. After this it again becomes a hollow way for 190 feet of the same width, but in some places attaining the height of 110 feet, or even more; and there is a very beautiful arched aqueduct passing from one side to the other at a height which now is lessened to 50 feet. From the excavation it continues a hollow way of 17 feet 6 inches wide and 41 high, for 750 feet; where are some tombs excavated in the rock at the top, and close on the N. side of the aperture; which continues 125 feet farther to an opening in the rock on the S. side of the excavation, through which the water seems to have forced itself over a rocky, broken, and steep descent directly into the great basin. Hitherto the cut has had the general direction of W. ½ S., but it now sweeps gradually round to the northward, and after a course of 325 feet, it is crossed by an arch bearing some inscriptions scarcely to be made out. The remainder of the hollow is 427 feet long, 30 feet high, and 17 feet wide, still sweeping N. until it terminates abruptly at about 30 feet above the level of the sea: the bottom of the rest of the excavation is tolerably regular, but in this portion it is impeded by large masses of rock lying across it at intervals.

The termination is, as has been remarked already, abrupt and very imperfect; nor are there any marks of the steps mentioned by Polybius, or any other regular communication either from the excavation to the basin or even to its entrance at the piers; but as the rocks are broken away in the latter direction, and as the action of the water has carried away every thing like art in the former, it is by no means impossible that there may have been at one time a flight of steps from the flood-gate to the hollow way; and also a regular channel available at pleasure, descending from the broken aperture, so as to give a rush of water to clear out the port. The Romans were not accustomed to leave such works as this must have been in an imperfect state; such a fine basin as the inner port of Seleucia would naturally receive every kind of attention they could bestow upon what was most likely the winter harbour of their galleys; for the piers were so constructed as to form an outer harbour during the summer season. Opinions are divided about the object of this great work. Some have thought it was for defence, others that it was a mere road to avoid the hill when passing from the sea to the east side of the town, while some fancy it was intended expressly to clean out the mouth of the port. As a defensive object it must have been comparatively useless, since the cut or ditch may be passed along the hill above the two tunnels, and also by means of the bridges; added to which, there is not any flanking defence whatever, which seems to be necessary even if it were but a road, in which case an easier line could have been selected along the foot of the hill directly through the city. Pococke says it was a water-course, and gives as a proof—the native name "Garez," which I find in the Turkish dictionary explained "a subterraneous channel to convey water." Pococke adds, "the water formerly ran through it, but now it does not go that way unless after great floods. It is said that the Arabs, coming into these parts, turned the water to the north-west, where I saw it run by a sort of subterraneous passage. The stream also in some parts takes its old course, though stone walls were built, which are still standing, to turn it another way." There was a stream running through it when I visited these ruins in 1832, and the water then ran along the excavation as far as the broken place, through which it made its way towards the old basin precisely as it did in the time of Pococke: I am therefore inclined to believe, with him, that it must have been really a water-course in any case, but whether to clear the entrance of the port, or simply to collect the several springs and rivulets rising in the hills above the city, (so as to command a supply,) is not so easily determined. There are, however, remains of other water-courses cut in the hills in the direction of the houses. This city was besieged, stormed, and taken by Antiochus the Great. The environs of Seleucia would have been still more celebrated in modern times had it not been for the campaign in Russia. In 1811 Napoleon had prepared a fleet at Toulon which was to have disembarked a large force in this bay, and Mr. Vincent Germain was waiting at Antioch for the expected troops, which had in the mean time been marched to Russia instead of taking the route from Suweïdíyah to India. Mar'ash was to have been the centre of the operations, probably on account of the fine forests near that town; but as the Beïlán Mountains would have furnished plenty of fine timber close at hand, it is not likely that this great Captain would have gone to Mar'ash when 110 miles through Antioch and Aleppo would have placed him at Beles, 200 miles lower down the river; for, although the details of the proposed operations are not known, there is every reason to presume that Napoleon meant to carry his troops down the river Euphrates to Basrah, which was to have been made his first *place d'armes;* and when it is recollected that the fine harbour of Grane (Koreïn) is within one day's march, he would, in all likelihood, have pushed on thither immediately, with a view to the gradual acquisition of Abú Shehr, Bahreïn, Maskat, Kishm, and the other numerous ports on both sides of the Persian Gulf, by gradual movements along the coasts from one place to the other; the Russian campaign however put an end to this, and indeed all his other projects; and the line of the river Euphrates still belongs to our old ally Sultán Mahmúd.

XX.—*A Sketch of the Progress of Geography;—and of the Labours of the Royal Geographical Society, during the year* 1837-8. By Captain WASHINGTON, Royal Navy, Secretary.

EACH year as it passes away contributes its share to the general stock of geographical knowledge, and is marked by the gradual progress of maritime and inland discovery; nor will the past twelve months yield to any since the establishment of this Society, either in the advancement of discovery, or in the importance and extent of the subjects of geographical inquiry brought before us.

For the moment our attention is arrested by the recent intelligence of Arctic discovery, and cheerfully do we award to Messrs. DEASE and SIMPSON their full measure of credit for the triumph they have achieved on the northern coast of America. Discovery has in itself a charm that irresistibly attracts attention from all minor subjects of geographical research, and leads us away from the slowly acquired, yet important, details that are absolutely requisite towards obtaining an accurate knowledge of the surface of the globe which we inhabit.

Such details must, however, necessarily form the chief subject of the following pages; for, although it would be a far easier and far more grateful task, to touch only on the progress of discovery in the various quarters of the world, this would afford a very imperfect report of the works of many laborious men, and would entirely omit the account of the advancement of topographical knowledge during the year.

EUROPE.

In Europe, more especially, such a sketch would be nearly useless, as in this portion of the globe nothing but the most minute details can satisfy the demand of the present advanced state of science:—itineraries and route surveys are obsolete, we may hope, in nearly every kingdom in Europe, and in many of these an accurate geodetic survey is now in progress.

To commence, then, with our own country:—

BRITISH ISLES.—In England, four whole sheets (or sixteen sheets as now issued) of the "Ordnance Map," extending over about 3000 square miles, and comprising parts of the counties of Norfolk, Suffolk, and Denbigh, have been finished during the past year, thus completing seventy-five sheets of the 120 of which the whole map will be composed; the portion published includes all the southern and midland counties, and the greater part of Wales. The surveys for this map are made on the scale of two inches to a statute mile, and are engraved one-fourth of this size, on the scale of one inch to a statute mile, or $\frac{1}{63360}$ of nature; each whole sheet contains 800 square miles of country, and about four times this space is annually surveyed, and engraved in the course of the following year.

Few persons, on looking at this map, reflect on the great labour that must necessarily be bestowed upon it, to bring it to the degree of accuracy of detail and beauty of execution manifest in some of the later sheets; we may direct attention to that containing the mountain of Cader Idris in Merionethshire, as a specimen of drawing and engraving not commonly surpassed: the whole survey is conducted under the superintendence of Colonel Colby, R.E.; the construction and engraving are under the direction of Captain Robe, R.E.; and Lieutenant Bailey, R.E., conducts the field surveys.

A portion of the mountain of Snowdon, engraved in the relief style by Mr. Freebairn from a model and beautiful drawing executed by Mr. G. Carrington, ordnance surveyor, represents admirably the relief of the ground. About four square miles of the Undercliff in the Isle of Wight also have been recently modelled on the scale of three feet to a mile; and might be accurately represented on paper with very little trouble by the above mode of engraving.

Nine sheets of the geological map of England, by Mr. Delabeche, assisted by Mr. M'Lauchlan, ordnance surveyor, including the counties of Cornwall, Devon, and part of Somerset, on the same scale, are also completed.

In Scotland orders have at length been given to commence the new government survey; preparations are making for continuing, in the course of the present summer, the series of primary triangles carried over a great portion of that country in former years by General Mudge and Major Colby.

In the mean time, some few points near Arran, on the west coast, have been determined by Mr. Galbraith of Edinburgh during the past year.*

In Ireland, the "Townland Survey," on the scale of 6 inches to a statute mile (thirty-six times the size of that of England), or $\frac{1}{10560}$ of the natural scale, proceeds rapidly; the counties of Meath, Leitrim, Longford, and Sligo, contained in 160 sheets, have been completed during the past year, making fourteen counties and 530 sheets published.

A map of the whole of Ireland, in four sheets, on the scale of $12\frac{1}{2}$ inches to a degree, based on the Government Survey, carefully executed and well engraved by Messrs. Walker, is just published at Dublin.

In connexion with the Townland Survey, an admirable Me-

---

* Ed. Phil. Journal, April, 1838.

moir of the County of Londonderry (the first of the series) is just published, giving an account of the natural, artificial, and general state of the county; a work which may be taken as a model for all descriptions of counties, and does honour to Lieutenant Larcom, R.E., and to the officers who have superintended its execution.

A curious statistical map of Ireland has also been engraved by Mr. Gardner, from a design by Lieutenant Harness, R.E., showing, by a great variety of shade, but without colour, the density of population in the different parts of the island.

The labours of the hydrographic department, under the direction of Captain Beaufort, R.N., have kept pace with the land survey of the British Isles.

The surveys of the rivers Thames and Medway having met at the Nore Light, Captain Bullock has been employed on the sands between that anchorage and Margate: he has found that many changes of great importance have taken place in their shape, extent, and position, of which neither the Trinity House nor pilots had any knowledge. We understand that the Hydrographic Office intends publishing a comparative view of the present state of these banks with that shown by some charts of the last century.

The nautical survey of the east coast of England is nearly completed, and most of it has been published. The Scottish shore has also been examined by Captain Slater and his assistants as far as the gulf of Cromarty, and in no very long period the whole eastern board of the island from Dover to John O'Groat's house will be engraved. The Orkneys are likewise in the skilful and industrious hands of Mr. Thomas, R.N., and will, when finished, present the same singular and interesting features as the beautiful chart of the Shetlands, which issued from the Hydrographic Office in the course of the past year.

The survey of the North Sea, under Captain Hewett, R.N., is one of the most important naval surveys that has been undertaken by the Admiralty; and the most interesting sheet of it, the result of about 100,000 soundings, will be speedily published. Many of the old banks have been found to be erroneously placed, and wrongly sounded; while numbers of new banks, absolutely unknown before, have been discovered, and others have been doubled in length; one for instance, the *Falls*, is ten miles longer than it was supposed, and has in some places not more than four fathoms on it at low water, while its ridge-like form—the character of all the North Sea banks—is so narrow, that a vessel might well have a deep-water cast before and after crossing it, and remain ignorant of its existence.

The examination of the coast of North Wales, by Lieutenant

Sherringham, R.N., is completed, and has shown how much use may be made of Caernarvon Bay as a port of refuge, if the Trinity House could be induced to light the south-west entrance to the Strait of Menai. The survey is now continuing between Milford Haven and Cardigan Bay in South Wales, which will include that dangerous group of detached rocks which project from St. David's Head, and on which so many vessels have been wrecked.*

The survey of the large æstuary of the Solway Firth, three fourths filled with shallow sands, and yet possessing channels of great value and safety, is nearly concluded; it has already produced a complete revolution in the manner of navigating it, and in the disposition of the buoys and beacons. The inhabitants of its shores are anxious to improve the approaches by lighthouses judiciously placed; and the residents on the shores of the Clyde are eagerly looking to the northern progress of this surveying party, under Lieutenant Robinson, R.N., in order to develop and augment their commercial facilities.

The steam-vessel commanded by Captain Beechey, R.N., was appointed to the survey of the Irish Channel too late last year to have produced any great result; but when finished the survey will be invaluable to the extensive commerce which frequents this channel, by enabling the mariner, who is not too lazy to keep his lead going, to find his way without sight of land, sun, or stars.

The coast from Dublin round by the north to Sligo Bay is now finished, and most of that part of the survey has been already published.

The great internal and deep-water lakes of Neagh and Erne are finished, and the surveyors, Lieutenant Wolfe and his assistant, are now employed on the navigable lakes Ree and Derg, traversed by the Shannon.

Exact registers of the tides kept at the Ports of London, Liverpool, and Bristol, as well as at all the Royal Dockyards, will help to furnish more accurate *data* wherewith to elucidate the theory of tides. The valuable researches of Messrs. Whewell and Lubbock on this subject, published in the Philosophical Transactions, entitle them to the gratitude of all physical geographers; but voyagers and travellers, in whatever part of the maritime world they may chance to be, can materially assist these eminent men in their researches by accurately registering the TIMES OF HIGH WATER daily, for a whole lunation, and the heights if convenient: the former is the principal object.

FRANCE.—The great topographical map of France, in 258 sheets, on the scale of $\frac{1}{80000}$, or $\frac{9}{10}$ of an inch to a geographical

* The writer is indebted to the Hydrographer to the Admiralty for much of the above information.

mile, publishing at the Dépôt de la Guerre, under the direction of General Pelet, has added 12 sheets during the past year, chiefly of the north-eastern provinces, making 54 sheets already published since the year 1833. The second part of the new geometrical description of France, by Colonels Puissant and Corabœuf, containing the results of the geodetic operations on which the survey is based, is completed, and will form part of the eighth volume of the "Mémorial du Dépôt de la Guerre." The Dépôt is also occupied with a map of the department of the Seine, in 9 sheets, on a scale of $\frac{1}{40000}$, or $1\frac{8}{10}$ inches to a geographical mile; and with a topographical map of Guyenne, in 54 sheets, which is nearly finished.

The *Pilote Français,* published under the direction of Admiral Hamelin, Chef du Dépôt de la Marine, ably seconded by MM. Beautemps-Beaupré and Daussy, containing the survey of all the coasts of France, advances rapidly to its termination. This noble work, of which that nation may justly boast, begun in 1816, will probably be completed during the present year.* The Dépôt de la Marine, besides other works which will be mentioned in their place, have also published the Voyage of the *Thétis* and *Espérance,* executed in 1825 and 1826, under the orders of M. de Bougainville, a copy of which, with their accustomed liberality, they have presented to the Library of the Geographical Society.

Other valuable publications, by men well known for their geographical works, are in progress in Paris, but there is one far too important to seamen not to claim distinct notice, namely, a table of "*Positions Géographiques,*" by M. Daussy, published annually in the Appendix to the Connaissance des Tems; this is revised every year, all recent determinations are added to it, and where any discrepancy appears the subject is discussed: that the results arrived at may not be always infallible, and that the decisions may be occasionally arbitrary,† does not prevent the work being very useful, and we sincerely hope M. Daussy may have leisure to continue carefully his "*additions,*" convinced that a more acceptable service to geography cannot be made.

Sweden and Norway.—Besides Colonel Forsell's map of Scandinavia in eight sheets, we learn that he has constructed a continuation of it in six sheets for the northern part of the country, but is not sufficiently satisfied with its accuracy to have it engraved.‡ In Norway they are now occupied in surveying the coast from Trondheim to the North Cape; but some time must elapse before we can look for a detailed map of this country.

* Notice des Travaux de la Société de Géographie en 1837, par M. Noel Desvergers, p. 274.

† See p. 240.

‡ Col. Forsell's Letter.

Geography will probably benefit by the visit to Lapland of M. Gaimard and his companions during the present summer.

The Society is indebted to the Foreign Office for an account of five Norwegians of Hammerfest, left during the winter of 1836 on one of the "Thousand Isles" near the south-eastern extremity of Spitzbergen, affording another proof of the endurance of man under circumstances of no common privation.*

ICELAND.—M. Marmier's *Lettres* give the account of the late expedition to this island, and the ninth *livraison* of the *Voyage en Islande*, by MM. Gaimard and Robert has just been published, at Paris, by the Dépôt de la Marine, and liberally presented to the Library of the Geographical Society.

BELGIUM.—The spirited director of the geographical establishment at Brusselles has lately published a geometrical plan of the city by M. Craan, in four sheets, on the scale of $\frac{1}{2500}$; two sheets of the map of the environs of Brussels, in nine sheets; and the first sheet of a cadastral map of Belgium, besides some statistical works.

PRUSSIA.—Engelhardt's map of Prussia, in twenty-four sheets, seems to be the best published; besides this there are various special maps of portions of this kingdom on a large scale. Berlin may be considered the head-quarters of geography in Germany; its Geographical Society has just held its fifth anniversary: Ritter here writes his admirable *Erd-kunde;* Berghaus publishes his *Annalen;* his *Atlas von Asien;* and in the present year he has added a *Physikalische Atlas* to his former labours.

SAXONY.—The great topographical map of this kingdom is publishing at Dresden under the direction of Colonel Oberreit, in twenty sheets, on the scale of $\frac{1}{57600}$, or 1·25 inch to a geographical mile; it is founded on a survey executed by the engineers between 1781-1805, and gives the height of all places above the sea. Five sheets are already published; the representation of the mountains is according to scale, and the whole is admirably engraved—the general effect perhaps is rather too dark.

The Statistical Society of Saxony, under Dr. Schlieben, steadily continues its labours.

HESSE DARMSTADT.—Twenty-five sheets of a map of this territory, to be contained in fifty sheets, on the scale of $\frac{1}{50000}$, from an accurate survey by M. Von Eckhardt, of Darmstadt, are now published. The mountains are represented according to the plan of Major Lehman.

FRANKFORT.—A map of the territory of Frankfort and its immediate neighbourhood, on the scale of $\frac{1}{75000}$, and based on trigonometrical points, has just been published by M. Ravenstein; who has also completed a portion of a beautiful model of the environs

* MS. in Library.

of the city and of the Taunas mountains, which accurately represents the relief of that picturesque country; it is on the horizontal scale of $\frac{1}{30000}$, and that of $\frac{1}{5000}$ for the elevation.

It is proposed to make a model of the Rheinlande from Mainz to Bonn, including the vallies of the Main and Lahn: to judge of it by the specimen, kindly presented by M. Ravenstein to the Geographical Society, it will be a very valuable work.

Several communications of much interest have been read before the Geographical Society at Frankfort during the past year, of which we may notice that on the Bolivian province Chiquitos, by the president, Dr. Kriegk; on the great coal fields of England, by M. Meidinger; and an ethnographical sketch of Old Palestine, by M. Seligmann.*

Baden and Wurtemberg.—Twenty-four sheets of the great topographical map of Wurtemberg, in thirty-eight sheets, are published; a map also of this kingdom, and of the duchy of Baden, in twelve sheets, on the scale of $\frac{1}{200000}$, or about three miles to an inch, has been finished during the past year by M. Woerl at Freiburg, and liberally presented to the Geographical Society. The map is remarkable in having the rivers, roads, and boundaries stamped upon it in colour. A model, in relief, also, of these countries, has been executed by M. C. Rath, at Heilbronn.

Switzerland.—The map of this country by Colonel Dufour, we learn from M. Chaix, is in progress; as also an orography, or description and height of mountains, within a circuit of twenty-five leagues around Geneva. MM. Woerl and Heck have also published a map of Switzerland in twenty sheets, on the scale of $\frac{1}{200000}$; and we learn that M. von Michaelis is just appointed to survey the canton of Aargau.

Bavaria.—Eighty sheets of the government topographical map of this country, to be completed in 103 sheets, have been published.

Austria.—The maps of the great government survey of this country, under the direction of Colonel Skribaneck, of the Bureau Topographique, engraved on the scale of $\frac{1}{144000}$, will consist of Austria (Ober and Nieder) in forty-seven sheets; the Tyrol and Vorarlberg, in twenty-four sheets; Salzburg, &c. in fifteen sheets; Styria and Illyria, in thirty-seven sheets; the first sheets appeared in 1811, and about one-third of the whole are now published: the survey of Moravia is finished, but of Bohemia, Croatia, and Gallicia no survey exists.†

Hungary.—The first seven sheets of a map of Hungary, in nine sheets, by Professor Schedius, at Pesth, are lately published; a government survey is also commenced. The rapidly-

* M. Ravenstein's Letter. † Baron C. Hügel's Letter.

increasing communication down the Danube will doubtless shortly lead to a better acquaintance with this fine country, as also with the adjoining states of Servia, Bosnia, and Montenegro,* which the pen of Vuk Stephanowich has shown to be well worth cultivating.

ITALY.—Thirty-six sheets of the great Austrian map of the Lombard-Venetian kingdom, in 43 sheets, are published. This map, drawn on the scale of $\frac{1}{28800}$, and engraved on the scale of $\frac{1}{86400}$, or ·83 of an inch to a geographical mile, is constructed on a grand triangulation, carried on from a base measured near Somma, on the banks of the Ticino;—it is engraved in a good bold style, and the glaciers are very well represented—the heights of the mountains are not generally marked.

SARDINIA.—Three volumes of Casali's Dictionary of this monarchy are finished, comprising the article *Casale;* and Zuccagni Orlandini's Corografia is not yet carried beyond the kingdom of Sardinia.

PARMA, PIACENZA.—Molossi's Geographical Dictionary of these Duchies is nearly completed; an analysis of this work has been kindly presented to the Library of the Society by our active correspondent Count Grăberg af Hemsö, at Florence.

At Bologna a Geographical Journal is about to be published by Count Ranuzzi, aided, it is said, by Chevalier Balbi: we trust that the authors will not content themselves with reproducing articles from other Geographical Journals, but give us an original account of much that Italy possesses of interest to the geographer, viz. the old maps of the Pizzicagni, at Modena; of Fra Mauro, at Venice; of the Portulani, of Benincasa, &c. &c.

TUSCANY.—Repetti's Geographical Dictionary of the Grand Duchy has reached the word *Livorno.* The last fasciculus contains an excellent article on the *Littorale Toscano.* This tract of land, about 203 miles in length, from the river Magra to the torrent of the Chiarone, the limit of the Roman territory, is divided into six basins, the three southernmost are usually called the Maremma; the detailed description of them embraces many points of interest in comparative geography.†

The 5th part of Count Serristori's Statistica d'Italia gives a full account of the statistics of Tuscany; the 6th fasciculus reveals, for the first time, we believe, the statistics of the Pontifical states.

NAPLES.—The survey of this kingdom, the execution of which is under the direction of Colonel Visconti, has been connected by a careful triangulation with Northern Italy. The drawings are made on the scale of $\frac{1}{20000}$, and will be published on the scale of $\frac{1}{80000}$,

* Montenegro und die Montenegrini. Stuttgart, 1837.
† Letter of Count Grăberg af Hemsö.

or $\frac{9}{10}$ of an inch to a geographical mile. The survey of the northern frontier was completed in 1836, and also part of the eastern coast, from Ascoli to Otranto. In the mean time a military map of the kingdom, on a smaller scale, viz. $\frac{1}{320000}$, was begun this spring, and will be advanced as fast as is consistent with such a work.

The 'Annali Civili delle Due Sicilie,' liberally presented by Colonel Visconti to the Library of the Society, contain much geographical information, with the measurements of many heights of mountains, &c.; as also does the 'Viaggio alla Meta,' &c., by Signor Capocci, Director of the Observatory at Cape di Monte, to whom and to Colonel Visconti we are indebted for much of the above information.

GREECE.—The map of Northern Greece, made by the French officers prior to the year 1836, is preparing for publication, in four sheets, at the Dépôt de la Guerre; a few officers are still employed surveying the country.* A paper on the Minoa of Megara, by Mr. Spratt, R.N., clears up a point of much interest in classical geography, and is highly creditable to this young officer employed on the survey of the shores of the Mediterranean.

TURKEY.—Weiss's general map of this empire, in twenty-one sheets, published in 1829, is the best that exists; portions of the country which have been the seat of war have been more accurately examined; and especially the valuable series of eighty-nine astronomical points in Wallachia, Bulgaria, and Roumelia, lately determined by officers in the Russian service, Captains Wrontschenko, Ortenberg, and Von Essen, and made public in June, 1837;† still no complete survey exists, and an intelligent traveller may here find sufficient occupation, and even discoveries to make, as has lately been shown by M. Ami Boué in his account of this country.

RUSSIA.—Thirty sheets of the great topographical map of this vast empire, to be completed in fifty-nine sheets, on the scale of $\frac{1}{420000}$, or about six miles to an inch, are now published. Much information relating to the Ural range is afforded us in the work of M. Rose; and we learn that M. Fedorow, the astronomer, lately returned to St. Petersburgh, has measured trigonometrically several of the peaks in the more northern part of this range, and found them to rise between 8000 and 9000 feet above the sea. M. Helmersen has also given a description of the Ural Mountains, and M. Hamel a note on the level of Moskow, the river Moskwa and the Oka, above the sea.‡

CAUCASIA.—At the southern extreme of the vast Russian empire, on the limits of Europe and Asia, the long contested

* Notice Annuelle, par M. Noel Desvergers, p. 277.
† Bulletin de l'Académie de St. Petersburg, vol. ii. p. 210.
‡ Bulletin de l'Académie de St. Petersburg, vol. ii. p. 300.

question in physical geography of the depression of the Caspian Sea has been finally set at rest by the trigonometric levelling, from Novo Tcherkask, near the mouth of the River Don, by Stavropol to Kisliar on the Caspian; whence it results that the level of the latter is 101 feet lower than that of the Black Sea—a remarkable coincidence with the result obtained barometrically by Meyer and Lenz in 1835.* The decision of this question is highly creditable to Messrs. Parrot and Struve, who proposed the expedition, and to Messrs. G. Fuss, Sabler, and Sawitch, who have carried it into effect at the expense of the Russian government. Much light must also be thrown on the topography of the country, 400 miles in extent, through which the levelling was carried, and of its immediate vicinity; the details, however, are not yet made public.

M. Koppen has just published a much improved map, in four sheets, of the southern part of the Crimea. M. Eichwald, also, has inserted all the latest researches in his map of the Caucasian provinces, which illustrates the second part of his journey to the Caspian recently published; and lastly, we have the account by M. Dubois de Montpéreux, of Neufchatel, of his journey through the Crimea and the Caucasus, which he seems to have carefully examined both as a geologist and historian, as well as a geographer.

ASIA.

The vast continent of Asia still offers to travellers a fertile field for geographical research, and throughout the greater portion of it we can only look for a few astronomical points, well fixed and careful itineraries, to enable us to correct our maps. Of the actual state of our knowledge of this country, the admirable work of Professor Ritter, of Berlin, *Erdkunde von Asien,* offers an able summary; and next to this, Berghaus' *Atlas von Asien,* which has now reached its fifth part, claims notice as the best work on Asia in general.

SIBERIA.—Commencing at the north, we have just received the second volume of M. Erman's journey, describing his route from Tobolsk, by Irkutsk, Kiakhta, and Yakutsk, to Okotsk; the description of the Lake Baikal, of the valley of the Lena, of the Aldan Mountains, and his remarks on the origin and manners of the inhabitants of the little known countries of Northern Asia will be read with great interest.

Farther to the south, we have, at length, the account of the journey of MM. Humboldt, Ehrenberg, and Rose, in 1829, who travelled along the eastern foot of the Ural range as far as Bogoslovsk in 60° N.; thence they returned to Tobolsk, crossed the

* Bull. de l'Acad. de St. Petersburg, vol. i. p. 2.

great steppe of Barabinska to Bernaul, and the Altai Mountains, whose highest peak, *Bieluka*, we learn from M. Gebler, reaches 11,000 feet above the sea. Proceeding thence to Baty, a Chinese military post on the banks of the Irtysh, they returned across the steppe of Ishim to the southern range of the Ural, and thence to Astrakan on the Caspian Sea.

Accompanying the narrative is an improved map of part of Northern Asia, between the limits of 51° and 60° N. lat., and 47° and 69° E. long., on the scale of $\frac{1}{2000000}$, or two inches and a quarter to a degree, which forms a valuable addition to the topography of that region.

M. Baer has communicated to the Society a notice on the frozen soil at Yakutsk, whence it appears that the frost there penetrates to the depth of 380 feet: M. Erman has corroborated this circumstance, and states that the mean temperature at Yakutsk being — 6° of Réaumur, even a greater depth of frozen ground might have been expected.

KAMCHATKA.—A map of this remarkable peninsula, showing the position and giving the heights of its series of volcanoes or *sopki*, has just been completed by M. Erman. Admiral Lütké also, in his excellent *Voyage autour du Monde*, has given an account of thirteen, and the heights of many, of these volcanoes: he has also determined the chief points of the eastern coast of Kamchatka, and of the country of the Koriaks and the Chukchis, to the north-eastern extremity of Asia.

NOVAIA ZEMLIA.—A very excellent report upon all former voyages to these inhospitable shores has been presented by M. Baer to the Academy of Sciences at St. Petersburg,* from which we learn that the islands may be divided in half, longitudinally, and the whole of the eastern portion must be erased from our maps. M. Baer has himself visited the islands during the past summer, but the ice prevented any further examination of the northern shores.

TIBET.—The journey of *Shi fa hian*, a Buddhist priest and pilgrim, although made at the close of the fourth century, is too remarkable not to be noticed here. The work is entitled *Foe Koue Ki*, a Chinese MS., which has been translated by the eminent orientalists Rémusat, Klaproth, and Landresse, and gives the account of the pilgrim's visit to the chief seats of his religion during a space of fifteen years. From China *Shi fa hian* travelled to the north-west and west, across the great desert of Shamo to the lake Lob; thence to Khoten, and by the Hindú Kúsh, probably to Peshâwar, across the affluents of the Indus, by Benáres to Tamlúk, where he embarked for Ceylon and Java, and thence

* Bull. de l'Acad. de St. Petersburg, vol. ii., p. 138.

returned to his own country. This work shows that the names of places throughout India were Sanskrit; that the Pálí language, its immediate derivative, was studied from Ceylon to Khoten; that the Buddhist religion was then flourishing in the great desert and upper course of the Indus; and that in the fourth century the Hindús made voyages in vessels, containing 200 persons, from Bengal to Ceylon, Java, and China.

Much information is contained in the notes of the learned commentators of this work; and Professor Wilson, from his knowledge of Sanskrit, has thrown still more light on the subject in his analysis of the work, read before the Asiatic Society of London.

HINDU'STA'N.—The travels of Moorcroft and Trebeck in the Himálayan provinces of Hindústán and the Panjáb, in Ládákh, Kashmír, Kábul, and Bokhárá, very ably edited by Professor Wilson, contain a great addition to our knowledge of this region: the work is accompanied by a map by Mr. John Arrowsmith, very carefully compiled from the best sources. The narrative of the journey of Victor Jacquemont in India, publishing at Paris, has reached the 17th Number. The Society is indebted to Captain Pemberton for a detailed account of the district of Manípúr, on the north-east frontier of India, accompanied by a series of maps; the itinerary of two Burmese embassies from Ava to Pekin in 1823 and 1833, by Major Burney;* and in the west of India Captain Wade's account of the river Setlej from Ludhiyánah to Mithan'kót,† have been published in the admirably conducted Journal of the Asiatic Society of Bengal. Thirty-six sheets of the great trigonometrical map of India, to be completed in 150, on the scale of four miles to an inch, are already published, and the survey, under the direction of Colonel Everest, E.I.C. Artillery, is now extending through the country north of the river Khrishnah.

BOMBAY.—The Geographical Society at Bombay proceeds steadily in its useful task of collecting and printing papers relative to India: many reports of high interest are contained in the two published volumes of its proceedings: it has also contributed to our Journal during the past year an account of the bore in the Gulf of Cambay, by Lieutenant Ethersey. This officer, as well as many others of the Indian navy, are employed on the nautical survey of the coasts and rivers of the western portion of India. The mission under Captain Burnes will doubtless furnish every information necessary to complete the geography of the main stream of the Indus from Attock to the sea, as well as of the adjacent provinces.‡ The survey of the mouths of the Indus by Lieutenant Carless, I.N., begun in 1836, has been continued

* Journal of the Asiatic Society of Bengal, Nos. 66 and 67.

† Ibid., No. 63, p. 169. ‡ Report of Bombay Geographical Society, 1837.

during the past year: the Kedíwárí branch has been surveyed from its mouth to its outlet; the Hajámrí branch was completed the year before, and we have now a trigonometrical survey of all the *open* mouths and branches, with a portion of the main rivers and the most dangerous part of the coast.

The surveys of the coast of Kat'híwár and Gulf of Cambay have been completed by Lieutenants Whitelock and Ethersey; the examination of the latter has extended as far south as Surat, including the mouths of the rivers Sábarmatí, M'hai, Dhandar, and Narbadá, which he has examined as far as the city of B'haróch.

To explore the regions west of the Indus an enterprising officer started from Bhúj Kach'h in January, 1837. He crossed the Thar in an almost direct line to Khaïrpúr, and arrived safely at Shikarpúr in his progress towards Belúchistán. The Thar, it is said, was found by this officer to present a less desolate aspect than is generally believed. Water was abundant on the line of his route, and the country was observed to be traversed by ranges of low hills, composed of sandstone, and thickly wooded.*

The chain of the Maldives has been completely surveyed by Captain Moresby, who is now engaged on a similar survey of the Chagos Archipelago. This officer will afterwards proceed to the examination of the bank *Saya de Malha,* situated about 300 miles south-east of the Seychelles; and probably also will survey the great and little *Basses* lying off the south-east point of Ceylon.† While on the subject of Ceylon, we must not omit to notice the history of this island by Joaõ Ribeiro, which has been published in the memoirs of the Academy of Sciences at Lisbon by their indefatigable secretary the Counsellor Macedo, from a rare and complete MS. It forms the fifth volume of the collection of *Memorias, &c., das nações ultramarinhas,* liberally presented to the Society by M. Macedo.

ARABIA.—The details of Lieutenant Wellsted's journey into the interior of 'Omán have been published during the past year; and the narrative proves even more attractive than was anticipated from the abstract of geographical information so promptly furnished to our Journal, and thereby doubling its value. We trust that some of the officers of the Indian navy now engaged in the survey of the southern coast of Arabia will profit by Mr. Wellsted's example, and endeavour to penetrate into the inland provinces of Hadramaut, Mahrah, &c. Of the survey of the Gulf of Akabah, at the head of the Red Sea, we have an excellent memoir by Lieut. Carless, I.N.; also a full description of Kishm and the islands at the entrance of the Persian Gulf, by Lieut. Whitelock, I.N.

* Report of Bombay Geographical Society, 1837. † Ibid.

PERSIA.—The comparative geography of Susiana or Khuzistán, a subject which has been discussed by D'Anville, Rennell, Vincent, Von Hammer, and partially by Kinneir, seems to be on the point of being cleared up, partly by the labours of the Euphrates Expedition, during which the rivers Karún, Bahamishír, and Kornah, were partially examined and much new topographical information obtained; but more especially by the researches of Major Rawlinson, who, we understand, has devoted some years to the elucidation of this difficult subject—difficult, chiefly for want of correct topographical information. It would seem that the canal and river at Shúster have hitherto been confounded; that the name Abzál is erroneously applied to the Desfúl river; that the river Shapúr is omitted in our maps, besides many other points in the positive geography of Khuzistán that we trust ere long to see set right.

The Society is indebted to the kindness of Her Majesty's Secretary of State for Foreign Affairs for the communication of Major D'Arcy Todd's two separate routes through the provinces of Tálish, Gilán, and Mázanderán; and to Mr. Taylor Thomson for an account of the first ascent of the Peak of Demávend by an European, when its height was ascertained barometrically to be 15,000 feet above the sea, and 11,000 above the level of the plain of Tehrán.

KURDISTA'N.—In this country, so difficult of access, we have Colonel Shiel's journey from Tabríz to Ván, round the southern shores of the lake to Bítlís, to Se'rt, then along the western bank of the Tigris to Jezireh, Erbil, and Suleïmániyeh, during which much useful geographical and other information was obtained. Mr. Ainsworth, too, on his homeward route from Bagdad by Mosúl, in the course of the last year, visited Nimrúd (the Larissa of Xenophon), Nineveh, Nisibin in Sinjár, Márdin, and by Diyar Bekr reached Constantinople. In this journey he observed barometrically the elevation of the various mountain ranges, and noted the dip and succession of their strata. From the expedition under Mr. Ainsworth, shortly about to proceed to Kurdistán, we hope to reap a rich harvest in comparative and physical geography.

ASIA MINOR.—In this classic land of travels we, at length, begin to collect accurate geographic information; and Mr. W. I. Hamilton in his late journeys throughout the country, from Kars and the ruins of Anni on the east, to Smyrna on the west, has set an example, which we trust will be followed by all future travellers. By his various routes he has connected many important places with the north coast; has obtained observations for latitude at more than 40 separate stations; has kept an exact itinerary by which the whole of his journey is being mapped on the scale of an inch to a mile; has fixed the site of several ancient places, and described remains and ruins which had escaped all former travellers; has given the best geological account hitherto furnished of this singularly elevated table-land; and, lastly, by a well-conducted series of barometrical observations, he has measured the height above the sea of its various plains and mountain ranges, and has completed his work by successfully carrying his barometer to the summit of Mount Argæus, and thereby being the first to prove that this mountain reaches the great elevation of 13,000 feet above the sea. Further to the east the astronomical positions determined by M. Fedorow in Ararat, and 14 points fixed by Captain Birdin of the Russian army in Caucasus and Asia Minor, have been examined by M. Struve and made public during the past year.* Of the routes followed by Captain Camille Callier, and M. Texier in 1833-4, nothing but a brief outline has yet been made public; we learn that the map of the former on the scale of $\frac{1}{500000}$, or about eight inches and a half to the degree, has been commenced at Paris.†

Returning to the west, the survey of the coast of Anatolia under Lieut. Graves, R.N., assisted by Lieut. Brock, will shortly be united to that of Ḳaramania by Captain Beaufort; it is now advancing into the deep gulfs near Cos, Halicarnassus, and Cnidus, where all is classic ground, and every day throws fresh light on ancient history; their researches also show not only where new marts may be opened to European commerce, but new paths for the progress of civilization, industry, and knowledge.

We are indebted to Mr. Brooke for an account of the little-known island and gulf of Symi which he visited in his yacht last year, and we learn that this spirited yacht-sailor is just on the point of starting to examine some of the numerous islands in the Indian Archipelago.

Syria.—Thanks to the expedition under Colonel Chesney and Mr. Ainsworth, we are here on known ground. In spite of climate, delays, and difficulties that would have deterred any but such resolute officers and men as composed this party, materials for a correct map of a large portion of Northern Syria have been collected; a line of levels has been carried across from the Mediterranean to the Euphrates; Northern Mesopotamia has been explored, and the survey of the great river Euphrates from Bírehjík to its outlet in the Persian Gulf, a distance, including windings, of about 1200 miles, has been effected. We have thus a certain base line extending in a north-west and south-east direction through 7° of latitude and 12° of longitude, from Iskenderún to Baṣrah, whence all future travellers may confidently take

* Bull. de l'Acad. de St. Petersburg, vol. ii., p. 219.

† Notice Ann. par M. Noel Desvergers, p. 280.

their departure when exploring either to the northward or southward of this line, and remember with gratitude what they owe to their countryman Lieut. Murphy, of the Royal Engineers, whose life fell a sacrifice to his exertions, and the climate, at Basrah.

A summary of the Euphrates Expedition, giving an outline of its most important features, has been published in this Journal,* while the more detailed narrative of its various incidents is preparing to meet the anxiety which the public naturally feels in an enterprise which was, in many points of view, one of the most interesting that ever left the British shores.

It were hardly necessary to add that the Royal premium "for the promotion of geographical science and discovery" has been awarded to Colonel Chesney, R.A., as leader of this expedition. Some valuable notes on the comparative geography of the Cilician and Syrian Gates by Mr. Ainsworth, and a memoir on the Bay of Antioch, and on the ruins of Seleucia Pieria by Colonel Chesney, clear up many of our former difficulties in reading the ancient accounts of the celebrated battle of Issus.

PALESTINE.—A remarkable point in physical geography has been here brought to light by the praiseworthy researches of Mr. G. Moore and Mr. Beek, who during the spring of the past year examined part of the shores of the Dead Sea, and carried a line of soundings across it; in the course of their examination they found, by the temperature of boiling water, that the level of the Dead Sea was 500 feet *below* the level of the Mediterranean. The truth of their experiments has been proved by the results obtained barometrically by Professor Schubert of Munich, who has lately found the lake of Tiberias to be 500 feet, and the Dead Sea to be 598 feet *below* the Mediterranean.† This is so remarkable a fact that we hope some traveller, supplied with good instruments, will soon visit Palestine to determine the level of this inland sea, and then explore the country southward along the valleys of El Ghor and El Arabah, a direct distance of only 100 miles, to the Gulf of Akabah, at the head of the Red Sea.

AFRICA.

The compact and unbroken mass of this great continent, where no deep gulf or navigable river of easy access opens a road into the interior for commerce and civilization, seems to offer an insurmountable barrier to the progress of discovery; nor can we say that geographical investigation has advanced more than a few steps beyond its coasts during the past year.

NORTHERN AFRICA.—Some few of the notes made by our

* Vol. vii. p. 411

† Allgemeine Zeitung, Nos. 83 and 109.

lamented countryman, Davidson, while in the city of Marocco and at Wad-Nún, in addition to those published last year,* have reached England. We had confidently hoped ere this to have heard of the arrival of Abú Bekr, his travelling companion, at Tumbúktú, but at the date of the last dispatches the return cafilah had not reached Mogador.

A valuable paper on the names of places in the empire of Marocco, written in the Arabic character, has been contributed to the Journal by our zealous honorary member, Count Gråberg de Hemsö. When we consider the barbarous orthography that disgraces all our maps of Africa, and of the East, we look upon this attempt to introduce a system of spelling according to a fixed standard as one of the most useful works that has yet appeared in our pages. If the Geographical Societies of Paris, of Berlin, of Frankfort, and of Bombay, could be induced to adopt some general standard for the orthography of Arabic, Turkish, and Persian names, perhaps all geographers, even if they might not fully approve of it, would for utility's sake adopt it.

The chart of the western coast of Marocco from Cape Spartel to Cape Bojador, from the survey of the late Lieutenant Arlett, R.N., are just published, at the Hydrographic Office, on the scale of eight inches to a degree; while to the southward, as far as Cape Verd, being chiefly the sea-coast of the Sahara, or great desert, the charts are engraved on one-fourth of the above scale, or four inches to a degree. Captain Vidal, R.N., during the past year has completed the survey of the Canary Islands.

ALGIERS.—The rapidly growing intercourse between France and her recent conquest in Africa promises to make us much better acquainted with this territory. An excellent account of the route pursued by the French army in its advance upon Ḳosṭantinah from Bonah by Sir Grenville Temple and his companion, M. de Falbe, has been kindly communicated to the Geographical Journal.† In this valuable report great attention has been paid to the Arabic orthography in the names of places, which so usually disfigure our maps; the sites of ancient cities are noticed, the physical geography of the country is well described, and a series of barometrical measurements gives us the height of thirty-one points, among which we learn that the city of Ḳosṭanṭinah stands 2900 feet above the sea. In addition to this, M. Dureau de la Malle has, in an able essay,‡ examined the ancient writers on the subject of Numidia, and shown where it is desirable to make researches, in order to throw light on its comparative geography.

A full report also has been given to the French government on the present state and resources of the newly-acquired province.

* Journal, vol. vii. p. 144. † Vol. viii. pp. 39 and 124.
‡ Province de Constantine, p. 125.

The officers of engineers have made a detailed plan of the city of Kostantinah; some of the exploring parties have pushed forward to the south-west, while others have explored a more direct road from the capital to the sea-coast at Storah. M. Dureau de la Malle has very opportunely published the travels of Peyssonnel and Desfontaines in the regencies of Tunis and Algiers, in the years 1784-6.* Colonel Harding, R.E., has communicated to the Society his journal† of a route to Kostantinah in 1820; and lastly, the Dépôt de la Marine has published the beautiful charts of the north coast of Africa from Cape Serrát in 9° 12′ E., to the Ja'farín Islands in 2° 26′ W., a direct distance of more than 600 miles.

At Carthage, Sir G. Temple and M. de Falbe are actively employed in exploring the ruins of the ancient city, and have been successful in bringing some inscriptions to light; the latter officer has lately made an excursion into the interior of the Beylik, in order to fix positions of some few places.

EGYPT.—Mr. Wilkinson's admirable work on the 'Manners and Customs of the Ancient Egyptians,' and Mr. Hoskins' 'Visit to the Great Oasis in the Libyan Desert,' are the two principal works on Egypt during the past year.

ABYSSINIA.—A letter from M. J. Russegger, chief of the Austrian mining expedition, dated Roserres, in lat. 12° 3′ N., in the country of Fasoglo, on the south-western frontier of Abyssinia, Dec. 19, 1837, gives us the most recent and best account of the progress of discovery in this quarter.‡ From a former journey into the country of the Nuba negroes, south of Kordofan, M. Russegger has reason to believe that the existence of the Jebel Kúmri, or Mountains of the Moon, in the position in which they are marked in our maps, is very problematical; and, with respect to the Bahr-el-Abiad, or White River, he believes that it flows in a direction contrary to that assigned in our maps, for it rises in the country of the Galla and Shangalla, and flows parallel to the Bahr-el-Azrek, or Blue River. M. Russegger also mentions many discoveries in mineralogy; and that he was shortly about to start on his second journey to the south, with the hope of reaching Fadassi, in the country of the Galla, whither no European has yet advanced. We heartily wish success to this spirited traveller.

We still anxiously expect the account of Dr. Rüppell's late journeys in this country; in the mean time, the travels of MM. Tamisier and Combes, mentioned in our last year's 'Sketch,' have been published at Paris. The survey of the north-eastern

* 2 vols. 8vo. Paris, May, 1838.

† MS. in library.

‡ Athenæum, No. 549. A periodical in which the most recent and accurate geographical information will usually be found.

extreme of Africa, by the officers of the Indian navy, is advancing to Cape Gardafui, but the details have not reached this country.

Western Africa.—Traversing the continent in this parallel we reach the Bight of Benin, to which the survey of the Áshántí coast will shortly extend. This tedious undertaking is now drawing to a close, and will then be of equal utility to the fair traders and to the anti-slavery cruisers. It is fortunately in the hands of such a man as Captain Vidal, R.N., who has steadily devoted himself during a long period of ill health to complete this unpopular work, and to connect with it a minute examination of the Canary Islands.

This gigantic survey, embracing the east and west coasts of Africa, from the Isthmus of Suez [Suweis] round by the Cape of Good Hope to the Pillars of Hercules, may be said to have been drawn and coloured with drops of blood. Twice did Captain Owen change his whole crew and officers; those accomplished surveyors, Captains Boteler and Skyring, fell a sacrifice during its progress, and now, in the hour of conclusion, the crews of the Etna and Raven have all but shared the same fate.

An account of Messrs. Oldfield and Becroft's ascent of the Old Calabar River will be found in the Geographical Journal, and we learn that Mr. Becroft has just returned from his fourth ascent of the River Quorra [Kawára], during which he found the natives very peaceable and well disposed to trade.

During the past year, the narrative of Mr. M'Gregor Laird, and Mr. Oldfield's ascent of this great river, in 1833, has been made public. It contains the full details of their voyage, related in a simple, straightforward manner, with a useful appendix by Mr. Laird, on the best means of establishing a trade with the interior of these countries, and the probable chance of success.

South Africa.—The important journey from which Captain Alexander is just returned, to the north-west of the Cape colony, through the country of the Great Namáquas, Boschmans, and Hill Dámaras, offers a route of 1500 miles, which has never appeared in any former map of Southern Africa. Crossing the Orange River at about 100 miles from its mouth, the traveller proceeded to the north-east as far as Nabís, the last missionary station; thence to the north-west, crossing the 'Oup or Great Fish River, in the parallel of 27° nearly; then, inclining to the northwards, he traversed many of its western tributaries, and reached the Kei Kaap or Great Flat, over which he and his party journeyed as far as the Kópumnaas or Bull's Mouth Pass, a remarkable defile, in the parallel of 24°, winding, for a distance of 40 miles, between mountains of from 2000 to 3000 feet elevation; he then emerged on the Great Desert of Tans, on the verge of the tropic, where the whole party nearly perished for want of water. Overcoming this difficulty, they reached the banks of the River Kúisíp, and followed its western course to the sea, at Walvisch Bay, in lat. 22° 52½', being the first Europeans that had reached it by land from the Cape of Good Hope. Quitting the sea-coast Capt. Alexander ascended the bed of the Kúisíp River in an east direction about 200 miles, to a large village named Ni-ais, inhabited by Dámaras of the Hills, situated in a fertile, well-watered country, abounding in cattle. From this point the expedition turned to the southward, crossed many of the western affluents of the Great Fish River, reached the Orange River at about 40 miles from its mouth, and returned to the Cape of Good Hope, after an absence of one year and ten days, bringing with them a large collection in natural history.

As before mentioned, this journey fills up a great blank in the south-western portion of Africa, and much credit is due to Captain Alexander for his perseverance in carrying through an expedition which throughout was exposed to very great privations and hardships. The information brought home also is highly important as establishing the fact of a station in the interior 200 miles from the coast, situated in a fertile country, inhabited by friendly natives, easily accessible by a few days' journey from Walvisch Bay, and in every respect affording all we could desire as a starting point to enable future travellers to traverse the continent of Africa in the parallel of 23°, and thus connect Captain Alexander's discoveries on the west with the labours of Dr. Andrew Smith, who reached the tropic in the year 1835, only 300 miles farther to the east.

The enterprising spirit of the frontier boors of the Cape colony, their roving habits, and circumstances resulting from the late Káfir war, have combined to cause a tide of emigration to set towards the north-east in a manner unknown in the annals of colonial history. A body of 5000 farmers, including their families, with cattle and sheep to a very great amount, have left the colony with the intention of establishing themselves wherever they can find a suitable place beyond its boundary. At first they pushed their exploring parties within a few days' journey of Delagoa Bay; but, from the latest accounts, they seem to have established themselves about the parallel of Port Natal, to the west of the Quathlamba range of mountains, through which, it appears, they have found a passage, and sent a large body of their delegates to the settlement at Port Natal. This irruption must shortly lead to a far better acquaintance with the geography of this portion of Africa.

AMERICA.

Traversing the Atlantic Ocean to the western hemisphere, we naturally follow the course of Captain Back in H.M.S. Terror, and, after reading his vivid description of the extraordinary circumstances that befel him and his crew in Hudson's Straits, and their providential escape from a ten months' imprisonment in the ice, we must feel thankful for their preservation, rather than regret that, on this occasion, the physical obstacles which were opposed to their progress were insurmountable; yet, although the main object was not accomplished, the voyage was far from being barren in results; and the examination of 120 miles of the north-eastern shore of Southampton Island, from Cape Comfort to Sea-horse Point, now for the first time laid down on our charts by Lieutenant Owen Stanley, is a proof that, even when drifted along at the mercy of a field of ice, an experienced observer will turn his opportunities to advantage: also, the very fact of the continued drift to the eastward is a strong argument in favour of the existence of the passage which the Rosses, Parry, Franklin, and Back, have successively gone in search of.

As if to compensate for the want of success in this attempt, our enterprising countrymen, Messrs. Dease and Simpson, two officers of the Hudson Bay Company's service, have succeeded, during the past summer, in tracing the hitherto unexplored country between Point Barrow and Franklin's Farthest; and thus a continuous line, of 60 degrees in extent, of the northern coast of America, from Point Turnagain, on the east, to Behring's Straits, on the west, has been explored by British hardihood and perseverance. The simple narrative of the gentlemen who conducted this expedition, so honourable to British commercial enterprise, has been promptly communicated to the Society by the governors of the Hudson Bay Company; and, from the excellent instructions drawn up by the resident governor Simpson, and from the perseverance and energy of their officers, we have every hope that Messrs. Dease and Simpson may succeed in tracing the 220 miles of still unexplored coast between Point Turnagain and Captain James Ross's Farthest, and thus accomplish the glorious task of determining the northern limits of the great continent of America.

The recent publication of the work named 'ANTIQUITATES AMERICANÆ,' which does great honour to the Royal Society of Northern Antiquaries at Copenhagen, has proved the discovery of America by the Scandinavians or Northmen in the tenth century. As such a subject must deeply interest all those who occupy themselves with ancient discoveries in geography and history, and as from the nature of the work its circulation will be extremely limited, it has been considered that an abstract of the historical evidence would appropriately find its place in our Journal.* From these documents it appears, that in the year 1003 the coasts of Connecticut and New York were examined, and probably also those of New Jersey, Delaware, and Maryland.

CANADA.—The charts resulting from the survey of the river

* Vol. viii. p. 114.

St. Lawrence from Quebec to its mouth, by Captain Bayfield, R.N., were published last year; and the surveyors are now completing the gulf of St. Lawrence, as well as the upper river from Quebec to Montreal. All that has been surveyed has been given to the public in upwards of forty sheets.

NEWFOUNDLAND.—The Dépôt de la Marine has published sailing directions for the eastern coast of this island, from Cape Bonavista northwards to Cape Norman, at the entrance of the straits of Belleisle, drawn up by Captain Lavaud; but the portion recently examined by this officer extends only from Cape St. John, in lat. 49° 57½′, the limit of the French fisheries on this coast, to Cape Norman in 51° 37′ N.*

UNITED STATES.—Of the progress of the survey under Mr. Hassler we have no recent accounts. Of the more western portions of this country, and especially of its natural history, we have a description in the work of Prince Maximilian Wied of Neuwied, whose travels along the banks of the Missouri, to the foot of the Rocky Mountains, are now publishing at Coblenz. Two numbers are on sale; and, to judge of the illustrations by some of the spirited original sketches which Mr. Charles Bodmer, a German artist, who accompanied the prince, has obligingly exhibited at the Geographical Society, it will add greatly to our acquaintance with the features of the fast-perishing Indian races.

The American Portfolio, now publishing in London, is also in this respect a highly valuable work; the portraits of the native chiefs are beautifully drawn and coloured, and the biographical sketches well and simply written.

We learn that Long's Peak, in lat. 40° N., long. 106° W. nearly, one of the range of the Rocky Mountains, has lately been measured, and found to reach 15,000 feet above the sea, but we have not been able to meet with any account of the measurement.

No detailed accounts of the operations of the survey of the western coast of America, under Captain Belcher, R.N., have been received, but there is good reason to believe that it is proceeding with energy.

This officer has lately visited the Russian port of Sitka, where there is an excellent arsenal well found in naval and military stores; and where he was received with every mark of friendly assistance. On his voyage thither he obtained a measurement of Mount St. Elias, which was also measured by Malaspina, in the voyage of the Atrevida, and found it to reach 15,000 feet above the sea.

SOUTH AMERICA.—Besides his admirable *Examen Critique*, Baron Humboldt has lately published a valuable notice, *Sur quelques Points Importans de la Géographie de la Guyane*, in which he briefly reviews all that was formerly known and has

* Annales Maritimes, Dec. 1837, p. 1231.

lately been discovered in this extensive tract of country north of the Amazons, chiefly with the object of keeping in view the bases on which the maps rest of the northern portion of South America; for, as he truly remarks, nothing is more important in astronomical geography than obtaining accurately such *data* and the knowledge of observations upon which they depend.

An interesting narrative, by Baron Humboldt, of two attempts to ascend Chimborazo in 1802, has appeared in No. 51 of the Edinburgh Philosophical Journal.

During the past year Mr. Schomburgk, our traveller in British Guayana, has explored the rivers Corentyn and Berbice, a detailed report of which appears in vol. vii. of the Journal. Since his return, this persevering traveller has again ascended the Essequibo to its junction with the river Rupunúny; following the upward course of the latter he ascended its tributaries, the Rewa and Quitaro, and on the 17th Nov. 1837, the date of his last letters, he had reached the latitude of 2° 31′ N., some miles beyond his extreme point on the former expedition, and was then about to start to explore the mountain range of the Sierra Acaray, and to connect, if possible, his observations with those of Baron Humboldt at Esmeralda on the Oronoco, already connected with Guayaquil on the Pacific; and thereby to determine the true breadth of the American continent, between Guayaquil and Demerara on the Atlantic ocean.

That enterprising traveller, Mr. Hillhouse, has also explored more than two hundred miles of the river Cuyuny.

Buenos Ayres.—Don Pedro de Angelis continues his publication of official documents relative to the history and geography of the provinces of La Plata. The fifth volume complete, and part of the sixth, have appeared since last year, and it is probable that the whole work will not be less than ten volumes.

In the fifth volume an account is given of the several missionary settlements formerly dependent on the College de Propaganda Fide of Tarija, situated between 18° 40′, and 23° 15′ S. lat., and important in a geographical point of view, as commanding the navigation of the two great rivers, the Rio Grande or Guapey of Santa Cruz de la Sierra, which runs into the Madeira, and the Pilcomayo, supposed to be navigable to the Paraguay. The account is dated 1800, and is drawn up by the Prefect of the Department for the information of the Governor of Potosi: it gives the geographical position of twenty-one points.

In the same volume is an interesting collection of records of expeditions undertaken by the Spaniards in the last century, to explore the Pampas to the south of Buenos Ayres, and to examine the coast of Patagonia. In these papers there is much information quite new to us respecting those parts, and the settlements formed on the coast of Patagonia about the year 1779, and afterwards abandoned by the Spaniards.

The sixth volume contains a table, never before published, of positions fixed by Malaspina, on the northern and southern shores of La Plata, in his voyage with the Atrevida and Descubierta.

A diary, by Viedma, of the expedition to form the settlement at San Julian's, on the coast of Patagonia, in 1778.

A survey of the frontier towns and forts of the province of Buenos Ayres, by Don Felix de Azara, in 1796, determining by observation the positions of all the principal towns in that province; and a variety of papers connected with the advancement of the frontiers of that state, into the territory formerly possessed by the Indians to the south; all which will be of great utility in assisting the geographer to fix more correctly the physical features of a considerable portion of the South American continent as yet very imperfectly laid down in our maps.*

Señor de Angelis deserves the highest praise for the zeal and ability with which, under very difficult circumstances, as we are told, he labours to make this collection as complete as possible. It is of equal importance to the historian and to the geographer, and we can well recommend it to all those who are interested in South America.

Thirty-two livraisons of M. Alcide d'Orbigny's beautiful work on the natural history of the republic of La Plata and the Banda Oriental are now published; in Mr. Scarlett's recent Travels in South America will be found some new information relative to Paytá, and a full statement of the proposed plan for steam navigation from Chile to Panama.

We anxiously look for the detailed account of the survey of the coasts of Patagonia, Chile, and Peru, by Captain Fitz Roy, in H.M.S. Beagle. The map of South America to illustrate the work, by Mr. John Arrowsmith, just ready, is compiled from the most authentic sources, and contains all that is known of that country. We are glad to see that the geographical positions determined in the voyage of the Beagle have been republished in German and in French; as, however, in the *Additions* to the *Connaissance des Tems* for 1840, Captain Fitz Roy's longitude of Iquique, in Peru, is rejected as being seven or eight miles in error, it is necessary to state that its position, as given in this Journal, vol. vi. p. 338, depends upon three independent chronometric measurements between Valparaiso and Lima; and we therefore venture to caution navigators against trusting to any other longitude than that determined in the late voyage of the Beagle.

* The writer is indebted to Sir Woodbine Parish for the chief part of the information relating to Buenos Ayres.

Australia.

New South Wales.—A summary of the last journey of Major Mitchell, surveyor-general of this colony, with a map of his discoveries in the country named by him Australia Felix, was published in vol. vii. of our Journal; the detailed narrative of this and of Major Mitchell's two former expeditions into the interior will shortly be made public. A map of the colony of New South Wales, by Major Mitchell, on the scale of eight inches to a degree, drawn and *engraved* at Sydney, represents, better than any other we have seen, the great physical features that mark the face of the country.

At Port Phillip a town named Melbourne has been founded at the north-eastern angle of the bay, and from the latest accounts is rapidly increasing in population and in flocks and herds, as the country is found admirably adapted for pasturage.

South Coast.—The colony established on the east side of St. Vincent's Gulf have laid the foundation of a town called Adelaide, and cheering accounts of its prospects and of the nature of the surrounding country have been received; an exploring party have made their way a short distance to the northward, and also across to the eastward through the Mount Lofty range as far as Lake Alexandrina; and while we write [May 19] the news has reached London that the river Murray has been found to discharge itself into the sea in Encounter Bay by a wide navigable mouth, four fathoms deep at the entrance, some miles to the eastward of Lake Alexandrina, through the eastern portion of which lake the river appears to hold its course.

North-West Coast.—The expedition, under Lieutenants Grey and Lushington, for exploring this most promising portion of the coast of Australia, sailed from the Cape on the 5th October direct for its destination, since which no accounts have been received; up to that time all the party were sanguine of success, and had taken the most prudent measures to ensure it. Their intention was to convey the most useful animals, plants, and seeds from Brazil, the Cape, and Timor, in order to spread them in the interior, and thus to benefit the country by every means in their power. Most sincerely do we wish them success. The nautical part of the expedition, under command of Captain Wickham, R.N., would first visit Swan River, and then probably proceed to examine Dampier's Archipelago.

North Coast.—The expedition to establish a settlement at Port Essington, on the north coast of Australia, sailed from England in the beginning of the year. It consisted of two vessels commanded by Sir Gordon Bremer and Lieutenant Owen Stanley, from whose experience as a surveyor may be expected much new hydrographical information.

An excellent map of the whole of Australia, in two sheets, on the scale of two inches to a degree, just completed by Mr. John Arrowsmith, is admirably engraved, and contains all the most recent discoveries in this quarter.

A paper, well worthy of serious consideration, on the geographical nomenclature of Australia, by Captain Vetch, R.E., is published in this volume of our Journal. It is to be hoped that all geographers, at least, will exert their influence in rescuing this country from the barbarous nomenclature that is daily gaining ground in it; and which, if not checked at once, will become so firmly established, that it will be nearly impossible to eradicate it.

POLYNESIA.

Parts of the western shores of New Guinea, and of the Archipelago of the Moluccas, have been examined by MM. Kolff and Modera, officers of the Dutch navy, and an analysis of both their voyages will be found in the Journal of the past year.

The Hydrographic Office has recently published a chart of the Arafúra Sea, a book of sailing directions, compiled by Mr. Windsor Earl from all the documents that were available: it is a great improvement upon all former charts of this sea, to which the name Arafúra has been given, as the numerous islands within its limits are chiefly inhabited by people of the Arafúra race.

The Dépôt de la Marine has just published a chart of the Indian Ocean, and also of the China Sea, compiled by M. Daussy; and the first volume of Horsburgh's Indian Directory, translated by Captain Prédour.

Mr. F. D. Bennett has given us some new information respecting the Marquesas and other groups in the Pacific Ocean, and has corrected the position of the Lobos, Guadalupe, and Rotch's Island.* The Actæon has fixed the position of a group of three islands discovered by Mr. Ebrill in the Dangerous Archipelago;† and Mr. Williams‡ affords us much new information respecting the little-known group of the Navigator's Isles. But we trust that the expedition which has just sailed from the United States will, ere long, accurately determine the positions of the numerous groups and islands in the Pacific Ocean.

The omission of any notice of the geographical distribution of animals and plants will strike the most cursory reader of this Sketch; but we can only refer to the names of Richardson, Swainson, and Watson, in our own country—of Meyen, Canstein, C. L. Buonaparte, and doubtless many others in different parts of Europe—and plead ignorance and want of time to search for information on the subject. Yet, on one point we must ven-

* Journal, vol. vii. p. 211. † Ibid. p. 455. ‡ Missionary Enterprises.

ture to add a few words, not with the most remote idea of doing justice to the subject, but with the hope of inducing those who take an interest in geography to turn their attention to one of the most important branches of strictly geographical inquiry.

ETHNOGRAPHY.—In attempting a brief outline of the recent progress of ethnographical research, it may be remarked that some new light has been thrown on the ethnography of Europe, and on the history and mutual relations of the different races of people who inhabit this part of the world, by researches into the intimate structure and affinities of their languages. Such researches have been conducted of late in a far more accurate and analytical manner than formerly. The Deutsche Grammatik of Jacob Grimm may be instanced as the most complete and elaborate analysis ever published of any one class of languages. This work has lately been completed by the appearance of the fourth and last volume; and the object of the distinguished writer has been to display the relations of all the dialects of the Teutonic speech with each other and with other languages of Europe and Asia.*

The history of the Indo-German languages has also been elucidated by the late works of Professor Bopp, who in his '*Vergleichungs-Grammatik*' has completed the task which he commenced some years ago. Dr. Prichard's work on the 'Origin of the Celtic Nations' is the first in which the question whether the Celtic dialects belong to the class of Indo-European languages was fully examined; and, in the opinion of all those who have considered the evidence which he has adduced, the question has been completely solved. A similar result has been obtained by M. Adolph Pictet, of Geneva, in a treatise more recently published on the same question.

The antiquities of the ancient Iberi, whose descendants speak the Biscayan or Basque in Spain and France, has been elucidated by Baron W. von Humboldt, in a work which is a model for researches of this description.† On the relations of the Basque to the American languages, some valuable observations are to be found in the late work of M. Abbadie.‡

The northern languages, and the history of the nations in the north of Europe and Asia, are subjects which have been investigated by Danish, Swedish, Finnish, and Russian philologers and antiquarians, whose works are less read in this country than they deserve to be. Professor Rask, of Copenhagen, was one of the most distinguished of these writers, and his premature death has been a great calamity to this department of literature. His work

* The writer is indebted to an author who has proved himself to be intimately acquainted with the subject of ethnography, for the notes from which the chief part of the above information is extracted.

† Prüfung der Untersuchungen über die Urbewohner Hispaniens, von W. von Humboldt. 4to. Berlin.

‡ Etudes Grammaticales de la langue Euskarienne, par M. d'Abbadie. Paris.

on the northern languages contains much valuable information on the history not only of the Scandinavian idioms, but also on those of the Finnish and Lapponic races.*

The lights reflected from different objects in this field of research have been condensed by Dr. Müller in his treatise entitled "Der Ugrische Volksstamm." In this work the author makes geography the groundwork of his researches, and follows the method pointed out by Professor Ritter in his admirable "*Erdkunde.*" The volume already published comprises an extensive survey of the Uralian countries and their primitive inhabitants. A view of Caucasus and the Altaic chain is to be the subject of a succeeding volume. The author agrees with Klaproth and other late writers in reprehending the application of the term Caucasian to the principal races of Europe and southern Asia. The true Caucasian nations constitute in fact a very peculiar group, most of them being entirely distinct from the nations termed by Eichhorn "Semitic." The Caucasian tribes, as it will perhaps be hereafter generally acknowledged, betray much greater signs of affinity, though this is but a remote one, with the Uralian nations and with others in the north of Asia. Erman, in his "*Reise um die Erde,*" has collected information which tends to elucidate the history and relations of many of these widely-scattered tribes. His researches are particularly interesting in one point of view by the indications of ancient affinity, or at least of early intercourse, which he has detected between the inhabitants of the north-western coast of America and the northern Asiatics. The forms of words, in his short collection of Ostiak words, and particularly the termination of nouns of number in *all,* remind us of the old Aztec, in which (as well as in the vocabularies of the tribes who dwell in the north-western tracts of America, in the ancient Aztlan, and in the path of emigration said to have been followed by the roving Toltecas) this combination of elements occurs in almost every third or fourth word.

AMERICA.—The ethnography of the New World, and the history of its languages, occupy the attention of distinguished writers in the United States. M. Duponceau, the venerable President of the American Philosophical Academy, has excited the interest of his countrymen to these researches, and Mr. Pickering, of Boston, has devoted his talents and the leisure which his active profession allows him to the task of preserving an impress of the thoughts and feelings of races of men who appear destined soon to become extinct.† The publication of a memoir on the Othomi language by Naxera, under the auspices of M. Duponceau, affords the first

* Rask über die alten Nordischen Sprachen.

† See Mr. Pickering's notes and publication of Rasles' Dictionary of the Abnaki Language, in the Memoirs of the American Academy of Arts and Sciences, new series, vol. i. 4to. Cambridge, 1833.

instance, and that a very curious one, of the discovery of a monosyllabic language in the heart of America. Till now the exclusive possession of monosyllabic languages had formed one of the strongest features of contrast by which the eastern Asiatics were distinguished from the inhabitants of the New World. A fact of some interest is the late invention of a new character adapted to the language of the Cherokees by a native of that race, educated in one of the colleges of the United States. In this character, of which the inventor has procured type, he publishes for the use of his countrymen a newspaper, well composed and edited, partly in English and partly in the native idiom of the red-skinned tribe.

Polynesia.—Of the ethnography of the insular natives of the southern ocean, the first sketch was drawn by Forster, the companion of Cook. By this eminent naturalist the natives of all the groups of isles of the southern seas from Madagascar to Easter Island were referred to two races, which have been termed the Papuan and the Polynesian. The opinion of Forster was adopted by Marsden, and has been followed by many writers; but late discoveries, and particularly those which have been occasioned by the late French expeditions sent out under Captains De Freycinet, D'Urville, and Duperrey, have proved that Forster's conclusions were, as might be expected, far too general, and that they require great modification. It seems to result from these later investigations that the aborigines, or at least the oldest known inhabitants of the larger islands in the Austral seas, as of Borneo, Celebes, New Guinea, New Britain, and New Ireland, are not Papuas or woolly-haired races, but tribes termed in many places Alfúrùs or Arafúras, who have straight hair, and who in physical character resemble the Australians.*

The Australians often are considered to be a branch of this division of mankind, and it becomes a question of great interest how far they may hereafter be identified with the Arafúra tribes by an examination of their languages. For such a comparison resources are yet wanting. By the missionary Threlkeld materials have been collected which afford access to the language of one Australian tribe, and these may hereafter furnish an opportunity of comparing this with other dialects. It is very desirable that this example should be imitated. On the southern coast and in the interior of New Guinea it is very probable that idioms akin to those of the Australian savages may be discovered. The northern coast of that great island is occupied by Papua, or black and woolly-haired tribes, who issuing from some unknown centre have spread themselves over the coasts of most of the larger, and have gained entire possession of some of the smaller, groups in the Indian and Pacific oceans. The era of their migration cannot

* In the first vol. of Dr. Prichard's Researches are given drawings of the skulls of Arafúras, who are there described as a distinct section of the human species.

be ascertained, but many circumstances seem to countenance the opinion of M. Lesson that it was long posterior to the dispersion of the Arafúras over the same countries, and even to that of the Polynesians. M. Lesson has attempted to distinguish several separate races among the tribes who are commonly included under that collective name, but the marks of discrimination on which he relies are deserving of less confidence than those which are founded on the affinity and diversity of language.* To this last circumstance M. D'Urville has directed his attention with greater success than any other writer among the late editors of the French voyages to the Pacific. His two volumes of philology connected with the narrative of the voyage of the Astrolabe, contain some valuable additions to the stock of information previously acquired, and the results which arise from a comparison of copious vocabularies are deduced with great ability.† They coincide, in their general tendency, with information conveyed in a later work, the publication of which forms an era in the history of Polynesian ethnography. We allude to the work of Wilhelm von Humboldt,‡ which treats professedly on the Kawi language in the isle of Java, a work of multifarious erudition, and one which displays a rare union of the powers of philosophical investigation with the clear discernment and comprehensive judgment of the statesman and man of the world.

In this work, not the Kawi language in particular, but the theory of language in general, has been elucidated; light also has been thrown on the history of the Oceanic nations in a comprehensive view; the revolutions have been traced which changed the old Polynesian inhabitants of Java into a comparatively refined people, speaking a language altered to, and greatly modified by, the sacred idiom of the Brahmins; and an insight has been opened into a still more remarkable connexion between the original basis of language among the Malaya islanders and the natives of the Indian continent. With respect to the population of the Philippines and of Madagascar, both D'Urville and Humboldt agree in one opinion, which is contrary to that generally prevalent in this country. It seems that the aboriginal people of the last-mentioned islands, the Ovahs, are so nearly allied to the race in Sumatra (who are the primitive Malays), and both nations to the Tagalas of the Philippines, that all of them may be almost termed one people, and considered as speaking proximate dialects of one language, whereas, it has been stated as the opinion of Mr. Craw-

* See a variety of Memoirs, by M. Lesson, in the Annales des Sciences Naturelles, and the first and second volumes of his admirable Supplements to Buffon's Hist. Naturelle.

† Voyage de l'Astrolabe, Partie Philologique, par le Capitaine Dumont d'Urville. Paris.

‡ Uber die Kawi-Sprache auf der Insel Java, &c. Von W. von Humboldt. Berlin, 1836.

ford, that the Polynesians of Madagascar are only accidental colonists of the shores of that island, and that the central parts were inhabited by people of a widely-different character, both physical and moral.

Africa.—On this, the most difficult region of ethnography, inasmuch as it is the least explored, we have just received the second volume of Dr. Prichard's admirable work, containing Researches into the Physical Ethnography of the African races, in which the author has, in a masterly and comprehensive manner, classified all the existing materials that could be obtained by diligent research, and fairly stated the conclusions resulting from them. In treating his subject, Dr. Prichard divides the population of Africa into different groups of nations, associated by analogies in their languages more or less pronounced; and, where information fails for such a principle of arrangement, by general features of resemblance physical and moral. Some of the principal groups are—1. The Atlantic nations, including all the Libyan tribes who speak dialects of the Berber, and, among these, the ancient Guanches. The author has compared the latest accounts of these tribes as given by Rozel, who accompanied the French expedition, and the information derived from a Notice on the Empire of Marocco in vol. i. of this Journal. 2. The nations of Senegambia, who inhabit the elevated plateau whence the great rivers of Western Africa take their rise. He has collected information from various quarters illustrative of the history of the Mandingos and the Fúlahs. 3. The maritime nations of Guinea, whose languages form one group of African idioms. 4. The nations of Sudán, first described by Ibn Batuta and Leo Africanus, differing in physical character and in their languages from the preceding. 5. Abyssinian, or straight-haired races, in the east of Africa, including numerous native African tribes extended over a great part of the continent, but chiefly inhabiting highlands. The Habesh of Tigre are distinguished from all these as an Asiatic colony. The author enters into an historical inquiry respecting the era of their colonisation, and connects them with the Arabs of Hamyar, or the Homerites of the Byzantine writers. In pp. 152, 153, he brings arguments to prove that the Abyssinian alphabet was derived from the Hamyarites, and not invented by the first Christian missionaries at Axum, as Michaelis and other biblical writers have supposed; and here we may observe that it has been discovered by Lieutenant Wellsted that such was precisely the fact.

The letters of the Abyssinians, or characters nearly resembling them, have been used of old by the Hamyarites, and this discovery, anticipated from a survey of historical probabilities, made by a British naval officer, has put an end to a controversy long agitated among the European literati, as to the era and manner in which the Abyssinians came to be possessed of the art of writing. Bruce strenuously asserted the high antiquity of the Ghíz alphabet, but he was opposed by the majority of learned men in Germany. 6. A sixth class are the Nuba races, an extensive group of nations, whose native region, the cradle of their national existence, is the northern slope of the great table-land of Habesk, and the banks of the great rivers which descend from it. This group contains races in which the negro type becomes gradually evanescent, till it fades away in the Berábera of the upper Nile, thus affording one instance of facts of which Dr. Prichard is in quest, viz. of the transition from the character of the negro races into those which bélong to other divisions of mankind. In this argument he takes great advantage of the observations of Dr. Rüppell. To the last-mentioned group the author adds the ancient Egyptians, whose history, and that of their three dialects, he has carefully traced. He seems in some way to connect them with the Nubian group, though he considers them as quite distinct from the Berábera. 7. The South African nations, meaning all those tribes who display in their languages any sort of relation to the Kaffirs, occupy, in the development of their history, a considerable part of this volume, and in this part of it the author has collected all the information within reach relative to African philology. This he thinks sufficient to support him in laying down the genuine African languages as a distinct class of human idioms, all related to each other by certain grammatical peculiarities, of which the idioms of Congo, of Kaffirland, and of the Mozambique country partake, and which may even be traced in the dialects of the old Egyptian. Whether the data on which these conclusions rest will be thought adequate by the learned remains to be discovered: the subject is worthy of investigation. As far as the South African nations, as above distinguished, are concerned, there seems to be sufficient evidence of a real affinity of language, such as denotes either consanguinity in the races to which they belong, or a very ancient intercourse and intimate connexion. 8. The last group are the Hottentots. The author concludes his work with general observations on the physical history of the African nations, and a discussion of the question which refers to their alleged inferiority in intellectual power.

In conclusion, it may be observed, that all travellers in new countries can materially assist the researches of the eminent men whose works have been mentioned, by collecting accurate vocabularies of the language or dialect spoken in the regions they may traverse. Surely, too, when its importance is so manifest, this will not be thought too great a task to be undertaken, and vigorously prosecuted, by all who are concerned in the promotion of the most interesting branch of geographical inquiry.

XXI.—*Narrative of a Journey from Mokhá to Ṣan'á by the Taríḳ-esh-Shám, or Northern Route, in July and August,* 1836. By Mr. Charles J. Cruttenden, Indian Navy.

During the time that the East India Company's surveying brig *Palinurus* of the Indian Navy was employed at Mokhá in making a plan of the roadstead, Dr. Hulton, the surgeon of the vessel, and myself, took advantage of the opportunity thus offered to endeavour to penetrate as far as Ṣan'á. We had previously made the attempt from 'Aden, on the south coast of Arabia, but, owing to the suspicious temper of the Sulṭán, were unable to effect our purpose.

Lieutenant Wellsted, I.N., has already, I understand, laid before the Geographical Society copies of inscriptions found at Naḳabu-l-Ḥajar,* and the Royal Asiatic Society have, I believe, published those found at Ḥiṣn Ghoráb, on the southern coast of Arabia, and our pleasure was, of course, great in finding the very same characters in Ṣan'á. We were the more surprised at this discovery, as Niebuhr says he could hear of none at that place, though the buildings on which we found these were, by the accounts of the townspeople, at least seventy years old.

My original intention was merely to offer to the Geographical Society the copies of the inscriptions, and an account of the place where they were found. As, however, I understand that a narrative of my journey to Ṣan'á may be acceptable, I have given a few rough notes made during our progress across the mountains, which may serve to show the nature of the country, and the principal towns.

The inscriptions were found in the neighbourhood of the most ancient part of Ṣan'á, near the foot of Dár-el-Ṭawáshí, or the "Abode of the Eunuchs:" it is also sometimes called Bakhírí. It is the eastern extremity of the town, and in former days was the part appropriated to the Jews. The letters are about 2½ inches long; three of the inscriptions are in relief, and the fourth is cut into the marble.

These inscriptions are exactly in the same character as those found at Ḥíṣn Ghoráb, about 70 miles to the westward of Makullah, on an almost insulated rock, on the south coast—as those at Nakabu-l-Hajar, in 14° 30′ N. long. 46° E.—and as those of Naḳhl Mayúk and Ḳoṣaïr, about 70 miles to the eastward of Makullah—but the two latter are said to be in red paint.

Several of the principal merchants in Ṣan'á assured us that these stones had all been brought from Máreb, which was about

* Journal, vol. vii. p. 20.

two long days' journey distant; and, in reply to our questions, told us that it was less expensive to bring these ready-cut stones from Máreb than to prepare them in Ṣan'á. The fact of Máreb being still called by the natives "Arḍ-es-Sabà" (*i.e.* the land of Sheba),* leads to the supposition that this might be the ancient Saba, but of this hereafter.

We determined to adopt the native costume as being better adapted for travelling, and, through the kind assistance of a wealthy Persian merchant at Mokhá, we found no difficulty in procuring a guide and eight mules to take us to Ṣan'á. We were particularly anxious to take the eastern road, or "Ṭaríḳ el Yemen," by Ta'ez and Dhamár," but the intestine feuds of the Bedawí sheïkhs in that part rendered travelling impracticable, and indeed this road has been blocked up for eleven years.

July 13, 1836.—We quitted Mokhá at sunset, as, the hot season having commenced, we were compelled to cross Tehámeh by night, and travelled along the shores of the Red Sea in a northerly direction, about two miles from the beach. Our party consisted of Dr. Hulton, myself, two servants, who also acted as interpreters, and four muleteers, all well armed, as even in Tehámeh travelling by night is not considered altogether safe.

In that tract the caravan-seraïs, or, as they are here called, the "Mekhàyehs," † are generally divided into several small apartments, each sufficiently furnished with rude wooden bedsteads, and small three-legged stools for coffee, &c. In the evening the inhabitants of these several chambers have their beds carried outside into the square court-yard, which forms the centre of the building, and placed in the open air, the weather being too oppressive to admit of any one's sleeping under cover.

The dews at night fall like rain, but, if the precaution is taken of covering the face with a light linen cloth, no evil effects result from the exposure, owing to the absence of trees; and the sensation of perfect refreshment that is experienced on rising is indescribable.

The "Muḳawwí," ‡ or coffee-house-keeper, supplies his guests morning and evening with curdled milk and a coarse cake of Jowárí § bread, which weighs about a pound, and is commonly called by the natives by the name of "one man's share." The whole cost of supper, breakfast, and a night's lodging, did not exceed a third of a dollar, or 1*s.* 6*d.*, and there are stated prices

* Máreb was the capital of Sabá. "Saba cujus metropolis Márab. Mariaba: Strabo. Plin. &c., paulo plụs triduo distans à Sanaa." Golius in Alphergan, p. 86.

† Properly Meḥáyí, plural of Maḥyá; a living-place.

‡ The furnisher of refreshments. This word is not derived from Ḳahweh, coffee.

§ Sorghum vulgare, called *dhurrah* by the Arabs.

for every article of food. At this time grain had risen greatly in price, owing to the continued drought that had prevailed for nearly four years.

The scenes of misery and wretchedness which we witnessed on our passage across Tehámeh were dreadful. In Mokhá it was no uncommon thing to see dead bodies lying unheeded in the streets, victims of famine, and this, added to the grinding tyranny and brutal oppression of the Egyptian troops, rendered the condition of these poor people almost insupportable.

Our first stage was about fifteen miles to the small village of Ruweïs,* where we halted till three o'clock on the following day. Thermometer in shade at 1 P.M. 96° Fahr.

July 14.—We pursued our route for about twenty miles, gradually diverging from the shore as far as Múshij, or Maushij,† a large village, containing perhaps eight hundred inhabitants, and celebrated for the quantity of "yásmín," or jessamine, which grows there; its flower, stripped of its stalks, and strung upon thread, is daily carried to Mokhá, where it is eagerly purchased by the women as ornaments for their hair. In each thicket of jessamine there is a well of pure and sweet water, so that these bowers form a very delightful retreat during the intense heat of the day.

The mosque at Múshij is noted as being the favourite resort of the Imám 'Alí, son-in-law of Moḥammed, who is believed to descend nightly in an invisible shape, and perform his devotions.

Múshij is under the dominion of Sheïkh Ḥuseïn bin Yaḥyá, whose character is so extraordinary, that it deserves to be more particularly noticed.

This chief owns the tract of land about three miles in length, by fifteen average breadth, extending from the back of Mokhá as far as Múshij, including the mountains bordering on Tehámeh. His influence with the neighbouring tribes of Bedawís is so great, that Moḥammed 'Alí, the Páshá of Egypt, is glad to purchase his alliance, or, more properly speaking, his neutrality, at the rate of eight or ten thousand dollars a-year. He is absolute in his own territory, and his commands are obeyed with the blindest devotion. His nominal residence is at the fortress of Ḥeïs, situated in a deep ravine eighteen miles N.E. of Múshij, but it is never certainly known where he is. He is held in such dread by the peasantry, that they scarcely mention his name but in a whisper, and he bears the highest character for impartial but strict justice. For several years his alliance has been courted by the Imám of Ṣan'á, who is naturally anxious to secure his co-

* Little Head. † Also called Maushid, Niebuhr's Descr., p. 224.

operation in defending the country from an invasion of the Turks; but the sheïkh persisted in remaining neuter, till the continued encroachments of the Governor of Mokhá roused him. He then at once renounced all friendship with Mohammed 'Alí, and, in company with Sídí Kásim, the exiled uncle of the Imám of Ṣan'á, invested the fortress of Ta'ez, where there were two regiments of Egyptian infantry, and which, as we understood, he succeeded in reducing.

The latest accounts we had, stated that he had furnished the exiled Sídí Kásim with troops; and that, in company with the tribes of Do Mohammed and Do Ḥuseïn, whose territories lie in the neighbourhood of Dhamár and Ebb, Sídí Kásim had dethroned 'Alí ben 'Abd-allah el Manṣúr, the reigning Imám, and established himself at the head of the government. Thermometer in shade at 1 P.M., 95° Fahr.

July 15.—Our next stage, of eight hours in a north north-east direction, to the hamlet of Sherjah, was of little interest, the country being nothing but an arid sandy plain, covered with a coarse kind of grass and stunted bushes, here and there intersected by the dry bed of a mountain-torrent. Thermometer in shade at 1 P.M., 95°.

July 16.—From Sherjah to Zebíd, six hours and a half, in the same direction, the country presented a better appearance, being in many places carefully cultivated. This valley is mentioned by Niebuhr as the "largest and most fruitful in the whole of Tehámeh;" and in a prosperous season it certainly would deserve that appellation. Four years of continued drought had, however, completely burned up the soil, and the husbandman could not but despond when he had placed the grain in the ground, and saw no prospects of a return for his labour.

Wádí Zebíd is in many places covered with a thick brushwood of tamarisk, which affords shelter to numerous wild guinea-fowl. We shot several, and found them quite as palatable as the domestic birds bred in England.

We did not reach Zebíd till midnight; and, the gates of the city being closed, we were obliged to search for accommodations in the suburbs, which, after some difficulty, we found.

Zebíd is a city of moderate size, not quite so large as Mokhá. It had a peculiarly gloomy appearance, owing to the dark colour of the bricks with which the houses are built, and the ruinous state of many of them. It is, I believe, considered as being the most ancient town in Tehámeh. The Arabs have a tradition that it has been three times washed away by floods, with the exception of the Mesjid el Jámi', or principal mosque, which certainly wears a venerable appearance.

That edifice is very large, and has an octagonal menàreh, which is ornamented with a light net-work of stone, giving it a very elegant appearance. The interior is the same as in other mosques, and consists of one large room, with the ḳiblah pointing out the direction of Mekkah, and several small adjoining oratories branching off in different directions, containing the tombs of deceased "welís," or saints. The "súḳ," or market, is remarkably well arranged, and divided into three compartments for fish, flesh, and vegetables. The supplies are ample for the garrison of 700 men, and the inhabitants, who may amount to 7000 persons.

With the exception of a few rusty 24-lb. carronades at the principal gate Zebíd is destitute of artillery. Like many other Arab towns, the number of tombs and mosques outside the walls render it practicable to march an invading force within 100 yards of the town without the slightest exposure. So long as the Arabs have no artillery the Egyptian troops are safe in Zebíd. In that case the town is well adapted for defence, the walls being high and crowned with numerous towers with loopholes for musketry. The town is well supplied with water; indeed, on our return, we found the road impassable from the overflowing of a mountain-torrent.

In many parts of Arabia, and particularly in Socotra, the marks of very heavy mountain-torrents are evident, but with one exception I never witnessed the effects of one at the time; this was on our return from Ṣan'á, when, crossing a broad mountain-stream, a sudden rush of water took place that carried a donkey on which I was mounted off his legs, and drowned him before we could succeed in catching him, nor did I land without much difficulty; ten minutes afterwards the stream was impassable, and the remainder of our party that had not crossed it were compelled to wait patiently till it had subsided. This must account for the numerous rivers that we see marked in old maps of Arabia; and in no place would a person be so easily deceived regarding their permanency as in Tehámeh, where the water always takes its old bed, and where the ravines are always clothed with thick underwood.

I have, in the small map, traced the river at Zebíd as it appeared on our return, but it must not be supposed that it is thus throughout the year. Four years, as I before stated, had elapsed without a drop of rain, and many persons may have been to Zebíd in the interim and found everything dried up, as we did on our first arrival. With the exception of that of Zebíd, all the streams I have laid down are perennial, generally of small size, but the best proof we have of their duration is the fact of their having many fish. Therm. in shade at 1 P.M. 96°.

July 17.—We left Zebíd in the afternoon, and after a very

long and fatiguing stage of nine hours and a half, or nearly thirty miles, in a north-easterly direction, we reached the city of Beït el Fakíh,* an hour after midnight. The country we had passed over was, if anything, more arid and barren than any we had hitherto seen, and we heard with much satisfaction the assurances of our guides that this was our last stage in the plains of Tehámeh, and that the following night would see us, "In shá āllah!" (please God), across the Turkish frontier and in the dominions of the Imám of Ṣan'á.

We found Beït el Fakíh a large town of 8000 persons, with a citadel of some strength in the centre of it. The town itself was unwalled, and consisted generally of a large kind of house, built partly of brick and partly of mud, and roofed with branches of the date-tree. It is the frontier-town of the Egyptian government, and as such is of some importance, it being the emporium for all the coffee that comes from the interior. The principal articles of trade in Zebíd and Beït el Fakíh are piece-goods from India, consisting chiefly of coarse blue and white cloth, English shawls, which are in great request, spices from Java, and sugar from Mauritius, which are bartered for money, wax, gums, and frankincense, and a small quantity of coffee that the neighbouring Bedawís bring down in preference to sending it to the Ṣan'á market. Indian Bamans are the principal merchants in the place; they are very numerous, but they have to pay a very heavy tax to the governor, and one of them declared, with tears in his eyes, that they could not make near so much profit as in India under the government of the English. A heavy duty is here levied upon all káfilahs (caravans) of coffee that arrive from Ṣan'á on their way to Hodeïdah, or Mokhá, and so vexatious are the continual demands upon the Ṣan'á merchants that it will end, in all probability, in their carrying their coffee to 'Aden, more especially as it is now under the English flag. The distance is nearly the same, and we frequently heard while in Ṣan'á that the merchants contemplated changing the route, if practicable; though of course, when this was said, they knew nothing of the treaty since formed by the Bombay government with the Sultán of 'Aden.

Beït el Fakíh is, without exception, the hottest town we found in Tehámeh, the thermometer being at noon 102° in the shade and 141° in the sun. The land between the city and the sea is higher than that on which the town stands, which prevents the sea breeze from cooling the air. In fact, we found the wind so heated in its passage across the dry sandy plain that lies between Hodeïdah and Beït el Fakíh, that it was less oppressive when it was calm. The fort is governed by a bimbáshí,† or sub-captain,

* Lawyer's house.

† Bíng-báshí: *i. e.*, head of 1000; a Turkish title.

and the garrison consists of about 500 men. The governor was extremely civil, but evidently thought we were going to the court of the Imám, to set on foot a treaty regarding the expulsion of the Turks from Yemen. He pointed out to us the numberless dangers of the road, and strongly advised us not to go; finding, however, his remonstrances of no avail, he no longer endeavoured to make us relinquish the attempt, and allowed us to depart with a good grace.

July 18.—We left the city at 6 P.M. and travelled in a north-east direction for eight hours immediately towards the mountains, the base of which we had for the last two days been skirting. After an alarm of robbers, on passing through some thick underwood at the foot of the hills, which made us alight and prepare for an attack, but which proved groundless, we reached a pass, and, crossing over a low shoulder of the mountain, descended, by a densely-wooded ravine, into the beautiful valley of Senníf.* Dark as it was, it was evident that the scene was changed. Tall, majestic elm-trees, mingled with the wide-spreading tamarind, and forming a natural avenue, met our view. The bubbling of a brook was heard, and the sound of our footsteps was lost in the grass. To us, who for six days had been travelling in a comparative desert, the change was delightful in the extreme.

July 19.—At 1 A.M. we reached the village of Senníf, and were soon established in a comfortable serái, or meḥáyé, with a temperature comparatively so much lower that we were glad to sleep under cover. The village was very full, owing to its being the day of the "súk," or market, and we in consequence could not procure beds till an hour before daylight, when we retired to rest much fatigued.

On rising in the forenoon we found Senníf a large village, built entirely of conical straw huts, with the exception of the sheïkh's house, which was a large barn-like building. Its population may be 1000 persons. The sheïkh sent us a fat sheep as a present, and offered us every assistance in his power. We found not only the appearance of the country but the dress of the natives totally altered; the men wearing conical straw hats and the women wide blue cotton trousers drawn tight round the ankle, their head-dress consisting of a handkerchief profusely ornamented with steel chains; they were also fairer than the peasantry of Tehámeh, owing probably to the greater coolness of the atmosphere in the mountains. In Socotra we found the Bedawís of Jebel Hajjíyeh much fairer than the town-Arabs, and attributed it to the same cause.

The valley of Senníf has the shape of a horse-shoe, and is as carefully cultivated as the lands of India; wheat, jowárí (Sorghum

* This cannot be the Sennef of Niebuhr, Voy. i. 334.

vulgare), and barley, flourished in great luxuriance, with several small patches of Indian corn, and some indigo.

Senníf is a market-town, and is also called "Súḳ el Jum'ah," or Friday-Market. There are seven market-towns between this place and Ṣan'á, in each of which the market is held on a different day of the week, and they are a night's journey distant from each other. The sheïkhs of the different villages levy a tax upon all merchandise, and take the merchants under their protection for the time.

We here first saw the Bedawís of the mountains; they are very slightly but elegantly formed, and their average height is five feet six inches; their colour is lighter than that of the Bedawís on the southern coast, and they have long, black, curling hair. The dress of the higher classes among them consists of a blue frock or shirt, with very wide sleeves, bound tight round the waist by the belt of their yambé' or dagger, and no sash, or, as it is termed in India, "kamar-band." The dagger is different from any other that I have seen, being much longer and nearly straight. Their turban is of blue cloth, with several folds of cotton of the same colour bound round it, the Bedawí disdaining to wear the straw hat used by the cultivators of the soil. They carry a short sword with a very broad, spoon-shaped point, if I may use the term, and a long matchlock. When on horseback they carry a very long spear, having a tuft of horse-hair close to the steel head. They appear to be very quick in taking offence, but their quarrels seldom last long. I have seen a man deliberately draw his sword and endeavour to cut down another with whom he was disputing, nothing but the folds of his turban saving his life, and I have been surprised to see the very same men quietly smoking their pipes together on the evening after the quarrel. We found them inquisitive, but not impertinently so. They would collect round us when we halted and listen to our accounts of "Wiláyah,"* or England, or to what they infinitely preferred, the musical box which we had with us. Some, indeed, after hearing the box for a minute or two, would exclaim, "'Audhá Billah min Sheïtán e rajím!" †—"God preserve us from the power of the devil!"—and walk away, but they were generally laughed at for their folly. They all expressed the utmost detestation of the Turks, or "El Aḥmarán" (the red men), as they designated them, and laughed at the idea of their endeavouring to penetrate into the interior through the intricate mountain-passes.

Our party was here increased by the addition of the leader of a large ḳáfilah, which was awaiting us about twenty miles further

* The (foreign) country; hence Wiláyetí, corrupted by the Bengalese into Biláïtí.

† "'Audhán billahi min Sheïtan er rajím," lit. (I fly) for refuge to God from Satan the stoned.

on, and two Ṣan'á merchants, mounted on two very beautiful Abyssinian mules. Therm. in shade at 1 P.M. 88°.

July 20.—As the intricacy of the passes would not admit of our journeying by night, we left Senníf at daylight and proceeded through a very romantic valley called Wádí Koleibah, on our way to Hajír, which was to be our halting-place for the night.

As we gradually ascended, the scenery hourly became more striking and magnificent. The hills were thickly clothed with wood, and we recognised several trees that we had formerly seen in the Jebel Ḥajjíyeh of Socotra. The villages became more numerous, and, the sides of the mountains being in their natural state too steep to admit of grain being cultivated, they are cut away so as to form terraces, which in many places gives them the appearance of an immense amphitheatre. The hamlets are generally built of loose stones with flat mud roofs, and, perched upon overhanging rocks as they generally are, they add considerably to the romantic beauty of the scene. After a halt of an hour during the hottest part of the day at one of these villages called Abú Ḳirsh, as its latitude, we again pursued our way up a steep ravine where we had to dismount. We here observed many large trees, one in particular, of a spongy nature, the stem about two feet six inches in diameter, and the leaves very large and of a leathery texture. It is called by the natives the "Ṭolak-tree," (Ficus Bengalensis), and is generally covered with the nests of the "baia,"* a small kind of sparrow. I have seen upwards of 300 nests upon one tree. They are of a pear shape, having a long funnel-like aperture at the base, and the interior divided into two compartments, one for the male and the other for the female and her progeny.

Partridges (the red-legged species) and Guinea-fowl are plentiful, though wild, and we also observed the jungle-cock of India.

A very fatiguing ascent of three hours brought us to a large fortified seráï, or, to use the mountain term, "simsĕrah," on the ridge of a mountain, and commanding the pass on both sides, and this we found was our halting-place for the night. The simsĕrah† of Ḥajír was a large square building about forty feet high. Round the interior were two stories of cells, and the central space was appropriated to the beasts of burden. We found the temperature here 79°, which, to us, who had not forgotten the heat of Beït el Faḳíh, was very low, and we were

* Baïah?

† Not found in this sense in the dictionaries, but Simsár, "a broker," or "valuer;" one who assists strangers in making bargains, gives a clue to the meaning of samsareh or simsereh, "the place of brokerage."

glad to creep into our cells, though we soon discovered that we were not the sole occupants of them.

On another ridge immediately above Hajír is a fortress of considerable strength belonging to the Bení Dhobeïbí * tribe, though nominally one of the frontier garrisons of the Imám. We found, however, that of late years his authority has been much curtailed, and the Bedawí tribes now levy an arbitrary tax upon all ḳáfilahs of merchandise, whether going or returning, that pass through their territories, in return for which they furnish them with a guard. We here found a large ḳáfilah of goods from Ḥodeïdah, bound to Ṣan'á, and an escort of about thirty men from the fortress to attend them, accompanied by Sheïkh Ghází Naijí in person. We joined their party, and whether they did not think it worth while to demand a tax from us, or whether they were in fear of the Imám, we knew not; they received us civilly, and said we were welcome to join them. Hajír is about 1200 feet above the sea.

July 21.—We started at sunrise, and, descending the ravine on the east-north-east side of Hajír, pursued our way through a broad and well-cultivated valley, gradually increasing in width as we approached Ḥuṣún† Diḳarah or Daḳrah, a very strong hill-fort on a conical-shaped mountain, belonging to the same tribe. The valley called Wádí Seïhán here opened out into a broad plain, increasing in size till it was lost in Tehámeh, some miles to the northward of the parallel of Ḥodeïdah.

The mountains on the north side of this plain are known as Jebel Ḥarráz, and on the other side they are called Jebel Burra'. On both these mountains are coffee-plantations, but those on Jebel Burra' are small and insignificant, while Jebel Harráz produces very superior coffee. It is likewise celebrated for the variety and abundance of its fruit.

In many places in the vicinity of our road we found large enclosures of several kinds of grain, but were cautioned not to straggle from the main body of the party, as a small tribe called the Bení Khórah, who reside in the ravines bordering on Jebel Ḥarráz, were in the habit of waylaying any unfortunate straggler, and, contrary to the usual custom of Arab banditti, murdering their captives. A party had been seen in the valley that morning, and our escort accordingly received an additional reinforcement from Ḥuṣún Daḳrah. This dreaded part of the plain is known as 'Khubt ibn Deran,‡ and we were shown several graves which

* Probably the Bení Doleïbí.—Niebuhr, Descr. p. 248.
† Huṣún, the plural, is commonly used by the Arabs for the singular, Hiṣn.
‡ Probably the Khobt Deṛham (Drachm-plain) of Niebuhr.—Descr. 149.

remained as monuments of the cruelty and ferocity of these miscreants.

We now crossed over an undulating country for several miles, much more barren than that we had left; and, shortly after fording a broad stream that runs down Wádí Ṣeïhán, we reached the village of Samfúr at noon.

We here found another large ḳáfilah from Ṣan'á bound to Mokhá with coffee, and, therefore, our guards left us to take back this party, and made us over to the charge of Sheïkh el Jerádí, who was to see us safe as far as Mofḥaḳ, or two days' journey towards Ṣan'á. The village of Samfúr may contain about 20 houses or huts. Therm. 82° in shade.

July 22.—On leaving Samfúr the next morning, we found several people by the road-side with baskets of fruit from Ḥarráz, amongst which we saw the peach and apricot, several kinds of grapes, walnuts, and a small species of pear, like the stone-pear of England.

The Ḥarráz mountains are at least 1500 feet high from the plain on which they stand, and by our estimate about 3000 above the level of the sea. They are apparently composed of a species of trap-rock. They afford a residence to many tribes of Arabs, who are nominally under the dominion of the Imám, but, like all the other tribes, pay no tribute beyond the tax which is levied upon their coffee as it passes through the gates of Ṣan'á. The coffee of Harráz is very superior, but not of the best kind, the trees from 'Uddeïn, the "'Uddeïní," being much larger than the others, and about twelve feet high.

The valley now became much narrower, in many places not exceeding twenty yards in width, while the mountains on either side rose to the height of 1200 or 1400 feet above the plain, thickly wooded to within 200 feet of their summit, where they presented a barren sheet of grey limestone rock. Under a huge mass which had fallen and completely blocked up the valley, we found a coffee-house and two or three small huts. Understanding that there was a coffee-plantation in the neighbourhood, and of the very best quality, we gladly availed ourselves of the suggestion of Sheïkh el Jerádí, and halted there for the day. A scrambling walk over the before-mentioned rock, by means of steps cut in it, brought us to the coffee-plantation of Dórah.* It was small, perhaps not covering half an acre, with an embankment of stone round it to prevent the soil from being washed away.

The coffee-plant is usually found growing on the side of any

* Caffee-plantage Eddóra. Niebuhr, Reise, i. 433.

valley or other sheltered situation, the soil which has been gradually washed down from the surrounding heights being that which forms its support. This is afforded by the decomposition of a kind of clay-stone, slightly porphyritic, which is found irregularly disposed in company with a kind of trap-rock, among which, as we approached Ṣan'á, basalt is found to predominate. The clay-stone is only found in the more elevated districts, but the detritus finds a ready way into the lower tract by the numerous and steep gorges that are visible in various directions.

As it is thrown up on one side of the valley, it is there carefully protected by stone walls, so as to present the appearance of terraces.* The plant requires a moist soil, though I believe much rain is not desirable. It is always found growing in the greatest luxuriance when there is a spring in its vicinity; for in those plantations where water is scarce the plant looks dry and withered. The bean is gathered twice a-year, and we were told that one of the Dórah trees, though of the smallest quality, ought to produce in the two crops at least ten pounds, or a Ṣan'á "maund."†

We found the fig, plantain, orange, citron, and a little indigo, growing among the coffee. A stream of water from a neighbouring spring was drawn through the garden, and we were told that the roots of each plant were regularly watered every morning and evening. The plant is said to live six years; three of which are requisite for bringing the tree to perfection, for three it bears, and then dies and is rooted up. Therm. in shade 75°.

July 23.—The following morning, accompanied by the sheïkh and his party, we left Dórah and took the road to Mofḥak,‡ travelling for nine hours in an east half north direction. The country was more open and highly cultivated. Barley seemed to be the principal grain, though there were many enclosures of Indian corn and wheat. In this stage particularly, we observed many of the trees found in Socotra. I had (when Captain Haines of the *Palinurus* was employed in surveying that island) accompanied Lieut. Wellsted, and was on the island for nearly two months; a great part of which time was spent among the rocky ranges of Jebel Hajjíyeh. At that time, Dr. Hulton, my present companion, was of opinion that they were peculiar to the island, but I have since seen them on the mountains at the back of

* See Niebuhr, Voy. I. Pl. lxiii, lxv.

† Man, pronounced maun by the Bengalese, whence the English "maund," an Indian measure introduced by merchants, as ḳírsh, in the plural ḳurúsh, from the Türkish ghurúsh, is derived from the German grosch, and used as the name of the piastre.

‡ Niebuhr, Descr. p. 250, Voy.

Ḍhofár, and also on the hills of Yemen, especially the dragons'-blood-tree, and the lubán or frankincense-tree, which we at first thought was not to be found in Socotra, but, on comparing the sabhúr tree* of Socotra with a specimen of the lubán that I procured on the southern Arabian coast at Ḥàsek, they were found to be exactly the same.

The valley of Dórah, through which we travelled in an E.N.E. direction, opened, after three or four miles, upon a large plain, in the midst of which was the village of Ṣeïḥàn.† The country was the same as that we had hitherto passed through, though not so mountainous. At three we reached the village of Mofḥaḳ, and found good quarters in a simsereh. This village of 50 huts is situate on the crest of an oblong hill, about 300 feet high, the sides of which are too steep for any beast of burden to ascend. It presents the appearance of an immense fort, and with a little care might be rendered impregnable. We here found another plantation of coffee of the 'Uddeïní sort. The trees were about twelve feet in height; but, owing to a scarcity of water in the immediate neighbourhood, looked sickly and faded.

Our Arab guards here left us, as no further danger was to be apprehended. A piece of white cloth and a little gunpowder made the old sheïkh quite happy, and we parted good friends. The lat. of Mofḥaḳ, deduced from a mer. alt. of the sun taken seven miles west of the village, is 15° 8′ N.; therm. in shade 74°.

July 24.—On the following morning we made a short stage of four hours, in an E.N.E. direction, to the village of El Hudheïn,‡ and here we sent on a courier to Ṣan'á with a letter to one of the principal merchants, which our Mokhá friend, Ḥájí 'Abd-er-Rasul, had kindly provided us with. Thermometer in shade, 73°; temperature of a spring, 64° (Fahr.).

In the immediate vicinity of El Hudheïn are several villages, and the inhabitants of these, hearing that two white men had arrived, crowded round us to beg we would tell them if any rain was coming. For some days the appearance of the weather had been threatening, and we therefore told them that rain would come, and they departed, quite satisfied that we knew all about it.

July 25.—At 6·30 started for Motteneh, distant eight hours and a half, in a general E.N.E. direction. At 3 m. we passed the village of Súḳ-el-Khamís; at 7 m. reached the village of Bowán, where we found a neat stone bridge thrown across the stream that flows to the Wádí Ṣeïhán; at 13 m. passed Yazil, a hamlet of about 30 houses. On leaving El Hudheïn we ascended gradually for about two hours, when we reached the ridge of the mountains; and from the summit a most magnificent view burst upon

* Subbúr?

† Sehân, Niebuhr, Voy. i. 432. ‡ Hadeïn, Niebuhr, Voy. i. 431.

us. The hills formed an immense circle, like the crater of a huge volcano, and the sides of which, from the top to the bottom, were cut regularly into terraces. I counted upwards of 150 in uninterrupted succession; and the *tout-ensemble* was most extraordinary. At the bottom of this basin ran a small stream,* which, from the height at which we were, looked like a silver thread.

Small hamlets, each with its little white mosque, were scattered over the sides of the mountains, and added greatly to the beauty of the scene.

We skirted the edge of this natural amphitheatre, and shortly afterwards reached a long table-land, very barren and stony, that extended to the village of Motteneh.

We had now attained our extreme elevation, and I do not think we were less than 5000 feet above the level of the sea; and, as this was the last stage before we reached Ṣan'á, I will here introduce a few hasty notes made by Dr. Hulton before illness rendered him unequal to any exertion:—

"The hills in the neighbourhood of Senníf are not high, and seem to be composed of a species of trap-rock of various kinds. Hills of the same kind prevail as far as Samfúr, where the compound becomes more crystalline, and partakes of the character of granite. Here the hills assume a remarkably varied appearance. Stupendous masses are heaped one on the other to an immense height, and others have rolled down of such dimensions as almost to obstruct the road through the valley.

"Near Mofḥaḳ this rock disappears, and a mixed kind then prevailed, with a large proportion of hornblende, aluminous matter, and quartz.

"From El Hudheïn the clay predominates; and from this bed it would seem that the chief part of the soil deposited in the valleys is washed down. After surmounting the lofty hills beyond Khamís,† the country is less mountainous, and appears to be more of a volcanic nature, large masses of cellular trap and scoriæ lying scattered on the plain."

During our stay at Motteneh we had a most terrific storm of thunder, lightning, and rain. It lasted about three hours, and washed down many of the huts in the village. We were told that it was the *first rain that had fallen for three years!* and the event was celebrated by a kind of festival. Therm. in shade, at 8 P.M., 72°; at 2 A.M., 53°. Pop. of vill. 250 persons.

July 26.—On the following day we left Motteneh at seven A.M., and continued over a table-land, in a N.E. direction; at 10 h. we passed the village of Lalwá, and at noon reached that of

---

* The Wádí Ṣeïḥán, according to Niebuhr's map.

† Súḳ-el-Khamís (Thursday market). Niebuhr, Voy. i. 431.

Assúr, seated at the eastern verge of this plateau, and saw the city and beautiful valley of Ṣan'á* stretched before us.

My first sensation was that of disappointment; but it soon gave way to a more pleasant feeling. We quickly descended about 1200 feet into the valley, and at three o'clock entered the suburbs of the city by the "Báb Ḳá'-el Yahúdí,"† or the Jews' Gate.

The first thing that struck us on entering the city was the width of the streets and their cleanliness. This, however, did not last long, as we became more acquainted with the town. The suburb of Bír el 'Azab, through which we were passing, is now walled round, although it was not so in Niebuhr's time. We saw no guns on the fortifications.

Proceeding through the Jewish town, we were met by Ismá'íl Waláni, our Ṣan'á friend, to whose care we were consigned. He received us very cordially, and conducted us to a very neat house in the Bír el 'Ázab, where preparations had been made to receive us. We were now told that our sudden arrival had caused great uneasiness to the imám, as he imagined we were nothing more than Turkish spies. We hastened to convince Ismá'íl of the error, and were shortly after waited upon by the vizier, Moḥammed Sa'dí, who had a long conversation with us, and, having ascertained that we were British officers, welcomed us formally, in the imàm's name, to Ṣan'á.

The etiquette of not allowing strangers to receive visits, which existed in Niebuhr's time, is still in force: we were therefore carefully locked up for the remainder of the day; but everything we could wish for was brought to us, and, with the exception of not being allowed to walk about the town, we did as we pleased. On the following day the vizier again called, and told us that the imám had provided another house for us in his own gardens, and accordingly we were conducted thither. Shortly after our arrival we were summoned to attend the Imám in his palace of Bustán el Metwokkil.‡

We found his highness in a large saloon, very splendidly furnished,§ sitting on a raised throne of richly-carved wood, with a silk canopy overhanging it. By his side stood his uncle Seyyid Moḥammed, or the Seïf el Khalífah.‖ The Imám wore a white turban, with a skull-cap of cloth of gold, a rich crimson silk robe reaching to his ankles, and a Kashmír shawl. His dagger, which was

---

* Capital of Yemen, or Arabia Felix.

† Báb-el-ḳá'el Yahúdí, i. e. Jews' Plain Gates. Yahúdí, also called 'Oseïr, was a separate town in Niebhur's time. Descr. p. 232.

‡ Properly Mutewakkil. El Mutewakkil billah (he who trusts in God) is a title assumed by the Imáms of Ṣan'á.

§ Luxury seems to have crept in since Niebhur's time. See Voy. i. pl. lix.

‖ The khalif's sword.

quite a blaze of jewels, had a gold hilt and scabbard. He received us kindly, ordered two chairs to be brought in for us, and accepted the presents we had to offer him. He repeatedly asked if we were not French, and seemed pleased at our declaring ourselves to be English. Dr. Hulton prescribed for some imaginary pain that he complained of, and we were then told that our audience was at an end.

A present of five sheep, some wax candles, and a bale of Persian tobacco, followed us; and we were besieged on all sides for a "Bakhshísh," or present of money, by the eunuchs who brought them. We were permitted to walk through the gardens, where we found many English fruits in great luxuriance; but there had been a great scarcity of rain here, as well as on the mountains, and everything looked dry and withered.

A large fountain played in front of our house, overshadowed by a huge vine that almost broke under its load of fruit. Walnuts, figs, and plums were plentiful; and the trees were the resort of a great number of black monkeys, which I suppose had been settled there for the Imám's amusement. We went out in the evening; and, though the people crowded about us in great numbers, no insult was offered, and we were allowed to do as we pleased. As, however, a daily journal of our movements here would be tedious, I will endeavour to describe the city and the court of the Imám as as we found them in 1836. In Ṣan'á we adopted the English costume, as it was already known there, and, having proclaimed ourselves English officers, we thought it best to appear in our uniform.

The Imám of Ṣan'á has two large palaces with extensive gardens adjoining; the whole walled round and fortified. The first and largest is called Bustán el Sulṭán, or the Garden of the Sulṭán; the other, which is the most ancient, Bustán el Metwokkil.* They are built of hewn stone, plastered over with a grey-coloured mortar, having the windows and cornices of a bright white colour, which gives the house a very light and airy appearance. Fountains appear to be indispensable in the houses at Ṣan'á, and in the Bustán el Metwokkil there are several. The Imám has a stud of very fine horses that are always piqueted in front of the palace. They come from the desert of Jóf, to the north of Ṣan'á, and for the first four years of their life rarely taste anything but dates and milk. They are larger than the "Nejdí" breed, but I believe are not considered as inferior to them in symmetry or speed.

At daylight every morning the levee takes place, and by eleven

---

* Mr. Cruttenden has kindly presented drawings of these two palaces to the library of the Society. The architecture appears Saracenic; simple and elegant, though with a mixture of circular and pointed arches, and not loaded with superfluous ornaments.—ED.

o'clock the Imám is no longer visible. 'Alí Manṣúr, who was the reigning prince during our stay (though since deposed), was much addicted to drinking spirits, and in fact was rarely sober after mid-day. He was a young man, born of a Nubian mother, and with a peculiarly disagreeable expression of countenance, owing to a cast in his right eye. As a Zeïdí in faith, the use of tobacco was forbidden to him; but it can hardly be supposed that a man who would disregard one precept of the Ḳorán would hesitate to break another. His days were consumed in smoking and drinking with the lowest of his servants, who plundered him in all directions. We were twice invited to join him in his house during these hours of recreation, but were too much disgusted to repeat the visit.

On Friday the Imám goes in state to the mosque, and the procession we witnessed was very splendid. Troops were called into the town to assist at the ceremony, and during the time of the procession the city gates were, as usual, closed. About fifty armed Bedawís formed the commencement of the caval-cade. They walked six abreast, and sang in chorus. The principal people of the town followed, mounted on horseback, each carrying a long spear with a small pennon. The Imám next followed on a splendid white charger, and very superbly dressed. He held in his hand a long spear with a silver head, having the shaft gilt. His left hand rested on the shoulder of a confidential eunuch, and two grooms led his horse. A very magnificent canopy, much like an umbrella in form, was carried over his head, having the fringe ornamented with silver bells.

The Seïf el Khalífah came next, having a canopy held over his head likewise, but smaller and less costly. The commander of the troops and the Imám's relations and principal officers followed, and about 100 more Bedawís closed the procession.

On reaching the square in front of the palace, the footmen ranged themselves round it, and the Imám, followed by his nearest relations, galloped repeatedly round the square, brandishing his spear, and making a feint of attacking the nearest horseman. After this had lasted some minutes, the imám stood still in the centre of the square, and the people rushed from every quarter to kiss his knees. He then retired towards the palace, and as he passed under the archway, a gun was fired to give notice that the ceremony was at an end.

The government of Ṣan'á under an Imám is, we were told, to be dated from the time that the Turks in the reign of Suleïmán, commonly styled "the Magnificent," were driven from that part of the country about 210 years since. The greater part of the fortifications in the old city of Ṣan'á were built by them, and there are the remains of a noble aqueduct yet existing, said to be their work. The first Imám was Ḳásim Abú Mohammed, a "sheríf" or descendant of the Prophet, of the family of Imám Hádi—Abú Mohammed's son took the name of "Metwokkil Allah," God's Vicegerent*—and this is now common to all the Imáms, as well as that of "Manṣúr,"† or Conqueror. The Arabs of Ṣan'á have a tradition that a descendant of the Imám shall assume the name of "El Hádí," the Regenerator,‡ and carry every thing before him, when, having taken the name of "El Mahadí," and converted all nations to the religion of "Eslam," the world will come to an end.

Sídí Ḳásim, uncle of the reigning Imám, who was banished during our stay, all his property having been confiscated, soon after our departure assumed the title of El Hádí, and so worked upon the superstitious fears of the Arabs, that he was enabled, through their assistance, to dethrone the Imám, and, under the name of "El Mahadí," assume the government himself.

The valley of San'á lies about 4000 feet above the sea; it is from 6 to 9 miles broad, extending northward as far as the eye can reach; it is bounded on the east by low table-land and a mountain called Jebel Nagam, rising about 1500 feet above the plain: to the west it is bounded by the table-land of Asúr and Lúlúwah, about 1200 feet in height; while to the south, at 7 miles' distance, it contracts into a narrow valley called Tarík el Yemen.

The population of Ṣan'á is great, perhaps 40,000, and I should say that in the four towns in the valley, viz., Ṣan'á, Ródah'§ Wádí Dhar, and Jeráf, there are not less than 70,000 people. The old city of Ṣan'á is walled round, and, including Bír el Azab, is 5½ miles in circuit: it has some guns, but in a very bad condition. The houses are large, and the windows of those of the higher classes are of beautiful stained glass. A handsome stone bridge is thrown across the principal street, as in wet weather a stream of water runs down it. The streets are narrow, though broader than those of Mokhá and Zebíd. Great hospitality was shown us on entering their houses; we were always pressed to stay, and never allowed to go without taking a cup of coffee, or rather of an infusion of the coffee husk called "Ḳeshr;" for, strange to say, though in the heart of the coffee country, coffee is never taken as a beverage, being considered as too heating. The infusion of the husk is very palatable; and we found it much more refreshing, and nearly as powerful a stimulant as the infusion of the bean itself.

The merchants form the principal body of men in the town. They are generally wealthy, and live in good style.

The Banians are also numerous, but they are compelled, like

* "The Truster in God."
† "Aided by God;" thence "Victorious."
‡ "The Guide, or Leader."
§ Rauḍhah or Rauḍah *i. e.*, Garden.

the Jews, to conceal what they really possess, and however wealthy they may be, to put on an outward show of abject poverty.

The principal trade is of course in coffee, but the San'á merchants are so fearful of trusting their goods to the Turkish government, that they prefer filling their warehouses with it in Ṣan'á to sending it to Mokhá.

The whole cost of transporting a camel-load of coffee from Ṣan'á to Mokhá is forty-four dollars, upon which the merchant clears a profit of three dollars and a half. It is brought into the Ṣan'á market in the months of December and January, from the surrounding districts. The nearest place to Ṣan'á where the coffee grows is Ḥaffásh, about a short day's journey south-east of Ṣan'á. Attempts were made by the last Imám to cultivate the plant in his own garden, but without success, owing, it appears, to the cold. The varieties of coffee are "Sharjí" (the best), "Uddeïní," "Maṭarí," "Ḥarrází," "Ḥabbat," "Ḥaïmí," and "Shírází;" of these the Sharjí and Ḥabbat are the smallest and best. Ḳeshr (husk), being more in demand at Ṣan'á, obtains a higher price. The best is the 'Anezí (Ḥabbat), and is sold at twelve dollars for the hundred pounds; the inferior sorts at four, five, and six dollars for the same quantity. The imports are principally piece-goods and Persian tobacco, with dates from Tehámeh; and a great quantity of thread, or rather twist, for weaving. The import duties in Ṣan'á are very light, and indeed almost nominal. Glass is in great request, and the demand is supplied from Egypt. Very magnificent silks and velvets are exposed for sale in the bázár, as well as spices, sugar, &c. &c.

The principal artisans are the Jews; these amount to about 3000 persons, and live in a quarter of the city appropriated to them. As infidels, they are exposed to many exactions and repeated insults. Each man pays twenty-five komásís per month, or about *a dollar* per year, for permission to reside in the city. A sheïkh is appointed, who is responsible for the regular payment of this impost, and of the heavy taxes that are likewise laid upon their vines, gardens, &c. They subsist chiefly by the sale of silver ornaments, gunpowder, spirituous liquors, and many by working as common artisans, such as shoemakers, &c. The mosques in Ṣan'á, about twenty in number, are very splendid, many of them having their domes gilt, particularly those containing the tombs of the Imáms. The baths are also very good, and on the same plan as those in Egypt: they are a favourite resort of the merchants, who meet here to discuss the state of trade and the news of the day, over their cup of ḳeshr and their never-failing ḥuḳḳah. The Ṣan'ánís are very much addicted to chewing the leaf of a tree which they call "ḳát." It appears by their account to exhilarate and produce appetite; it also causes great thirst, and if taken in large quantities, will bring on spasms. It is the never-failing accompaniment to the breakfast or dinner; and, from long use, appears to be indispensable. The wealthy merchants on week-days generally ride a very fine-spirited kind of donkey, much larger than the English ass, and very strong and fleet. These donkeys are similar to those from Baḥreïn, in the Persian gulf. On Fridays, all who can afford it, appear on horseback to join the procession of the Imám to the Mesjid el Jámi'.

About five miles N.N.W. of Ṣan'á is the town of Róḍah, which is much cleaner and neater than the capital, being the residence of nearly all the merchants, who retire to their country-houses after the business of the day is over. The gardens at Róḍah and Wádí Ḏhár, another town five miles to the west of the former, are very fine, and the vineyards beautiful. The vines are trained over a trellicework about four feet from the ground, and are very extensive: the best kind of grapes are "el Beḍá'," a small white grape without a stone, and the "'Ayún," a large black one very richly flavoured. Peaches, apricots, plums, &c., were abundant, and indeed formed the principal food of the lower classes. The town is well watered by several small streams: on the banks of one of these I saw the only gum Arabic tree which we had seen during the journey: there was an immense quantity of gum dropping from and incrusted on the bark, but it appeared to be considered as of no value by the town-Arabs. Half way between Róḍah and Ṣan'á is the town of Jeráf, built much in the same style as the former places, and furnishing vegetables chiefly, for the Ṣan'á market. Each of these towns is governed by an Amír, who levies the government taxes in the name of the Imám.

The climate of Ṣan'á is too dry to be healthy: there is rarely any dew at night, and the wind produces a feverish feeling in the hands and face. In average seasons, rain falls three times a year, i. e., in January, in very small quantities, and frequently not at all; in June for about eight days: by this time the seed is sown and the cultivator looks forward to this season with great anxiety. Lastly, it falls in the latter end of July, when it is in the greatest abundance: a few farmers defer sowing till this period; but this is not commonly done when they can reasonably expect rain in June. In July the wind from the south-east prevails during the day, but, declining in force during the afternoon, it is met by a current of air from the north-west, and the two strata of clouds meeting in different states of electricity, thunder, lightning, and rain are the result. Thus it is that rain is never known in these regions at this period without being accompanied by vivid lightning and loud peals of thunder.

For nearly four years preceding our visit they had hardly had a drop of rain, which was the cause of incalculable distress to the people in the vicinity. This long drought had added considerably to the general unhealthiness of the country. The people on our first arrival were dying at the rate of 150 per day, from the effects of a malignant fever which was raging with great violence, and which generally carried off the sufferer in four days. The famine was also dreadful here,* and dead bodies were seen in every corner of the streets awaiting the compassion of some one to afford them the means of burial. Many of the wealthy merchants fed a stated number daily, and boasted loudly of their charity. On inquiry, however, we found that the food furnished consisted merely of refuse grapes, such as were literally of no use. The Imám, too much engaged in his favourite amusements of smoking and drinking, thought little of the distress which his people were enduring; and though the incessant funeral chant, as the bodies passed under his windows on their way to the burying-ground, ought to have roused him from his gross sensuality, he heeded it not: the consequences fell heavy upon him. In a month from that time he was dethroned, insulted, and immured in a dungeon, while his uncle, who supplanted him, wisely endeavoured to secure the affections of the people by relieving their distresses with a liberal hand.

During our walks through the city we discovered the accompanying inscriptions,† and forthwith copied them on the spot. On close inquiry we found that the stones had been brought from Máreb, about two days' journey distant to the N.E.,‡ and that there were many more to be found there. The longest inscription was on a slab of white marble, and, when we saw it, served to cover a hole in the roof of a mosque. A bribe of a dollar had a magical

* To give an idea of the great scarcity experienced in 1836, and at the same time of the usual fertility of the soil in this part of Arabia Felix, I subjoin an account given to me by one of the principal Banian merchants in the town. Wheat, which in time of plenty is sold at 1*s.* 6*d.* the gaddá of 54lbs., was now selling at 3 dollars or 13*s.* 6*d.* the gaddá. Barley, usually sold at 6*d.*, was now 8*s.* the gaddá. Jowárí, usually 9*d.*, now 9*s.* the gaddá. Beans, usually at 8*d.*, now 12*s.* the gaddá. Ghí, or clarified butter, usually sold at 4*d.*, now 1*s.* per lb.

† That these inscriptions are specimens of the *Musnad* or ancient Ḥimyarí character, will hardly be doubted by those who have read M. Rœdiger's learned paper in the *Zeitschrift für die Kunde des Morgenlandes* (Gœtt. 1837, 8vo., p. 332), though the corresponding alphabet given by him from an Arabic MS. may be fairly set down as one of the many literary frauds with which the half-learned in Asia have endeavoured to raise their reputation at the expense of truth; such, for example, are '*Ancient Alphabets*' and the late Mr. Price's interpretation of the *Persepolitan Inscriptions*, from a Persian MS. The striking resemblance, at first sight, between these Himyarí characters and those on the ancient Bactrian coins decyphered by Mr. Prinsep in the *Calcutta Asiatic Journal*, did not escape Mr. Cruttenden's notice; and it is possible that a further acquaintance with these characters will prove their identity: if so, the happy conjectures of Dr. Lepsius, in his tract on *Indian palæography*, will receive an extraordinary and unexpected confirmation.—F. S.

Professor Gesenius, to whom all our inscriptions have been sent, has, we understand, decyphered the words, "King of the Himyarites."

‡ Between seventy and eighty miles, according to Niebuhr's calculation.

effect on the scruples of a servant, and the stone was brought to our lodgings that night to be copied, and carefully replaced before daylight. The Jewish workers in gold assured us that frequently square gold coins were brought to Ṣan'á by the shepherds of Máreb for sale, and a Banian merchant corroborated this account, adding also that jewels, particularly pearls, are found there usually, after heavy rains, when the people closely search the water-courses, and generally find something to repay them for their trouble. In the Imám's garden I one day found a marble head, and on inquiry learned that this also came from Máreb: to my great mortification I was told that the figure arrived at Ṣan'á perfect, but was immediately broken by order of the Imám, as a relic of ancient idolatry, and I was unable to find the other parts of the statue: the head, however, I secured, after some demur on the part of the gardener, and brought it to England; it is the only specimen of sculpture that I have ever observed in Arabia Felix, and as such I esteemed it a curiosity.

Hearing so much about Máreb we forthwith determined to go there, and for that purpose sent for a sheïkh who lived in that part of the country. The Imám, however, now became jealous of our proceedings, and for six days we were not allowed to leave our house in the garden. Our agent Ismá'il also gave us strong hints that we had better go at once, as the people in Ṣan'á were becoming troublesome, and the imám attributed it to the "Christians." At this time the uncle of the Imám, Sídí Ḳásim got timely warning that his head was in danger, and escaped that night, accompanied by about forty of the Imám's servants and guards, mounted on his best horses. We now found that we could do nothing: our garden-door was always locked, and what with disappointment and anxiety, my companion's health began to decline, and he soon was in a state that precluded all possibility of travelling.

After nearly three weeks of severe illness, Dr. Hulton thought himself strong enough to return to the vessel. As travelling on horseback was out of the question, I constructed a kind of palanquin out of an old litter belonging to the Imám, and procured twelve men to carry it.

I also went to take leave of the Imám, who sent us two fine horses and some shawls, as a parting present. On the 20th of August therefore we left Ṣan'á, having been resident in the town nearly a month.

The thermometer during our stay reached 75° as the highest and 55° as its lowest point, between the 26th of July and the 20th of August.

The crowd collected in the streets on the occasion of our departure was immense, but so far from insulting us, many appeared

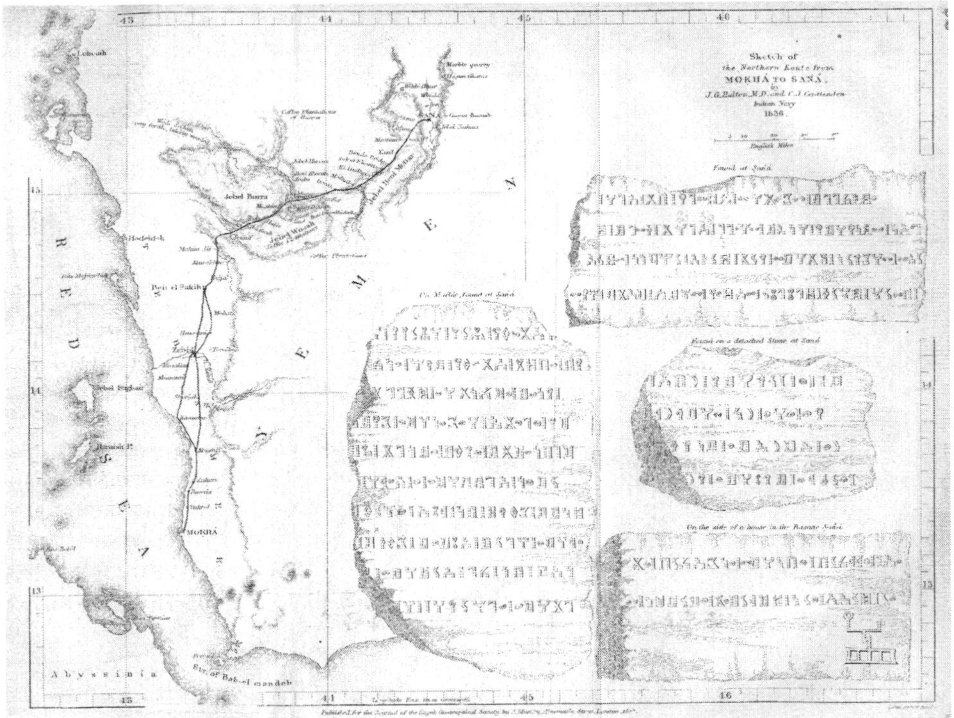
Sketch of
the Northern Route from
MOKHÁ TO SANÁ,
by
J. G. Hulton, M.D. and C. J. Cruttenden
Indian Navy
1836.
English Miles
Found at Saná
On Marble, found at Saná
Found on a detached Stone at Saná
On the side of a house in the Bazaar Saná
SANÁ
Jebel Bura
Hodeidah
Beit el Fakih
Zebid
MOKHÁ
Jebel Sugbair
R E D S E A
Y E M E N
Abyssinia
Str. of Bab-el-mandeb
Marble quarry
Coffee Plantations

to commiserate the emaciated appearance of Dr. Hulton. All our friends assembled to see us depart, and accompanied us to the gates of the city, when we bid them farewell, and pursued our way slowly to Mateneh. Our journey to the sea was of course slower than it had been upwards, but on our arrival at Senníf my friend Dr. Hulton was sufficiently strong to travel on the back of a donkey; and in fourteen days from our leaving Ṣan'á we reached the gates of Mokhá.

Nothing worth notice occurred during our way down, except the demand of a few dollars made by the Arabs as toll for the passage of the escort across the Wádí Seïhán, and the improved appearance of the country, owing to the great fall of rain that had taken place during our residence at Ṣan'á. We found the ship still anchored at Mokhá, and all our friends very uneasy at our protracted absence. I hoped that the return to the sea would restore Dr. Hulton's health, but it was too late, and he died very shortly after he reached the ship.

In closing this brief Memoir of our journey into Yemen, I can only regret that the task has not fallen into abler hands than mine: I am well aware that an account drawn up by a man of such general attainments, and especially in geology, as Dr. Hulton, would have been far more valuable and satisfactory.

XXII.—*On a New Construction of a Map of a portion of Western Africa, showing the possibility of the Rivers Yeú and Chadda being the Outlet of the Lake Chad.* By Captain W. Allen, Royal Navy. Read 25th June, 1838.

The little knowledge we have gleaned with immense sacrifice, in the interior of Africa, appears still more scanty until brought together and combined; and the isolated relations of individual travellers sometimes contain facts apparently unimportant, and at first sight at variance with others; whereas a careful examination of all the data will frequently reconcile conflicting statements, and by bringing one traveller to the assistance of another, will enable us to throw considerable light on this interesting subject.

Among the desiderata there are two very important points, namely, the verification and connexion of the various geographical positions, and the still unsolved problem of the nature of the lake Chad, as to whether it be a *still water* or have an outlet. These I propose to make the subject of the following paper:—

In the examination of the longitudes of former travellers, it is unfortunate that there are but few data on which to work; and I am obliged to prove the errors of others, from the assumed correctness of my own positions. Having, however, generally been in a vessel, I had instruments and facilities for astronomical observations, which Captain Clapperton, on whom we depend, could not in his land journeys possess.

From the simplicity of the operation, I conclude his latitudes were all by observation, therefore I shall not interfere with them. But as he expressly says in his first journey from Kukah to Sakatù, that his longitudes are by dead reckoning,* and as I have reason to believe that his others were so, I shall take the liberty of considering them all in the same light.

Although we did not visit the same places, and there are consequently no direct means of comparison between our routes, there is fortunately a sort of connecting point at Kulfu, where he gives the bearings and distance of Rabba, at which place I had very good chronometric observations. He says the latter is three days south of the former; therefore carrying back his distances from the latitude of Kulfu, on the meridian of Rabba, we shall have Bussah in longitude 4° 42′ E.

As this differs 1° 29′ from the position of that place on Clapperton's chart, that quantity may be adopted as a correction to all his longitudes. But a portion of this, namely, 51′, arises from the erroneous position which he assumed for Badagry,† from which place he took his departure; therefore the remainder has arisen, probably from his having over-estimated his distances, an error into which all travellers are liable to fall, and which must increase in proportion to the difficulties and fatigues of the journey. If this principle be admitted, it will be necessary to apply a second correction, in the ratio of the excess of the distance, between the known points Badagry and Bussah. However, as it seems unreasonable to suppose that he always committed an error of equal amount, I would propose to take a mean between the *constant* and the *increasing* corrections. Then the longitudes will be as follow:—

Bussah	4°	42′ E.
Womba	5	49
Guari	6	26
Zaria	7	4
Bebeji	7	37
Kano	7	44
Sakatù	4	38
Katagúm	9	16
Kukah	12	34

It is impossible to make use of the distances given in itineraries, unless we have some data by which to estimate a day's march. This will vary according to circumstances, and there must be a considerable difference between a Ḳàfilah journey and that of a

* Denham and Clapperton's 2nd edition, vol. ii. pp. 220 and 251.

† Properly Badaghí.—*Vide* Robertson, notes on Africa, p. 283.

messenger. When the distance is given between cities not very remote, and having from their importance frequent communication, I think the latter ought to be calculated on, and might be taken at twenty-five miles. By adopting this in the distance between Rabba and Kulfu, it would place the latter more to the north, and within the distance of two days of Yáuri, as given by the Landers; otherwise two days would bring Yáuri too much south, and too near Bussah, to agree with the time they took in passing between those cities.

The distance from Bussah to Rabba, the Landers say is seven days by land and four by water. I presume the land journey is through Kulfu.

Both at Sakatù and Yáuri, the distance between these cities is said to be five days. This, if we take the messenger-day of twenty-five miles, will agree very nearly with the position of the former city, by Clapperton's latitude and his corrected longitude, and gives 10·8 miles per day for the distance to Kano, which is, perhaps, enough for horizontal distance, considering that the route is difficult and circuitous. This journey between Sakatù and Kano was performed three times, each in about seventeen days.

On his first expedition, travelling between Bornù and Sakatù, Clapperton takes his departure from a lunar observation* at Kuka; but he neither gives the data, nor mentions his having taken such an observation at that place. Now, as in coming westward, he falls short of the position I have assigned to Kano, by his longitude corrected from Bussah, either he or I must be wrong; but as the estimated distance generally exceeds the true instead of falling short, the presumption is that the lunar observation places Kukah too far to the eastward. It is therefore better, perhaps, to reject it altogether, and assume a position for Kukah, by carrying back his itinerary from the corrected longitude of Kano.

Thus I leave Clapperton's map untouched, with the exception of bringing his longitudes a little more to the westward. The case is very different, however, with respect to Lander's itinerary between Kano and Danrorah, after the unfortunate death of his master. This has been so erroneously laid down, that the latter place falls within thirty-five miles of Fandah, from which he was informed that he was then distant twelve or thirteen days.† This route must, therefore, be altered entirely, which the data enable me to do with considerable precision; and it will be checked by the reported distance of Danrorah from Fandah, and of Jakobah from Dagbòh.

---

* Denham and Clapperton, vol. ii. p. 251.

† I cannot ascertain how or by whom this itinerary was laid down, but the position of Fandah was not at that time ascertained.

Lander went from Bebeji to Kuttup, and returned to Zaria; thus making a triangle, of which the first and last places at the base are known;* and Kuttup, which is three days short of Danrorah, is at the apex. I have very carefully estimated the distances on the routes to and from Kuttup, from the hours which he gives, leaving out of the question his vague courses, and I find, after making proper allowance for winding, they are about 120 miles each: these, therefore, form the legs of an isosceles triangle, and their intersection points out the position of Kuttup. To the remainder of the journey of three days to Danrorah, I have given an analogous direction, which agrees with Lander's reported distance of Fandah, and mine between Dagbòh and Jakobah: Danrorah is half a day from the latter place, which lies probably in lat. 9° 30′ N. and long. 9° 26′ E.

This corrected map of Lander's journey, viewed conjointly with that of Clapperton, which it nearly joins, reveals a remarkable and highly important feature; namely, the almost uninterrupted continuance of a valley between the lake Chad and Jakobah, in the neighbourhood of which city flows the river Chadda. This leads me to the second part of my subject, the nature of the lake Chad. I approach it with great timidity and caution, as I am aware that very different views of the subject have been hitherto entertained by many able geographers.

Major Rennell was of opinion, that not only the waters which descend from the mountains of the moon, but the outpourings of the mighty Niger, Jalibà, or Kwara, may be evaporated, from the low and extensive morass in which he supposed them to terminate. The hypothesis of Major Rennell, for whose sagacity I entertain the highest veneration, would never have been proposed but for the imperfection of the materials on which he had to work: and the discovery of the outlet of the Kwara or Jalibà completely subverted it, and proved at the same time how vague were the ideas of the Arabian geographers respecting the interior of Africa.

My rejection of that hypothesis therefore will not be deemed presumptuous, especially as so able a geographer as M. Reichard, of Lobenstein,† had, as early as 1802, assigned very strong reasons for supposing that the outlet of the Jalibà would be found in the Bight of Benin. The discovery of the large lake or inland sea by Denham and Clapperton seems at first to support this theory, especially if it has, as the former believed, no outlet; but by giving an approximation of the size of the lake, Denham has shown how small that reservoir is, compared with the volume of water discharged into it. The Caspian, for instance, receiving

* By Clapperton's latitude and longitude.
† Von Zach, Monatliche Correspondenz, May, 1802.

the drainings of a comparatively low range of mountains, has a surface of more than 10° by 3° in breadth. The Chad is only about 3° in length, and receives the deluge of tropical rains, from a range 25° in extent, and doubtless higher than the snow-capped mountains of Abyssinia, since the principal branch of the Nile takes its rise in the eastern side of the same lofty range, the Kumri,* or mountains of the moon, or as Sultàn Bellò says, *ninety-nine* mountains all of them beginning with the letter F.

It is true the lake Chad is not the only receptacle; we have reports of the lake Fitri, and there are, perhaps, many others: but we may presume from the alluvial nature of the whole of that vast tract, that they are merely inundations of the low land; and even the permanent lakes are but slight depressions of soil, and may have no great depth. In the next place, we have now positive proof of the existence of a large river, the Chadda, flowing from the east, and uniting with the Kwara from the west, which it certainly rivals, if it does not surpass it in magnificence; having much greater average breadth, although I believe the depth in general to be less. This magnitude implies a proportionate length of course, and if it do not pass through the lake Chad, it must have its sources far away to the east, in some of the *ninety-nine* mountains. The Mendefi range seen by Denham is not sufficiently remote or lofty to be the parent of the *whole* of such a stream; yet it would lie right in its course, especially if the slope on the southern side be as extensive as that on the north, which Denham found to be 100 miles. The mountain may, however, be volcanic, of which there is very great probability:† in that case it may rise abruptly on the south side, from the alluvial or river level, and allow the Chadda to sweep round its base. This course indeed Denham gives it, from the accounts of the natives, but he joins it with the Shari, thus making the stream flow in opposite directions at the same time.

Again, Denham assures us that the water is fresh; he says frequently "it is very sweet and pleasant."‡ Although instances have been given of lakes without outlets being sweet, I doubt if these be well established; and the contrary is certainly the received opinion.

Lastly, the river Chadda is very clear compared with the

---

* Properly Jibàlu-l-Ḳumrạ, *i. e.*, the Blue Mountains, though the original name was doubtless Jibàlu-l-Ḳamarì, "the lunar mountains," as the Arabs derived their knowledge of this part of Africa from Ptolemy, who calls the whole range "the mountains of the moon."—(Geogr., p. 115.) S.

† The corrected position of the highest point of the Mendefi range falls exactly on the line of the volcanic elevations in the Atlantic of St. Helena, Anno-Bom, St. Thomas's, Fernando Po, and Camaroons Mountain, an extent of more than 1500 miles.

‡ Denham and Clapperton, vol. i. p. 64.

Kwara. This is rarely the case with mountain streams; but as with the Rhône, the Rhine, &c., &c., it is always the consequence of having passed through a lake, where the almost stagnant waters have time and space to deposit the débris from the mountainous regions, with which they were charged. Indeed the Chadda was so clear, that the men drank its water from alongside, after letting it stand a little while; it was remarkably "sweet and pleasant." While in the Kwara we drank the water which had been suffered to cool and settle in the boiler. The Chadda began to rise a little earlier in the season than the Kwara, but this must be the case whether it pass through the lake or not, since it must have its feeders in the Kumri mountains, many degrees to the south of the sources of its rival, and therefore must begin to fill a little sooner and faster, but not much, otherwise the rivers would have different levels at their confluence.

If the fact mentioned by Mr. Laird, that the water of the Chadda is colder than that of the Kwara, was the result of comparative experiment, and not depending on the fallacious evidence of the senses, I cannot admit it to be a proof that the whole stream is direct from the mountains, without having been delayed in its course by being spread into a lake.

I made experiments in the Kwara only, but my self-registering thermometer failed, as the quicksilver passed the index every time. This however was sufficient to prove that the water at the bottom was at a much lower temperature than at the surface.

These are the circumstances which have led me to the belief of the identity of the Chad and the Chadda, or more explicitly, of the lake and river. I shall, I trust, bring forward some strong evidence in my support, and will honestly state the arguments on the opposite side of the question, as they appear principally in Denham's assertion, and the implied belief of Clapperton. They however were pre-occupied with one idea, and thus easily satisfied themselves that they had proved it. I thought at the outset of this inquiry, that I should be able merely to bring forward some plausible reasons in support of my opinion, but as I advance, I find the evidence crowding so closely on me, that I hope to be able to prove, not only that the alluvial or river level can be traced from the lake Chad, nearly to the part of the Chadda which I was able to reach; but that with a few breaks, I can show the continuity of the river for nearly the whole of that distance.

The similarity of the names has led others, as well as myself, to look to the Shari* of Denham as the outlet of the lake, notwith-

* Shárí, if Arabic, would signify glittering, an epithet often applied to a large body of water; but the name may be a Bornuese word, and it may have been incorrectly expressed.—S.

standing his assertion to the contrary, and I confess I began this inquiry, with the hope of finding in his description some circumstances from which I might draw a fair presumption of the possibility of his having been mistaken in the direction of the current of that river. Those who know the difficulties under which travellers in general labour, and how truly they are multiplied an hundred-fold to the African traveller, will not be surprised at my imagining him to have fallen into such an error. For myself, if I succeed in proving my proposition, I shall consider that I have made a greater discovery in Africa within my closet, than while personally labouring on the spot.

The consequence, however, of an attentive reading of the account, which I had in some measure forgotten, was, that I very soon perceived the necessity of abandoning all idea of drawing off the water of the lake by that channel. Nothing can be more conclusive or circumstantial than Denham's description. He saw the river and crossed it at several points, and not only assures us that the current runs past Loggùn and Showi with considerable rapidity *towards* the lake, but having gone *with* the stream in a canoe as far as its entrance to "that sea of fresh water," he speaks of the difficulty of paddling *against* the current on his way back.* He also describes the water as discharging itself by several branches, and that it consequently forms a *Delta*.

We have also a confirmation of the northerly course of this river, from the account received at Sakatù by Clapperton; who was told that it is only four feet deep above Loggùn, before it is joined by the Asha, which comes from the east through Bagermi.† The Shari doubtless has its origin in the Mendefi range to the south-west.

Having thus disposed of the Shari, I feared that I should be obliged to give up the tribute of the ninety-nine mountains to the effect of the sun's rays, very much wondering nevertheless, how he would dispose of such a vast body of water, if once raised into vapour, except by letting it fall again on the surface of the lake in the shape of dews; and every one knows the dampness of the African climate, which causes the finest steel to rust in a very short time.

However, a new light broke in upon me in the course of my investigation. Soon after the first arrival of the travellers on the borders of the lake, they came, says Denham, "to a very considerable stream called the Yeú; in some parts more than fifty yards wide, with a fine hard sandy bottom, and banks nearly perpendicular; and with a strong current running three miles and a half to the eastward. As I expected, every one of the Arabs said this was the Nile, and that it ran into the great water the

* Denham, vol. ii. p. 6 and 7. † Clapperton, vol. i. p. 30.

Chad. It is sometimes double the width and a great deal deeper. The air from a running stream of CLEAR water, and the freshness it imparted to all around, was such a relief, after a march through sandy deserts, that both man and beast were in a manner renovated by its effects. The men, and even the women, bathed and washed, and the negroes swam all the horses." The principal feature in this description is the *clearness* of the water. It struck me very forcibly as an anomaly, although he says also that the Shari was clear at its embouchure; but this is incompatible with the formation of a Delta, and of the fact which he relates of its bearing on its bosom floating islands, &c., and I supposed this must be the outlet I sought for; upon very slight grounds it may be thought, but I must request that judgment may be suspended, while I lead up to my hypothesis, by referring to the authority of travellers from all parts of the continent, and in all times, up to our own enterprising countrymen, who have trodden the very ground. If I were to quote all the writers who throw light on this interesting question, I should swell my paper unnecessarily, since it has been already done by far abler hands. It will be sufficient for my purpose if I observe, that the accounts of the ancients of a great central river, traversing Africa from the west, were confounded with those of the Arabian geographers of the middle ages, who all spoke of a large river flowing from the east. They were considered as identical; but the ocular testimony of Mungo Park having been received as confirmation of that of Herodotus, it was immediately concluded that Africa could not be traversed otherwise than from west to east; and the Arabians were thrown into discredit. Deceived by the generic terms applied to both the rivers of the negroes,—by the transmission of the description through the natives, who doubtless consider the direction of rivers with reference to their own journeys, and independently of the current,—and by the universal belief of the communication with some great central lake, these accounts have been considered as relating to one and the same river, until the discovery of an intermediate outlet has shown the possibility of the continent's being traversed by streams from opposite sources.

Justice, however, has not been sufficiently done to the Arabian writers; and I hope to be able to prove, that their account of the eastern *Nil el 'Abìd*, or river of the Negroes, is nearer to the truth than has been generally imagined.

Idrísí,* Abulfeda, Leo Africanus, &c., all speak of the westerly course of this river. The former† says that it forms an island at *all* times of the year, 300 miles long by 150 in breadth, which is

* Jaubert's Idrísí, p. 13.

† Jaubert's Idrísí, p. 18; also Major Rennell, in the Proceedings of the African Society.

inundated during the rainy season; consequently, the whole of that tract is alluvial, and agrees remarkably well with Clapperton's description of it.*

They all show the drainage of the western side of the mountains, where the Nile of Egypt has its source, by a large river flowing to the westward, and communicating with the river of Tumbuktù; though their reports are vitiated by having put forth their opinion that there was also communication with the river of Egypt. Still they may have been misunderstood from the application of the generic term Nil to the three streams.

Modern discovery has proved them in the main to be right. While Mr. Lucas was collecting information in Africa from the Sherif Imhammed, a native of Fezzan, relative to the countries of Sudán, Bornù, and Kashnah, with which he was said to be perfectly well acquainted, information equally valuable was obtained in England from Ben Ali, a native of Marocco, of the same countries, which he also had visited. As they agree in the most important particulars, they may be taken as very good authority.

In speaking of the great river which in Arabic is sometimes called Nil el Kibir, or the Great Nile, and sometimes Nil el 'Abíd, or the Nile of the Slaves, they say,† "Its rise and termination are unknown, but its course is *from east to west.* So great is the rapidity with which it traverses the empire of Kashnah that no vessel can ascend its stream; and such is the want of skill, or such the absence of commercial inducements among the inhabitants of its borders, that even with the current neither boats nor vessels are seen to navigate.‡ In one place, indeed, the traveller finds accommodation for the passage of himself and his goods, but even then, though the ferrymen, by the indulgence of the Sultan of Kashnah, are exempted from all taxes, the boat which conveys the merchandise is nothing more than an ill-constructed raft, &c." "The depth of the stream, which is more than 100 miles to the south of the city of Kashna, the capital of the empire of that name, is estimated at twenty-three or twenty-four feet English. Its width is such that at the island of Gongo,§ where the ferrymen reside, the sound of the loudest voice from the northern shore is scarcely heard."

"Having passed the stream, the face of the country, and with it, the mode of travelling are changed. High mountains and

---

* Denham and Clapperton, vol. i. p. 81.

† Mr. Lucas's communications, Proceedings of the African Association, vol. i. p. 123.

‡ This is precisely the case on the Chadda at the present day; but Denham says that in the former Sultan's time there were many canoes trafficking on the river Yeú near Kabuhari.

§ Gongo or Gondo means island in the Haussa language.

narrow valleys, extensive woods and miry roads, succeed to the vast plains and sandy soils of the Zahara* and its neighbouring kingdoms. No camels are found; their place is supplied by small horses, asses, and mules." † The country of Kashnah, properly speaking, "is bounded on the north by the mountains of Eyré,‡ and by one of the districts of the great Zahara; on the south by the (great river) Niger, and on the east by the empire of Bornù and the kingdom of Zamphara.§ The capital is five days north of the Niger."||

Hornemann's notices are pretty correct, except that he considers that there is but one stream, having in all its course a current running in the same direction as that of the river of Tumbuktù, or from west to east. He says,¶ "The river which was seen by Mr. Park waters Nyffe and Cabi, where it is called Gulbi, and in Haussa Gaora, (Kwara), runs eastward into these districts of Bornù, where it takes the name of Zad; all these names mean the *great water*. It is a very large river. The breadth of the Zad was given me for one mile, others said two, but in the rainy season the breadth is said to be a day's journey (*i. e.* eight hours). The Bidumas** always keep in the middle of the stream. They are a very savage, heathenish nation."

Hornemann's distance from Kashnah to Bornù, 330 miles, agrees very well with Clapperton's.†† His supposition of the easterly current may be explained from his considering it to be a continuation of the river of Tumbuktù; and as I said before, from the probability that his native informant referred to the trending of its banks in the same direction as his journey, rather than to the course of its current.

In the Journal of the Arts and Sciences for 1823, an account is given by an officer at Sierra Leone (Major Laing), of the journey of a Mohammedan named Mohammed Misrah,‡‡ from Alexandria through Súdán and the countries in question. His testimony is of a varied nature, making both for and against my hypothesis, although it can have but little weight either way.

* Ṣaḥrá, *i. e.*, Desert.—S.

† Lucas's information, Proc. Afr. Ass., vol. i. p. 118. Idrísí, p. 15, says they have camels in the country of Lemlem.

‡ Eyres, *i. e.*, Aurâs, the mountains on the south side of Fezzan.

§ Zanfárah of Idrísí.—S.

|| The sheriff does not describe the western limit, but it is evident that it ought to be Zamfara, and the word is misplaced, as, according to Clapperton, Zamfra, of which Zirmi is the capital, lies to the west of Kashnah.

¶ Hornemann's notices, Proc. Afr. Ass., vol. ii., p. 201.

** Bidoomahs of Denham and Clapperton.

†† Clapperton was thirty-six days one way and thirty-seven the other between Kuka and Kano; Idrísí gives thirty-six days as the distance between Kauga and Ghanah. But in countries where the frequent wars cause not only the destruction of cities but the depopulation of whole districts, names are probably of little duration, or may be as migratory as the inhabitants.

‡‡ Probably Misrí, *i. e.*, Egyptian.

He says, that he had conversed with people who had travelled from Nufi to Sennár, along the banks of a very broad river, which frequently overflows its banks, inundating the country to an immense extent; but he positively believes this river to be a link of communication between the Niger and the Nile, calling it the Bahr el 'Abìd. He asserts, on the other hand, that he has made the circuit of the lake Fitri, which he declares has no outlet, although it receives a large river 400 yards wide at its embouchure. He says, at the distance of ten days south from Bornù, the Jalibà flows through the Kafir* country, where it is known by another name; that there is no river of any consequence within that distance of Bornù, although he must have passed the Asha, the Shari, and subsequently the Yeú. His accounts of distances prove that his memory was not always faithful. From Bornù Biriní to Kano, he gives only ten days; Idrísí gives thirty-six, as well as Clapperton, and Hornemann says 330 miles. Mohammed's testimony is therefore entitled to little credence in the details, although it is evident he had travelled through the countries he describes. There is something like evidence to be gleaned from Sultàn Bellò's almost unintelligible account of the countries of Súdán, given in the Appendix to Clapperton's journal.† The river Baku is described as larger than the Kwara. The relative positions of the rivers are, the Kwara in the west, the Kaduna in the centre, and the Baku or Jaku‡ to the east. The people of Nifi are said to have come originally from Kashnah, and their prince from Atághér. "He first conquered the territory of Bení, from the river called Bakú to that which is named Kaduna, &c.; he then embarked on the Kwara, &c." Mention is made of an eastern Kwara, and again, "the river of Kwara runs through mountains &c., and issues from the mountains of the moon, &c." This is a very consistent description of a river, such as I have attempted to trace from the lake Chad: the river Chadda in fact, which like the Baku, is larger than the Kwara.

Thus we have very strong presumptive evidence from the Arabian geographers and from native travellers, of the existence of an immense river, running through the countries of Bornù, Wankarah, and Haussa, &c., "100 miles south of the empire of Kashnah;" with almost unvarying assertions of its course being from the east towards the west, and having its sources in the same mountains which gave birth to the Bahr el Abyad, its origin being clearly pointed out by Mr. Brown as the Misselád. In Sudán it is not a mere mountain torrent, but a mighty stream, at the common river or alluvial level, which is proved by its overflowing its banks for more than 300 miles. If this stream flows

* Kafir, *i. e.*, Infidel. † Pages 339, 340. ‡ The Gükwa of Clapperton (?).

from the east, it cannot terminate in Wankarah, since, according to Idrísí, it retires to its bed after the inundations, and is a river at all times.* Neither if it flow *towards* the east, can it have its sources in the mountains to the west, which have but little elevation, and could not furnish such a stream. Still less can it flow in the region of the south, distinct from the Chadda, by which that region is known to be occupied. It only remains to examine the more important information which we derive from those who have gone over the very ground, who have travelled on the banks, and who have crossed and recrossed the river in question, at many points of its course between Bornù and Haussa, and even to the basin of the river Chadda, and if by their means, leaving out for the present their opinion of the direction of the current, I can trace the continuity of the river level, I think no one will deny the identity of the streams.

In the month of December Dr. Oudney and Captain Clapperton commenced a journey into Sudán from Kukah, having joined a káfilah of twenty-seven Arabs and fifty Bornuese.

The low country was still inundated by the overflowing of the Yeú; although the river had fallen about six feet.† In what Clapperton calls temporary rivers, formed by the overflowing of the Yeú during the wet season, there was still a considerable body of water, running at the rate of two or three knots.‡ We may imagine the main stream to be of great magnitude, when it required a whole day, and considerable difficulty, for the káfilah to cross what is described as an inferior or "temporary" branch. It must have been deep, as they were obliged at Dammasak to cross on rafts,§ made of bundles of reeds laid on two poles; and the camels swam over. They continued "*winding* along the banks of the river, or occasionally cutting off a bend by a cross-path." Near the lake Tumbum, the country to the S.W. as far as the eye could reach, was a dismal swamp.|| Skirting this the travellers arrived at Bede-guna, near the swollen river they had crossed on rafts at Dammasak. The country S.E. and S.W.¶ was an entire swamp. The lake Zumbrum is twelve miles S.S.W. from Bede-guna. From this place they continued to travel over a *level* country, sometimes by the borders of the Yeú; which at one place is** said to be 150 yards wide, but nearly dry. "Its waters were dull and sluggish as far as we observed; and during the middle of the dry season, the naked channel and a few pools

* Idrísí, p. 18.
† Denham and Clapperton, vol. ii. p. 188.
‡ He does not say in which direction.
§ As described by Imhammed, Proc. Afr. Ass., vol. i. p. 123.
|| Ibid., vol. ii. p. 199.
¶ In this direction Idrísí places the town of Gharbíl at the foot of a mountain and on the bank of the *Nile*.
** Ibid., vol. ii. p. 210.

of water, sometimes far apart, are all that remain of the river." "This may be taken as a fair average breadth *downwards* as far as the lake, where however the depth seemed considerably increased."* This is within a quarter of a mile of the city of Katagúm.

After leaving this, the road was along a large swamp to the south, and a branch was twice passed which Clapperton calls the Shashum, a narrow stream falling into the Yeú: I imagine it, however, to be another sub-division of its waters. By Clapperton's map a large arm branches off at Old Birní, and passes Bede-guna; where it is said not to dry up throughout the whole year.† Another arm branches to the south at Katagúm; thus completely answering to the description of Idrísí, that the river surrounds and intersects the country of Wankarah.

A day's journey further, Clapperton says, "Since leaving the wells of Beere-Kashifery on the southern borders of the great desert, we had not met with rocks, or even pebbles; till now, the very channels of the rivers being destitute of stones; the whole country consisting of soft alluvial clay."‡ He now (at Katungwa) saw a range of low rocky hills stretching nearly south-west, called in the Haussa language Dushi, or the rocks. Zangeia is situated near the extremity of this range. "The prospect to the south is bounded by high blue mountains." After this our traveller found the country diversified by hill and dale. A range of hills south-west, called Dull, about 600 or 700 feet high, running nearly north and south. Some rocky streams, the Girkwa and Sokwa, rising in these mountains, are passed before arriving at Kano; which I conjecture from the context to be on an elevated plain. After this place, Zangeia, he passed over a range of hills or mountains separating the basin of the Yeú from those of Guber and Guari, where it is not to our purpose to follow him; thus far, however, he had been travelling on an uninterrupted alluvial level, overflowed and intersected in many places by a large river. It is fortunate that very near this spot, Lander passed in an endeavour to reach Fandah, after the death of his master; and on his return to Zaria, from within half a day of Jakobah, he also traced an extensive alluvial level, occupied by a large river or rivers. This level or valley is clearly defined by the mountains of Zaria and Guari on the west, and on the east by the blue mountains seen by Clapperton and Lander. But to explain this clearly, it will be better to commence with his itinerary at Bebeji;§ which I have before mentioned as a known starting point. After passing the hills which Clapperton in his first journey calls Dushi, a generic term, and in his second Nora, on the third day he

* Denham and Clapperton, vol. ii. p. 222.
† Vol. ii. p. 203. ‡ Vol. ii. p. 231. § Page 288.

crossed a large river, and two small ones, running westerly, near Karifo.* His route then was at the base of a long range of mountains stretching southerly, the blue mountains which Clapperton saw to the south at Zangeia. He marched continually over rocky and hilly ground until he reached Kuttup, crossing several large streams † running to the westward, one of which he calls the Kudúnia. After leaving Kuttup,‡ he crossed a large river, the Rari, flowing to the south-east, then traversing the rocky ridge on the opposite side of the valley, he arrived at Danrorah. From one of these heights, he had the prospect of an extensive plain, of several days' journey before him. Here he was obliged to turn back, but he obtained a sort of connecting point, for he was told that the city of Jakobah was at the foot of a large hill, which he saw at the distance of half a day to the eastward.§ On his return he does not mention the passage of the river Rari again, but he must have crossed it as he re-visited Kuttup. This city he describes as situated in a beautiful plain. From thence he took a route more to the westward, and at a regular ferry he recrossed the Kudúnia, or a river so much broader, and deeper, and more rapid, than he saw before, that he had great difficulty in passing it.|| Continuing to travel in a *plain,* he crossed and recrossed a large "*noble*" river running to the south;¶ and after passing the town of Eggebee, the situation of which in a beautiful and *extensive plain* he describes with very glowing colours, he arrives at the city of Zaria, which is situated on the north-east side of the mountain pass from Nufi; the range which separates the valley of Wankarah** from that of Guari and Koton Kora. On the west a part of the same elevation he also saw at Accoran.††

Here is the alluvial level, clearly traced in the return from Kuttup all the way to Eggebee, where the plain, extending "many miles," may easily be supposed to reach Karifo,‡‡ from which it cannot be very *many miles.* The course of a "noble, deep, broad, and rapid river," has also been marked out; having been crossed and re-crossed in such a manner as to leave no doubt of its identity. Since they flow in the same alluvial level I presume there will be no difficulty in joining the "westerly streams," crossed near Karifo with the "noble" southerly river,

---

* Clapperton's Second Expedition, p. 289.
† These must be tributaries from the Mandara range.
‡ Page 296.
§ Page 297. This resembles Idrísí's description of Mellel, in the Lemlem country. Lander says Jakobah is the capital of the Yem Yem country.
|| Page 299.
¶ Pages 302, 303, &c., Second Edition, vol. ii. p. 140.
** Page 149.
†† Page 301.
‡‡ Carifo in the map annexed to Clapperton's Travels.

two miles east of Eggebee, with the "large southerly river" requiring a ferry, a quarter of a mile west of Makami, with the "deep, broad, and rapid" Kudúnia,* and with the "large" south-easterly river Rari.

It now only remains to unite the Yeú or Shashum with the Karifo stream, towards which it is directed in Clapperton's map, and the Chadda with the Rari. If the contiguity of such large rivers, apparently without the intervention of high land, be not thought sufficient evidence, I can bring the report of many natives, independently of each other, given at various places, but nearly all within the tract of country of which we have been treating.

1st. Captain Clapperton was informed, both at Bedeguna and Katagúm, on the Yeú, that this river had its rise among rocky hills in the country of Boushi, near Jakobah and Adamawa.†

2nd. At Girkwa,‡ he was told by a black sheríf that he would cross on the following day a river communicating between the Kwara and the Yeú.§ The captain did not believe this report, probably from some mis-statement of it.

3rd. Lander was told at Danrorah, by his servant Mohammed, that "a river called Shar or Sharra, deriving its source from a lake Chad, flows about half a mile from Jakobah; and that canoes can be paddled from the said lake to the Niger at any season of the year."‖

4th. Lander at the ferry above Bussah, when making some allusion to the river,—"A Felatah came forward and made the extraordinary assertion, that instead of running to Fandah it took a turn to the eastward, and disembogued itself in the lake Chad in Bornú.¶

5th. Mr. Laird made many inquiries at Fandah about the course of the Shari. "The answers invariably were, that it came from Lake Chad; and one man, a native of Kukah, offered to take me up there in twelve days, without changing the canoe."**

6th. I was myself informed at Dagbóh, on the Chadda, "that Jakobah was seven days higher up the river; and the city of Kukah, in Bornú, is said to be only six days, for canoes, above Jakobah, on the same river."

7th. Lander told me at the same place that he had received information from natives, which I did not hear, that one could go

* I allude to the second river which he calls by this name; I conceive the first to be a mountain stream, falling into the large river. The name is common to other streams, and to towns. Clapperton mentions one of each; and Lander, having heard of a river of that name when he descended the Kwara, thought it was the same he saw in Wankarah, and they have been made to join in the map of his route The butter-tree is called in Haussa Maï Kadania.

† Den. and Clapp. 8vo, pp. 202 and 222.

‡ Qu. Jaku of Bello. See Appendix to Clapp.'s Journal, p. 329.

§ Id. p. 236.

‖ Clapp. 4to, p. 297.

¶ Lander's Voyage, ii. 97.

** Laird and Oldfield, i, 232.

up the Chadda, from Jakobah on "*one water*" to the great water of Chad. Indeed he affirmed that he had information from various quarters that this noble river is the outlet of that immense lake.

8*th*. Lander says that at Wawaw "the king's head drummer, a Nufi man, stated, in answer to our inquiries, that the Chadda, Shari, flows into the Niger at Fạndah; and a regular intercourse is kept up with the natives on its banks, for the purposes of trade, by means of very large canoes. The sheikh, he said, resided very near the Chadda, which in Bornù spreads into a large body of water.

Lastly. He was told that the Kudúnia falls into the Niger near Fạndah; that is, the river which he had crossed so frequently in the Boushi country, which passes near Jakobah, which is the same as the Chadda, according to my information, and which comes from a lake Chad, according to Mohammed, Lander's servant.

I will now bring forward, and endeavour to combat, Denham's evidence on the other side of the question; no easy task, it would appear at first sight, since he is very circumstantial, and evidently intends to record his opinion that the lake has no outlet. But, as I said before, the difficulties he had to contend with must be taken into consideration, especially as he had to do with a very winding river, and therefore might easily be mistaken in its course. On the 24th of May, he came to the Yeú at Lada, 70 miles from Kuka,* where he says, "The river here makes a bend resembling the letter S; the water is extremely shallow, and a dry path over the bed of the river appeared close to our halting-place, although the banks were high, and capable of containing a very large stream. I walked out, following the *easterly* course of the stream, in search of game; but within four hundred yards of the river the ground was so choked with high grass and prickly underwood that I was obliged to take a path more inland." He passed a great many small lakes in the vicinity of the river; one was very deep, containing hippopotami in great numbers. From the ruins of the walls of old Birni "we obtained a sight of the river Gambarú,† running nearly east, notwithstanding its *windings*, and only a few miles distant." Again—"We came to the river, which is here a very noble stream, nearly a quarter of a mile in breadth, and situated between two high banks thickly overgrown with jungle, bushes, and bamboo. We endeavoured to ascertain if there was any current; but the water appeared *perfectly stationary*. Omar Gana, however, and the shouas who had accompanied us, were unanimous in declaring that after the rains a very strong current from west to east constantly flowed."

* Denham, i. 206. † Den. vol. i. 212.

Again—"We followed the course of the river to the eastward (which he in another place calls the Yeú), nearly three miles. There being no pathway, we were obliged to break through the high grass, trees, and thickly-scattered bamboo, which made it a fatiguing excursion; and after all we could only get a sight of the water by following the tracks of elephants, and other animals, whose ponderous bodies beat down everything before them. At length we came to an open dry shoal of sand, the bed of the river extending more than two hundred yards; we were told that the stream was here again called the Yeú." At another place, "We proceeded *winding* with the river. In several places we had the banks clear of trees, and covered with verdure* for some hundreds of yards, and the stream nearly as broad as the Thames at Richmond." Again—"Crossing the Yeú at a *dry spot,* we came to the banks of a large water, called Dammasak,† about five miles distant from the ford."

On their return to Tripoli in August they again crossed the Yeú, "now a considerable stream full of water, and running *towards* the Chad at the rate of three miles an hour." The Arabs, again, appeared to have a different opinion respecting it, calling it the Nile, which Denham supposed meant the river of Egypt. He says, again, "Bellal accompanied me down the river about nine miles, where, increasing in width about 100 yards, it flows into the Chad, with a strong and *deep*‡ current of water."

At the eastern end of the lake he had very vague accounts from the natives and from Barca Gana; the amount of which is, that the Bahr-el-Ghazál, which once received the waters of the Chad, and was a day's journey broad, is now dried up, and they knew of no other river communicating with the lake: that Fitri had a stream running out of it,§ and was not like the Chad, which every one knew was a *still water*.

Having now quoted all that Denham says respecting the nature of the lake, and the course of the river Yeú, I would merely ask whether it is not only possible, but very probable, that he might be mistaken in the current of a river which is described as very

---

* Qu. African verdure ten feet high (?).

† Clapperton found the water of Dammasak so deep, broad, and rapid that it required a whole day to pass it on rafts. This could not, therefore, be a still water, as one might infer from Denham's description. I mention this discrepancy to show how easily travellers may fall into error in that country. One account must be wrong. Denham's was in the beginning of the rainy, and Clapperton far on in the dry season.

‡ Rivers at their discharge are usually shallow. The depth is, at all events, incompatible with the strength of current which might have been in consequence difficult to ascertain.—A.

§ This is in opposition to the account of Major Laing's informant (see p. 299, line 7).

winding and difficult of approach, except by paths made by wild animals in the prickly bushes on its banks; a river that is in some places narrow, and in some a broad and noble stream, sometimes nearly dry, sometimes deep enough for camels to swim over, requiring rafts, as at Dammasak and two or three places mentioned by Lander, corroborating the testimony of Imhammed's "ill-constructed rafts," and deep enough for the huge hippopotamus to conceal himself at the bottom; a river, in fact, which in some places is said to be so sluggish as not to have any perceptible current, at others to run at the rate of three and a half knots, and which at one place, Dammasak, is described in an opposite manner by the two travellers? This perhaps is the strongest evidence of the difficulty of coming at the truth in that country. These arguments do not appear to me sufficient to overturn the reasons which I have given for believing that the waters of the lake Chad, collected from the lofty Kumri mountains, flow by the river Yeú through Wanḳarah and Boushi, to join the Chadda near Jakobah.

A strong presumptive evidence that this river flows *from* the lake may be also drawn from the description which is meant to serve the opposite purpose. Thus the farther it is from the lake the more important it appears, which may be in consequence of its having received an affluent from the Eyres (Aurás) mountains to the north* and the Mendefi range to the south; and certes, the increase in volume of a river is *below* its affluents. Again, the characteristics of the outlet of a lake may be said to be clearness, depth, rapidity, and singleness; those of the discharge of a large river are generally turbid, shallow, sluggish, and divided waters. And it is hardly probable that the Yeú would reunite the branches by which it surrounds Wanḳarah, in order to discharge itself by a single embouchure, in an alluvial country of its own formation, although it must be collected to pass between the Nora or Dushi hills, and the blue mountains which Clapperton saw to the south at Zangeia.

Of course it cannot be imagined that a river larger than the Kwara, which has been navigated to within seven days of Jakobah, can derive all its waters from the lake. It is most probable that near this city, where it takes a sweep to the westward, it receives another and a very large affluent from the Mendefi mountains. Indeed, a native informed Denham that having passed that range, he crossed a large river flowing between two high mountains, although his account of twenty days south would lead him far away from where Adamawa is supposed to be.

---

* The existence of such an affluent is noticed by Idrísí, p. 19.

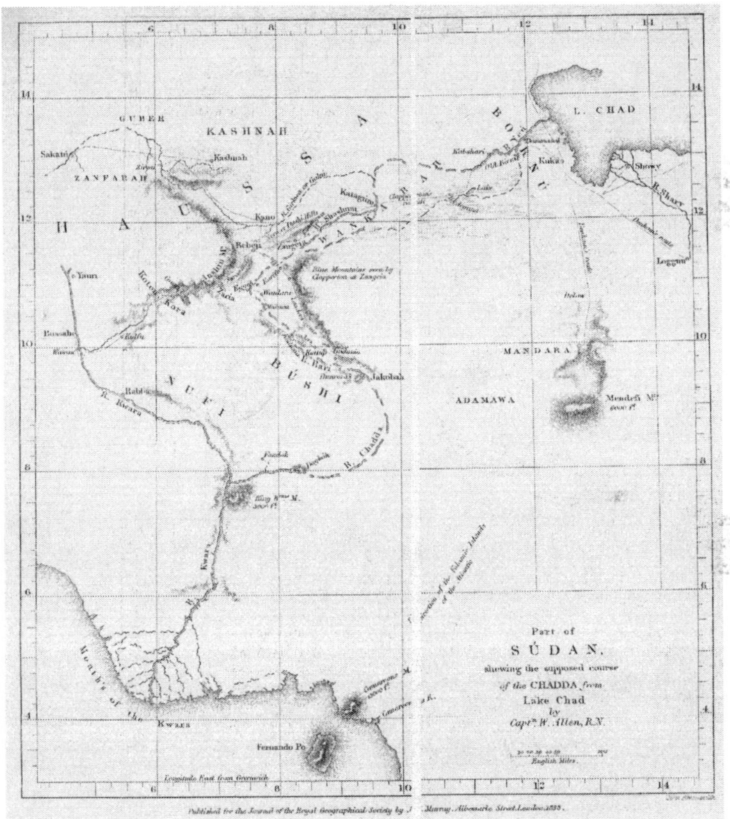

Published for the Journal of the Royal Geographical Society by J. Murray, Albemarle Street, London 1833.

Thus, if this hypothesis be proved, there is an uninterrupted navigation to the very centre of Africa; to the foot almost of the long-sought mountains of the moon, since Mr. Laird found two fathoms in the Chadda in the lowest part of the dry season: the natives and Idrísí say it is navigable for canoes throughout the year. Clapperton found some branches in Wankarah deep, where the water had already subsided six feet; and in addition to the length of the Chad, there is probably a chain of communicating lakes to the eastward. A great deal might be said on the importance of this easy access; but as I have already extended this paper beyond what I anticipated, I will conclude by putting the case in a very simple form. Clapperton was told that the Yeú had its origin near Jakobah, and Lander was told near Jakobah that the Chadda, which flows half a mile from that city, had its origin in *a* lake Chad. He had also crossed and recrossed a large river flowing between these points. If these reports be true, and they were distinct and independent of each other, they must relate to one and the same river.

* * * * *

Subjoined are the positions of the principal stations at which I obtained astronomical observations on the banks of the Kwara and Chadda in 1835:—

	Lat. N. ° ′	Long. E. ° ′	Variation. ° ′
Cape Nun	4 21	5 53	
Eggabóh	5 28	6 25	
Okóh	5 54	6 38	
Adda Mugu	6 30	—	
Attáh	7 6	—	
Stirling	7 49	7 6*	19 51W.
Kattam Kárafi	8 7	—	
Fandah	8 12	7 42	
Egga Island	8 43	6 42	
Rabba	9 13	6 26	20 36W.

* The mean of many observations, with an artificial horizon, for longitude by three chronometers, and a lunar distance, gave the longitude of Stirling 7° 3′ E.; by Jupiter's Satellites, 7° 14′ E.; Idem, second observation, 7° 16′ E.

XXIII.—*Notes on the Country lying between the Meridians of 55° and 64° East, and embracing a section of the Elburz Mountains in Northern Khorásán.* By James Baillie Fraser, Esq.

The great range of Elburz, which skirts the southern shore of the Caspian Sea, and pursues a course nearly east and west from the plains of Mogám to the highlands of Mergháb (improperly called Moorghâb), and the Házárah country, spreads out into greater width between the meridians of 55° and 64° east; that is, from Gúrgán* of Asterábád to Mergháb, and presents to the desert on the north a slope which has obtained the name of *Atak*,† literally the *skirt* (of the mountains), in the same way as *Dáman*, a province of Afghánistán, signifies the skirt of the hills sloping down to the banks of the Indus.

This tract, formerly one of the richest districts belonging to the sovereigns of these countries, and still covered with ruins of great cities, and vestiges of former prosperity, is now nearly a wilderness, or but sparsely occupied by tribes of Turkománs, who live partly on the produce of their flocks and herds, but much also on the proceeds of that plunder which they gain in inroads and pillaging parties systematically made by them upon their Persian neighbours.

The mountainous tract above alluded to, irregular in shape, and spreading out in some parts to a breadth of full 200 miles, is divided longitudinally into ridges, separating valleys which communicate with each other by passes or defiles, and form an extensive aggregate of districts, many of which are well inhabited and cultivated.

These districts may be conveniently arranged in several lines running from west to east, as they are divided by the principal intervening ridges, nearly as follows. Upon the skirts of the most southern ridge are found those of Meï-omíd, 'Abbás-ābád, Mazínān, Sabzawār, Níshápúr, and Sheríf-ábad. Close behind these, but separated from them by ridges of greater or less altitude, lie Jájerm and Sang-khást, Nárdín, Jaghetáï or Juveïn, Bām, and Miyán-ābād; Isferáïn, Ṣúfi-abād, Báwer, Sultán-Meïdān or Mádan, forming what is locally called the *Serd-viláyet*, or Cold Country, being that between the first-mentioned line of districts and the great central line which comes next to be described. This includes the districts of Armúdlí and Semalghán; the plains of Bujnurd, Shírwán, Khabúshān, Rádkān, Chinárán, Ṭús, and Mash-hed, which are all continuous with, or opening into each other. At the western extremity of this line are found some of

---

* Or Jurján.—S.

† A Turkish word, pronounced *etek* at Constantinople; it is exactly the *sinus* of the Romans.—S.

the passes to the lower country of Gurgán, hereafter to be noticed; while the eastern end opens into a nearly continuous valley which reaches to Herát.

A chain of mountains, unequal in height as in breadth, confines the great line of valleys just described, on the north-east, and among these lie irregularly disposed a number of valleys and elevated plains, each forming a district in itself separated by or intermingled with lofty and craggy ridges. The north-eastern base of this mountainous tract which bounds the desert runs nearly N.N.W. from the point at Ak-derbend,* where it is pierced by the Mash-hed river, to Paráú, another point to the N.W., where that part of the range terminates, and the mountains turning back at an angle, probably acute, towards the south, preserve a very irregular and indented line. Paráú, as we are informed by persons acquainted with the country, is a mountain bounding on the north a valley, the bed of a fine stream, and known by the same name; and from its extremity, we are told, a narrow chain extends to the lesser Balkán, 120 miles. Thence to the greater Balkán is said to be a distance of ten farsakhs,† or forty miles; and thence again to *O'k-kóh,*‡ or the Arrowhill, the distance is called twenty farsakhs, or eighty miles, from which, according to our informant, Khívah is not more than five days' journey. From O'k-kóh to Kuruk súfí, where the Atrak falls into the sea, was said by the same person to be thirty farsakhs. These distances, though given by a person well acquainted with the country, are so uncertain that they cannot be relied on; insomuch, that no attempt has been made to use, or to reconcile them with other distances or bearings in the accompanying sketch.§

The following are a few of the best-known districts in this range of mountains. To the eastward lies Kelát-Nádirí,|| which is the last place of importance in that direction. Then, opening to the Atak, are found *Deregez, Sheúrek, Derbādān, Afghāz,* a fine wooded valley beneath a hill of the same name; *Geristán Fírízeh* close on the verge of the Turkománs, *Germáú,*¶ in the same vicinity, *Gífhān, Rāz, Kúsh-Kháneh,* or the Hawk's Nest,** *Bálkání, Kolpílí, Thikmaván, Zerzáú.* These are all said to be fine districts, with plenty of villages, and some of them wooded. Further removed from the mountains, and upon their skirts, lie the wider districts of Tĭhejen, watered by that river, and inhabited by the Turkomán tribes; Abíverd was once a fine district, but is now abandoned, as are Dirum and Nisá. N.W. of these lie

* White Defile (Turkish).—S.

† Or farsang, the parasanga of the Greeks.—S.

‡ Turkish name.—S.

§ Preserved in the library.—Ed.

|| Nadir's kelát, being his favourite stronghold.—S.

¶ Germ-áú or Germ-áv, *i.e.* warm water, a provincialism for Germ-áb.—S.

** Kúsh in pure Turkish signifies only bird.—S.

Akhāl, inhabited by Tekeh Turkománs, a well-watered sloping plain; *Archaman, Burmah, Bāmeh, Kizil Rebát,** *Shehrul-islám,* described as a place of extensive ruins built of fire-burned bricks, and *Paráú,* which appears to have been a place of pilgrimage, where there are ruined shrines to 'Alí, &c.

More in the mountains, and north of the last-mentioned range, lie the more mountainous districts, which are too numerous to be detailed, many of them consisting of little more than a single village. On the way from Deregez we passed through Meïheú-Kokan, T'hverí, Kasmān, Daúlat-Kháneh, Ráwir, Shehrazān, Karácheh, and Sukíyeh, to Shírwān. On our left we had the fine valley of Kal'ah Yúsuf Khán.† On the right were Gerenheh and Inshá,‡ with others, which give their waters to the Atrak.

From Shírwān we passed through Dereh Shúr§ and Hazár-jeríb|| to Bujnurd, having Gérmeh Khán upon our right, while on the left the mountain of 'Alátágh¶ separated us from Sang-khást. Proceeding westward, we descended into the Mónah valley, passed through part of Inshá (another district), and leaving the valley of the Atrak to the right entered Semalghán, from whence passing through the desert places of *Zerd, Kohistán, Julgehi-Kelát-há,* and *Kalpúsh,* with *Múchíh* on our right, and *Armudlí*** on the left, to Deheneh-zoh, the pass by which we descended into Gurgān.††

The mountains facing westward, from Paráú to Senker, are furrowed deeply with ravines, each of which gives access by a pass, more or less steep and abrupt, to the plain that intervenes between the base of the mountains and the Caspian sea. Of these valleys or passes the principal are the *Paráú,* the *Chander,* the *Sunt,* and the *'Serisheú.* These all pour their streams into the Atrak, of which they form a large proportion; and their course is necessarily much to the south of west. A ridge of hills, running a long way into the plain, and named Marábeh, divides the Atrak on the south from the Ṣáráṣú,‡‡ which joins it at the terminating point of the range. Through this valley runs one pass. The next is that of Deheneh-zóh, by which we descended. Next comes Deheneh Ḳúzlú,§§ which leads from a fine julgeh (green valley) Kará Shokh; then Dehneh Eshеklí,|||| which leads from the plain of Armúdlí; then Ḳúzlú Dehneh and Jul-

---

* Red-hostel. † Lord Joseph's castle.—S. ‡ Please God!—S.

§ Salt valley.—S. || Thousand acres.—S.

¶ 'Alá ṭágh (Turk.), beautiful mountain.—S.

** Pronounced Armút-lí, *i. e.* Pear-place.—S.

†† Called by the Arabs, who substitute *j* for *g*, Jurján, the ancient capital of the country, of which a complete account, drawn from Asiatic geographers and historians, is given by the Baron de Sacy in the Mines d'Orient, i. 338.—S.

‡‡ Ṣárí sú (Turk.), *i. e.* yellow water?—S.

§§ Walnut-tree mouth (T.)—S. |||| Ass's mouth (T.)—S.

geh, which leads from Ḳalpúsh. The mountain of Seráú intervenes between Ḳúzlú and Esheklí Dehenehs; and Níl kóh,* a mountain celebrated in the Sháh Námeh, stands at the mouth of the latter.

But the mountainous tract which we have been describing becomes more peculiarly important from being the separating boundary between the fierce Turkomán tribes and the northern districts of Persia, which have been desolated by these marauders; and as the passes by which they reach the upper country all lead through this barrier, it becomes a matter of interest to ascertain the number and character of these several lines of approach.

From the best information that could be obtained upon the spot, the following are the principal passes, from the Atak to Upper Khorásán, beginning from the eastern end of the tract in question:

1*st, Puli-Khátún,* or the Ladies' Bridge. I am not sure whether a separate pass has obtained this name from the bridge so called, which is thrown, as I believe, over the Mash-hed river, or whether the name merely designates a particular bridge in one of the succeeding passes. It is by this way that plundering parties destined for Turbat, Ḥaïderí,† and the road towards Ghóriyàn, and Herát, ascend the hills.

2*nd, Aḳ-derbend.*

3*rd, Muzderán.*—I do not know whether these are one and the same, or different passes, but they are both on the road from Mash-hed to Serakhs. Burnes, in coming from that place, passed *Derbend* at forty-eight miles from Serakhs, and calls it the same as Muzderān. Derbend (or the gate) is a common term for such passes, and this may mean Aḳ-derbend, or the white gate. On the other hand, Burnes says nothing of the Mash-hed river, which certainly bursts through the southern continuation of the Ḳel'at mountains at Aḳ-derbend, from which it seems probable that he ascended by a separate pass, namely, that of Muzderán, leaving the Mash-hed river and the Aḳ-derbend pass to his left. There is also a doubt as to the precise point where the Mash-hed river joins the T'hejen.

The scenery of these passes, whether one and the same, or different, is described as picturesque, though not rich in wood. The ascent from the Mash-hed side is comparatively equable; but the corresponding descent towards Serakhs, on the N.E. side, is more precipitous.

4*th, Kelát-Nádirí,*‡ a pass to the Atak, lies through the inclosed valley of Kelát; but independently of its natural obstacles, which are great, no one can pass through that valley unless with the consent of the chief, who resides in this natural stronghold.

5*th, Chapishlú.*§—To the west of Kelát. It is steep and rugged,

* Blue mount (P.)—S.
† Haïder's tomb.
‡ Nádir's castle.
§ Blindman's-buff-Place.—S.

and with great difficulty practicable for guns; but I am not in possession of any further particulars regarding it.

6th, Alláhu-Ákber, or *Nardewán Nádirí* (the ladder of Nádir). —This leads directly from the valley of Deregez, and then by T'háverí or other roads into the valley of Khabúshán. It was improved by Nádir Sháh to facilitate the passage of artillery, and is defended by three bastions. Though steep and rugged on the south side of the mountain, the road down to the Deregez valley is tolerably good and easy. I saw heavy guns dragged both up and down this pass.

7th, Tuĭnjí.—A very steep ascent from the south, impracticable as it now stands, for the passage of guns—a long, tortuous and steep, but less difficult descent into the Deregez valley. It is connected with the pass of Zírek, which traverses the central ridge, and is sharp but not difficult.

*8th, Zohi-Gerenheh.**—A fine romantic pass; difficultly practicable for guns, and very defensible. It crosses the most northern ridge; and its lower extremity is defended by a bastion with guns mounted, which commands the water.

9th, Zohi-Deringeh.— A very romantic defile like the last, difficult though practicable for guns. There is a fine spring on each side of the *Kot'hal*† (or ascent and descent). One principal branch of the river Atrak takes its rise at Inshá, a point in this defile, and the stream must be repeatedly crossed in threading it. There is a *derbend* or rocky pass like a gateway at its lower extremity.

These are the principal passes from the Atrak to the upper country; but there are a variety of others of less importance and more difficulty, several of which lead directly to Deregez, from vallies opening at Rádĭkán, Chinárán, and Khabúshán, &c.; others again lead from various points, but join the passes already enumerated. To the west of Bujnurd the hills become lower and more open, so that the paths by which the Turkománs can invade the upper country are by far more numerous. There are four or five which lead into the Mónah valley alone; several others which lead to Semalghán, and others still more easy, that crossing the Atrak and its valley, lead to Jájerm, Meï-omeïd, 'Abbás-ábád, &c. Between Deregez and Kelát-Nádirí there are also some pathways of great difficulty and scarcely practicable for horsemen.

Itinerary with bearings and distances from Sabzawár to Mashhed by the *Serd-viláyet*, or upper (cold) country.

May 30th and 31st, 1834.—From Sabzawár to Káhríz-degez, nine hours and a half, chiefly by night. First half over a plain, then through a defile among low hills, and a pass called Chehel

* The word *Zoh* signifies a defile.

† Kutal or Kútal, a mountain-pass.—S. Kut'hal in northern Persia.

Pehlewán or the Forty Champions. The village, a miserable ruinous place on a plain twenty miles broad by sixty or seventy long, commencing at the mountain and district of Jaghetái on the west, and issuing out on the plain of Níshápúr on the east; it is bounded on the north by the hills of Bám and Miyán-ábád.

May 31st and June 1st.—On to Maskún, a small village, and then by night through the 'Alí Zengí pass, and up a grassy valley with a stream to 'Abdu-llah-gáú; distance about forty miles from Káhríz-degez, and direction about N. 25 E., on the whole time eleven hours, through hills of moderate height covered with grass, with plains among them of various extent. 'Abdu-llah-gáú is a thriving village of a thieving tribe, in the plain or valley of Sulṭán-Meïdán, and within two miles of that town, now ruined.

Bearings from Káhríz-degez:—Village of Maskún of route N. 30 E. 12 miles. Direction of route, N. 35 E. 15 miles; of 'Abdullah-gáú, N. 25 E. 30 miles; of Ṣúfí-ábád (not seen) N. 5 W. 15 miles; of Bám (not seen) N. 28 miles; of Miyán-ábád (not seen) N. 25 W. Peak of Jájerm-hill, pointed out, probably an error, N. 42 W.

June 2nd.—From 'Abdu-llah-gáú to Nádir-tepeh,* to a ruined caravanseray, where we halted for breakfast; all the way through green valleys and cultivated land, with moderate pasture-hills on either hand. Then entered the long valley of Khabúshán, Chinpárán, &c., nearly in the same direction, and after eleven hours and a half, of which three-quarters in halts, reached the small village of Ḳal'ah Seyyed-hà, distance about 42 miles.

June 3rd.—On to Gunáh-ábád;† along the same plain, partly cultivated, partly meadows, to Gunáh-ábád; generally fine grazing-ground, but the country ruined by wars and plundering parties; distance about 24 miles.

June 4th.—On to Mash-hed,‡ passing through Khushmítí, and along a fine plain, with plenty of grass and cultivation; distance 28 miles.

From Mash-hed to Deregez; and from thence, by Bujnurd, to Gurgān.

June 18th and 19th.—To Rádkán, as formerly, through the plain of Mash-hed, by Ṭús, and through the meadow of Chinárán.

June 20th.—From Rádkán to Muchulún, about 14 miles, passing through the thriving village of Súhún, and all along the skirts of the hills that bound the valley on the N.E.

June 21.—To Yestanjí, 10 miles—a miserable village; direction from Rádkán, distant 24 miles, N. 58 W.; rising among hills on the right to Chamani-Naucheh; and then on same direction, crossing a small range to Ḳal'ah Yúsuf-khán—passed

* Nádir's-hill. (T.)—S. † Sin-abode.—S.

‡ Mash-hed Imám Riẓá, *i. e.*, the place of the Imám Riẓá's martyrdom.—S.

many pretty valleys, with much verdure. Kal'ah Yúsuf-khán* is a fine village, in a green, well-cultivated, and pleasant valley, through which runs the *Yaëk*,† in a full stream, which joins that which flows by Khabúshán.

On nearly same direction, ascending the narrow valley, or defile of the Yaëk, to T'háverí—a knot of three or four villages, in a small plain, surrounded by well-cultivated hills.

June 22.—On from T'háverí, to the pass of Zík,‡ about six miles; then descend into the pretty green valley of Bosmeyú. Direction of Kal'ah Yúsuf-khán, from top of Zírek-pass S. 20 to 25 W. On to top of Túïnjí § pass—a very steep zig-zag path—up the face of a precipitous cliff. From top of Túïnjí to Mohammed-ábád of Deregez, 15 miles distant N., 50 E. On descending, first very precipitously;—then through a tortuous defile, for full twelve miles;—then, turning sharp to the left, up the bank of the ravine, we entered the valley of Deregez, and rode in a northerly direction for full 16 miles to the Persian camp—no bearings obtained.

July 1.—Bearings from *Kizil-kótal*, a height overlooking the Atrak, about 8 miles N. E.:—Abíverd—due east 12 miles. *Bábájuk*, *Mír*, and another village, N.N.E. about 8 or 9 miles. General direction of the hill-skirts and the hills themselves, from Kelát Nádirí, and probably from Serakhs—so far as can be seen and judged—probably to Nisá Dirún, Mehineh, &c., for 60 miles at least, about N.N.W. and S.S.E. Nothing to be seen looking on the Atak but an apparently boundless plain, brown, arid, and dusty—without a vestige of habitations, except the villages mentioned.

July 2.—Mohammed-ábád, we crossed the plain or valley of Deregez, to the foot of the pass of Alláhu-Akber ||—which ascended: road good—broad, and ascent, for the most part, not abrupt. No bearings taken, but direction by general observation somewhat to the westward of south.

July 3.—Descend Alláhu-Akber pass—very rough and precipitous—but the heavy guns were taken down this way; and on, about 4 miles in all, to the Serdár's ¶ camp, in the hollow of Meïheú-kokân, a rich green valley, with a fine stream—up which we marched, and crossed the intervening high ground, to a descent nearly as lofty and rugged as the Allahú-Akber pass, by which we dropped into the valley of T'háverí again.

July 4.—On, in a westerly course, up the T'háverí valley—a fine rich green *julgeh*—and through those of Kasmán, Daulat-Kháneh, Ráver, Shehrizán, to Káracheh.** No bearings, but by occasional glances at the compass direction very little to north

* Joseph Khan's Castle.—S. † Yáïk, or Yáyik (T.) *i. e.* broad, wide.—S. ‡ Zírek (P.) swift.—S. § Túyinjí (T.) watchman.—S. || 'God is Great.' ¶ Commander's. (P.)—S. ** Black Well. (P.)—S.

of west. The whole way, a succession of the richest and greenest meadows, watered by fine streams.

July 5.—Quitted the line of julgehs, or natural meadows, by a pass to the left, and rose high among a maze of hillocks—the skirts of the northern hills of the Khabúshān and Shírwán Valley, to Sukhí, about 18 miles, the road traversing a tract of arid heights;—then 6 miles more, to Shírwán, along a plain—rich, but a good deal burned up. Shírwán, a small town, once containing 3000 houses and fortified, now a ruin almost empty, and the wall broken down in every part.

July 6.—From Shírwán to Bujnurd, over a series of craggy heights to the Kotal of Alakhahband—from whence, on left, the fine village and district of Zauran—into which descended. Further on, to right, saw the pass of Ḳal'ah-zoh, which is said to lead directly to the Atak without any bad descent, and is accordingly a great route of the Turkomans;—next turned up a narrow and winding ravine, called the Dereh Shúr (salt-valley), which is the boundary between the lands of Khabúshán and Bujnurd. At the head of this Dereh we came to the plain of Hezaijeríb, formerly a notable place of Turkoman resort, but now cultivated. A pass, called the Aḳ-kotal, leads from this plain over a succession of broken ground, into the valley of Bujnurd.

July 12.—Onwards 7 miles, to Lenger Village, about N.N.W.

July 13.—On 4 miles, to Akhór pass,* from Bujnurd, about N. 75 to 80 W.—a high hill, overlooking the Mónah Valley, and hills intervening between it and the Atrak.

Bearings from thence:—Kal'ah-khán of Sémalghán, distance 14 miles S. 75 W. General direction of course of Atrak, W. 25 N. Mónah Valley lies E. and W. Akhal on the Atrak, as pointed out by the guide, N.E. Mehineh, N. 25 E. Paráú, N. 25 W.

On our right was a high hill, *Kará-búlák*,† with a village, Karáchäï, at its foot; Germeh-khán, a julgeh or valley, N. and N.E. of Bujnurd, furnishes a stream to the Atrak. The water from Shírwán comes round it on the N., and falls into the Mónah in a fine full stream. That from Shírín-Dereh, a small dell among the maze of hillocks and hills on the opposite side, between the valley and the Atrak, enters the valley nearly opposite. The waters of Kúsh-kháneh and Germ-ab (or Germáú), which run in a channel not in sight, join the Atrak (into which the Mónah has fallen) a good way below; and we hear of no other stream of consequence on that side, till those of the Sunt, Chander, Paráú, &c., which came from their respective hills and valleys.

On the left, the water from Sémalghán joins the Mónah—then

* Stable. (T.)—S.

† Black Town. (T.)—S.

that of Inshá, joined by that of Aïràn—both rising in the mountain of Ḳurḳud; then, lower down, the streams from the Khálbásh-julgeh, and from Múchíh.

The mountains N. of the Atrak appear to be a most singular mass of dry, brown, scorched hillocks and rocky ridges, heaped on each other, and ribbed and furrowed in the most extraordinary manner, but exhibiting not the smallest particle of either vegetable or animal life. A few dark spots, it is true, on nearer inspection, upon those nearest the valley, denote bushes of the cedar or thorn,* common to such arid regions; but in the three or four days' journey across these mountains, I could hear of no habitable spot, and scarcely of a fresh-water spring. The Sunt Mountain is the highest to the north of the Mónah and Atrak valley, as Ḳurḳud is on the south.

Onwards, descending into the Mónah valley, and through several Turkomán villages, to Ḳal'ah-khán of Sémalghán.

July 15.—Onwards, across a desert, dry track of irregular ground, skirting Ḳurḳud, to the site of a village called Zerd,† long ago ruined by the Turkománs, on a pretty streamlet, when breakfasted. Then, in same direction, to the Mohák valley, and turned up to Kahistàn,‡ another ruined village, where we bivouacked: it is on the Inshá stream, and fairly whithin the centre of the Ḳurḳud mountains.

July 16.—Onwards, among heights and ravines, the skirts of Ḳurḳud, down to a deep broad hollow, *Derehi-kelát-há*, or "The Valley of the Forts," leading into the Khálbásh valley or plain, between which and that of Armúdlí intervenes a long low ridge, the tail of a part of Ḳurḳud. This valley, irregularly varied with high grounds, is all barren and arid. Onwards, among hillocks on the S.E. side of the Múchíh valley. The Atrak here, within about four farsakhs or sixteen miles, makes a bend to the northward;—turn up through fine corn-land of the Góklán Turkománs,§ and descend through the Deheneh-zoh—a precipitous and long pass through a deep forest, to the camp of Maḥmúd Walí Khán Góklán;—general course from Bivouac at Ḳohistán to this camp, nearly W. ½ N., from 48 to 50 miles.

July 18.—5 farsakhs, or 20 miles farther to Haïder Khán's camp, course about S.W. by S.—a perfect plain, covered with the withered and burnt-up vegetation of the spring, where not eaten bare by the flocks of the Góklán Turkománs.‖

* Juniper? (S.) † Yellow. (P.)—S. ‡ Mountain Track.—S.

§ Or kóklen. Klaproth; Asia Polyglotta, p. 217.—S.

‖ The spelling of the names is in some places doubtful, as it is very difficult to ascertain which, *z*, *s*, *k*, or *g*, is required, unless the word has been written down by a learned native. Some of the names ending in ún should probably be terminated by án, as that syllable, when final, has now for nearly three centuries been usually pronounced ún by the Persians.—S.

XXIV.—*Notes on a Journey from Belize to Guatemala, and return by the River Polochic in* 1834. By E. Legh Page, Esq.

After having passed a week at Belize, I embarked on the 10th of June, 1834, with Mr. Chatfield, appointed His late Majesty's Consul to Guatemala, and sailed for the Rio Dulce, about 100 miles to the southward. On the following evening we crossed the bar at its mouth, where we found only five feet nine inches water; and on the morning of the 12th proceeded up the river, or rather the outlet of the Golfo Dulce, which winds for about 23 miles in a general S.W. direction, as far as the fort of San Felipe, situated on the western shore of the gulf. Off the bar we were joined by Captain R. Owen, of Her Majesty's Ship *Thunder*, who with some of his officers took advantage of this opportunity to survey the gulf.

The entrance* to this stream is so narrow and so overhung with verdure, as to be almost hidden, until the very jib-boom of the vessel penetrates it: each side is bounded by hills rising almost perpendicularly to the height of 300 feet, clothed with impenetrable forests, and tenanted solely by monkeys and other wild animals; and with parrots of the brightest plumage. The heat was suffocating, as it did not proceed alone from the direct rays of the sun, but from the trees meeting above and thereby excluding the free air; this however had its advantage, for the sea-breeze is drawn into it, as through a funnel, otherwise the river could hardly be navigated by sailing-vessels. As we proceeded slowly, fresh beauties of nature burst upon the sight; occasionally a white cliff stood boldly prominent. At these spots the trees grew less densely, and the sun darted his fiery rays upon us; again the trees would meet aloft, and then the shadow and the gloom were a relief from the glare and heat of a tropical sun. Numbers of gaudy butterflies sported amongst the flowers that struggled, in gay colours, through the underwood; bright-coated lizards would dart along some overhanging bough; the monkeys had retired to the forest's deeper shade, and only showed themselves to peep at their new visitors, and grin and chatter, and then screaming to drive each other wantonly away.

At about 8 miles from the sea the stream expands to a small lake 9 miles long by 2 broad, called by the Spaniards the golfete or "little gulf," which narrows again for 6 miles, till we arrived at sunset off the Castillo or Fort San Felipe:† we here entered

---

* West point of entrance is in lat. 15° 50′ N., long. 88° 46′ 20′ W. from Owen's Survey.—Ed.

† In 15° 38′ N., 89° 1′ 45″ W.

the Golfo Dulce and steered across for Izabal, where we anchored early in the morning of the 13th.

Izabal* is a village of about forty huts, three houses, and a cabildo or court-house, situated on the southern shore of the lake, about 16 miles SW. by S. of San Felipe. From the hill behind the village is a beautiful view over the whole lake or gulf, bounded to the north by the mountains of Santa Cruz,† while to the S.E. is the lofty summit of Mico,‡ usually wrapt in clouds.

Having made the requisite preparations for our journey to the city of Guatemala, we started on the morning of the 15th of June, and travelling to the S.E. we shortly commenced the ascent of the mountain of Mico, by a slippery muddy road, or rather path: it took us six hours to reach the Cumbre or summit of the pass, whence we descended to the Rancho del Mico, composed of a few huts, and pushed on to Quirigua, a hamlet of about a dozen huts, which we reached at four o'clock, having been ten hours on the road, during which we had only accomplished six leagues of distance.

June 16.—Started at daylight, the road leading in a S.W. direction, occasionally over verdant plains, at times covered with gigantic trees and rank luxurious vegetation. At three leagues we commenced rather a precipitous descent, which brought us to the village of Encuentro, or 'Junction' of the rivers Motagua, or Rio Grande, and the Managua—a village of about 200 inhabitants, situated on the banks of the former river, which sweeps past at a rapid rate, and gives animation to a scene otherwise miserable enough. A fish called bobo, which is caught in abundance in this river, is much esteemed. We swam our mules over with some difficulty, and crossed ourselves in a flat-bottomed boat, called a bongo by the natives.

Continuing our journey we immediately commenced a steep ascent, the scenery gradually changing till we got to the region of the stately pine which stood in groups on the very edge of the precipice—the path led along the banks of a deep ravine, and was sufficiently dangerous.

Farther on the country assumed the appearance of an English park, then a picturesque ravine, or a flat valley interspersed with lofty trees.

Crossing several streams flowing to the N.W. to join the river Motagua, at nine leagues we reached Gualan, a village of fifty houses, with a plaza or square—the houses of one story and

---

* Izabal lies in 15° 24′ 20″ N., 89° 9′ 53″ W.—Variation in June 1834—7° 45′ East.—Ed.

† 3420 feet above the sea.

‡ Its summit 2835 feet high.—Ed.

usually with a colonnade. We were detained here two days for want of mules. This was the rainy season, which lasts from April to November; and, although the mornings were fine, at three P. M. the clouds gathered on the wooded hill around, lightnings flashed, and torrents of rain poured down for about three hours, when the sky cleared, and all was bright and calm as a summer's evening in England.

June 19.—Left Gualan at sunrise: our journey lay in a S.W. direction, through a country rich in all the beauties of woodland scenery. At noon reached the village of San Pablo, a small but pretty assemblage of cane-built huts. We here turned off nearly south, and, after crossing a small tributary stream flowing to the N.W. to join the Motagua, we reached the town, or city as it is called, of Zacapá at two P.M., after a journey of eight leagues and a half.

Zacapá is a place of some extent; the streets regular; a plaza, on one side of which is a handsome church, on the other a court-house and gaol were erecting in a substantial manner. The people seemed much better clad than any we had yet seen. The population may amount to 5000 persons.*

June 20.—Half-an-hour to the westward of the town crossed in a ferry-boat a broad river, flowing to the northward, at a spot called Comantan, consisting of a few huts; hence our road continued in a W. by S. direction, through a beautifully-diversified country, and across several streams, to Chimalapa, a distance of seven leagues. Chimalapa is a village of about ninety huts, with the ruins of a church, from the top of which is a beautiful view of the windings of the noble river Motagua, which here flows through a rich and varied country, spreading fertility around those who, alas! hardly know how to profit by it. We here measured an acacia tree whose stem was four yards, and its luxuriant foliage 120 yards in circumference.

June 21.—At one league to the westward forded the river Chimalapa, which flows N. W. to the Motagua, and continued in a W. by S. direction, along the southern bank of the latter river, although at some distance; passed large herds of cattle; much more cultivation; fields of maize, &c. At six leagues turned off to the S. W., and at nine leagues reached the village of Guastatoya, a small but pretty hamlet, situated in a beautiful valley: the cottages had pointed roofs.

On the following morning continued our journey through a rich and fertile country. At five leagues we stopped at a farm to feed the mules: we here observed that all the people were more or less afflicted with goitre, or guëguïcho, as it is called, said to arise

* About ten miles south-east of Zacapá is the town of Chiquimula de la Sierra, a place of much trade, and containing about 6000 inhabitants.

from some property in the water they drink. At nine leagues reached El Puente, or "the bridge," of stone, with two arches, over the adjacent stream, a tributary to the Motagua. This is a miserable collection of huts, where it was difficult to get anything to eat, as this road is only travelled during the rainy season.

June 23.—Started early for the capital, about ten leagues in a S.W. direction. For several hours we were without food, as not a hut could afford bread or anything else eatable. Passed through some fine scenery and saw a beautiful fall of water: at length, at a sudden turn of the road, we came in sight of the city of Guatemala, and, being at a considerable elevation, we looked down upon its whitewashed houses and churches: descended into the valley, crossed a neat stone bridge over the Rio de las Vacas, and again ascending reached the gates of the city.

The city of Guatemala la Nueva, or "the new," to distinguish it from "la Antigua," or the old capital, has been too often visited to need any description here. I will merely state that it lies in a valley or plain, about fifteen miles long by ten wide, and at an elevation, according to Haefkens, of nearly 5000 feet above the sea.* Its population has been usually stated at from 35,000 to 40,000 persons. The market is abundantly supplied with meat, fruit, vegetables, and fish from the far-famed lake of Amatitan.

The distant view from the city across the plain to the west is bounded by the three lofty mountains rising from 7000 to 8000 feet above its level, known by the names of Volcan de Agua, or Water Volcano, a perfect cone, clothed with perpetual verdure to its summit; the Volcan de Fuego, or Fire Volcano, generally emitting a white smoke from one of its peaks; and the Volcan de Pacaya: and nothing can exceed the boldness and beauty of the effect produced by these noble objects when seen against a cloudless sky, shortly after sunset, as the sun's last rays still gild their lofty peaks, while all below lies buried in a mass of shadow.

Aug. 22.—Anxious to form a nearer acquaintance with these great physical features of the landscape, I quitted the capital with two friends, and we directed our steps across the plain towards the old city of Guatemala, distant about eight leagues, in a general W.S.W. direction. The road led at first through plantations of maize on one hand, and on the other some remarkable mounds or tumuli, supposed to have been burial-places of the Indians before the invasion of the Spaniards. After two leagues across the plain we descended a deep ravine, and reached the Indian vil-

* It is not on record that either the elevation of this plain or of the neighbouring volcanoes has ever been measured, and it would be a good service rendered to geography if some traveller would obtain a series of barometrical observations in the city of Guatemala, and on the Volcan de Agua.—Ed.

lage of Mixco, built upon the declivity of a hill: here are a plaza and small church; the population is said to be 4000; the costume of the Indians is remarkably handsome. Passing through the village we commenced a steep ascent by a rugged road;—from the summit the view of Guatemala, the wide-spreading plain, and the village of Mixco, at our feet was very picturesque. We now crossed some fields of luxuriant grass, and abounding in wild flowers, whose delicious perfume was very grateful, and reached successively the villages of San Ignacio and San Lucas: two leagues beyond a sharp descent brought us to La Antigua, or ancient city of Guatemala.

The ruins of this town are seated in a rich valley, nearly surrounded by mountains, among which the volcanoes are proudly conspicuous; its mouldering ruins presenting a strong contrast to the rife and luxuriant vegetation everywhere around. I need hardly mention that this city was abandoned in consequence of the numerous disasters that had befallen it from its foundation to the fatal earthquake of 1773. It now, however, contains a population of about 12,000. In the two days we remained here we visited all the neighbouring villages; and the Ciudad Vieja, or site of the original capital, when founded by Alvarado in 1527, and which was overwhelmed by an eruption of water from the volcano in 1541; it lies about a league to the south of La Antigua. About one league beyond are the villages of Jocotenango and San Miguel de las Dueñas, near to which we ascended to the brow of a hill, and had a magnificent view of the Pacific ocean.

Aug. 25.—At 1 P.M. we set out from the old city for the summit of the Volcan de Agua. At 3 we reached the village of Santa Maria, where we halted to collect our party, which consisted of M. Lüdert of St. Petersburg, Mr. Croskey of Philadelphia, and myself, two servants, and sixteen Indians. At 5 set forward again, and for an hour traversed a country studded with trees, and covered with luxuriant grass, among which the bright scarlet star of the native single dahlia was conspicuous; by degrees the road became narrow and steep, till we reached a part of the mountain called La Cruz, from a cross being erected there. We here were obliged to abandon our mules, and after a short halt we proceeded onwards by torch-light, scrambling through the rank grass and dense underwood with great difficulty and fatigue for another hour, when we bivouacked for the night by the side of blazing fires.

At 5 on the following morning we again started on our ascent of the mountain, which became more difficult at every step, partly owing to the slipperiness of the long wet grass. We now entered the region of pines—some noble trees swung their branches to the sweeping wind with a solemn and mournful sound—others lay prostrate on the ground, their bleached and broken stems blasted by the lightning or broken by the fury of the storm.

At sunrise we saw a vast sea of clouds floating far beneath the elevation we had gained: an Indian here tracked a tiger for some distance from a spring, but we saw nothing of it.

Two hours of this fatiguing ascent brought us at 7 o'clock to the summit, and we gladly descended into the apparent crater to escape the cold and cutting winds, where, by the aid of a good fire and a substantial breakfast, we soon recruited our strength.

The crater, or hollow space at the summit of the Volcan de Agua, is from forty to fifty yards in depth, and about 150 in diameter—the sides and bottom are strewed with masses of rock, apparently showing the effects of boiling water or of fire, and between them spring up bushes and trees. After some time spent here, we climbed to the crest of the crater, or summit of the mountain, which is stated to be 6800 feet above the old city of Guatemala at its base, or 12,620 feet above the level of the sea,* and from this elevated point we had a glorious view in every direction: immediately around us, and almost at our feet, the villages and numerous haciendas of the valley—the cities of Old and New Guatemala—the village and lake of Amatitan—more distant the province of Suchiltepec, and the range of volcanoes stretching away to the northward, even to the distant plains of Chiapa; to the north the mountains of Vera Paz and Belize; to the eastward the district of Sonsonate and the province of San Salvador, while to the N.E. and to the south the view is bounded by the Atlantic and Pacific Oceans, the former at a distance of 140 miles, its nearest point; the latter only 30 miles distant, or apparently almost at our feet.

To our astonishment, while here we suddenly encountered two large wild oxen: the Indians by shouting tried to frighten them over into the crater; they however escaped: we saluted them with our pistol-balls, and they retreated much to our satisfaction, yet not so much so to the hungry Indians, who would gladly have captured them.

After leaving an inscription to commemorate our visit, we started on our descent of the mountain: this was doubly fatiguing and difficult—and we proceeded with a cord round the waist, each being held by an Indian before and behind us. After reaching a certain distance flowers began to bloom on all sides; I recognised the dahlia, lupin, and a species of large poppy; we also gathered some fine strawberries: on our arrival at the "Cross" we were completely exhausted; but after a short rest mounted our mules

* Journal, vol. vi. p. 123. This elevation is probably not far wrong, but we do not know on what measurement it rests. The same authority gives the height of the village of Santa Maria 6920 feet above the sea.—Ed.

and travelled down to the village of Santa Maria; and from thence in two hours reached our old quarters in the ancient capital.

On the following morning we set out on a visit to the lake of Amatitan, about four leagues to the S.E. of La Antigua; the road lay through a fertile and luxuriant country. At two leagues we reached the farm of Barcena, prettily surrounded by wood. At three leagues we came to a pass in the limestone rock, so narrow that two mules could scarcely move abreast: on emerging from it at a considerable elevation we had a fine view of the city of Guatemala in the distance; we then commenced a steep descent by a winding path, and after four hours' ride reached Amatitan.

This is a small village of about 100 houses, at the western end of a small lake of the same name: the lake itself is a fine sheet of water, about three leagues in length, from N.N.E. to S.S.W. by one in breadth, bounded by lofty hills, wooded to their summit. It affords an abundant supply of fish called moharras, which are excellent eating, and are sent in great numbers to the capital. The shores also supply salt.

Aug. 28.—We started in a N.E. direction on our return to the new city of Guatemala: from a rising ground, shortly after leaving Amatitan, we had a fine view of the lake and of the river Michtatoya, which winds its course on the eastern side of the lake, and flows towards the Pacific. At two leagues we passed through a small dirty village called Villa Nueva; and at four leagues halted to examine a large sugar estate named Villa Lobos. We were here shown some of the largest caterpillars I ever saw, from five to six inches in length: from hence we crossed the plain to Guatemala la Nueva.

After a stay of three months in the capital of Guatemala, during the whole of which time I had met with the greatest hospitality—had enjoyed the most delightful climate in the world—with a luxuriant soil—affording some of the most beautiful productions of nature—with a situation as lovely, and with scenery as picturesque, as any country can boast—it was with great regret that I found that circumstances, which it were needless here to relate, obliged me to return to England. I decided therefore upon travelling to the coast by a route seldom chosen, and never, as far as I am aware, described; namely, by the river Polochic.

In company, then, with my former travelling companions, I quitted the city of Guatemala before daylight on the morning of the 14th September, and journeying in a northerly direction, by a good road, skirted on the left by a deep ravine, or baranca, over which we occasionally passed on a succession of naturally-formed bridges. The scenery was bold and beautiful: the ravine seemed to have been caused by an earthquake rending the mountain, yet leaving at intervals arches or passes formed by masses of the rock, tumbled as if by accident together, yet forming a bridge, strong and perfect, as if built by man: the sides of the ravine were steep, almost to perpendicular, and some hundred feet in depth: large flocks of parrots rose screaming from the depths below, their plumage brightning in the rays of the rising sun: at times, in the deep gloom of the valley beneath, we could distinguish the fire of the Indian burning his charcoal for the market of that fair and hospitable city we were leaving.

After about two hours' ride we crossed a well-built bridge, and entered the village of Chinauta, which consists of a collection of about sixty huts and two churches. The river Chinauta is a small clear stream, which falls into the Rio Grande de Motagua, and owing to its serpentine course we crossed it not less than fourteen times during the day's journey. Quitting Chinauta, we passed successively the villages of San Antonio, San Rafael, and the farm of Carizal. Continuing along a good road we came to the Vuelta Grande (a 'sudden turn,' almost at right angles), and from an eminence had our farewell view of the city of Guatemala, the white churches of which were still visible.

We next passed the village of Jiji, consisting of about ten huts, and a beautiful ravine, called the Baranca de la Negra, and finally arrived at half-past three at the farm and sugar-mill called Trapiche Grande, about nine leagues from Guatemala. Having in the course of the day made a considerable descent, we found ourselves in a sultry climate compared with the temperate one of the capital:—the difference in temperature in one short day's journey in this country is very striking.

Sept. 15.—Started at six A.M., and shortly after halted to examine the remains of an ancient Indian fortification. It presented a succession of ruined breastworks of stone, but was so concealed by brushwood that we could make but little of it. We proceeded through a beautiful valley called the Vega de los Plátanos and a hamlet named Buena Ventura, a few huts embosomed in wood; had occasional views of the Rio Grande, after which the road became steep, stony, and bad: descending through a thickly-wooded country, we arrived at the Rio Grande, on the shady banks of which we were detained an hour in swimming our mules, and ferrying ourselves over. The river is here a noble and rapid stream. After halting for breakfast, we proceeded to the Aguas Calientes, or "hot springs." The face of the country was here completely changed: the trees grew distant from each other, and large masses of rock were scattered about in wild confusion: the water issues from the ground almost at a boiling heat, with a strong sulphureous smell. I endeavoured to force my mule to step into the stream, but he would not; and on dipping my finger in I was glad to withdraw it. About one hundred yards farther on we came to a similar spring, but emitting a stronger sulphureous vapour. We continued by a rough broken road over the Llano Grande, or "great plain:" passed the small village of Texas, and halted at the Trapiche, or sugar-mill of Llano Grande, a pretty spot, surrounded by groves of orange and lemon trees. We started again at five P.M. by a good road, reached the estate of Choacus, and began the ascent of the mountain of Choacus by moonlight. By a steep and rugged road we reached a forest of pines, which cast their gloomy shadows across the path, while below us the bright fires of the sugar-mill offered a wild and uncommon scene.

Three hours brought us to the summit, which I should estimate roughly at 2000 feet above the plain; and descending on the northern side at ten P.M. we came in sight of the moon-lit plain of Salamá. After breaking all our horses' cruppers from the steepness of the descent, we reached a small plain, then ascended a second series of the heights of Choncono, and at eleven P.M. arrived at Salamá—distance estimated twelve leagues—and took up our quarters at the hospitable house of Don José Soria, one of the few remaining old Spaniards, courteous, kind, and well-bred; a real hidalgo.

The town, or as it is called, the city of Salamá, is the residence of the chief of the department of Vera Paz, although Coban, about twelve leagues farther north, is its capital: the town may contain from 5000 to 6000 inhabitants, of which a few are merchants, about one-sixth ladinos, and the rest Indians. Here is a plaza, a cathedral, a cabildo, and some respectable shops; it is surrounded by farms, and is situated in a luxuriant valley.*

After one day's detention here for mules we quitted Salamá at 3 on the morning of the 17th, travelling in a north-easterly direction—by a rough road we crossed the hills of Cachil and Quililá, and at 9 o'clock arrived at Purulá, a small Indian village about five leagues from Salamá. After two hours' halt we commenced the ascent of the mountain of Purulá; the road at first good, and the scenery beautiful: heavy rain with thunder and lightning continued throughout the ascent: for some distance the path led along the banks of a river, after which it became, if possible, worse than that of the Mico mountain. At the end of three hours we reached the summit, which may be estimated at 1500

---

* At San Jeronimo, about three leagues S.E. of Salamá, is a large estate belonging to Mr. Marshall Bennett of Belize, which formerly belonged to the Dominicans. In the village are about 2000 ladinos. Sugar, cochineal, and cattle, are the chief products of the estate.

feet above the plain: the first part of the descent is bad, but it afterwards becomes good, opening on all sides upon romantic views. On the right the river Polochic winds its course through a fine plain, while directly before us, surrounded by orange groves, was the peaceful village of San Miguel Tucurú, which we reached at 6 P.M. Here is a poor church, a poorer cabildo; a plaza, and some mud huts; a population perhaps of 500 persons. We could not obtain bread to eat, and with difficulty any thing else.

Sept. 18.—Started at 6 A.M., and by a steep descent reached the river Polochic, which is here a rapid stream, foaming over a rugged bed: at intervals are small islands clothed in the richest verdure: the road continues along the banks of the river skirted on the left by a rich plain, crossed by numerous streams in their course to swell the torrent of the river. At 9 reached the large farm of Rachguizch on the right of the road; and at noon the little hamlet of Chemaguin in a luxuriant valley surrounded by well wooded hills. After two hours' halt we pushed forward with fresh horses, and shortly ascended a hill, whence we had a view of the plain of Polochic and the distant glittering water of the Lago Dulce: we descended through a thickly-wooded country, and soon came to the banks of the Polochic, which is here rapid and much widened, and crossed it by a singular suspension bridge, called the Hamaca, slung, as it were, from one side to the other across the stream—an ingeniously contrived and well constructed bridge, formed of the strong rope-like stems of the *bejucos* or parasitical plants, wattled together and attached to the trees on either side: the *lianes* or binds are annually renewed:*—by this means we safely crossed the torrent, and the horses having swam over, by a short steep ascent we reached a good level road through groves of the Manaca palm, bending gracefully like a large feather; the platanilla with its strings of coral-red flowers, and the mojan: these woods extend for many miles, and afford a cover for numerous wild animals which infest this unfrequented part of the country.

At 5 P.M. we reached the village of Teleman, having rode about twelve miles in a general E.N.E. direction.

The Embarcadero de Teleman is a small hamlet of about a dozen huts, situated on the northern bank of the river Polochic, at its highest navigable point; the heat and the musquitos proved to us

* Mr. Page's Journal, which he has kindly presented to the Society, and from which these extracts are made, contains a picturesque sketch of this curious bridge. We learn from Mr. Fletcher, who has several times travelled this road, that the bridge is about 100 yards long, by about 2 wide, and is suspended about 3 yards above the stream in the rainy season, when alone it is required.—Ed.

that we had reached the low lands, and the limit of our land journey.

Sept. 19.—At noon we embarked in a piragua or bongo, a flat-bottomed boat, about thirty feet long, manned by a patron or coxswain and three sailors, and turning a bend of the river lost sight of Teleman, and glided rapidly down the stream.

The Polochic is a fine river, which rises in the mountains of Purulá, in the district of Coban, in lat. 15° ½ N., long. 90° ½ W. nearly, whence it holds a general E.S.E. course for a direct distance of about seventy geographical miles, and falls into the Golfo Dulce* at its south-western angle. It is navigable only as far as Teleman, and thence to the gulf its course is extremely winding.

We passed on the right bank the site of Pueblo Viejo, or the "old town," where a small tributary falls into the river. Farther on to the north rose majestically in the distance the wooded mountain of Cajabon.

Upon a rock overhanging the water we observed two beautiful rose-coloured storks, posted like sentinels; whilst numerous juoacaromeyas or large macaws, parrots, king-fishers, wild pheasants, and turkeys, flew across the stream; passed the Boca Nueva, the outlet of a stream of some extent, on the right, and a few wretched huts to shelter the Indians when gathering sarsaparilla, which abounds near all the rivers in this vicinity. We could just see through the gloom of the evening the mouth of the Cajabon† river on the north shore. Owing to rain, mosquitos, &c., we passed a very uncomfortable night, and the following morning through the haze we distinguished the gulf and hills behind Izabal, where we arrived at 1 P.M. The distance from Guatemala to Teleman is about 45 leagues, from Teleman to Izabal about 30, making about 75 leagues as the whole distance by this route.

---

* The *Golfo Dulce*, from the late survey by Captain Richard Owen and Messrs. Smith and Lawrance, officers of the *Thunder*, is found to be 24 geographical miles in length, by about 10 miles in breadth; the average depth of water is from 6 to 8 fathoms, with a bottom of bluish clay.—Ed.

† It is about 2 miles up on the banks of the river Cajabon that the site of the town of New Liverpool has been selected by the Vera Paz British Agricultural Company.

XXV.—*Memoir to accompany the Survey of the Delta of the Indus, in* 1837, *by* Lieut. T. G. Carless, Indian Navy. Communicated by the Geographical Society of Bombay.

About fifty miles from the sea, the river Indus, a few years ago, divided into two grand arms, the Bagár and the Sátá: all its mouths and branches were then open, and some were navigable for vessels of considerable size. Extensive changes have since taken place; and these have been carried to such an extent, that the course of the river in the Delta has been entirely altered and most of its branches destroyed. During the dry season, no communication now exists between the Bagár and the main stream, a sand-bank having accumulated at the confluence, which fills up its bed for a distance of several miles, and is five or six feet above the level of the water. In all the branches diverging from it, the water is salt for the greater part of the year, and they are then merely inlets of the sea. The Sátá, or eastern arm, as it is called in the late maps, pursues the same course to the ocean as the great river from which it is supplied, and is, in fact, a continuation of it. In every part it preserves a similar magnitude; and for the last ten or twelve years it has been, as it is now, the principal channel of the Indus. In its passage to the sea it receives many local appellations, but is best known, near the coast, as the Wanyaní. Of the five branches it formerly sent off, the Mall and Mutní, discharged by the Mall and Káher mouths, are completely choked at the part where they leave the parent-stream, their beds being now nearly level with the adjacent country and partially overrun with jungle. The other three, the Hajámarí, K'hédíwárí, and Titíyah, pursue a westerly course to the sea, and are the only branches that are now favoured by the fresh water in the dry season. The two first empty themselves by the mouths of the same name, and possess navigable channels into the main river, but the latter is merely a narrow and shallow creek that conveys a very small portion of water to the Richel mouth. Above the Delta, the Indus throws off two more branches, the Pinyarí and Feleïlí, which are rivers only during the inundation; after it has subsided, they dry up in many places for miles, and in others form a series of shallow, stagnant pools that have no connexion: they are, besides, closed by bands or dams thrown across them above the the sea-port towns in order to retain all the fresh water that passes into them for agricultural purposes.

The Indus formerly reached the sea through eleven large mouths, but three now suffice, in the dry season, to discharge its waters. Of these, the Pítí, the Pintiyání, Juwah, and Richel, belong to the Bagár; and the Hajámarí, K'hédíwárí, Kúkewárí Káher, and Mall, to the Sátá; the Sír and Kórí are entrances to the Pinyárí and Feleïlí branches, and complete the number. Besides these there are many small mouths, some still open, and others that are partially choked; but, as it would only tend to confuse, I shall not name them. At present, the Kúkewárí, which gives egress to the waters of the Wanyaní, or main river, is the grand embouchure of the Indus: in the late maps it is called the Górá, but erroneously so, for that mouth was deserted by the stream some years ago, and its site is now occupied by an extensive swamp, intersected by several small creeks. There are only two other mouths, the Hajámarí and K'hédíwárí, that discharge fresh water in the dry season. Some of the branches that have been partly destroyed receive a small supply by means of canals cut for the purpose; and are said to be fresh in the upper part, but at all their mouths the water is salt.

From the foregoing sketch of the state of the Indus in the Delta, it will be perceived that only two of its large branches have at present a communication with the main stream in the dry season. A trigonometrical survey of these and of the main river, from its mouth to the Títíyah, a distance of thirty-five miles, has recently been completed, and I shall now proceed to describe them, with the part of the coast that has been examined, and such other facts, connected with the Delta and its rivers, as have fallen under my observation in the course of the survey.

Between the eastern and western mouths of the Indus, the sea-coast* of the Delta runs nearly in a straight direction to the N.W., about 125 miles: the shore is low and flat throughout, and, at high tide, overflowed to a considerable distance inland. With the exception of a few spots covered with jungle, which serve as marks for indicating the mouths of some of the rivers, it is destitute of trees or shrubs; and nothing is seen for many miles but a succession of dreary swamps. Wherever these occur, the land is scarcely discernible two miles from the shore; but, at those parts where there are bushes, it is visible from the deck of a small vessel about double that distance. Between the Sír and Korí mouths it is overspread with low mangrove-jungle running far into the sea; and, from the former, a bare uninhabited marsh, without a single bush or other object to relieve the eye, extends up to the Mall mouth, a distance of thirty-five miles: it is intersected by four large salt-water creeks, or rather inlets, the Kajah, Rúdah, Warí, and Khaï, which run far into the land, and are probably portions of destroyed rivers. Above the Mall all the mouths of the Sátá reach the sea within a space of twenty miles; and, at this part, the

* In the charts now in use it is laid down above half a degree too far to the eastward, and the same error will be found in every part that exists at the Hajámarí and Korí mouths, where the longitude has been ascertained by numerous observations. The former is in 67° 25′ 21″ east, and the latter in 68° 30′ east of Greenwich.

coast presents a peculiarity of formation not observable elsewhere. From the more elevated plains further inland, it runs off in a number of long spits, or tongues, of considerable breadth, which, from the rounded form they have assumed, bear the appearance of having been thrown up by a violent rush of water. They are composed of the soil deposited by the inundation; and, being always submerged at high tide, have not become firm, like the tracts that are only overflowed at intervals. The highest part is covered with a kind of long thorny grass, and they are separated from each other by creeks leading into the different rivers, which, with their numerous ramifications, form an excellent water-communication throughout the districts on this part of the coast. About the mouths of the Sátá, tamarisk-jungle is seen for the first time near the sea: there is a large patch on the north point of the Richel mouth; and, although not very high, it serves as a mark for distinguishing that entrance from the others. On a coast so devoid of objects, and so extensively flooded at times, it is often difficult to discover the mouths of the different rivers; and the pilots, many of whom have passed their lives in navigating it, are frequently a long time before they can find them.

The broad alluvial bank, projecting everywhere from the coast, extends from the bay of Karáchí to the N.W. extremity of Kach'h, and in width varies considerably. Off the mouths of the Sátá, where numerous shoals and flats have been cast up by the greater strength of the tide, it runs out in some places five or six miles, and is dry for a distance of twenty miles along the shore. At the Kórí mouth it is of a similar breadth, but only dries here and there in small patches. In other parts the outer edge is only two miles from the land, sometimes less; and there is a depth of water on it which, from three fathoms, decreases gradually as you approach the shore to three or four feet. That part of the bank lying off the mouths of the Sátá, which extends farther out and has attained a more advanced stage of formation than any other, has been thrown up partly by the strong current from the rivers, and partly by the heavy sea of the S.W. monsoon: it covers an area of not less than sixty square miles; and, with the exception of the narrow channels leading across, it dries at half-ebb. The interior part is composed of soft clay, mixed more or less with sand, and, near the shore, grass is seen springing up over the surface in large patches. Farther out, extensive sand-flats have been formed, which have risen to such a height that they are not covered in many places at the highest tides: they are smooth and firm, and a few blades of grass have already made their appearance on the most elevated parts. Being thrown up by the sea they are steep along the outer edge, frequently rising almost abruptly from a depth of two or three fathoms; and they abound with a species of long, thin worm, which serves as an article of food for the swarms of aquatic birds that frequent the coast. From the report of the natives, the whole of this portion of the bank is rapidly rising above the surface of the sea; and, before many years have elapsed, it will probably be added to the land. Outside the bank the bottom is very soft, but on it, except in a few spots, and in all the channels leading across it, extremely hard, being composed of sand mixed with a very small quantity of clay. Very little mud is found in any of the rivers, except the Kórí; it being apparently all carried out of them by the strong tides, and not deposited until at such a distance from the shore as to be out of their influence.

The tides are everywhere extremely irregular: between the Sír and Mall mouths there is a feeble current setting constantly to the E.S.E. along the shore, and the flood or ebb can only be distinguished by the rise or fall of the water, which is only four feet. No rivers are discharged on this part of the coast, and the sea is very slightly discoloured in consequence. Off the mouths of the Sátá the ebb runs out of the different channels in a broad muddy stream, direct from the land, but quickly loses its strength as it leaves the bank. Here the water is fresh at low tide 5 or 6 miles from the shore, and the rise and fall increases to 10 feet: in other parts, where the small mouths are numerous, the tides change their direction almost every hour, and are scarcely felt 2 miles outside them.

From the beginning of October to the middle of March the Sind coast is navigated by the boats of the country without difficulty or danger. The soundings are everywhere a sufficient guide; and, in general, decrease so gradually that no danger is to be apprehended in approaching it. The only shoal of any consequence is the great bank off the mouths of the Sátá, which projects beyond the line of the direct route to the northern rivers. From the dangerous appearance of the breakers during the sea-breeze the native boatmen have a great dread of approaching it, and at night always stand out into eleven or twelve fathoms before they venture to pass it. During the fine season, land and sea breezes generally prevail, with cold clear weather; but the wind sometimes blows very fresh from the N.E., and the atmosphere is obscured by clouds of fine dust many miles from the land. The season terminates long before it does on the Malabar coast, and the navigation then becomes very dangerous. Early in February, westerly winds set in with considerable violence, and, for the first fortnight, the weather is always very tempestuous. Strong gales, also, sometimes blow from the same quarter; and there is a heavy, tumultuous sea running, which breaks across the mouths of all the rivers. Short intervals of moderate weather occur afterwards until the middle of March; but, after that date, the Indus may be considered closed for the season. A few of the most adventurous of the Dutch boatmen sometimes navigate the coast, and enter the river as late as the end of April; but it is considered a very hazardous undertaking, and their vessels are frequently wrecked in attempting it.

The fisheries carried on along the coast give employment to about 100 vessels, and are a profitable source of commerce to the inhabitants of the seaport towns. Maghribí, situated upon the Sír river, sends about 30 boats, and Karáchí nearly double that number. The dingys, or sea-boats employed, seldom exceed 10 or 15 tons in burden, and, from their construction, are well adapted for navigating the coasts or entering the shallow mouths of the rivers. They have great beam in the centre, and the bow and stern, which are precisely similar in form, are very sharp; some are decked over, but others have merely a light platform abaft for the convenience of steering, and they are rigged with a single lateen sail. The upper part of the stem is always carved and profusely ornamented with coloured beads, shells, and other trifles: this is done by the boatmen from a superstitious belief that it propitiates the saints and secures success in fishing. These boats sail very fast in smooth water, and, from their buoyancy, are enabled to run into the shallow mouths through the breakers without sustaining the least injury or inconvenience. The best fishing-ground is about the great bank, and along the coast immediately below it: in the fine season the sea about those spots is covered with buoys, and clusters of boats are seen at anchor, waiting for the turn of tide to examine their nets. Besides sharks, which abound, several kinds of large fish are obtained; two of these, called by the natives Ringán and Sírí, are varieties of the cod, and are valuable for their sounds, which, with shark-fins, are exported to Bombay for the China-market. Other kinds, such as the Sír-fish, cavalho, and red snapper, are taken in such quantities that only the best are preserved for the purposes of commerce, as the supply would otherwise exceed the demand. Every evening the boatmen proceed to the nearest river and send on shore all those that are not required, for they imagine that if they were thrown overboard it would scare the others from the coast. At the spots frequented by them for this purpose, the shore is covered with fishes in every state of putrefaction, and there is such an intolerable stench, that it is almost impossible to remain in their vicinity. Most of the men engaged in the fishery belong to Karáchí, and are rather superior, in point of intelligence and appearance, to the other inhabitants of the sea-coast.

The large boats employed in the trade to Maskat, and the ports on the Malabar coast, are from 30 to 50 tons in burden, and sail principally from Karáchí: they are built in the same form as the fishing-dingys, but are disfigured by a very high poop, and have their sides raised several feet above the gunwale with mats and bamboos. There is not a greater number than 30 or 40 of these sea-going vessels in Sind, and none of them are armed. A few of the largest belong to the merchants at Vikkar and D'hárájá, and ascend the rivers on which those towns are situated without difficulty.

Hajámarí.—The general course of the Hajámarí, or, as it is called in the upper part, Síyahan river, is W. S.W., and its length forty miles. It quits the parent-stream about 22 miles from the sea, and is the most winding of all its branches: in some parts you do not advance in a direct line more than one mile in three, and the reaches turn back so directly upon each other that nothing but a narrow neck of land, scarcely 100 yards broad is left between them. Its mouth opens like a funnel, and, with the exception of that part where the stream takes its course along the right bank, is occupied by a broad flat, partially covered with water: this forms a continuation of the bank, everywhere extending from the coast, which is here rather more than a mile in breadth. The best and largest channel leading across it runs in a N.N.E. direction towards the north point of the river, and is 600 yards wide: at the entrance there are heavy breakers on either side, and at high water no greater depth is found on the bar than thirteen feet. Besides this channel there is another, 300 or 400 yards above it; but it is extremely shallow, and can only be used by the smallest boats, when the weather is moderate. About this mouth the land is extremely low, and entirely destitute of trees or other remarkable * objects by which it may be known.

A short distance above the entrance the river has a width of 550 yards, which decreases gradually as you proceed upwards, until it contracts to a narrow stream, not more than 50 yards broad. Bander Vikkar, its port, is situated 20 miles from the sea. Below that town the channel occupies about half the stream, and, with a few exceptions, crosses from side to side at the middle of each reach: the deepest part is generally found close to the steep banks, and the shallowest wherever they rise gradually from the water. The soundings are very irregular, and there is frequently a depth of 6 or 7 fathoms, but at high tide there is nowhere less than 12 feet, until you arrive at an extensive bar or flat 19 miles from its mouth, on which not more than 7½ feet are found. This is the only obstacle that exists below Vikkar,

* A beacon has been lately erected, on the north point of the river, that can be seen ten miles from the land, and buoys are to be placed in the channel leading into it, as soon as they can be got ready.

and the large boats arriving from Kach'h and Guzerát are frequently detained until the height of the springs before they can cross it. Immediately above Vikkar the river runs in a succession of reaches, broader in the middle than at the ends, where they contract so much as to leave a passage barely 80 yards wide; here it is alternately deep at the angles and shallow wherever it widens. On the flats the depth varies from 3 to 8 feet, but they are full of holes, and a continuous channel of more than 5 nowhere exists. Near its junction with the main stream, its breadth for many miles rarely exceeds 50 yards, and in the dry season, when there is not more than 3 feet water on any of the shoals, it is easily forded in six or eight different places: at this part it has shrunk within its former limits considerably, and appears like a large creek, wandering through the deserted bed of a river; the old banks are visible throughout, and at least 300 yards apart.

Quitting the main river in a direction nearly opposite to the course of its current, the water is prevented in a great measure from entering the Hajámarí, and the quantity that finds an outlet through it is very small. In the upper part the current is extremely feeble in consequence, but below Vikkar the tides, influenced by the ocean, are much stronger; even there, however, they are not rapid, for their velocity is scarcely ever greater than 3 miles an hour, and is generally much less. At Vikkar the flood is irregular in its approach, and 10 miles above it ceases to be felt altogether. At the mouth of the river there is a rise and fall of 8 feet, and it diminishes gradually in ascending to 9 inches.

At high water vessels drawing 9 feet can pass into this branch without difficulty, but none of a greater draught than 7 can ascend it as high as Vikkar. The boats that navigate it throughout in the dry season do not draw more than 2 feet 8 inches.

In its course to the sea the Hajámarí sends off several creeks, which connect it during the inundation with the branches of the Bagár. Of these the Chagalú quits it about 5 miles below Vikkar, and the K'hédíwárí, once a broad and deep river, a few miles farther down: midway between them there is also a large canal called Bórá-ke-wá, constructed some years ago by a wealthy Hindú, whose name it bears. After the swell has subsided they all partially dry up, and the only communication that then exists between this river and the western branches is by a salt-water creek, close to its mouth, called the Baghá. At high tide it affords a passage to the largest dúndís, and by this route they are enabled to enter the Richel, and proceed up that river into the Bagár, which is said to be accessible to vessels drawing

15 or 16 feet. From that arm they pass through a creek into the Gahrah, a small stream that leads to the harbour of Karáchí. On the eastern side the Ráhpúrah creek leaves the Hajámarí 5 miles above its mouth, and after pursuing an irregular course to the S.S.E., falls into the lower part of the K'hédíwárí branch; it is never filled until the night-tide has run to its greatest height, and even then is so narrow and shallow that it can only be used by the smallest boats. Opposite the village of Juwah, a few miles higher up, a large canal, called Jahlú, also affords a communication with the K'hédíwárí: it joins that stream near its confluence with the main river, is navigable for boats of some size, and connected by a small branch with the Ráhpúrah.

About the mouth of the Hajámarí the land from a short distance appears dry and covered with the richest verdure, but, on a closer inspection, is found to be one vast swamp, thinly spread with a species of coarse thorny grass, mixed with rushes, and intersected in every direction by small creeks. Above these plains, which extend about 5 miles from the coast, and are partially overflowed by the spring tides, the jungle commences, and cultivation is first seen. The country, however, is still very swampy; for although the banks of the river are sufficiently high to keep out the water, it obtains access to the interior through the numerous canals cut for the purpose of irrigation, and converts each field into a deep marsh, only passable on the narrow bank of clods raised to prevent its escape. About this part the jungle is composed entirely of low mangrove bushes, but a short distance higher up the mangrove disappears and gives place to the tamarisk. Below Vikkar it is thinly spread over the face of the country in large patches, and in some places approaches to the verge of the river, but generally keeps some distance from it. The intervening spaces present to the view a succession of grassy plains or rice-grounds, with here and there thickets of tall reeds and sedges. About the dreary tracts of the sea-coast there are few villages, but, as you ascend the river, they are passed in quick succession; and numerous flocks and herds are seen roaming about in every direction, which give an appearance of animation to the scene not observable in any other part of the country. Above Vikkar the tamarisk attains a greater height, and the jungle increases in density to such a degree, as to be almost impervious. With the exception of a few low points covered with reeds, the banks of the river are so thickly fringed that it is scarcely possible to pass along them; and the boatmen, when tracking their Dúndís, often find it difficult to proceed. Very few villages are visible from the river, but they are thickly scattered along the banks a short distance inland.

Above and below Vikkar, where extensive tracts are met with

clear of jungle, a large portion of the land is under cultivation, but above that town the spots cleared for tillage are small and confined to the immediate vicinity of the villages. Nearly all the flats that now occupy the bed of the river are made available for agricultural purposes; and these, although not so productive as other localities, are preferred by the indolent natives on account of the little labour they require. Rice is almost the only kind of grain produced in this part of the Delta: barley and moong are cultivated, but the quantity grown is so small that it scarcely merits notice. From every part of the Hajámarí numerous canals have been cut to convey the water to the fields in the interior, but only those below Vikkar are filled in the dry season. Very few water-wheels are in use, and they are employed in the irrigation of spots devoted to the cultivation of esculent vegetables.

Amongst the seaports of Sind, Bander-Vikkar ranks next in importance to Karáchí: the port takes its name from a small village in the vicinity, but the town is called Bárí Górá, after the old Górá river, which formerly flowed past it to the southward. It is the only town in the Síyáhan district, and belongs to Mír Naṣír Khán, the brother of the reigning Amír, who obtains a revenue from it of rather more than a lac of rupees, derived from his share of the produce of the soil, and a duty of 10 per cent. on the trade. It contains about 200 houses, constructed of reeds and grass plastered with mud, and, including the contiguous villages, has a population of about 1200 persons, composed of Hindús, Jaṭs, and a small number of Belúches. At the few shops it possesses nothing is procurable but rice, coarse cloth, tobacco, sweatmeats, and some trifling articles of domestic use: a strong spirit extracted from the sugar-cane, which is in great request amongst the inhabitants, may also be had. Mutton is exposed for sale three days in the week, and butter, eggs, and fowls, which are both cheap and good, are supplied from the neighbouring villages. Vegetables can only be procured in small quantities, and are very inferior, consisting of egg-plant, carrots, radishes, garlic, onions, and sweet potatoes.

Opposite the town the river, although not more than 170 yards wide, is deep, there being four and five fathoms close to the bank; it affords in consequence every facility for the discharge or shipment of cargoes; and the largest boats are enabled to lie close to the storehouses for that purpose. Besides numerous dingys or sea-going vessels, fifty or sixty flat-bottomed boats, of all sizes, from 5 to 50 tons in burthen, will generally be found here in the fine season.

Although the appearance of Bander-Vikkar is wretched beyond description, it possesses nevertheless a considerable trade, having become within the last few years a depôt for the greater part of the foreign and internal commerce of the Delta, which is entirely owing to its situation on the only river that now affords a communication by water with the upper parts of the country. As the water forsook the branches leading to the old ports D'hárájá and Sháh-bander, the trade naturally flowed into this channel, where it has remained stationary for some years; but as the river is fast filling up at its confluence it will probably be diverted before long to other places. If it does not become choked, and a commerce is ever established by the Indus, Bander-Vikkar may naturally be expected to rise into considerable importance. The average number of boats that annually sail from it with full cargoes is about 400; of these ten are sent from Bombay, three from Demán, three from Maskat, nine or ten from the Makrán coast, and the remainder from Kach'h and Guzerát: in size they vary considerably, but it will not be far from the truth if 20 tons is assumed as the medium tonnage. The value of the trade is as follows:—

EXPORTS.

Rice, 8000 karwárs	£24,000
Ghee, 2000 maunds	1,600
Gáír, 3000 maunds	900
Grindstones, about	50
	£26,550

IMPORTS.

From Bombay.			
		Brought forward	£6,850
English cloths, about	£1,200	Copper, 300 maunds	1,800
Sugar, 200 bags	600	Lead, about	150
Iron, 1000 maunds	500		
Malabar Coast.			
Pepper, 200 bags	700	Timber	1,000
Coir, coarse cloths, cocoa-nuts, &c.	350		
Maskat.			
Dates, about 6000 maunds	1,200	Dried fruits, pomegranate rinds, &c.	250
Slaves, about 100	800		
Kach'h.			
Cotton and coarse cloths	1,500		
Carried over	£6,850		£10,050

For some years past no route has presented such facilities to the natives for the conveyance of merchandise to the upper parts of this country as that by the Hajámarí branch, but in consequence of the heavy duties exacted at different places on the main river they have never been able to avail themselves of it. With the exception of a small quantity of timber none of the goods imported to Vikkar ever find their way above the Delta, and these consist merely of such articles as are required for the most pressing wants of its inhabitants. Nothing is received from Upper Sind but a little indigo; wheat, although grown there in large quantities, being generally procurable at a much cheaper rate from Kach'h and the Makrán coast. The most valuable part of the trade has for many years passed into Sind by the seaport of Karáchí, from which place it is conveyed to Tálá and Haïder-ábád on camels. The amount of the imports levied along this route is about 12 per cent., but by the river it frequently exceeds 35, and is never less.

Along the course of the Hajámarí the beds of several deserted streams are still visible, which formerly connected it with the main river and other branches. One of them, the old Górá, flowed past Vikkar to the southward, and a few years ago was navigable for large vessels. Some idea may be formed of the rapidity with which the branches of the Indus fill up, when I mention, that in 1831 Captain Burnes passed through it on his way to Haïder-ábád, and that when I visited it, four years after, the greater part of its bed was level with the plain. Hitherto it has been considered a matter of doubt whether large vessels ever navigated this celebrated river or any of its branches; but the recent discovery of one on the banks of the Hajámarí, about two miles above Vikkar, has placed the point beyond dispute: she lies half embedded in the ground, about 150 yards from the river, and is at least 400 tons in burthen; her hull, which is of the old-fashioned build and pierced for fourteen guns, is almost entire; and she is said to have belonged to the fleet of the Kalórá princes, the former rulers of Sind. One of the men employed during the survey as a pilot, had served on board of her in his youth, and by his account the remains of others, of a much larger size, were visible until very lately in different parts of the Delta.

On examining the Hajámarí again, since the survey was completed, I found that considerable alterations had occurred throughout its course. In many places the channels have shifted from one side to the other, and in general become shallower; new shoals have risen above the surface, and there is scarcely a part of the river where the depth is the same as it was before. At the mouth it is less in some places by 5 feet, and in strong westerly winds the water now breaks right across it. At the N. point the combined effect of the S.W. monsoon and the inundation is exhibited. Formerly the bank was steep, and there was a bank of 7 or 8 feet close to it, but now it shelves gradually off and projects 50 yards farther into the stream, a flat of hard sand, so firm that in walking over it the foot leaves no impression, having accumulated along it for a distance of 400 yards up the river. Large masses of clay have been torn from the bank by the violence of the current and carried out upon this shoal, where they have become firmly fixed as the sand was thrown up, and from a short distance now appear like detached rocks protruding through the surface. When this branch was first surveyed a narrow channel of 12 feet led from the main river into it, but on examining it again not more than 5 could be found in the deepest part; a shoal has also accumulated just before it, and this has been thrown up in such a position that it prevents the water in a still greater degree from turning into it. The most extraordinary change however is observable in the second reach from the junction, where the old steep banks of the river are 300 yards apart, and which is about a mile in length. Previous to the last inundation, the stream, 50 yards wide, took its course along the RIGHT bank, and the remaining space was filled up by a piece of land, in such an advanced stage of formation that it had been brought under cultivation: now the stream flows close along the LEFT bank, and the flat has shifted its position to the opposite side.

From the direction in which the Hajámarí leaves the parent stream, very little water passes into it, and the current has not sufficient velocity to prevent the soil brought down from settling in its bed; it is fast filling up at the junction in consequence, and before long will become impassable from the same cause that has operated in closing the Bagár and other branches. This, however, might perhaps be prevented. About half a mile below the part where it is thrown off, the main river turns with a sharp bend to the southward, and the whole body of water is precipitated against the right bank, which is very steep, with considerable force; if a canal were cut from the angle in the same line as the current runs down upon it, and about 1500 yards in length, it would reach one of the windings of the Hajámarí, and, from the volume of water that would be thrown into it by this means, with the increased rapidity of the tide, it might in a few years regain its former magnitude. The advantages that would be gained by a work of this nature are obvious, the Hajámari mouth being accessible to larger vessels and more easily navigated than any other now favoured by the fresh water. It is impossible to calculate what alterations would take place in the lower part of the main river by turning a portion of its waters into another channel, but I do not think it would be affected either in its course, depth, or magnitude. Of one thing, however, we may be assured, the Indus can scarcely be in a worse state at its mouths than it is at present, and any change must be for the better.

K'hédíwárí.—This river leaves the main stream 16 miles below the confluence of the Hajámarí, and, after pursuing an irregular course to the W.S.W., reaches the sea 3 miles below the entrance of that river, in latitude 24° 6′ 30″ N. The mouth of this small branch is approached by a broad channel that crosses the upper or northern extremity of the great bank nearly in a straight line; it is 4 miles long, from 600 to 800 yards broad, and at high tide has in most places a depth of 16 or 18 feet. The shallowest spot is an open space of some extent about half-way up it, where several channels, some of them partly destroyed, diverge across the bank to the southward and eastward; here there is not a greater depth than 10 feet. In strong westerly winds the sea breaks across the mouth of this channel, which is not sheltered in the slightest degree from the swell, and even in the fine season the breakers are heavy on the shoals that bound it, whenever the sea-breeze blows rather stronger than usual. Last year there was a bar of 8 feet stretching across it near the mouth of the river, but it has since been entirely removed by the inundation, and there is now 16 and 18 feet in the same spot.

The K'hédíwárí mouth is 650 yards wide, but at low tide not more than half that breadth is occupied by the water, the remainder being filled up by a mud flat projecting from the left bank. On the south side, the land, which is extremely depressed, extends from the more elevated part higher up the river in a broad spit or tongue of soft mud, covered with grass, and at high water is submerged for several miles. The opposite bank, although considerably higher, is also overflowed on the springs when the flood has attained its height; and the entrance then presents an appearance so totally different from what it does at other times that it is difficult to recognise it as the same place: all signs of a river disappear, and nothing is seen but a broad sheet of water, dotted here and there with herds of cattle, and bounded in the distance by a low indistinct line of banks and villages.

From its mouth the river runs nearly east in one long, straight reach for a distance of 3 miles, and then receives the water of a small stream called the Ad'hiyárí. At the point of junction its width is not more than 200 yards, but further down it increases to 500, and at high tide there is nowhere less than 13 feet in the deep channel. On the left bank, a muddy swamp, intersected by numerous small creeks that convey a small portion of fresh water to some of the half-destroyed channels of the great bank, extends from the mouth of the river to the Ad-'hiyárí: it produces nothing but coarse grass or rushes; and, as before remarked, is submerged at high water. On the opposite side the country, although very swampy, is everywhere under cultivation, and scored by innumerable canals cut to the fields in the interior. Besides these, there is also the Ráhpúrah creek, which connects this branch with the Hajámarí.

About the junction of the Ad'hiyárí the river winds in a succession of short reaches to the northward, then takes an easterly direction for about 2 miles, and, turning with a sharp bend to the southward, joins the main river 8 miles above its mouth.

Throughout this part its average breadth does not exceed 160 yards, and it presents precisely the same features as are observable in the other branches: the banks are alternately steep and shelving; and the deep channel, in which there is often 3 or 4 fathoms, always runs along the former. At the sharp bend, where the river turns suddenly to the southward, a narrow bar extends from the low projecting point across the stream, in a lateral direction, to the opposite bank; and on this, in ordinary tides, there is not a greater depth than 5 feet. At its confluence, this branch, like all the others now open, is extremely shallow—a broad sand-bank having been thrown up along the right bank of the main river which reaches across it.

The Ad'hiyárí leaves the parent stream a short distance below the K'hédíwárí, and enters that branch after pursuing a course to this W.S.W. for about $2\frac{1}{2}$ miles; in its passage it is joined on the right bank by a stream of similar magnitude called the Músáwarí, which is also an offset from the main river. In the narrowest part the Ad'hiyárí has a breadth of 35 yards, its banks are steep on both sides, and at high water the depth is never less than 9 feet. In the K'hédíwárí the tides are regular at night, but during the day the ebb runs 8 hours and the flood 4; the velocity of the former is seldom greater than $2\frac{1}{2}$ miles an-hour, and the strength of the latter varies in different parts of the river from $\frac{1}{2}$ a mile to $1\frac{1}{2}$. In the Ad'hiyárí the tide rises 4 feet on the springs, which increases to 6 at the mouth of the river, and to 10 at the entrance of the channel leading across the great bank towards it.

The district through which the K'hédíwárí flows is entirely destitute of trees; near the main river a few small tamarisk-bushes are seen, and this is the only part where they are met with. The soil is in general favourable for the production of rice, and, as the land is clear of wood, a large portion of it has been brought under cultivation; in some places, however, it is extremely sandy, and yields nothing but rushes or reeds. Along the lower part of the river, the few villages visible are situated some distance inland, but above the junction of the Ad'hiyárí they are thickly scattered along the banks and throughout the adjacent country. At Béló Kúkéwárí, the largest, the hákim and principal zemíndárs of the district reside.

From the report of the natives this small branch has been gradually increasing in magnitude for some years; the last inundation swept away the bar at its mouth, and, besides removing several shoals in the lower part, deepened it considerably. In consequence of this favourable change, it was navigated by large boats for the first time this year, and, as it was found that grain could be procured at a cheaper rate than in other parts of the Delta, they will probably resort to it in greater numbers next season: I am, however, inclined to think it will experience, before long, the same fate as the other rivers, and become choked at its confluence. At present it possesses many advantages over the Hajámarí, the only other branch now open, and, in affording access to the main river, presents equal, if not superior facilities, to the grand embouchure itself. From the sea the distance, through the Ad'hiyárí creek, is only 10 miles, and at high tide there is not less than 9 feet in the shallowest part; all the difficulties of the navigation will disappear when a few buoys are laid down, which is about to be done, and by this route a steamer drawing 6 or 7 feet might enter the main river in two hours, at a spot 7 miles above its mouth.

From Haïder-ábád the Indus pursues a S.S.W. direction to the ocean, and, with the exception of two sharp bends, one at the part where the Pinyárí quits it, and the other near the confluence of the Hajámarí branch, its course is rather direct. The distance in a straight line is 90 miles, but, by the windings of the stream, about 102. In the Delta it receives many names; below the Hajámarí it is known as the Wanyaní, and near the sea is sometimes called the Manijá, but this name is only applied to it by a few of the natives residing on its banks, and is not generally recognised in other parts of the country. From the Hajámarí to the deserted bed of the Naïr, it receives the appellation of Pópét; and above that is called successively the Múgrah and Sátá, from having once formed a part of those now abandoned rivers. The width of its mouth, the Kúkéwárí, is nearly a mile, but at low water a mud flat on the right bank contracts it to 770 yards. On the full and change of the moon, the tide rises 1½ foot higher than usual, and then, like all the other mouths, it can scarcely be distinguished even from a short distance, the country on both sides being inundated for several miles.

When I first examined the Kúkéwárí mouth there were three channels leading from it across the great bank outside, which here projects 5 miles from the land; and that running to the N.W., in a line parallel to the shore, towards the K'hédíwárí entrance, was by far the best: the last inundation destroyed all but one, and this, from the greater body of water discharged through it, has attained more than double the width it then had, and is rather deeper. It takes a course across the bank to the S.W. in the same line as the river above, and preserves the same breadth, 1100 yards, until near the sea, where it gradually widens to 1½ mile: the greater part of this broad space, however, is occupied by an extensive shoal, with only 7 or 8 feet on it at high tide; and the channel becomes divided by it into two smaller ones; these again have shoals at their mouths, and by these continual divisions the great channel has no less than four different openings by which it may be entered. The best of these small channels is 500 yards broad at the entrance, but about a mile above contracts to 130, and it has a depth varying from 12 feet to 3½ fathoms; it is defended from the swell that usually accompanies the sea-breeze by the projecting point of a shoal, and in this respect possesses a great advantage over the others, which are unprotected. Above the large shoal the depth of water in the great channel is 10 and 12 feet, and it increases as you approach the mouth of the river to 4 fathoms, but there is one spot where there is not more than 9 feet right across. In the dry season the current does not run with such velocity as might be expected: during the months of January, February, and March, it never exceeded 3½ miles an-hour, and was frequently much less: the flood varied in strength from a quarter to one mile an-hour. Outside, the sea rises 10 feet on the springs, but this quickly decreases on the bank to 6, and at the mouth of the river to 5: two days after the change of the moon there was a further rise of 1½ foot; but this, which is always the case in the night, only happened twice in the day, and the country was then flooded for miles. At the entrance of the channel leading to the Kúkéwárí mouth, the navigation is certainly intricate, but a few buoys or marks would render it sufficiently easy for steamers drawing 6 or 7 feet, and I do not think they would experience any difficulty in entering the main river by it.

About the Kúkéwárí mouth extensive alterations have occurred since it was examined last year. The following description of it is taken from the first report published by order of the Bombay government, and, when compared with the account that has been given of its state at present, shows the extraordinary changes the mouths of the Indus annually undergo:—"The broad bank that has accumulated before the Kúkéwárí mouth projects 5 miles from the land, and is intersected by three channels which give egress to the waters of the main river; two of them cross the bank nearly in the same direction as the course of the river that supplies them, but the other, turning to the N.W., runs some distance in a line parallel with the shore, and, after uniting with the channel of the K'hédíwárí, reaches the sea about 5 miles to the southward of the Hajámarí mouth. The latter, which is the best of the three, is between 300 and 400 yards broad, and in the shallowest part has a depth of 9 feet. At the lowest state of the tide the central parts of the bank are elevated 12 feet above the level of the sea, and the beds of the two southern channels from 3 to 4: the vast body of water issuing from the main river rushes through them with great impetuosity, and with a noise that in calm weather is heard some distance: they then form what may be termed rapids, and on the extreme edge of the bank terminate in a fall of about 12 inches." Of these two channels only one, as before remarked, now exists, the other being choked at the point of separation: the lower part of it, however, has not been destroyed, and there is still a small quantity of water forced into it over the bar of sand that has been thrown up. The northern channel, which was 5 miles long, and formerly connected it with the K'hédíwárí mouth, has been entirely obliterated, with all the branches it sent off towards the sea; not the slightest trace of them now remains. These changes have been caused by the rapid current of the inundation throwing up a bar at the point where the channels are separated at an early period of its rise; the water being thus prevented from passing into them, they were subjected, as the swell attained its height, to the action of a feebler current setting off-shore across them from the depressed part of the coast between the Kúkéwárí and K'hédíwárí mouths, and quickly filled up. The closing of two of the old outlets by throwing the whole body of water into the channel leading direct from the river, which was formerly the worst of the three, has increased its depth and breadth considerably, and, its bed being now much lower than it was, the rapid described in the first report with the fall at its mouth has disappeared.

Near the sea the Indus attains a magnitude at high tide conformable to the idea generally entertained of it; but at other times, when the fall of water has uncovered the shelving flats and contracted the width of its bed, it appears inferior in point of size to most of the other large rivers of the Eastern hemisphere. As before stated, it has been surveyed trigonometrically up to the Títíyah, a small branch thrown off 35 miles from its mouth. In this part its average breadth is about 700 yards, but in some places it is much broader, and in others contracted to half that width by shoals projecting from the old banks far into its bed. The greatest and least breadth is found abreast of the Mutní branch, where from 1060 yards it decreases to 305; in three or four places besides it is 1900 yards wide, and there are few reaches where it does not diminish in some part to 350 or 400. Throughout this portion of the river the banks in general are alternately steep and shelving, and the former in most cases indicate the side on which the deep channel lies; in a few of the reaches the high bank is formed of several ledges, and then it has a narrow ridge of shallow water running along it.

In general, half the bed of the stream is occupied by a dry flat, extending from the old bank, and the shallowest part of the channel is always found at those spots where you cross between two of them from one side to the other. There are also a few detached shoals, with from 3 to 6 feet on them in the centre of the river. At the bends of the reaches the depth of water is always great, there being frequently 6, 8, and even 10 fathoms; but the crossing-places, between the flats, are much shallower; at these spots it varies from 13 to 18 feet. Opposite Míyáhan, a large fishing-village, one mile above the junction of the K'hédíwárí, a very narrow ridge extends in a lateral direction across the channel from one flat to the other; and there is another close to the Síyáhan branch, which projects from the right bank about a mile down the centre of the reach, and connects a line of small shoals. On both these bars there is 9 and 10 feet at high tide, and they are the shallowest spots in this section of the river. Abreast of the old Mutní branch the bed of the stream is rather foul, but one of the channels has from 10 to 16 feet in it, and, although narrow, could easily be navigated.

The strength of the current varies from $1\frac{1}{2}$ to $2\frac{1}{4}$ miles an hour, and the latter is about the average velocity: it nowhere exceeds $3\frac{1}{2}$ miles, and only flows at that rate in two or three places, where the water is confined to a narrow channel at the bend of a reach. This part of the river was surveyed in February and March, when it is at its lowest state, and the comparative weakness of the current must be attributed to the diminished quantity of water it discharges in those months. Along the steep banks, especially where they are much indented by the separation of large masses, it sometimes runs into small eddies, but they seldom extend farther into the stream than 10 or 12 yards. It is only at these places that it is necessary to pay particular attention to the boats in tracking, for, the water being thrown off from every small projecting point of the bank, if the track-rope is not brought down from the mast-head to the stem before it strikes the bow, they are shot out suddenly into the middle of the stream, and carried down the river. At the junction of the Títíyah, and opposite Panj-gajer, a large village, three miles below the Síyáhan branch, the eddies are upon a larger scale, and afford some insight into the manner in which the shoals are formed that eventually effect a change in the course of the river. At the first-named spot the current strikes against the steep bank, a little above the Títíyah,* and is thrown off it, nearly at a right angle, down the middle of the stream: half a mile below that branch it turns into a small bay or bight, and runs up past it, at the rate of about a mile an-hour. This upward current prevails over a space of 180 yards broad in the deepest part of the river, and at the point where it turns into the bay the formation of a shoal has already commenced. A short distance below it there is another similar bight and eddy. The flood does not ascend the main river higher than Anní, a village ten miles from its mouth,

* Accompanying the MS. of Mr. Carless is a plan of this part of the Indus, on the scale of 4 inches to a mile, showing the alterations in its course near the Títíyah branch, and the direction of the eddies.—Ed.

nor even to this distance on the neaps; but the influence of the ocean-tides is perceptible as high as the Títíyah, where there is a mean daily rise and fall of 4 inches. The velocity of the flood varies in different places, but never exceeds a mile an-hour, nor does it continue to flow longer than 2 or, at the utmost, 3 hours.

The country about the mouth of the main river is submerged at high tide for some distance, and continues swampy as high as Bétrí, a village four miles from the sea, at the head of the first reach. On the left bank the land, being free from bushes or reeds, and the soil good, is under cultivation to a considerable extent, and affords also excellent pasturage to large herds of buffaloes. Numerous canals, that are filled by every tide, conduct the water to the fields; and there is also a navigable creek of some size about 2 miles above the entrance of the river, called the Babúlú, which leads into the lower part of the old Górá branch, and affords a communication by the Gumbrah, another large creek, with the Káher mouth. At Bétrí the Mutní river is not more than 3 miles distant, and boats can be seen in it very distinctly. Bétrí was formerly a very large town, but, from some connection with the alterations of the stream, was gradually deserted, and now does not contain above a dozen scattered huts. Near it there is a small lagoon 8 or 10 feet deep, supplied with water by cuts from the main river, which forms a kind of reservoir for the numerous canals that irrigate the grassy plains and rice-grounds in the vicinity. Along one side the bank is very steep, and, as this peculiarity could only have been produced by a strong current, it is evidently a portion of what was once a connecting branch between the Mutní and the old Górá river. On the right bank of the main river nothing is seen for some miles but a dense mass of reeds, occupying the site of an extensive morass that was formed some years ago when the stream deviated from its course more to the eastward. At the lower part of it a large navigable creek is thrown off which traverses the swampy land of the coast in a westerly direction, and reaches the sea by a small channel that joins the one leading from the K'hédíwárí mouth near its entrance. On the other side of the river, opposite the lower part of the morass, the Babúlú creek leads to the old Górá mouth, situated 2 miles below the Kúkéwárí: it has still a width of 850 yards, but is so choked with mud as to be impassable at low tide even for small boats. Above the thicket of reeds the country becomes more elevated, and is bare of trees up to the junction of the K'hédíwárí, where a few small tamarisk-bushes are first perceived. Below that branch there are few villages in the vicinity of the river, but along the course of the Babúlú creek they are numerous.

About the K'hédíwarí the tamarisk-thicket is thin and intermixed with patches of reeds; but as you advance it quickly becomes dense, and the bábul * (*mimosa Arabica*) appears in large quantities. In this part the river has a most desolate and melancholy appearance, for, the country being overrun to the edge of the banks, very few villages or fields are seen from it. The first village of any size is Míyáhan, situated on the right bank, about 10 miles from the sea, and inhabited entirely by fishermen. It is the principal station of the river fishery; and the man who farms the tax on it resides there. He pays 1600 rupees per annum to the Amírs, and is entitled to one-third of the quantity caught. About Míyáhan the soil contains a large portion of sand, and is so loose that in some places it is thrown up by the wind in heaps 8 or 10 feet high. Amongst the cultivation barley appears in small quantities; and there are a few spots cleared for grain and esculent vegetables. At Anní, a small village about a mile higher up, two ferry-boats are kept constantly in readiness to convey passengers across the river. Near this village there is a dry bare flat, extending some distance down the stream, in the form of a long spit, which has a deep but narrow inlet between it and the steep bank: this is the favourite haunt of four large alligators; and it was here we saw these monsters for the first time on the Indus. They were of enormous size, at least 25 feet long, and of that species with a long pointed snout. The Sindian boatmen told us they were much smaller but more rapacious higher up the river, and that the large ones, being very sluggish, rarely attack a man. As evidence of this we saw the stream in the vicinity covered with fishermen floating fearlessly past them on pots, with their bodies half immersed in the water.

Thirteen miles from the mouth of the river the destroyed bed of the Mutní branch is seen: the space filled up forms a triangle, each side about a mile and a half long, and in most places is overrun with bushes. There is still a small creek flowing through the centre of the plain, amongst the irregular furrows left in the bed, but it is separated from the main river at low water by a broad bar of sand; and as there is not a greater rise of tide even on the springs than 2 feet, none but the very smallest boats can pass into it. The Mall branch, which is about three miles higher up, appears to have been as large as the Mutní, and is still more completely choked, the land having attained a more advanced stage of formation. The greater part of its bed is now covered with jungle or under cultivation, and the remainder is occupied by heaps of loose sand, with pools of water between them that have no connexion. At the point where the banks contract to the

* Bábul, from the Sanskrit Barbura, is a generic name for the Acacia, or Mimos tribe.

breadth preserved by the lower part of the river it is 700 yards wide, and it is dry for several miles farther down. The cause of the changes that have taken place in these two branches, and also of the foul state of the main river in their vicinity, is apparent in the nature of the soil, which about this part of the Delta is extremely loose and sandy.

About a mile above the Mall a canal was cut two years ago, from the main river to that branch below the part that has been destroyed. It was 15 feet broad, but during the last inundation attained a width of 30 yards, and deepened so much that it is now navigable for small boats at high tide, even in the dry season. The natives frequently call it the New Mall, and before long it may perhaps become a new branch. Nearly opposite the canal is situated the village of 'Alí-ábád: it is of some size, and has a market for vegetables, which are produced in large quantities. A short distance above it a low flat projects into the river from the left bank. The natives point to this spot as the place where a large ship called the Fateḥ Jang, that carried 40 guns, was wrecked. About two years ago her upper works were visible above the water, and they succeeded in taking out her masts and some of the casks from her hold; but she has since sunk in the mud. About Panj-gajer, a large village near the Síyáhan branch, the country is more open than in other parts, and sugar-cane cultivation is first seen. Panj-gajer means five yards; and the village owes its name to the following circumstance. Ten or twelve years ago the chief of the district cut a canal near it to the Mall branch, which was 5 yards broad; the inundation increased its size and depth so much that it became a small river, but afterwards filled with sand, and dried up altogether. He then opened another 300 yards above it, which met with exactly the same fate, and he is now cutting a third through the bed of the first one. This gives some idea of the continual labour required to keep the canals open, and the fields in the interior properly irrigated.

About the Síyáhan branch none but small villages are seen until you arrive at Killán on the left bank, containing seventy or eighty houses, and inhabited chiefly by Hindús. About this part of the country the remains of several destroyed rivers are visible. I traced three, the Náïr, the Mugrah, and the Killán, several miles inland, and the natives told me there was another called the Sátá 10 or 12 miles higher up. The bed of the Náïr is almost level with the land and overrun with a dense mass of underwood: it is said to have been a river of considerable size fifteen years ago, flowing from the Bagár or Great Western arm, and to have been connected with the Síyáhan branch. The Mugrah is rather more than a mile above it, and is only partly filled up, there being numerous hollows still in its bed that have pools of water in them throughout the dry season: the natives say it was also thrown off from the Bagár, and has been in this state for the last fifteen years. About 3 miles inland a small branch, called the Ḥarám Kháneh, joined it to the Náïr, and the beds of other creeks are visible in every direction in the tract of country between them. The Killán is on the opposite side of the river, and in the centre of its bed there is now a large village named after it, surrounded with fields and groves; a short distance beyond the village it appears like a long straight glade cut through a forest, and at one spot there is a small lake left by the inundation, enclosed between high irregular banks overhung with trees placed in most picturesque groups, amongst which the mango, tamarind, and plantain are conspicuous. This river appears to have formed a connecting branch between the Mugrah and Sátá.

The Títíyah quits the Indus 35 miles from the sea and joins the Richél river near its mouth: it is not more than 30 yards broad and has not a greater depth in many places during the dry season than 1½ or 2 feet. There are two large villages near its confluence; one of them, Kóterí, is pleasantly situated in a grove of fruit-trees, and the other, Béman-jo-poro, is the residence of the Kárdár of the district. The main river has not as yet been surveyed above this small branch, but it has been examined by Lieut. Wood of the Indian Navy up to Ḥaïder-ábád, from whose reports the following description of that portion of it has been drawn up:—

Between the Títíyah and T'hat'hah shallows abound, and the navigation becomes intricate; in some spots the river is very foul, and wanders in many channels over the whole extent of its inundation bed: they are narrow and winding, and in general not more than one of them is navigable. In the bends of the reaches their depth is often very great, but in other parts they have not more than 9 or 10 feet in them; they also change their direction with great rapidity, for in two instances the fair channel, which ran close along the left bank when the Indus steamer passed up the river, was found two months afterwards to have shifted its position over to the other side. At the commencement of the dry season the current runs in some places at the rate of 5 miles an hour, and its average velocity is about 3½ miles. In this part of the river the banks are alternately shelving and steep, and marks of the alterations that have taken place in its course are everywhere visible: in some places they have evidently, at no distant period, formed shoals in its bed, and at others they are fast giving way before the strength of the current. The country is thinly populated, and in general covered with thick tamarisk-thicket, occasionally mixed with the bábul: few villages are met with, and the largest does not contain 100 houses. At Uplán, a few miles above the Títíyah, but on the other side of the river, there are two large canals leading one to Sháh-bander and the other to Maghribí: they are filled about the middle of July, and navigable for small boats to those towns, until the final fall of the river takes place in the last week of September. Abreast of T'hat'hah the banks of the river are a mile and a quarter apart, but the water does not occupy above one-third of this space, the remainder being filled up by an extensive dry shoal. The Bagár or Great Western arm was formerly thrown off 4 miles below that city, and the sand-bank that has been cast up at its confluence now forms the right bank of the river.

Above T'hat'hah the bed of the river in many places is full of small shoals, and the intricacy of the navigation increases; the banks are also frequently low on both sides, and no longer indicate, as in the lower part, where the deepest water is to be found. At the village of Kaddí its width is 980 yards, but at the angle of the reach in which it is situated, not more than 390: abreast of the Pinyárí it increases to 760, and at Trikal, a few miles below Ḥaïder-ábád, again contracts to 380. The strength of the current is considerably diminished, and there are very few places where its velocity is greater than 3 miles an hour. In this section of the river the depth of water has not yet been fully ascertained in every part, but in the lines of soundings taken across the shallows at certain distances, 7½ and 8 feet have always been found in some part. The only branch thrown off by the Indus in this part of its course is the Pinyárí, which quits it at Baná, a village half way between T'hat'hah and Ḥaïder-ábád, and in the dry season is not more than 40 yards wide: the stream does not extend many miles from its efflux, and its waters are clear and stagnant. Near Trikal a small creek communicates with the Feleïlí branch, but it is only filled by the inundation, and then insulates the land on which the capital is built. A short distance above T'hat'hah, and on the same side, the first hill is seen: it is low and of sandstone formation. From this place the Shikárgáhs or hunting-places of the Amírs extend on both sides of the river up to the capital; they are composed principally of large mimosa and tamarisk trees, and are so overrun with thick underwood as to be almost impervious. Most of the low sandy tracts between them are thinly covered with bushes, and very little cultivation is seen until you arrive at Ḥaïder-ábád, where there are a few fields and gardens. From that city a range of low table-hills, about 200 feet high, extends along the left bank to Trikal, 12 miles below it, and there is also a small group at Jérkh, 8 miles farther down.

The local divisions of the Indus, which have been entirely disregarded by those who have written about it, merit notice, for in some instances a knowledge of them will prove practically useful. The confusion apparent in many of the accounts descriptive of the general character and features of the river must be attributed to the authors not having paid sufficient attention to this subject, for, being ignorant of the causes that have led the natives to separate it into sections and distinguish each by an appropriate appellation, they have in many instances made the mistake of supposing that these names denoted distinct branches. Thus the Síyáhan has been described as flowing into the Hajámarí at Vikkar, whereas the Síyáhan and Hajámarí do not possess with regard to each other the relation of two separate branches, but are merely local appellations applied to different portions of the same river.

The terms employed to designate the different sections are often derived from some physical peculiarity, natural or artificial, or from some tradition referring to the locality, but they are most frequently named after the district through which they flow, or the principal tribe that occupies it: examples of all these different names are met with in Sáhú, Síyáhan, Pópét, and Mugrah. The first term, Sáhú, means "good," and is applied to that portion of the river so called, because the banks are composed of firmer materials and the stream is less liable to vary than in the sections immediately below it. Síyáhan is taken from the district through which that branch flows. Mugrah is derived from a Belúch tribe that formerly occupied the country about the old river of that name, and Pópét, which in the Sindian language means a parrot, from a tradition that in former times the boatmen always sacrificed one of these birds as an offering to a celebrated Pír or Saint on entering this division of the river, the navigation of which was dangerous and perplexing. The following table exhibits the different sections below Ḥaïder-ábád, with their limits and the derivation of their names, and will, perhaps, render this naturally complicated subject more intelligible:—

	Names.	Limits.	Derivation of Names.
Main trunk of the Indus.	Sáhó.	From a village close to Ḥaïder-ábád to the first range of hills above T'hat'hah, 40 miles.	From a word signifying "good."
Main trunk of the river, called in the late Maps the Sátá or Great Eastern arm.	Sátá.	From the latter point to the Títíyah branch, 28 miles.	From once having formed a part of the old Sáta river.
Ditto.	Mugrah.	From the Títíyah to the bed of the old Náïr river, 5¼ miles.	From having once formed a part of the Mugrah, which was called after a tribe of that name.
Ditto.	Pópét.	From the Náïr to the confluence of the Síyáhan, 4¼ miles.	From a tradition.
Ditto.	Wanyaní or Banyaní.	From the latter point to its mouth, 24 miles.	From the number of Hindús of the Banyan Caste resident on its banks.
Ḥajámarí branch.	Síyáhan.	From point of junction to Vikkar, 20 miles.	From the district through which it flows.
	Hajámarí.	From Vikkar to its mouth, 20 miles.	From a drunken Ḥajjam or barber having been drowned in attempting to swim across it.

Of the different sections of the main river, three of them, the Pópét, Mugrah, and Sátá, possess the same physical peculiarities, and a correct knowledge of them will be found in a practical point of view of some importance: it is in this part of the Indus that the destroying process is continually going on, which causes so much vacillation in its course amongst the subordinate branches in the Delta, and produces those frequent changes which render the navigation so uncertain. In the division above it (the Sáhó) this unsteady stream is confined to a permanent channel by the firmness of the soil, and the existence of a low, but in this particular important, range of hills, until it passes into the Sátá section, where the destructive operations commence. The causes that produce the frequent alterations observable throughout this part of the river are seen in the nature of the soil, and the velocity of the current at particular places. The soil is loose, light, and easily soluble, being composed of fine micaceous sand, mixed with a small quantity of clay; it is apparently distributed throughout the Delta in extensive patches, being met with at intervals along the whole course of the river below Ḥaïder-ábád, and in all the branches I have examined. Wherever it prevails, but especially in the three sections I have attempted to describe, the destruction of the banks continues without intermission, for in the dry season, when the level of the river is low, the current, in flowing along those that are high and steep, is constantly exerting itself to undermine their bases, which soon causes the upper portion to slide into the water: in this manner the navigable channel is often partially filled up and forced to shift its position by the large masses of soil that fall into it. This phenomenon is of such frequent occurrence that at night, when the range of sound is most extensive, as many as thirteen reports, produced by the precipitation of pieces of the bank into the water, have been counted in the short space of a minute. Dr. Heddle, in his valuable report on the Indus, when speaking of the effect this peculiar action has on the course of the river, relates a circumstance which shows the extent to which it sometimes takes place. At a village where he came-to for the night, a large portion of the bank, measuring 400 feet in length and nearly 100 in breadth, suddenly gave way a short distance a-head of his boat: the noise attending its fall resembled that of a large body of water rushing over a precipice, and the agitation of the river that followed caused the boat to roll as if in a heavy sea: a few huts were precipitated with it, and it was with difficulty the property in some of them was saved. On the following morning the part of the channel into which the bank had fallen, where prior to the accident there was a depth of 3 fathoms, was converted into a shoal partly above water, and the boat in getting under way was obliged to make a considerable circuit to clear it. When it is considered that so many reports, produced by similar causes, may be heard in the space of a minute, some idea may be formed of the rapidity and extent of the destructive action constantly going on in this part of the Indus; the alterations are in fact so frequent, and the position of the deep channel fluctuates to such a degree, that a pilot is of little use, further than by the knowledge he may possess of the character of the stream in the different sections, and of the extensive local changes that are continually taking place.

In the upper and lower parts of the Wanyaní section, where the banks are comparatively permanent, the soil is a stiff clay, but in the centre, about the old Mall and Mutní branches, a large tract occurs of the same sandy incoherent kind that is met with in the Mugrah and Pópét divisions; and similar physical peculiarities are observed in the character of the river.

Below Ḥaïder-ábád the navigation of the Indus is very intricate in some places, but, with proper precautions, unattended with danger or any great risk to property. Unaided by steam it will always be extremely tedious, and even with that power applied to it the time required for ascending the river will not be shortened so much as is anticipated; for it is evident, from the foul state of its bed and the strength of the current in many parts, that frequent delays will occur. From the continual alterations that take place in the channels it is often difficult to find and follow them; and the boatmen seldom possess a sufficient knowledge of their direction to act as pilots. In dropping down the river in the dry season they always have a small boat sounding a-head; but even with this precaution they often run aground, where a few months before there had been abundance of water, and much difficulty is at times experienced in getting into the deep channel again. From the prevalence of strong northerly winds in the dry season steamers are better adapted for navigating the Indus than sailing vessels; and it will be seen from the description that has been given of the main stream, that one drawing 6 or 7 feet might enter it by its own mouth, the Kúkéwarí, or by passing through the K'hédíwárí branch, and ascend it as high as the Títíyah without difficulty, even when it is in its lowest state. The mouths, it is true, undergo great changes annually, and this will always render them difficult of access; but it appears to me that, as they must always discharge the same body of water, the channels, although they may alter their position, will generally be of the same average depth. Although steamers of the draught I have mentioned may navigate the river up to the Títíyah, I do not think any that draw more than 4, or, at the utmost, 5 feet, will be able to ascend it to Ḥaïder-ábád in the dry season. In sounding across it in several places below that city, no greater depth was found than 7½ and 8½ feet; and this being the case, it is probable that a continuous channel does not exist of more than 5 or 6 feet; for it appears to be a peculiarity of this river to throw up narrow bars or ridges, stretching obliquely across it, from bank to bank. Two have been discovered in the main stream in a distance of 35 miles, and in a part that is both deeper and more free from shoals than any other: they are also met with in the K'hédíwárí, and in the Kórí, or eastern branch, they are both more numerous and on a much larger scale. Steamers built expressly for the navigation of the Indus ought to be constructed with flat bottoms; and if it is intended to employ them as tugs, they should possess a speed of at least 9 miles an hour.

On the banks of the river fuel is scarce, and, except in the shikár-gáhs or game-preserves, there is no large wood. In those spots it is abundant; and if permission could be obtained to take it away, no scarcity would be experienced, in the event of steamers being introduced, for some years. It is not, however, likely that the Amírs will grant any request that may be made to them on this point; for in all the late treaties there was nothing about which they showed so much anxiety as the preservation of these hunting forests. In the Delta the tamarisk is the only resource: it seldom attains a great height or thickness, and as a single bush affords but few billets; it requires considerable time to cut down a cargo. During the late trip of the Indus steamer the wood of the bábul (*mimosa Arabica*) was found to answer admirably, and that of the tare or mangrove was equally good: the latter, however, is scarce, the tree being only found near the sea. At present, if proper arrangements were made, a sufficient supply of fuel might, perhaps, be commanded to keep two small steamers constantly plying.

Sailing vessels ascending the river in the dry season, when the winds are blowing strong down it, can only proceed by tracking, and seldom make a greater progress than 12, or at the utmost 14 miles a-day. This method, although extremely tedious, is well suited to the peculiarities of the Indus, and, as now performed, admits of little or no improvement. The only danger is experienced in the bends of the river, where the current sometimes runs in strong eddies, and is continually exerting its influence to undermine the banks. If a boat is driven with force against the steep bank in these places, and a large mass becomes detached by the concussion, which sometimes happens, her destruction is almost inevitable.

The form of the dúndí, or flat-bottomed boat of the country, is well adapted to the navigation of the river, and there is no kind of vessel better calculated for the transport of goods. Many of the largest are 80 feet long, and 60 tons in burthen. They have no keel, and both the bow and stern, which are perfectly flat, rise from the water at an angle of about 30 degrees. They are very high abaft, where there is a small deck, and are generally steered with a long curved oar. The boatmen are very expert in handling it, and frequently propel the dúndí with it alone. These vessels have also a broad triangular-shaped rudder, which, as there is no sternpost, is hung over the slanting stern, and moved by ropes on each side. The masts are stepped on a plank secured at each end to the gunwale, and are supported entirely by ropes: the after-sail is square and very large; the foresail of a lateen shape; and in order to give them greater power they are usually hoisted abaft the mast. When laden these vessels do not draw more than 4 feet. The chief defect in their construction is the small power possessed by the rudder, which arises from the great breadth of the stern below the water-line. This might easily be remedied; but any great deviation from the general principle on which they are constructed would not be an improvement. From the scarcity of large trees, and the high price of teak plank, the workmen are obliged to use the small wood of the country in building their boats; and most of them are formed of innumerable pieces, fastened by bamboo pegs, nails being only employed to secure the knees and ribs. They are in consequence liable to many accidents that would not affect their safety if they were constructed of better materials; and in the lower part of the river there are about forty or fifty lost annually. Dúndís are hired at so much per karwár (about three-fourths of a ton), in proportion to the length of the trip; and the rates, as now charged, are as follow:

From Bander-Vikkar	to T'hat'hah .	$1\frac{3}{4}$	rupees per kurwar.
,, ,,	to Haïder-ábád	$2\frac{1}{2}$ to 3	,, ,
From Haïder-ábád	to Sihwan . .	$2\frac{1}{4}$ to 3	,, ,,
,, ,,	to Lárkhánah .	4 to 6	,, ,,
,, ,,	to Shikárpúr .	6 to $6\frac{1}{2}$	,, ,,

These rates are for ascending the stream in the dry season: the crew attached to each boat is never large enough to manage her properly, and extra men must always be hired, who are paid and found in provisions by the employer.

The jumtís, or state barges of the Amírs, are of the same form as the other flat-bottomed boats, and some of them are large and commodious. I saw one with four masts that measured 90 feet in length; and the natives told me there was another at Haïder-ábád 20 feet longer. These vessels are constructed entirely of teak, brought from the Malabar coast, and are much better built than any other description of boat on the Indus. They have generally two large open cabins, or rather pavilions, on deck, elaborately carved all over, and furnished on all sides with silk curtains. The foremost one, being considered the post of honour, is always occupied by the Amírs; and that aft, which is the largest, is devoted to the use of the chiefs and followers in attendance. In going down the river against the winds they are impelled by four or six oars, each of them so large as to require five men to pull it; and on these occasions they are always gaily decorated with flags and streamers innumerable.

The periodical rise of the Indus has been ascertained by Lieut. Wood, who was directed to remain in Sind for that purpose; and as his observations have already been published, I shall confine my remarks on it to the effect it has in the lower part of the Delta. At Ḥaïder-ábád it was found to be 15 feet 3 inches. Above the Delta the water only rises to the level of the banks when the swell has attained its height, but a short distance below T'hat'hah overflows them, although not to any great extent. At the confluence of the Síyáhan, 22 miles from the sea, where the banks are 11 feet high, the rise is 13 feet, and the whole country below this is inundated. It decreases towards the sea in proportion to the slope of the land, for at Vikkar it is only 6 feet above the highwater mark of the ocean tides; and the low flats on the coast, which are often flooded even in the dry season, are seldom covered to a greater depth than 2 feet. None of the permanent villages are removed during the continuance of the swell, but are merely defended from the water by a bank of earth thrown up around them, and the cattle are turned out to roam about in search of food. Throughout the Delta the water is retained, and the supply to the different fields regulated, by dykes thrown up along the banks of the rivers, which in some places are 6 feet high, and extend in a continuous line for a distance of 40 miles; they also serve the natives as a road. During the swell the flood-tide causes a heavy bore at the mouths of most of the rivers, but it quickly loses its force, and does not ascend them to a greater height than 6 or 7 miles.

Although the water of the Indus is extremely muddy in appearance the quantity of soil suspended in it is not so great as might be expected. When reduced to the consistence of stiff mud, or the same state in which it is deposited on the banks, I found it at the Kúkéwárí mouth to be 3 cubic inches in a cubic foot of water; but the mean of several experiments, made above the influence of the ocean tides, gave only 2½, and this is, I think, about the average quantity in the dry season. During the inundation the amount is much larger, being 4 cubic inches, which is no doubt caused by the greater strength of the current at that period.* On examining the soil obtained from the water, Dr. Heddle found it to be composed of argil, or fine clay, and carbonate of lime, with a quantity of mica in the form of fine sand. It also contains a portion of common salt, with carbonate of soda

* Taking 3 cubic inches as the average throughout the year, and the quantity of water discharged by the river, as given by Lieut. Wood, I find, by a rough calculation, that the Indus conveys to the sea annually 10,503,587,000 cubic feet of mud, and this would cover a space 8½ miles square to a depth of 4 feet.

and nitre. The water, besides the mud suspended, holds in solution a proportion of saline ingredients, principally common salt, carbonate of soda, and nitrate of potash; but the amount is not so great as to render it at all disagreeable to the palate. An idea prevails very generally amongst the natives of Sind, that the water of the Indus is unwholesome.

The changes that have occurred in the branches of the Delta within the last twenty years are most remarkable, and exhibit the inconstant character of this celebrated river in the strongest light. I was fortunately able to obtain some information respecting them; but as few of the natives possess any knowledge of the country beyond the immediate vicinity of their own villages, I could not in some instances trace the course of the destroyed rivers to my own satisfaction. About twenty years ago the main stream of the Indus flowed to the sea by the Bagár, and the Sátá or great eastern arm of the maps, which at present forms the lower part of it, had no existence:* this fact is asserted by all the natives, and is corroborated by the report of the gentlemen attached to Mr. Smith's mission, who found the river in 1809 emptying itself by the Pítí or westernmost mouth. At this period the Bagár threw off numerous branches, and some of them were navigable for large vessels. The Sátá was the first that quitted it below T'hat'hah: it was not very broad, and after pursuing the same course as at present for a few miles, turned more to the eastward, and traversed the country between the Mall and Pinyárí, in a S.S.E. line. Ten or 12 miles farther down, the Bagár sent off a larger branch, and this divided into two streams a short distance below its confluence: the Mugrah, which was connected with the Sátá, by the Killán river, was one of them, and the Náïr the other: the latter appears to have emptied itself into the Mall branch. Between the Sír and Mall mouths, a distance of 35 miles, no rivers are now discharged, but there are several salt-water creeks, or rather inlets, running 8 or 10 miles into the land. The Sátá and Mugrah must have reached the sea at this part of the coast, and these creeks are, in all probability, the remains of their mouths and lower parts. The account given by the natives of the changes that have occurred in the other branches of the Bagár is extremely vague and unsatisfactory: throughout the line of the Hajámarí portions of destroyed rivers are met with everywhere; but the clue afforded by them is so slight, that it is not sufficient to trace their entire courses with any degree of accuracy, and until farther information is obtained on the subject, I shall not attempt it.

From the report of the natives it appears that a very high inundation sometimes occurs in Sind, which invariably causes great

* In some old maps I obtained at Bombay, the Sátá is laid down as a small creek, and pursues quite a different course from what the main river does at present.

alterations in the lower part of the Indus: it is said to happen once in about half a century. About eighteen years ago* one of these floods came down; the river rose several feet above its usual height during the swell, and the strength of the current was much greater than in ordinary seasons: whole villages were swept away from the banks, and in many parts of the country the crops were completely destroyed. On this occasion the river altered so much about the part where the Sátá was thrown off, that a larger body of water than usual was forced into that stream, and it increased in size considerably; the change became greater every successive year, until at last the main river turned into the Sátá and abandoned the Bagár altogether. It did not, however, pursue the same course as that branch for many miles, but forced a passage for itself, nearly in a straight line, through several creeks, across the Mugrah and Näïr, into the lower part of the Górá river, and shortly after opened a new mouth, the present Kúkéwárí. Before this happened, many of the branches were navigable for large ships, and at an earlier period were frequented both by the Company's cruisers and merchant vessels. The rulers of Sind had also a fleet of fifteen ships stationed at Sháh-bander, which owes its name (the King's Port) to that circumstance, and it is mentioned in the histories of the country that they sometimes ascended the river as high as T'haṭ'hah. The line of route they pursued from the sea to Sháh-bander is accurately pointed out by the natives: they entered the Richél, the only accessible mouth, and passing into the Hajámarí, through what is now the Kejeïrí creek, ascended that river to a part about 10 miles above Vikkar, where it joined the Bagháná, or, as it is now called, the Mall, on which branch, but considerably lower down, Sháh-bander was situated; they could also pass into the Górá river from the Hajámarí, and navigate it down to Bétrí, then a large town. At this period the Richél mouth, which is now nearly closed by a sand-bank, had a depth of 4 fathoms, and there was a high beacon erected on the south point to facilitate the navigation: this, from its resemblance to a minaret, the natives called Múnara: no trace of it now remains, but its name has been retained in that of a village built near its site. Such are the alterations that have occurred in the lower part of the Indus within the last eighteen years.

Having completed the description of the rivers I have examined, and given some account of the changes they have undergone, I shall now proceed to offer a few general remarks on the Delta and its inhabitants. The country on the sea-coast is submerged at high tide for a distance of 3 or 4 miles, and continues

* It is almost impossible to obtain the exact date of any occurrence in Sind from the natives; but if their assertions are correct, this flood must have happened in 1819, the year in which the great earthquake was experienced in Kach'h.

swampy about 2 miles further in-land. These marshy tracts, which in most places are destitute of bushes, afford excellent pasturage for large herds of buffaloes, and on that account are considered by the peasantry as valuable property: notwithstanding their dreariness they often present a greater appearance of animation than is observed in more populous parts of the country, for every creek is full of boats, and men are seen in every direction cutting grass for the cattle of the distant villages. It is of a coarse thorny kind, that grows to the height of 16 or 18 inches, and is only obtained close to the sea, for a short distance inland the small rush springs up in such abundance that it is impossible to separate it: being impregnated with salt it is preferred by the cattle to the herbage of the upper parts of the country, and the natives say they thrive on it much better. The districts immediately above these swampy plains are the most productive in the Delta, and a great part of the land being free from jungle, which here only appears in patches, is easily brought under cultivation: they terminate about 12 miles from the sea, and the dense mass of jungle with which the whole of Upper Sind is overrun, commences. It is principally composed of the tamarisk-bush, mixed here and there with the bábul; but the latter in some places is met with in large quantities: the mangrove is only seen on the coast, where there are also a few saline shrubs of the same species as those so common on the shores of Arabia. The soil, in general, is composed of clay mixed more or less with sand, and contains a large quantity of salt, which frequently appears in a thin crust on the surface: in a few spots it is a stiff tenacious clay, that hardens when exposed to the sun, and turns white; but in the upper half of the Delta, where it contains a very large proportion of sand, it crumbles into a fine dust, and in strong winds rises in clouds that obscure the atmosphere and penetrate everything. The soil of Lower Sind is not, in fact, so rich as has been represented; for, although a large quantity of rice is produced, it is all of a very coarse inferior quality, and the esculent vegetables are small and tasteless. Much of the land lies waste and neglected, and in many places, spots that have been already cleared for tillage have been allowed to return to a state of nature: with a little labour they might again be rendered available for agricultural purposes, but the system pursued by the government and its agents towards the peasantry discourages every effort, and paralyses all exertion. Besides rice, which is of two kinds, white and red, barley, múng,* and bájerá,† are raised in the dry season by irrigating the fields from cuts to the river, and the sugar-cane and castor-oil plantain cultivated to some extent. There are a few fruit-trees in the upper part of the Delta, such as the mango, plantain, and tamarind, but the fruit is inferior.

* Phaseolus Mungo. † Holcus spicatus.

In the winter season the climate of the Delta is delightful, being cool, dry, and bracing; the temperature ranges from 45° to 82°. and during the day is most agreeable: fogs sometimes occur, but they are by no means prevalent, and quickly dissipate as the sun rises. In the summer months the heat is excessive, and less rain falls than might be expected. During the inundation the climate is very unhealthy: fevers, dysentery, and agues prevail, and all the inhabitants that reside constantly in the Delta have an appearance of premature old age, which is doubtless to be ascribed to this cause. None of the chiefs or wealthy land-holders remain there during the hot months, but repair to Haïder-ábád, and do not return to their estates until the water left by the swell has dried up. The dryness of the atmosphere may be seen in the annexed abstract of a meteorological table.

The population of the Delta is miscellaneous, being composed of Hindús and Jats, with a mixture of various tribes from the adjacent countries. Besides Belúches and Jókíyahs from the western side of the Indus, many parties are met with that have emigrated at different periods from Kach'h; and there are also a few villages inhabited by Patháns or people from Afghánistán, the Panjáb and Bikánír. The Belúches are generally fine-looking men, with a Jewish cast of countenance, aquiline nose, high forehead, and large expressive eyes: in stature they are somewhat above the middle height, and their make bespeaks more activity than muscular strength. They are a rapacious, vindictive, and dissolute race, and as far as I have been able to form an opinion of them, appear to possess few good qualities. The soldiers have a more martial appearance than most Asiatics, and are undoubtedly brave, but they would have no chance against regular troops, from the total absence of discipline amongst them. This tribe, which conquered Sind about sixty years ago, is composed of many small sub-divisions or rather families, but none claim or exercise any superiority over the rest, and they are found mixed together throughout the country; the largest and most powerful is the Laghari, whose chief resides some distance above Haïder-ábád. There is no peculiarity of physiognomy observable in the Jats to distinguish them as a distinct people: amongst them you see every cast of countenance that denotes ignorance and stupidity, and they are now perhaps the most miserable, superstitious, and degraded race in India. They are a branch from the Jats or Játs, a Hindú tribe that anciently occupied the country about the upper part of the Indus, and were converted to the Mohammedan faith on the subversion of the Hindú dynasty in Sind. The Jókíyahs belong to a Rájpút tribe that formerly governed the country; but in the Delta they are not numerous. The Hindús are of the Lowáná and Bhattiyá castes, and do not differ from those of British India: patient, persevering, and industrious, they are in possession of nearly all the trade of the country, and form by far the most wealthy class in it. All those following any occupation on the river are called Mowánás: the men that navigate the dúndís are generally natives of the Panjáb; but the name by which they are distinguished in Sind is unknown there. In enumerating the tribes of the Delta, the Sayyads and Fakírs, or religious mendicants, must not be omitted, since they form a large proportion of its population, and may be considered almost a distinct class. Their number is estimated by the Sindians at 100,000, which is perhaps somewhat exaggerated: it must, however, be considerable, for large parties are frequently encountered travelling about to collect alms, and there are many villages entirely inhabited by them. The Sayyads, or descendants of the Prophet, are generally shrewd and intelligent men, and there is no peculiarity about them either of dress or appearance by which they can be distinguished from the wealthiest class of Muselmán inhabitants. Most of them possess grants of land, and they also derive a considerable revenue from the offerings of the pious. These impostors had formerly great influence with the rulers of the country, but it is now on the wane, the present Amírs not being so bigoted and superstitious as their predecessors: it will be fortunate for the country when it ceases to be felt altogether, for they add by their exactions to the misery of the lower classes, without conferring the slightest benefit on them in return. The Fakírs pursue an erratic life, subsist entirely on the charity of individuals, and disfigure their bodies in the same manner as those in other parts of India: their principal place of resort is the shrine of Lál Sháh-báz, at Sihwan, where there are seldom less than five or six thousand collected, who receive their food daily from the temple.

The condition of the lower classes of the peasantry in Sind is truly wretched: unable at times to obtain a sufficiency of food and clothing for themselves, it is quite out of their power to provide for the wants of a wife and family, and they never marry. The consequence of this miserable state of existence may easily be imagined: theft is common, and many of the villages are full of public women of the lowest description, living in a state of hopeless poverty and disease. Very few fine-looking men are seen in the Delta, and the women are nowhere so destitute of personal charms: they are ugly and haggard even in youth, which is to be attributed to the hardships they undergo in early life, and the unhealthiness of the country during the swell. Smoking is universally indulged in to excess, and the strong spirit distilled from gaur* is in great request amongst all who can afford to purchase it. The lower orders use bang, an intoxicating and very deleterious drug obtained from hemp, in large quantities. In

* Molasses, or raw sugar.

most of the towns there are numerous dancing girls, and the only amusement of the inhabitants consists in smoking their capacious hukkahs, and drinking until they are intoxicated, whilst these women exhibit their indecent postures before them: the pleasures of the Sindian are, in fact, entirely sensual, and his rank in the scale of civilization is consequently very low.

Amongst the animals of Sind the camel, both from its size and utility, ranks first in importance: very few are seen near the sea-coast, but in the upper part of the Delta droves of forty or fifty are frequently passed: they are rather smaller and lighter in limb than those of Arabia, but from being better fed are much finer looking animals. The horned cattle do not differ from those of India. Horses of a diminutive breed are met with at every village in great numbers. In the woods wild hogs abound, and there is also an animal very common in the interior which, from the description, must be the elk. The country is infested with jackals, who prowl about day and night, and are very ravenous: on several occasions they have been known to attack men. A lynx and a leopard were seen, and tiger-cats three or four times; but none of these animals are numerous. Hares and deer are abundant everywhere. The dogs are large and ferocious; so much so, that it is dangerous for a stranger to approach them without being accompanied by some of the inhabitants.

It is almost impossible to convey an idea of the vast quantities of water-fowl that frequent the rivers and the swamps of the sea-coast: the latter are literally covered. The varieties of the wild duck are innumerable, and geese, pelicans, flamingoes, spoonbills, storks, cranes, royal and grey curlews, herons, snipes, with several other kinds, are seen in immense numbers. The Egyptian ibis is common. There is also a large bird about the size of a turkey, called the kalam, which only makes its appearance in the winter months, and is found both in wet and dry situations: I have seen the fields actually covered with them. In the thick woods of the upper part of the Delta partridges, quails, and plovers are equally numerous. Among the small birds are some with very rich plumage, which I think are unknown. The domestic fowl is remarkably fine, and generally of the kind with black bones and skin.

The fish obtained in the greatest quantity is the delicious pulla: it is of a most delicate flavour; but so rich from the quantity of oil it contains, that some people cannot eat it. The mullet is also abundant, and attains a large size. Besides these there are several other kinds that are common to most of the Indian rivers. Otters and turtles abound everywhere, and porpoises are seen as high as T'hat'hah. There are also great numbers of water-snakes of a large size in all the rivers.

Kórí.—The Kórí, or eastern branch of the Indus, was surveyed in 1833, and as it exhibits some physical peculiarities not observable in any of the other rivers, I shall conclude this paper with a short description of it. The Kórí, which separates Sind from Kach'h, once formed the lower part of the Feleïlí, and it also received the waters of a large branch, thrown off by the main river during the inundation near B'hkur: the beds of both these branches are now partially filled up throughout the whole line of their course, and the portion of water they receive during the swell is prevented from passing into the Kórí by bands that have been thrown across them by the Sindians. The alterations caused in this branch by the earthquake of 1819 increased its magnitude so much, that it became a small gulf or arm of the sea, and it now gives a better idea of a great river than any other branch of the Indus. At its mouth it is 6 miles wide, and the Sind coast, being very low, is not visible from the Kach'h side: it begins to contract at Kótásír and continues to do so up to Lak'hpat, a fortified town situated 39 miles from the sea, where it diminishes to a narrow stream, 200 yards wide, and is so shallow that, if the bottom were firm, it might be forded at low water in several places without difficulty. The broad bank fronting the coast of the Delta extends right across the entrance, and terminates a short distance below it on the shores of Kach'h: in many places the sand-banks are dry at low tide, and the sea outside them is very shallow, there not being a greater depth than 5 fathoms 8 miles from the land. Two channels, the Ad'hiyárú and Súr, lead out of the river across this mass of shoals; they are broad and deep, having a depth of 20 feet in the shallowest part, which is on the bar at their mouths. Above the part where these two channels separate, the river for some miles has a depth of 7 and 8 and in some places 14 fathoms, and there are no sand-banks until you arrive at Kótásír: near this town the bed of the stream is extremely foul, and vessels of any size are prevented from ascending higher by several shallow bars or ridges, that reach across it in a lateral direction from side to side. A short distance above these, extensive flats of soft mud occupy half the breadth of the river, and the channels being narrow and intricate, the difficulty of the navigation increases every mile as you advance towards Lak'hpat.

During the neaps the tides are very irregular; they run at a rate of from 2 to 3 miles, vary in duration from 4 to 8 hours, and rise sometimes 9 feet, but at others only 6. On the springs they are alternately weak and strong, and in the latter case, when they have a velocity of 6 miles, they continue to flow for a period of 8 hours: this is caused by the sudden influx of water from the ocean during the night, when they attain their greatest height. At Kótásír there is a rise and fall of 10 feet, and at the mouth of the river it increases to 13, but at Lak'hpat it does not exceed 4. The Kórí is navigable for vessels drawing 16 feet to within a short distance of Kótásír, but they could not proceed any higher on account of the bars stretching across the stream from side to side: even the country boats that frequent it, which seldom draw more than 6 or 7 feet, are obliged to remain at that town and send up their cargoes to Lak'hpat in dúndís.

About the mouth of this branch the land is low and swampy: on the Sind side it is overrun with a dense mass of stunted mangrove bushes and overflowed by every tide: the Kach'h shore for some miles is a mere ridge of loose sand, thrown up between the river and a broad tract of marsh land intersected with small creeks, that extends from Kótásír to Jakaú. None of these creeks have any communication with the Kórí; but the largest, which reaches the sea a few miles below the mouth of that river, affords a passage for the boats of the country to Jakaú, a seaport town of some importance in the N.W. extremity of Kach'h. Along that part of the sandy ridge exposed to the action of the sea, the beach is literally covered with heaps of clay balls, that have apparently been formed by the waves rolling the small pieces of clay, detached from the banks and thrown up here, on the smooth firm sand.

About the mouth of the river the country on the Sind side, up to Lak'hpat, is low and flat, and thinly covered with saline shrubs or the decayed trunks of bushes that have been destroyed by the salt water: this part of the Delta is evidently depressed below the level of the rest, which is to be attributed to the effects of the earthquake of 1819, and being from this cause partially flooded at times to a great extent, even in the dry season, is uninhabited. On the Kach'h side, the country above Kótásír presents to the view a confused mass of rugged broken hills, evidently of volcanic formation, which at a distance assume the appearance of moderately elevated table-land: here and there the small spurs sent off from them project in low promontories into the river, and the rocks extend under water half way across its bed: the only mountains seen are the J'hárah hills, situated 8 miles to the S.E. of Lak'hpat, which are about 1000 feet high.

From Lál Chettah, a small mosque 10 miles above Kótásír, the low hillocks on the Kach'h side run in a direct line up to Lak'hpat; but the river, receding gradually from them for some miles, and then turning suddenly towards that town, leaves an extensive tract of low land at their bases of alluvial formation: this is said to have sunk several feet in some places during the earthquake, and a small fort in the upper part, erected close to the river, the ruins of which are still visible, was overthrown. During the inundation it is now covered with water to some depth, and in the dry season salt is obtained in large quantities, from those parts that are below the level of the river. Near Lál Chettah, where the stream is 2½ miles wide, a large rock covered with oysters rises in the centre of the deep channel, and on the opposite side of the river there is a ruined fort, called Bastah Bander, which formerly belonged to the Ráós* of Kach'h, and was destroyed by the Sindians during their wars with those princes.

Kótásír is a small village and pagoda on the Kach'h side, situated about 16 miles from the sea, and is dependent on Naránsír,† a fortified town full of pagodas, about a mile farther inland: they are both celebrated places of pilgrimage of the Hindús, and are said to be of equal antiquity with the most ancient of the Hindú temples in other parts of India. The pagoda at Kótásír is erected on a small rocky eminence close to the river, and, like all these edifices, has been constructed with more solidity and strength than elegance: a broad terrace runs round it, defended by a low massive wall with embrasures, and in these several small guns are mounted. One side of the hillock on which it stands is bounded by the swamp that now occupies the site of the old bed of the river, and before it, a mud flat extends from the banks, about a hundred yards into the stream: across this flat a handsome stone causeway has been carried out to another but smaller temple, where there is a large tank built in the river for the convenience of the pilgrims when performing their ablutions.

The fortified town of Lak'hpat is built upon the edge of the elevated land, about a mile from the river, and incloses a space about 800 yards square, of which not more than a third is occupied by houses. It is of an irregular shape, and the walls are defended by numerous towers and bastions, with guns mounted on them of all sorts and sizes. Most of them are so old as to be entirely useless; but one, a long brass six-pounder, with the arms of Portugal engraved on it, deserves some notice, from the peculiarity of its construction. In the upper part of the breach there is a square piece cut out, about 2 feet long, and 8 inches wide, which admits of a small gun being placed inside the large one, with the muzzle projecting about a foot beyond the open part. It is furnished with a handle, and appears to have been contrived for loading with greater celerity and safety. Lak'hpat was built about thirty-five years ago, by Jem'ahdár Fateḥ Moḥammed, a celebrated Kach'h general, to defend the frontiers of the kingdom against the encroachments of the Sindians. It is now garrisoned by 50 Arabs and 150 native soldiers, and contains a population of about 5000 persons, composed principally of merchants and

---

* Princes or chieftains.

† Naráyan-sir?

Hindús, who have fled from Sind to escape the tyranny of the Amírs. About the town the country is barren and unproductive; and, from the quantity of shells found in the soil in most parts, has probably at a remote period been submerged. Kóterí, the landing-place on the Sind side, where there is a small custom-house and guard, is 4 miles below Lak'hpat, and numerous ferry-boats are constantly passing between them, full of men, cattle, and merchandise. From this station the goods are conveyed on camels to the eastern parts of the Delta, and distributed throughout the inhabited districts of the T'har.*

The effects of the earthquake that visited Kach'h in 1819, and laid most of its towns in ruins, are visible in every part of the Kórí. Opposite Kótásír the banks of the river on the Sind side are perpendicular for about three miles, and close along them there is a depth of 14 fathoms. In this part the land is of alluvial formation; but all the strata exposed to view in the face of the banks, with the exception of two or three of the upper ones that have been deposited since, are broken up in confused masses, and inclined to the horizon at an angle of 30 or 40 degrees. This is also the case throughout the tract of low land lying at the base of the hillocks between Lak'hpat and Lál Chettah. Previous to the earthquake, the river, instead of pursuing the course it now does, turned close round the rocky eminence on which Kótásír is built, and reached the sea between its present mouth and the Jakaú creek. Along this line the country is overflowed at high tide to a depth of 2 or 3 feet; and the old banks may still be traced. The alterations that have taken place in this part of the river are very extensive, and have evidently been produced by the sinking and upheaving of the ground during this awful convulsion of nature. The fact is attested by the remains of several boats, which are still visible, half buried in the soil that fills up the bed of the old river, and it is probable that to this cause the Kórí owes its present magnitude.

Abstract of the Meteorological Journal, in the Delta of the Indus, 1836-7:—

		Dec.	Jan.	Feb.
Thermometer (Fahr.)	Max.	82°	83°	82½°
	Mean	65	64	63
	Min.	49	50	49
Evaporation	Max.	·620	·520	·780
	Mean	·320	·330	·330
	Min.	·210	·155	·100

* Desert, from the Sanskrit St'hala.

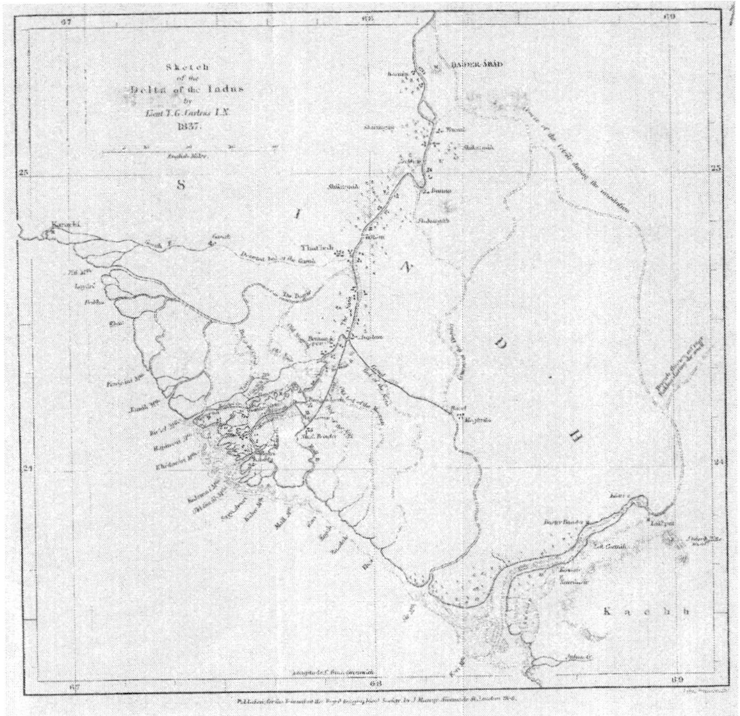
Sketch
of the
Delta of the Indus
by
Lieut. T. G. Carless I. N.
1837.
English Miles
S
I
N
D
H
Karachi
Thatteh
Hyder-Abad
Kachh

XXVI.—*Report on* Mr. Ainsworth's *Observations on the Magnetic Intensity of the Earth, made during a Journey across the Continent of Europe in* 1838. By Major Edward Sabine, R.A.

Mr. Ainsworth has accompanied the manuscript record of his observations by the following description of the instruments with which they were made:—

"The observations of the dip were made with a needle $4\frac{5}{10}$ inches in length, with a steel axle, supported by agate cylinders. This instrument, which had previously accompanied the Euphrates expedition, was made by Robinson of London. At the time of observing, it was always adjusted to the magnetic meridian by the compass, and also by taking a position at right angles to the vertical position of the needle.

"The observations on the intensity of the horizontal force were made with two pair of cylindrical needles, vibrated in small arcs, commencing with 20°, in an apparatus similar to M. Hansteen's. In these observations every care was taken to prevent errors from difference of strength in the supporting fibre of silk, in the distance of the needle from the bottom of the box, and in the equipoise of the cylinder as well as the level of the box itself. One pair of these needles, E 1 and E 2, had been previously used in the Euphrates expedition. The other pair, A1 and A2, were made for the Kurdistán expedition, by Mr. Robinson, and brought to a steady, although with regard to one another, an unequal force, by immersion in boiling water. Successive experiments made on the same and on different days at Westbourn Green, near London, gave irregular results before this immersion; but the force at the same place became afterwards pretty constant at the same temperatures. The needles were $3\frac{3}{10}$ inches in length."

On the 9th of September Mr. Ainsworth, being then at Constantinople, and having completed his series of intensity results across the continent of Europe, enclosed needles E 1 and E 2 in a packet addressed to me, under cover to Captain Beaufort, Hydrographer to the Admiralty, for the purpose of having observations repeated with them in England, and compared with those made before his departure, in order to examine the steadiness of their magnetism in the interval. They reached me at Tortington, in Sussex, on the 18th of October, and were immediately examined as follow:—

		h. m.		Degr.		Seconds.
E 1,	Oct. 18,	1 40 P.M.	Therm.	53.5	100 vibr. in	559. 5
	Oct. 20,	1 50 P.M.		64.0	. . .	560.83
				58.75		560. 2
		h. m.		Degr.		Seconds.
E 2,	Oct. 18,	2 32 P.M.	Therm.	53	100 vibr. in	562.83
	Oct. 19,	4 40 P.M.		56	. . .	563.2
				54.5		563.0

Mr. Ainsworth's journal contains the following record of observations made with these needles at Westbourn Green, near London, before his departure from England:—

		h. m.		Degr.		Seconds.
E 1,	May 30,	4 28 P.M.	Therm.	67	100 vibr. in	559.5
	May 31,	10 15 A.M.		64.5	. . .	561.5
				65.75		560.5

		h. m.		Degr.		Seconds.
E 2,	May 30,	4 15 P.M.	Therm.	67	100 vibr. in	561.0
	May 31,	10 28 A.M.		64.5	. . .	567.5
	June 1,	10 40 A.M.		65	. . .	566.5
				65.5		565.0

From the well-ascertained difference in the horizontal intensity at the two stations, Tortington and Westbourn Green, these needles ought each to have made the time of performing 100 vibrations about 4 seconds *less* at Tortington than at Westbourn Green; whereas, when corrected for the difference of temperature, the times at Tortington are nearly the same as those in London. So far there is an indication of a slight loss of force sustained during their continental journey. It is, however, so slight,—not exceeding probably the fluctuations of the horizontal intensity itself at different seasons, and on different days,—or at least, not exceeding the limits of the differences which *these needles* have occasionally shown at the same place on different days,—that I have not thought any correction on this account called for.

Table 1 contains the times of vibration of the four needles at the stations observed at by Mr. Ainsworth. Table 2 contains the conclusions drawn from them. The values of the intensity are expressed in relation to the force at Paris, as Paris has been taken as a fundamental station by previous observers in the same parts of the European continent; and Mr. Ainsworth's results are thus brought into immediate comparison with theirs. His observations at Paris were made in M. Arago's magnetic cabinet, which is the usual place of observation there. In computing the horizontal intensities by E 1 and E 2, the times of vibration at the different stations have been corrected to a mean temperature, employing the co-efficient ·00038 found for E 1 in reducing the results obtained in the Euphrates expedition. As E 1 and E 2 were made of the same steel, are of the same dimensions, and have so nearly the same magnetic strength, the same co-efficient has been considered applicable to both. In the final results, which occupy the last column of Table 2, and exhibit the total intensities deduced from their horizontal components by means of the observed dips, the observations with E 1 and E 2 have been alone employed, as these needles are the only ones which have received the necessary verification of their unchanged rates. But the horizontal intensities computed by A 1 and A 2 are also given in separate columns, both distinctly and combined, for the purpose of comparison with those of E 1 and E 2. We may infer from the comparison that A 1 and A 2 both altered their magnetic state considerably between London and Paris; but as, when their results are combined, the difference of the force at those stations is the same by A 1 and A 2, as by E 1 and E 2 (·932 to 1.000), we may also conclude that what the one lost in force the other gained. This is, I believe, a very common circumstance with needles when first put together in pairs: they undergo more or less alteration, causing their respective forces to approximate; and this change takes place very shortly after they have been put together, their differences being subsequently pretty constant. Such appears to have been the case with needles A 1 and A 2, which, after Paris, present a tolerably uniform relation to each other. They have probably, on the whole, lost a little force: we may infer this from the general comparison of their results with those of E 1 and E 2: but they show a regularity which gives promise of their being very useful needles in the hands of so diligent and careful an observer as Mr. Ainsworth.

In the journal of the observations at Isella, in the pass of the Simplon, it is remarked that the dip instrument had been rudely examined by the Custom-house officers at the Sardinian frontier, and that the dips in the different positions of the instrument were not so accordant as previously. Mr. Ainsworth complains of this also in a letter from Constantinople; and the injury which the needle appears to have received at that time is traceable throughout the subsequent observations.

TABLE 1.

Station.	Date.	Time of 100 horizontal vibrations.				Therm	
		E 1.	E 2.	A 1.	A 2.		
	1838.						
London . .	May 29 / June 7	560·5	565	367·25	385·5	66	Westbourn Green.
Paris . . .	June 20-21	541	546·5	357·75	369	68	Observatory.
Chalons . .	July 2	530·5	532	350·5	360·25	72	Near the Säone.
Geneva . .	July 10	522·5	525.5	346·75	358	74	Observatory.
Isella . . .	July 16	525	536·5	349·25	359	80	S. side of the Variola.*
Milan . .	July 20	517.5	527	345	356·25	82	Botanic Garden.
Verona . .	July 23	516	524	not obsd.	not obsd.	74	Convent Garden.
Venice . .	July 26	510·5	517·5	341·5	351·5	71	Botanic Garden.
Ober Laybach	July 29	510	513·5	340·75	350·5	69	Inn Garden.
Vienna . .	Aug. 8	521	526·5	348·75	357·75	71	Bn. Hügel's Garden.
Pest . . .	Aug. 14	517	525	344·25	353·75	70	Summit of Blocksberg.
Galatz . .	Aug. 27-28	493·5	499	330·75	339·5	68	B. Consul's Garden.
Constantinople	Sept. 10	466·5	456·5	307·75	319·0	71	B. Ambassador's Garden.
Tortington .	Oct. 18-20	560·1	565·3	..	..	56	In a plantation.

TABLE 2.

Station.	Hor^l. Intensity. Paris = 1·000.				Means.		Observed Dip.	Total Intensity. Paris = 1·348 E 1 and 2.
	E 1.	E 2.	A 1.	A 2.	E 1. and 2.	A 1. and 2.	° ′	
London	·930	·934	·949	·916	·932	·932	69 24	1·362
Paris	1·000	1·000	1·000	1·000	1·000	1·000	67 35	1·348
Chalons	1·043	1·042	1·042	1·049	1·043	1·045	65 25	1·289
Geneva	1·077	1·087	1·064	1·062	1·082	1·063	64 37	1·298
Isella	1·072	1·046	1·049	1·057	1·059	1·053	65 20	1.305
Milan	1·104	1·087	1·075	1·073	1·095	1·074	64 29	1·307
Verona	1·104	1·092	not obsd.	not obsd.	1·098	..	62 49	1·236
Venice	1·125	1·118	1·097	1·102	1·121	1·100	64 05	1·319
Ober Laybach	1·126	1·133	1·102	1·108	1·129	1·105	63 07	1·284
Vienna	1·080	1·079	1·052	1·064	1.080	1·058	64 16	1·280
Pest	1·097	1·084	1·080	1·088	1·090	1·084	63 15	1·245
Galatz	1·202	1·200	1·170	1·181	1·201	1·176	61 15	1·284
Constantinople	1·348	..	1·351	1·338	1·348	1·345	56 34	1·258
Tortington	·927	·925	..	..	·926	..	68 51	1·377

I have not drawn any result from the observations of E 2 at Constantinople. In every other instance the time of vibration of that needle is greater than that of E 1; but in this instance it is less. From this circumstance, and from the agreement of the results of E 1 with those of A 1 and A 2, it is probable that the observations with E 2 were on this occasion affected by some accidental irregularity.

I am able to supply the following comparisons of Mr. Ainsworth's results with those of other observers:—

Geneva,	Quetelet,	1830	1·292
	Ainsworth,	1838	1·298
Milan,	Humboldt,	1805	1·312
	Quetelet,	1830	1·294
	Ainsworth,	1838	1·307
Lohitsch,	Keilhau,	1826	1·314
Ober Laybach,	Ainsworth,	1838	1·284
Vienna,	Keilhau,	1826	1·325
	Ainsworth,	1838	1·280

The manuscript journal of Mr. Ainsworth's magnetic observations is deposited with the Geographical Society.

EDWARD SABINE.

Tortington, 23rd October, 1838.

ANALYSES, &c.

I.—*Alte Geographie des Kaspichen Meeres, des Kaukasus, und des Südlichen Rüsslands.* Von Dr. E. Eichwald. 8vo. Berlin. 1838.

[*On the Ancient Geography of the East Coast of the Caspian Sea, &c.*] Communicated by W. R. Hamilton, Esq., F.R.S.

The hydrography of the Caspian Sea generally has long been one of the great *desiderata* with geographers :—an equally interesting subject of inquiry is the examination of its eastern shore, and of the intervening tract of land between this sea and the Aral. From the time of Anthony Jenkinson in 1557—Christopher Burrough in 1579—Ssoimonoff in 1719—Peter Henry Bruce in 1723—Jonas Hanway, Woodroofe, and Elton in 1746—Gmelin in 1770—down to Muravieff in 1820—and Bassargin in 1826, we have no account of this vast inland sea, equal in extent to the whole kingdom of Spain, that can be relied upon for its accuracy. The author of the present work, although writing professedly on the *ancient* geography of the east coast of the Caspian, appears to have given the most detailed account of its *present* state that has yet been produced. In addition to his own journey* to that country, Dr. Eichwald has had the advantage of free access to all the documents in the Admiralty at St. Petersburg : it is, therefore, to be presumed, that his account may be received as the most correct hitherto published.

The first 200 pages of the work contain an abstract of all former accounts relating to this country, by the Greeks, Romans, Arabians, and modern Europeans—from Herodotus to Muravieff. They also give the detail of Colonel von Berg's expedition in 1825-6, to carry a line of levels between the Aral and the Caspian—from pages 192 to 197, the geologist will find an account of the structure of the steppe or table-land, more than 700 feet in height, which lies between those two seas—and, from pages 106 to 170, the hydrographer will find a brief statement of all

---

* See Reise, &c., von Dr. E. Eichwald.

charts of the Caspian, from that of Adam Olearius, in 1696, to the latest times—and the antiquary will discover some Arabic inscriptions from Derbend and Galatí, explained by Frœhn; and a Sanskrit inscription, recently found at Bákú, with an explanation by Bopp.

The following is Dr. Eichwald's description of the *present* state of the country, on the ancient geography of which his work professes to treat:—

The eastern coast of the Caspian Sea, with the exception of the projecting promontory of Tuk-karagan and a few gulfs, extends nearly in a direct line more than 10° in length from N. to S., between the parallels of 37 and 47 nearly. The small gulf of Aster-ábád forms its southern limit, whilst its northern boundary is marked by a very large bight at its N.E. extremity, which receives the waters of the Emba. The principal branch of this river expands before it reaches the sea, into a great many shallow basins, like lakes, the northern branch of it being nearly choked up with sand. All this part of the coast, as well as that adjoining to the N. and N.W., is extremely flat and shallow; as the large rivers, namely, the Aral, the Wolga, and the Tuck, which here empty themselves into the sea, are constantly bringing along with them a quantity of sand which is accumulated on the shores; consequently the whole of the northern part of the Caspian Sea, which is exceedingly low, together with the adjoining eastern coast, is so shallow, that for the distance of several miles from the shore, there is only a few feet depth of water, and an immense number of small sand-hills and banks of sand, make it difficult to land on any part of it. Similar sand-hills occupy also the shore itself, which extend to a considerable distance inland among the steppes; but it cannot be said that they form any connecting chains of hills.

At the very commencement of this Sinus Mortuus (Mertvoï Kúltuk), there rises a small chain of calcareous hills, called the Chink, which forms, as it were, the rampart of a remarkable high level, named Ustürt, which extends under the 45th parallel, between the Aral and Caspian Seas, with a breadth of about 160 miles. This high plain is so steep towards both those seas, that it rises above the Caspian 639 English feet, and its fall towards the side of the Aral is not less, though his last is 117 English above the level of the Caspian. This plain is never at a less elevation than 550 feet, and is in some places more than 727 feet above the level of the Caspian: it extends nearly in a direct line between the two seas, and it sinks so insensibly, that there is nowhere to be observed anything like a connected chain of hillocks, and it can only be considered in the light of one elevated plain.

The extreme headlands of this high plain—namely, the Aḳ-ṣaḳál, the Surak, the Ḳará-úl, and the Ḳará-táú, form so many small connected hill-tops round the bight of Tuk-ḳará-ṣú, which is the southern branch of the Sinus Mortuus. The summit of the Manghisláḳ and Tuk-karagan hills is still higher than these hill-tops; they form the projecting promontory of Tuk-karagan, and they rise nearly perpendicularly to the height of some hundred feet. This continuous range of hills incloses the whole coast from this spot to Alexander's Bay, and extends nearly straight from N. to S., with the exception of a small divergence to the E. They consist throughout of a recent tertiary calcareous formation, as will be more particularly described in the historical narrative of the journey. The depth of this coast is very remarkable; it is seldom so little as 6 fathoms, generally between 10 and 18; and this has been observed in Alexander's Bay.

This gulf is connected with a large bight, the entrance of which is formed by a precipitous rocky bank, from whence the bight widens considerably, and receives several rivers descending from the high plateau, namely, the Sirbásh, the Kichík and the Kumbenska.

Further east, and a little to the south, where we come upon the gulf of Kenderlin, the land shelves off, but as it is entirely surrounded by hillocks, it seems to receive no stream. Here, however, the hills of Kenderlin form again small chains connected with the larger hilly chain of the Karakhteh, and further inland they are quite lost in the elevated plateau. A coast stream, the Turakhth, here empties itself into the sea, between the Kenderlin and Ḳará-bóghá gulfs.

The narrow entrance into this supposed very deep gulf is confined by a number of rocks, amongst which are some dangerous eddies, and it is surrounded in all directions by a steep bank; along the whole of its coast there is but one inconsiderable stream, the Makranda. To the east it is bounded by some small hilly knolls, which extend from north to south; but these are less remarkable for their height, than for the great number of salt lakes, most of them very small, which are in the lower valleys.

Finally, under the 40° of N. lat. the gulf of Balḳán is bounded by the extreme points of this elevated plateau: here also the hills rise steep and precipitous from the bank, and present at the top porphyritic formations, which in remote times have broken through beds of granite, and which occur also round the gulf of Krasnovedsk, at the entrance of the bay of Balḳán, and in some of the islands in the bay; but farther to the east they are at a greater distance from the shore, and are more inland where the flat and elevated plateau prevails. Here, likewise, this plateau consists of a tertiary calcareous formation, which, towards the Balkán, crops out in single protuberances. The great and little Balkán, and two lofty and insulated eminences, being the extreme points of the Búlán-mountain, a continuation of the farthest part of the elevated plateau, and to which they are united by the chain of the Kurreh-mountain, and the Koshasúrí-mountain, form its southern limit, and shut in to the east the bay of Balkán; so that the Amu-daryá, whose old bed I followed up for 5½ miles,* could only fall into the bay between the great and little Balkán, whilst it wound round the southern point of the great Balkán, and thus emptied itself into the bay in a direction from south to north.

In consequence of the confined outlet of this old river the bay is continually more and more choked up with sand, and has scarcely a depth of a few feet, though this is somewhat more considerable at the entrance into it, about the island of Dógadán.

All this part of the coast is very much choked up with sand, and is very flat: the extreme points of the Ustürt, or the high plateau between the Caspian and the Aral, terminate at the end of the Krasnovodo-mountain, which to the north joins on to the chain of the great Balkán. This plateau again unites with the Mangojar hills, which are lost in the Aral chain; and we thus trace a continued high range of hill from the N.E. to the S.W., less and less elevated as it approaches the south, presenting at its extreme limit, near the Balkán Gulf, the appearance of a crater, and porphyritic formations raised by volcanic influence through granitic and calcareous hills. It is exactly here, where the low shores presented the least resistance, that the eruption must have taken place; and the same has been observed on the opposite or western shore of the Caspian, on the peninsula of Absheron,† whose igneous exhalations, naphtha-springs, and other volcanic outbreaks, are still in activity. Here also we come upon the lowest extremities of the great chain of Caucasus. Finally, we observe also similar porphyritic formations on the southern coast of the Caspian Sea, about the snow-capped volcano of the Demávend, which proceed eastwards by Herát through the Hindú-kush into the three gigantic mountain-ranges of central Asia, namely, the Himálaya, the Kwen-lun or Kulkun, and the Múztágh, or the mountains of ice.‡

One consequence of this great and extensive elevation on the west, south, and east coasts of the Caspian, has evidently been the great depression of western Asia, represented on the one side by

* See the Periplus of the Caspian Sea, p. 268, *sq.*

† Ibid. p. 176, *seq.*

‡ See the excellent description of the mountain-chains and volcanoes of Central Asia, by A. von Humboldt, in Poggendorff's *Annalen der Physik und Chemie,* 1830, p. 3.

the Caspian Sea itself, and on the other by the low level which stretches out far to the north, the interminable south Russian steppe, which is not indeed so deep as the Caspian, but which, as far as Saratov and Orenburg, is below the level of the ocean,* and which at the eastern slope of the Aral, which rises from it as an elevated mountain range, contains up to the sources of the Tobol so many salt lakes, evident proofs of a sea-coast in very ancient times.

The depth of this great depression of the old world increases in proportion as we approach the Caspian and the Aral seas: thus on the north-west coast there is a depth of more than 50 toises below the sea-level,† at the mouth of the Tuck; while on the north-west coast of the Aral, in the desert of Borsúk, on the eastern slope of the Ustürt, where it is lost in the Mongojar hills, there is a relative depth of 31 toises below the surface of the ocean.

The elevation of the Ustürt, which is more than 300 feet‡ above the level of the ocean, and which connects with the Aral range by the Mongojar hills, affords an easy and convincing proof that the supposition of an outflow from the Aral into the Caspian, as well as that of any river flowing into the latter from the eastward of the great steppe of the Ḳirghiz, are equally untenable. From the moment in which the Ustürt and the Mongojar extremity of the Ural were raised to their present elevation, at least during the great contemporaneous crater-like depression of western Asia, that is, since the first existence of the Caspian Sea, no river could possibly flow into that sea, either from the Aral or from the great steppe of the Ḳirghiz, or generally from central Asia.

But in the same manner as the lake of Aral has been formed by the Ṣíḥún (Jaxartes), and by the Jíḥún (Oxus, or Amú-daryá), and is at this moment supplied by them, other, and much smaller seas or lakes have been formed and supplied in the great Ḳirghiz steppe, to the east of it; thus the Aḳ-ṣaḳál lake has been formed by the waters of the Turghaï river, and two other lakes owe their existence to two other streams, the Sara-ṣú and the Chin.

The whole of the north part of the barren highlands of the east coast of the Caspian is inhabited by Ḳirghiz-Ḳozzáḳs, that to the south by Turkmáns and Khívahlís. These tribes wander about with their Aúls, and carry on the coasting-trade. A

* "Central Asia," by A. Von Humboldt.

† This must have been written before the completion of the late levelling from the Black Sea to the Caspian: from which it appears that the latter is 101 feet below the former; and if the relative difference of level of the Aral and Caspian be 117 feet, as here stated, the Aral will consequently lie about 14 feet *above*, instead of 190 feet *below*, the ocean.

‡ More than 500 feet.

large sandy steppe, called Sam, which is entirely barren, and without water, marks the commencement of the elevated plateau; further south is a large salt lake, called Jareb Guzken, which still further south has several other smaller ones in its vicinity.

The southern end of the east coast of the Caspian is devoid of any mountainous formation, is quite flat and sandy, and has evidently been once covered by the sea, from which it is continually retreating. Besides the embouchure of the old Amú-daryá, which has here, under the 39th parallel, a depth of several fathoms, a circumstance which denotes also the bed of a large river in very remote times, we have in the 37th parallel, the Atrak, which seems to rise out of a salt spring, and flows northerly from Jorján or Gurgán into the sea. As this whole extent of coast consists of a shifting sand, and is therefore extremely shallow, to a considerable distance from the shore, with a depth of no more than a very few feet, we might naturally expect to find on the coast only detached sand-hills, connected sometimes in small chains, which evidently formed in ancient times sand-banks in the sea. Among these may especially be noticed the Green, White, and Silver Hills; which last lie near the mouth of the Gurgán, which in the middle ages was an important emporium for the trade with India. Aster-ábád also, which doubtless once flourished in the south-eastern-most angle of the Caspian sea, may then have been a trading place of some consequence, though at present the river which washes its walls has a course of several miles to the west before it reaches the sea, and the bay, which is very much choked up with sand, is no longer able to admit the larger class of merchantmen.

The remainder of the first part of Dr. Eichwald's book is taken up with a dissertation on the ancient authorities for the tribes and peoples inhabiting the eastern coast of the Caspian. The second division is devoted to the ancient and comparative geography of Caucasus, and of south Russia. The work is accompanied by an ancient and modern map of Caucasus; the latter extending from 38° to 45° N. lat., and from 36° to 50° E. long., on the scale of 2½ inches to a degree, and contains the latest observations of the officers of the Imperial Russian Staff.

II.—*Observations Météorlogiques et Magnétiques faites dans l'étendue de l'Empire de Russie.* Redigées et publiées par A. T. Kupffer, Membre de l'Acad. de Sci. de St. Petersbourg, &c. 4to. 1837. Communicated by Colonel Jackson.

Independently of the reciprocal advantages derived by nations at peace with each other, they not unfrequently confer, by their joint efforts for the advancement of knowledge, a lasting benefit upon mankind; for while commercial interests are sacrificed by those unhappy differences which occasionally sunder the bands that unite them, the scientific truths they have discovered together remain secure from contingencies, imperishable and glorious monuments of the noblest impulse which can actuate nations.

Science, in its higher branches especially, is too frequently regarded as matter of curiosity rather than of utility, but surely we have seen enough of the important results obtained by the application of science, to hail every new discovery as the forerunner of material, tangible benefits. But even were it not so, is not truth lovely in itself? And can there be given to our comprehensive faculties a nobler direction than the investigation of those laws which, emanating from the Great First Cause, maintain the harmony of the universe? Certainly not; and when those laws more immediately affect the planet on which we live, they have a still more powerful claim on our attentive consideration. Of these laws, those which regulate the magnetism of the earth, that occult force, known only by its effects, are of peculiar interest, since by the application of this force we have been enabled to cross the trackless ocean, and draw together in social compact the widely-spread families of mankind.

The variations of the needle, its dip, and the different intensities of the magnetic influence in different places, are so many circumstances which could not fail to awaken attention, as being not only remarkable in themselves, but as materially affecting the value of an instrument to which we owe the discovery of a new world, and which is in daily use, whether for directing our progress to foreign shores, or for correctly delineating the countries we know.

To discover, therefore, the cause of the variations of the needle, and the laws which regulate those variations, so as to determine the exact value of the needle's indications, became a most important desideratum; and it was reserved for Humboldt to add to the other glories of an imperishable name, that of stimulating the leading nations of the world to an exact and systematic observation of the phenomena of terrestrial magnetism. His call has been responded to, and in the work before us we have a valuable proof of what may be effected by a system of observations properly methodised and carefully executed.

The name of M. Kupffer is too well known in connexion with some of the most important physical phenomena of the earth, lately discovered, to require any eulogium here. We shall, therefore, merely give a succinct notice of the work in question. This, as its title indicates, is not confined exclusively to magnetic observations, but embraces meteorology generally; indeed the connexion between the whole range of meteorologic phenomena is so close, they are so mutually modified each by every other, that the independent and unconnected observation of any one would be of little utility. Not only is it necessary to observe simultaneously all the meteorologic and magnetic phenomena at any one station, but similar observations must be made at various places, at the same moment, in the same manner, and with instruments reduced to a common standard of comparison.

This has been effected in Russia, over which extensive empire a great number of observatories have been established:—

"At some of these points," says M. Kupffer in his preface, "both meteorologic and magnetic observations are made; at others only meteorologic. A normal observatory has been founded at St. Petersburg, at the *Institut des Mines*, where a certain number of officers receive the practical instruction requisite to become observers in the establishments in the provinces; detailed instructions are communicated to them, in which a rigorous and uniform system is followed both in the distribution of the hours as well as in the choice of instruments and the method of observing. Eight times a day they observe the atmospheric pressure, the temperature of the air, the humidity of the atmosphere, and the quantity of water which falls either in the form of rain or snow. At some places they also observe, at the same hours, the magnetic dip and variation: the change in the variation is also observed, at certain times of the year, simultaneously with those set on foot in different parts of Europe. It is proposed also to add to these, observations on the temperature of the ground, on atmospheric electricity, and on the intensity of the force of terrestrial magnetism. In each observatory there are, besides the officer specially intrusted with its superintendence, two assistant observers, who live in the observatory, as a single observer would not be sufficient for such a task. These assistants have been formed at the normal observatory at St. Petersburg, where they have passed one or two years in making the same observations, so that they are quite equal to it. That the zeal of the observers may not slacken, they are employed only for a few years, and are from time to time relieved by others who have recently left the normal observatory. This arrangement facilitates the introduction occasionally of improvements in the system of observations."

From such an arrangement the most satisfactory results may be fairly anticipated, and the scientific world will be put in possession of them by successive publications.

The present volume, the first of the series, is divided into two parts; the first contains the necessary instructions for making the observations, and the second the observations made by Mr. Kupffer himself and his assistants at the observatory of St. Petersburg.

The instructions, drawn up by Mr. Kupffer, are divided into two sections: the first is devoted to thermometric, hygrometric, and barometric observations, the direction and strength of the winds, the quantity of rain and snow which falls, and the appearance of the heavens; while the second is exclusively appropriated to observations on the horary variations of the dip of the magnetic needle, the absolute dip, the variations of the declination, and the positive declination. These instructions have the great merit of being lucid, and they embrace all the necessary details except those minute cautions and directions which can only be communicated orally, while teaching the manipulation of the instruments themselves.

The second part of the volume contains, in tabular arrangement, the various meteorologic observations made daily at St. Petersburg, from the 1st July, 1835, to the 30th June, 1836, inclusive. The thermometer, hygrometer, and barometer were each registered eight times a-day; every two hours from 8 A.M. to 10 P.M. inclusive. Observations on the winds and weather were made at the same hours. The quantity of rain or snow fallen was observed twice a-day, at 8 in the morning and at 8 in the evening, excepting in cases of very great rains, when the quantity was observed as soon as they had ceased to fall, and immediately registered as to quantity and time, independently of the regular observations; the snow which fell was thawed. In summer the fall of rain was more frequently observed, on account of the evaporation.

In addition to his observations, Mr. Kupffer has given a set of psychrometric tables, which will be found eminently useful as abridging the calculation of the psychrometric formula. These tables refer to the mode of observing the moisture of the air by two thermometers, the bulb of the one being covered with a wetted muslin; and they are so arranged, that by entering them with the indications of the covered thermometer and the difference between the indications of the two thermometers, the tension of the aqueous vapour in the air is immediately found.

The magnetic observations were not begun till much later, and they are not contained in the present volume.

From the well-known zeal and ability of Mr. Kupffer, his scrupulous attention to exactitude, and unremitting patience, we receive with particular satisfaction the result of his personal observations as data upon which we can confidently rely; and the general direction of the meteorologic and magnetic observations in the Russian empire could not, we feel assured, be placed in better hands.

We cannot, however, close this brief notice without expressing our admiration of the zeal with which the Russian government encourages the investigations of mathematical and physical science. The munificent grants of money, and the assistance of every kind afforded by the emperor for scientific objects, are worthy of a great sovereign, and justly entitle him to the gratitude of all those who, like ourselves, cultivate knowledge as a means of happiness; and we sincerely hope that the joint labours of Russia and England in the field of meteorologic and magnetic research will, from the vast extent of their respective possessions, and the acknowledged ability of their observers, be soon rewarded by a rich harvest of important discoveries.

III.—*Reisebeschreibung nach Arabien und andern umliegenden Ländern.* Von Carsten Niebuhr. 3d band. Hamburg. 1837. 4to.

[*Carsten Niebuhr's Narrative of a Journey to Arabia and other contiguous Countries.* Third vol.] Communicated by the Rev. G. C. Renouard, M.R.A.S.

Of all the acts of royal munificence which have a just claim to the gratitude of mankind, none surely can rank higher than those which are calculated to advance knowledge and promote a friendly intercourse between distant nations. Much of the glory reflected on our own country, and on other European nations, in the most brilliant epochs of their history, is due to the patronage afforded by their sovereigns to expeditions despatched for the purpose of discovering unknown regions; but the moral and scientific value of most of those expeditions was tarnished by one deteriorating ingredient which entered too largely into the motives of their patrons. The lust of wealth and dominion was too plainly the leading principle in the councils of the monarchs who aided Columbus and Gama. Nor was it till the middle of the last century, that Denmark gave the noble example of sending out, at the national expense, a body of travellers for the sole purpose of pursuing scientific inquiries. Frederick V., probably at the suggestion of his excellent minister, Count von Bernstorff, and with the assistance of the celebrated Michaelis, to whom the selection of the persons to be employed was in a great measure left, despatched four young men duly prepared, on a mission to Constantinople, Egypt, and Arabia, for the purpose of obtaining accurate information respecting the geography, climate, and natural productions of those countries, as well as of the languages and condition, moral and physical, of their inhabitants, especially with a view to the elucidation of the Scriptures of the Old Testament. Having received their instructions at the close of 1760, the travellers set sail for Smyrna on the 7th of January, 1761, and Carsten Niebuhr, the only survivor, for his four companions, the draughtsman, physician, linguist, and naturalist, all sunk under the influence of the climate not long after their arrival in Arabia, reached Copenhagen at his return on the 20th of November, 1767. His description of Arabia, published in 1772; the first volume of his travels printed in 1774; the observations relative to botany and other parts of natural history, made by Professor Peter Forskål, together with plates illustrative of them, edited by Niebuhr, in 1775; and finally, the second volume of the travels in 1778, fully justified the choice of Michaelis; nor did any of the travellers appear more deserving of approbation than Niebuhr himself. Though the first two volumes comprehended the largest and most important part of his travels, yet a third still remained which was to give an account of his journey through Syria and Asia Minor in his return homewards. This volume, by which the work is completed, appears to have been ready for publication in 1779; but its appearance was retarded by various untoward circumstances, so that all hope of its publication seems to have been abandoned, till about two years ago, proposals for printing it were circulated by M. Perthes of Hamburg; and it is to the liberality of that bookseller, as well as to the affection of Niebuhr's family, particularly of his daughter, that the admirers of her father's works have at length the satisfaction of accompanying him to the end of his long pilgrimage.

The present volume is divided into five sections or chapters, containing the traveller's remarks on Aleppo; his voyage to Cyprus, and thence to Jaffa and Jerusalem; his observations on those places; his return to Aleppo, and journey from thence through Kóniyeh to Constantinople; geographical information collected in his passage through Asia Minor, and an abridged account of his route through Bulgaria, Vallachia, Moldavia, Poland, and Germany, to Denmark.

The last section has been very judiciously shortened by the editors, as much of the ground passed over is not only well known, but has been sufficiently described by later travellers; and as this portion of the author's narrative does not supply matter enough for a volume of nearly the same thickness as the preceding ones, there are added, in the form of an appendix, several valuable tracts relating to the nations and antiquities of Western

Asia and Abyssinia, drawn up by Niebuhr, together with the astronomical data of the celestial observations made during his journey. For the republication of all of these, particularly the latter, the literary world is much indebted to the Editors, as it is always desirable to have the works of an eminent writer collected, and few persons have an entire command of the various periodical publications through which these papers are scattered. That is particularly the case in this country, where the 'Monthly Astronomical Correspondence' of the late Baron von Zach is not easily met with. So little have the Editors of the present volume been influenced by any undue desire of augmenting its bulk, that no less than ten papers contributed by Niebuhr to the 'German Museum,' between the years 1781 and 1791, have been omitted, though most, if not all of them, would still be read with interest.

The description of Aleppo and Antioch, the account of their inhabitants and commerce, and the remarks on the antiquities and condition of Cyprus, contained in the first section, are very deserving of notice, particularly the observations on Citium * (pp. 21-25); and the amusing anecdotes respecting that extraordinary son of an extraordinary mother, Edward Wortley Montagu (p. 29). From Jerusalem, Niebuhr returned to Jaffa, and went on by sea to 'Accá (Acre) and Ṣaidá (Sidon), making an excursion from that place to Damascus, having to his surprise and satisfaction, met with, at least, one country in Asia in which an armed escort was not wanted (p. 82). From Ṣaidá he proceeded by sea to Ládhiḳiyyah (Laodicea), stopping by the way at Tripoli only long enough to take the altitude of some stars (p. 92), as he was anxious to meet M. van Masseyk, the Dutch Consul at Aleppo, at Ládhiḳiyyah, from whom he was sure of receiving much valuable information. There he was detained some weeks, waiting for the departure of a caravan, and had thus leisure to extend his inquiries respecting Syria, and to revise his plans and routes.

As his road northwards lay through Bayás and Adeneh, he not only passed through a tract little known, but was able to make a plan of that town, as well as of Ḳóniyah, the ancient Iconium. These plans, though not minutely exact, are valuable as expressing relative bearings; and in that respect the slight sketch of Jerusalem given by Niebuhr (Tab. iv. p. 52) may be of service when compared with the elaborate plan published by Mr. Catherwood. In his route from Ḳóniyah to Constantinople, Niebuhr

* Now called Chitium, which is pronounced Shiti by the modern Greeks. Niebuhr does not seem to have been aware that Shiti (Chiti) is identical with Chitium; or that the inscriptions copied by him (tab. iii. p. 22) are quite different from those published by Pococke (Descript. of the East, ii., tab. xxxii.), and interpreted by Gesenius (Scripturæ Phœniciæ Monumenta, p. 122, tab. 11, 12).

passed through two remarkable places, one of which has seldom been visited by other European travellers; viz., Kútáhiyah, the ancient Cotyæum, remarkable in modern times as the centre of the tract where the carpets are manufactured, which, for wear and colour, have not yet been equalled by our own: the other place is Afiyúm-ḳará-ḥiṣár (Opium-black-castle), so long erroneously supposed to be on the site of Apamea Cibotus, which stood near the confluence of the Marsyas and Meander. D'Anville (Geogr. Ancienne, ii. 53), with whom this error originated, seems in this, as in some other cases, to have been misled by an apparent resemblance between the ancient and modern names.

Niebuhr's observations on the government and condition of Moldavia and Vallachia, are still worth reading, notwithstanding the great political changes experienced in those principalities within the last few years, in consequence of the Sultán's ill-judged attempt to resist the overwhelming power of Russia. The peculiarities and affinities of the Vallachian language did not escape the notice of Niebuhr, and the information he gives will be new, at least, to most of his English readers, so little is known in this country respecting that remarkable monument of the Roman dominion in Dacia, and of the amalgamation of the Latin and Slavonian languages.

The astronomical data given in the appendix have long been known to all geographers who could make use of the Baron von Zach's work. They, with a few others found among M. Niebuhr's papers, are here reprinted with the corrections and reductions of M. von Zach and Professor Bürg, which greatly enhances their value, as they are thus made immediately applicable in practical geography. Of the remaining papers, which all have merit, the most remarkable are a dissertation on the Noṣaïrís, a singularly fanatical sect, who are numerous in the northern mountains of Syria (iii.); some observations on the exposed position of the temple at Jerusalem, and its liability to suffer from tempests (iv.); and memoranda respecting Abyssinia, which were collected at Caïro in 1762, about six years before Bruce visited that country (vi.). "This paper deserves to be compared," say the editors, Messrs. Gloyer and Olshausen, "with the author's review of the two first volumes of Bruce's Travels, printed in the German Museum for June, 1791." The illustrations of Xenophon's Anabasis (v.), drawn up during Niebuhr's stay at Brusa, have been in great measure superseded by Major Rennell's able work, and the more extensive observations of modern travellers, some of the most important of which, however, have not yet been published.

IV.—*Journal de la Navigation autour du Globe, de la frégate la Thétis et de la corvette l'Espérance, pendant les années 1824-6.* Par M. le Baron de Bougainville, Capitaine de Vaisseau. 2 vols. 4to. Paris, 1837. Publié par ordre du Roi.

During the course of the last twenty years five distinct naval expeditions sent out by the French government have circumnavigated the globe, and the narrative of their respective voyages, with a full account of all their observations in detail, illustrated by admirably engraved charts, and generally by some excellent drawings, have been liberally published, by order of the Minister of Marine, at the expense of the state. Such liberality must afford a cheering prospect to the navigator, when exposed to numerous privations in a tedious voyage round the globe; or to the surveyor struggling against the difficulties of a long-protracted examination of inhospitable shores, and cannot fail to stimulate officers employed on such expeditions to exert themselves to the utmost in the service of their country.

Of the expeditions above mentioned that of the *Uranie* and *Physicienne,* commanded by M. de Freycinet, took place in 1817-20, and brought home a number of facts relative to the physical history of the globe; the voyage of the *Coquille,* under the orders of M. Duperrey, occurred in 1822-5, and is distinguished for his examination of the little-known islands of New Zealand and the Caroline group; the expedition of the *Thétis* and *Espérance,* commanded by M. de Bougainville, took place in 1824-6; the voyage of the *Astrolabe,* under M. Dumont D'Urville, in 1826-9, is remarkable for his examination of the eastern coast of New Guinea, of New Zealand, and for the great attention paid to the Polynesian languages; and, lastly, that of the *Favorite,* Captain Laplace, noted for the examination of the Gulf of Tonkin, and of the Archipelagos of the Natunas and Anambas in the China Sea.

The narratives of four of these expeditions have been for some time before the public, but the detailed account of the voyage of the *Thétis,* although more than eleven years have elapsed since her return to Europe, has only recently issued from the press. The chief results of this voyage, with respect to hydrography and the collections in natural history, were of course immediately made known to the French government, and probably found their way into some periodicals at the time; but it cannot but be a subject of regret that the authorized narrative of the voyage, and of its results, should have been so long kept back from those who take an interest in the advancement of geography, and from the public at large.

In fulfilment of a similar mission to that intrusted to M. de Kergariou, in *la Cybèle,* in 1817, and to M. de Camper, in *la*

Cléopatre, in 1822—namely, to show the French flag in the Indian and China seas—the frigate *Thétis,* commanded by the Baron de Bougainville, quitted Brest on the 2nd March, 1824. In the course of her passage to the Canaries, where she anchored on the 13th March, the vessel had experienced a set of 2° 43′ to the eastward; on the 29th crossed the Line in 17° 40′ west of Greenwich; on the 26th April sighted the Cape of Good Hope; and on the 5th May, in lat. 39°, long. 37° 40′, the magnetic variation attained its maximum, namely 32° 12′ W., decreasing immediately afterwards as she steered to the eastward; on the 20th the *Thétis* anchored at the Isle of Bourbon, where she found her consort *l'Espérance,* Captain Camper, arrived from Rio de Janeiro.* After a stay of a few days the two vessels sailed in company, recrossed the equator on the 18th June, between 67° and 68° E. long., and on the same day the variation was at zero. On the 23rd sighted and fixed the position of the north part of the coast of Suadiva, the southernmost group of the Maldivas,† in lat. 0° 51′ 10″ N., 73° 20′ 40″ E. of Greenwich, making the channel of Adumati 58 miles wide.

June 25.—Made Adam's Peak, on the island of Ceylon, and on the 29th anchored at Pondichery. M. de Bougainville devotes a chapter to a description of this city, and contrasts well the monotonous uniformity of its *white* town with the animated scene presented by the junglers, the bazaars, and attractive dances of the Bayadères, &c., in the *black* town. He states the population of the five French establishments in India—viz. Pondichery, Chandernagor, Karikal, Mahé, and Yanaon, at 4000 whites and 132,000 blacks, and gives some other statistical details.‡

July 30.—Quitting Pondichery the *Thétis* traversed the Bay of Bengal, entered the Straits of Malacca, anchored for a few days off that town, and also at Sincapore, thence crossed the China Sea to Manila, where she arrived on the 17th September, and anchored in the port of Cavite. Two days previous, in lat. 13° N., long. 115° 20′ E., the variation of the compass was observed to be *nil,* with the ship's head East. At Manila the vessels remained three months, in order to refit, and where they found everything requisite for their equipment.

From Manila the *Thétis* went to Macao, on her passage thence

* There appears to be an error here of 10° of long. in M. de Bougainville's text not marked in the *errata.* See also vol. ii., p. 113, of Appendix.

† The whole of this group has lately been surveyed by Captain Moresby, of the Indian navy, and published on the scale of 4 miles to an inch. By his observations the north point of Suadiva is in 0·55° N. long., 73° 19′ 0″ E. of Greenwich, and the channel of Adumati 53 miles wide. The channel to the north of Adumati, in lat. 2° N., is 16 miles wide.

‡ There must be some mistake here, as the official returns in 1835 give 980 white, 1515 mixed, 165,241 blacks. Since the 1st of July, 1836, a fixed light, 97 English feet above the sea, has been shown at Pondichery, visible from 15 to 18 miles from the deck of a ship.

to Tourane examined the isle of Haï-nan, which forms the eastern side of the gulf of Tonkin. On the 13th Jan. 1825, anchored in the Bay of Tourane, in Cochin China. M. de Bougainville differs from Horsburgh and Crawford in considering this bay so excellent as they represent it; they probably knew it only from an imperfect survey, but it was afterwards thoroughly examined by M. de Kergariou and the officers of the *Cybèle*, in 1819. After about a month spent in this bay, in execution of a mission with which M. de Bougainville was charged to the emperor, as he is styled, of Cochin China, and which afforded an opportunity of seeing a little of the manners of this curious people, the *Thétis* sailed for Surabaya, on the north coast of Java; on her passage she examined, for the first time we believe, the Anambas islands,* a group extending 150 miles from north to south, and as much from east to west, lying about 100 miles from the eastern coast of the Malay peninsula. Some of the islands rise up in blocks of granite, with white marks, similar to those of the Pedra Branca, in the Strait of Malacca,—indeed their general structure is granitic, with coral reefs in all the sheltered spots. Here are said to be 50 islands; they are of moderate height, and in parts wooded; the group may contain from 1500 to 2000 inhabitants, apparently Malays, which M. de Bougainville agrees with Horsburgh in considering as pirates. The islands are called *Scianthan* by the natives, and the officers of the ships explored more than one excellent anchorage. The position of twenty-two of the principal islands or remarkable points were also accurately determined—extending between the parallels of 3° 27′ and 1° N. and 106° and 108° 21′ E. long.; high water at full and change, at 5h. 25m.; rise of tide 2 feet 4 inches; the current set N.N.W. and S.S.E., at its greatest strength 3 miles an-hour; the variation observed on different islands 1° 7′ E.

Proceeding to the southward, the *Thétis* crossed the Line for the third time since her departure from Brest; ran through the dangerous strait of Gaspar, reconnoitred the Carimon-Java islands, cleared the straits of Madura, and anchored on the 20th March off Surabaya, where the ships remained for six weeks, and unfortunately suffered much from dysentery. Quitting Java, at a short distance to the N.E. of Madura, they found the soundings change suddenly from 46 fathoms, mud, to 26 fathoms, coral—immediately after to 44, gravel, and then 47, blue mud as before. M. de Bougainville cautions navigators to beware of being lulled into security, as coral is commonly found in these seas rising from a bottom of sand. Steering through the strait of Allas, the shores of the island of Lombock were examined, and the road-

* Since re-examined in the voyage of the *Favorite*, and included, with many other corrections, in the excellent chart of the China Sea just published at the Dépôt de la Marine at Paris, by MM. Daussy and Wissocq.

stead of Tanjong-Louar or Peejow found to be the best on the coast, and far superior, in the opinion of M. de Bougainville, to that of Bally. Having cleared the S.E. point of Lombock, remarkable from its resemblance to the angle of a lofty bastion at the foot of which stands a column on a low base, the *Thétis* and *Espérance* steered to the southward, rounded successively the S.W. point of Australia and the southern extreme of Van Diemen's Land, and on the 1st July anchored in Port Jackson. M. de Bougainville remained here nearly three months, and devotes his 14th chapter to a description of the colony of New South Wales. On the 21st September the vessels sailed for the coast of Chile, and arrived at Valparaiso on the 23rd November, without having made any land in the passage of 6000 miles across the Great Ocean, between the parallels of 31½° and 37½°. In this passage the current set the vessels 163 miles to the north and 125 miles to the east—the variation of the compass increased from 8° 30′ to 13° E.; it began to decrease in 177° E. long., and reached its minimum 5°, in 123° W.: from this point it varied slightly between 7° and 8° as far as 105° W. long., when it was again 5°; it shortly after increased to 13° and 16°, and remained so till the ships' arrival at Valparaiso.

M. de Bougainville makes some very just remarks on the want of correct charts of the Great, or, as it is called, the Pacific Ocean. The difficulty of accomplishing such a task he fully admits, but surely, where so many lives and so much property are annually risked, it cannot be admitted as an excuse that all the *ingénieurs hydrographes* of this or any other country are fully employed in more urgent business, and have not leisure to attend to it; nor does M. de Bougainville admit as an excuse that we have not yet sufficiently correct *data* to construct such a chart. Let but a really competent person devote himself to the task, and it will be found that far more *data* are available than is generally imagined, and quite enough to warrant the immediate construction of a chart on a large scale in several sheets.*

After a stay of six weeks at Valparaiso, the vessels sailed on the 8th January, 1826, for Rio Janeiro, where they arrived on the 21st March, left again on the 8th April, and finally anchored at Brest on the 24th June, after an absence of two years and four months.

During the whole of the voyage great attention was paid to meteorologic, as well as to the more common magnetic observa-

---

* La Carte du Grand Océan, par MM. D'Urville et Lottin, published in 1833, and corrected to 1834, at the Dépôt de la Marine at Paris, is by far the most correct chart of this ocean. It has the merit also of distinguishing between the high and low islands; but being a general chart, in one sheet, it is necessarily on a small scale.

tions, as variation of the compass, &c.; the indications of the barometer, thermometer, and hydrometer were invariably registered four times in the 24 hours; and the appendix, besides some valuable notes, contains a list of 150 geographical positions determined on the voyage chiefly by M. Fabré and other officers, with a discussion of the observations on which they rest:—the position of the ship at noon each day is also given with the variation, current, winds, &c.—in short, a complete abstract of the ship's log-book—a more valuable document than at first might be imagined; but the late researches by Colonel Reed, R.E., into the law of storms has shown how extremely valuable are these minute meteorologic registers: from the too often despised log-books of merchant-ships he has educed the basis of a theory, which, when carried into practice, may be the means of saving property to a large amount, and the lives of numerous fellow-creatures. May this be an inducement to all who traverse the ocean to pay strict attention to such registers, assured that their labours will not be in vain!

The second volume of M. de Bougainville's work contains an itinerary from Santiago de Chile to Buenos Ayres, by M. de la Touanne, who has also enriched the work with many highly picturesque views of scenery in Cochin China, Philippine Islands, Java, &c., and some characteristic sketches of the natives of the Indian and China seas.

The narrative of the voyage of the *Thétis* is related in the simple style of a nautical journal. M. de Bougainville well observes that "he is a sailor by profession, and that he will not give up the sextant or the lead for the painter's pallet," and that it is to facts that he trusts to make his account useful; yet, although unprovided with a naturalist or hydrographer, an unusual omission in a French voyage, M. de Bougainville was enabled to give to the Jardin du Roi at Paris a fair collection of objects in natural history, and some living animals, and has shown himself by his observations a worthy successor of his celebrated father, who conducted, in 1766, the first voyage round the world made by the French marine, in the memorable expedition of *la Boudeuse* and *l'Etoile*.

One cannot but regret that M. de Bougainville should give place in his work (p. 361) to the story of the Master of a French ship, called the *Gustave*, having *discovered* a bay on the coast of Chile in June, 1836, to which he, as discoverer, gave the name of *Baie des Français;* whereas any one either at Valparaiso or Callao, or at any other intermediate port on the coast, could have told M. Langlois that the tenders of the *Beagle*, Captain FitzRoy, had not only entered but thoroughly surveyed the bay in question in June, 1835.

V.—*Der Ugrische Volksstamm, &c.* [*Researches upon the country between the Ural and the Caucasus, with reference to history, geography, and ethnography.*] By Ferd. Heinr. MUELLER, Ph. D., &c. Berlin, 1837. Communicated by Dr. I. C. PRICHARD, M.D., F.R.S.

THE object of research which the author has proposed to himself is the history of the great nomadic nations of Asia, whose movements have produced in various times, and particularly during the middle ages, a powerful impression on the eastern countries of Europe. The light already thrown on many subjects connected with this pursuit by Thunmann, and more lately by Julian von Klaproth, has encouraged the author to pursue their investigation by the study of the Byzantine historians in connexion with Arabian and other eastern writers. The present work, of which this is the first volume, is devoted to a preliminary survey of the region which has been the principal theatre, not of the military exploits of the nomadic tribes, but of their national development. The nations whom Dr. Müller has chiefly in view, are those belonging to the great Finnish race, by him termed Ugrische Völkerschaften, and to this stock he refers not only the Madjars, or Hungarians, but likewise the more celebrated Huns, as well as the Chasares, Bulgarians, and several tribes who have been by other writers derived from a Tartarian origin. Much information has been already collected with reference to this subject by the Russian academicians Pallas, Gmelin, Lehrberg, Frähn, Sjögren, and others; but the views adopted by some of these writers are erroneous in part, and none of them has fulfilled the design of our author, which is, as he says, "to give a survey of the theatre in which human nature has here developed itself, and of its relations of the region to the families of men that were formed within it."

In the course of his work Dr. Müller proposes to display the relations which the ancient races of the Uralian mountains bear to those of the Caucasus and the chain of Altai. He follows Klaproth in supposing the Ural to be the native seat of the race, to the history of which his inquiries are chiefly directed.

The first section of the present volume contains a geographical survey of the Ural mountain-system, which the author divides into three regions, the southern, middle, and northern Ural. The principal rivers sent forth by the southern Ural are surveyed and followed through their course; the countries watered by the Jaik and the Biela have frequently changed their population, and even their physical condition, and many regions now occupied by vast steppes were at one period the abode of an agricultural people. The plains between the southern extremity of the Ural and the Euxine were the great road by which the nomadic nations of Asia poured themselves into Europe. The middle Ural sends forth the rivers of Perm, viz. the Kama and the Chysova, and several considerable streams which flow westward into Siberia. There are two principal transits over the middle Ural, one of which is supposed by Lehrberg to have been the situation of the "Eiserne Pforte," or Iron Gates, mentioned in the old Russian annals, and by which it is said that the Novgorodians gained access to *Iugrien*, or the country of the Ugri. By Lehrberg, as well as by Karamsin, it is maintained that this country included a considerable tract of northern Siberia, and that a part of the Asiatic empire of Russia was in fact tributary, in the middle ages, to the republic of Novgorod. The northern Ural is the most inclement part of Europe. It is only in the best season of the year that the Vogouls and Samoiedes hunt in this region. Yet, though now correctly termed a desert, the northern Ural was in former times, according to Dr. Müller, inhabited. It formed a considerable part of the land of Iugrien, or Iugoria, mentioned in the ancient imperial title of the Russian Sovereigns. This, in the opinion of Lehrberg, Müller, and many other Russian antiquarians, was the native country of the Ugri, Hungarians, or Madjars, and of the more celebrated Huns.

Novaïa Zemlïa is only a continuation of the Uralian chain. Müller has collected in his work all the early notices of this country, and the coast near the Straits of Waigatz, and of the trade of the English and Dutch to the coasts of the White Sea.

The second section contains a survey of the *Alginskoi Syrt*, a broken and often interrupted series of high levels which connects the southern extremity of the Ural with the chain of Altai, and likewise of the great steppes inhabited by the Kirghiz Tatars.

The third section is devoted to a survey of the river-systems of the Irtish and Oby, and a geographical account of the countries watered by those streams. The aboriginal inhabitants of these countries are the Ostiaks and Samoiedes; the tracts between the two rivers are now occupied by a Tatar tribe, the Barabintzes, and at Beresov there are many Russian settlers.

The fourth section contains the physical geography of the countries to the westward of the Ural, and a survey of the great rivers which issue from that chain in a westerly direction, and which divide the northern countries of Europe; it also gives an outline of the ethnographical history of their inhabitants.

The fifth, and last section, comprises a geographical survey of the countries occupied by Finns and Laplanders, the western division of the group of nations derived from the Ugrian stock.

In the volume which is to follow the present, and which will conclude this part of the author's work, he promises to "survey the fruitful plains of Sarmatia, from the midst of which the great rivers flowing into the Euxine and Caspian will conduct him into the Alpine valleys of the Caucasus, where he will investigate in that region the primitive as well as the present abodes of the nations belonging to the Ugrian race.

VI.—*Report on the Eastern Frontiers of British India; Manipúr, Assam, Arracan, &c.* By Captain R. Boileau Pemberton, 44th N.I. Calcutta, 1836.

Of the several passes into the territories of Ava, through the mountain chain which forms the eastern frontiers of the British dominions in India, we had none but the most imperfect and unsatisfactory accounts previous to the late Burmese war, and of the existence of some we were wholly ignorant. The few from among our subjects who, in the pursuit of traffic, had ventured into the fastnesses of the mountains had scarcely ever penetrated beyond the first ranges which immediately overlook the low lands of Bengal and Kachar.

Of the countries beyond, or on the eastern side of this mountain barrier, our information was necessarily still more imperfect; and with the exception of that part of the Irawaddi river between Ava and Rangún, our knowledge had not advanced one step beyond the point it had attained when the clear, laborious, and accurate Buchanan withdrew from the field of investigation.

When, at the commencement of the Burmah war, our ignorance of the whole frontier became manifest—the impolicy, the shortsightedness of not having instituted, even by force if requisite, a proper examination of the mountain passes flashed upon the mind of even the most careless observer, and its lamentable effects were afterwards shown in an expenditure of life and treasure without parallel in the annals of Indian warfare. Another proof, if proof were wanting, to be added to those lately brought forward,* of the absolute necessity of a complete survey, to insure the good government of any country.

To remedy this evil officers were at various times despatched to examine these frontier districts, and the report before us, drawn up by one of the most able of these officers, and who was for some time joint-commissioner in Manipúr, combines, in a clear and masterly sketch, the results of these investigations in both a geographical, political, and statistical point of view.

As the work is printed at Calcutta, and probably very few copies are to be met with in Europe, a brief abstract of the physical geography of these countries is offered in the following pages.

The Report professes—1st, To give a general description of the great chain of mountains which running from the southern borders of the Assam valley, in lat. 26° 30′, extends to Cape Negrais, the extreme southern limit of our possessions in Arracan, in lat. 16° N., and forms a barrier on the east along the whole line of the Bengal Presidency, from one extremity to the other.

Secondly, To describe the nature of the passes and countries

* See Major Jervis's Address at the meeting of the British Association at Newcastle, in 1838, on the Present State and Progress of the trigonometrical survey in India.

by which this great mountain-chain has been penetrated, and which may be conveniently subdivided under three heads :—

1st. Those routes which extend from the frontier of the Silhet and Kachar districts, through the Manipúr territory to the Ningthí or Kyen-dwen river and central portions of the northern provinces of Ava.

2nd. Those lines of communication extending from the southern borders of the valley of Assam into Manipúr.

3rd. Those through the province of Arracan, by the several routes of Yalak, Aeng, and Tongo, to the towns of Shembegwen, Membú, and Prome, on the banks of the Irawáddi river.

Thirdly, To describe the countries extending east from the banks of the Ningthí river to the frontiers of China, as far as we have become acquainted with them from native information.

Fourthly, To endeavour to estimate the comparative value of the different passes from the British territories into Ava, &c.

And lastly, To describe the countries of Kachar, Jintíyah, and the Kossiyah hills, which have been recently annexed to the British Indian dominions.

The mountainous chain which forms our eastern frontier may be considered a ramification from that which, sweeping round the south-eastern border of the Assam valley, stretches nearly due west along the northern frontier of the Silhet district, and terminates at the great southern bend of the Bramahpútra, in long. 90° east. The point of divergence is situated between the twenty-sixth and twenty-seventh parallels of north latitude, and the ninety-third and ninety-fourth degrees of east longitude: from thence it runs south for about 60 miles, with a mean breadth of from 80 to 120 geographical miles. At the S.E. extremity of the Silhet and Kachar districts it trends in a S.W. direction to the twenty-third parallel, when its breadth is upwards of 180 geographical miles, from the sea on the west to the Ningthí river on the east; it thence runs full 360 geographical miles in a S.S.E. direction to the seventeenth parallel, where its breadth scarcely exceeds 20 miles, and from that point it again trends to the W. of S., and terminates in the rocky promontory of Negrais.

The loftiest points of this great chain are found at its northern extremity, on the confines of the Manipúr territory, where the peaks attain an elevation of from 8000 to 9000 feet above the sea, and average from 5000 to 6000. On the Kachar and Silhet frontier the measurements hitherto made give an altitude of from 2000 to 4000 feet; east of Tripurah and Chittagong they fluctuate between 2000 and 5000 feet. On the Arracan frontier the Table Mountain, in lat. 21° N., long. 93° E., is said to be upwards of 8000 feet above the level of the sea. From this it gradually declines, and may range between 3000 and 4000 feet at all those passes which have hitherto been, or are ever likely to be, resorted to for purposes of traffic or war.

The rivers which flow from this mountainous tract of country on the east and west, though numerous, are, with but few exceptions, of trifling importance. On the western side the seven principal are the Súrmah or Barak, the Gúmti, the Fenni, the Chittagong River, the Kuladain or River of Arracan, the Talak, and Aeng. On the eastern side the Man or Maïn River, the Yúma Kiyúng, the Manipúr River or Impan Turál, and the Maglung, which flows through the valley of Kubo from north to south.

Of the geological structure of this tract of country we know little: between Manipúr and Kachar a light brown sand-stone and a red ferruginous clay prevail on the lower heights; above these slate or shale is found. In the central ranges, west of Manipúr, limestone occurs. Crossing the Kubo valley we reach the Ungoching hills, where lignite coal is found, which has been traced to the opposite bank of the Ningthí, where it rises to the surface in abundance: the great central ridge, when crossed between Manipúr and Assam, is composed of hard grey slate, and on its northern face boulders of granite were found resting on the inferior heights.

The timber on these heights is various and plentiful. Oak, fir, teak, and bamboo, in all its varieties, from the most delicate and small to the most gigantic, abounds.

The report then describes sixteen of the tribes of people by which this range is inhabited. All the tribes around the Manipúr valley partake strongly of the characteristic features of the Tátar countenance, whereas the Kúkís, or southern tribes, resemble the Malay more than the former.

Manipur.—The present boundaries of this country are, on the west, the Jírí river to its confluence with the Barak, thence south to the mouth of the Chikú, which is a point of some political importance, as it marks the union of boundary of the states of Manipúr, Kachar, and Tripurah. From this point the southern limit of Manipúr is ill defined as far as the Impan Turál, when it follows the course of that river due south to the parallel of 23° 55′, then east to the foot of the chain of mountains which separates the valleys of Manipúr and Kubo. The eastern boundary-line runs north along the eastern foot of this range to the N.E. corner of the Lúhúpa country, in lat. 25° 5′, beyond which no exploring parties have hitherto been able to penetrate. On the north the great central ridge which, running from N.E. to S.W., separates Assam from Manipúr. The territory comprised within these limits occupies an area of 7000 square miles, of which a valley of 650 miles of rich alluvial soil constitutes the central portion; this valley rests at an elevation of 2500 feet above the sea, its extreme length is about 36 miles, with an average breadth of 18 miles. The range of mountains which forms the western barrier of this valley has nine celebrated peaks and ridges, whose elevation varies from 5790 to 8200 feet above the sea, chiefly of compact sandstone and slate. The loftiest peak of the eastern boundary of the valley does not exceed 6700 feet, and from this elevation they descend to 4900 feet above the sea. Many detached groups of hills of about 500 feet in height appear in various parts of the valley, while to the east of the ancient capital there is one range whose central peak, called Nongmuïching, is 2700 feet above the valley and 5200 above the sea.

The chief stream which waters this valley is the Khongba, or Impan Tural, which falls into the Ningthí. The surrounding mountains are covered with the noblest varieties of forest-trees, as cedar, oak, pine, redwood, ash, willow, &c. Limestone appears to form the substratum of the whole valley, over which is a great depth of alluvial deposit. Iron also is found, and salt abounds. Fruits, both of temperate and tropical climes, are plentiful. The climate is favourable. In short, from its situation, extent, climate, and products, Manipúr is a very valuable acquisition.

The Report, at p. 51, goes on to describe in detail the three different routes by which Silhet and Kachar are connected with Manipúr, and also those by which the latter is connected with the Burmese territory. Then the three routes from Manipúr into Assam.

ASSAM.—Having stated the boundaries of Upper, Lower, and Central Assam, the Report gives the area of the whole valley at 18,900 square miles, with a population of 602,500, and a revenue of 44,000*l.* sterling. The lines of communication between Assam and Kachar and Silhet are then described and their several advantages pointed out, and the general resources and capabilities of the country are displayed. The valley of Assam is rich in mineral treasures; nearly all the streams wash down particles of gold; iron is found under the Naga hills and elsewhere, and coal to the east of Rangpúr.

ARRACAN.—Although founded on rather imperfect data, the total area of this province may be stated at 16,500 square miles; its population in 1831 was 173,928, and the net revenue 32,630*l.* sterling. Of the land but little is under cultivation; but foreign commerce is much on the increase. Not less than twenty-two passes are reported to exist between Arracan and Ava, but of these only five are in use, and have been explored by different officers of the Indian army—the best of them is the pass of Aeng, so called from a village, of about 150 houses, of that name; its summit is 4664 feet above the sea.

The Report then completes the description of our eastern frontier from Assam to Arracan, nearly the whole of which is the result of personal examination by Captain Pemberton.

PONG.—The kingdom of Pong extends from the foot of the mountains forming the south-eastern boundary of the Assam valley, in lat. 27° N., down to the twenty-second parallel, and from the mountains which separate the Manipúr and Kubo valleys on the west, to the Chinese province of Yunan on the east. A chapter is devoted to a highly interesting account of this little known country. The next section of the Report, from p. 146 to 188, contains an estimate of the comparative value of the different passes from the British territories into Ava, and the author offers many suggestions which appear likely to facilitate the rendering them either lines of commercial intercourse or military operations.

KACHAR.—Captain Pemberton then proceeds at once to an historical, topographical, and statistical description of the whole province of Kachar, and after stating its boundaries, he estimates its area at 4200 square miles, of which more than half are considered to be of a fine rich plain. The population of Southern Kachar, by far the most valuable portion of the province, may be assumed at 50,000 persons.

JINTÍYAH.—This province, as well as that of Kachar, is divided into a northern, central, and southern district; the latter, which is the most important, has an area of 650 square miles, and a population of 150,000, and is described as composed of well-watered and fertile plains. The central or hilly portion contains an area of 2340 square miles, with a population of 20,000, and is a tract of undulating table-land elevated from four to five thousand feet above the sea.

KOSSIYAH HILLS.—This tract of country forms an irregular parallelogram, the length of which from north to south is about 70 miles, and its average breadth 50, giving an area of 3500 square miles, which consists of three portions of unequal breadth and diversified character—the first, or more northern, is a closely-wooded tract rising from the Assam valley, and stretching by a succession of gentle undulations for 20 miles, to the heights on which stands the village of Mopía, 2746 feet above the sea, and from which the northern crest of the more elevated central plateaux is seen resting at an elevation of between four and five thousand feet.

From Nangkláu, which stands on the northern crest to Musmáï, which is similarly situated on the southern verge of this elevated region, the direct distance is about 35 miles; and it is within these limits that the region is included, whose salubrity has been so much extolled by its friends, and so much questioned by its opponents. The elevation of this lofty region, imperfectly described by the name of "table-land," appears to vary from four to six thousand feet above the sea, which would give an annual mean temperature of from 59° to 65° Fahr., or from 19° to 13° lower than that of Calcutta, which is nearly 78°.

At such an elevation, and with such a temperature, the change from a residence in the plains to one on the hills is, during the hot season, the most delightful it is possible to conceive.

Viewed geologically, this tract consists of two portions separated by the Boga Paní, which flows between them from N.E. to S.W. The northern portion consists, almost exclusively, of granite, masses of which are seen protruding through the soil at every step, and large boulders are scattered over the surface. In the southern portion coal and iron are found. Nangkláu, Maïrang, and Charra Púnjí, have each been selected at different times as a *sanatarium* for our invalids in India, and the Report discusses at length their claims to salubrity: from the experiments of Mr. Cracroft, who instituted a series of observations to determine the actual quantity of rain that fell at Charra Púnjí, it appears that the total quantity in four months, between June and October, was 225 inches! The peculiar position of Charra would perhaps warrant the expectation of a great amount of rain; but this exceeded all conjectures; yet there appears no reason to doubt the result. From observations in Arracan, we learn that 197 inches of rain fell between the months of June and October, 1825, which seems the only parallel on record.

Captain Pemberton concludes his excellent Report in the following words:—

"On our own frontier, we have seen a spirit of enterprise awakened, which, in Arracan, has been already productive of a most extraordinary revival, and increase of commercial pursuits; and in Assam, the same ardour is rapidly diminishing the space which separates her numerous tribes from the industrious and energetic inhabitants of Yunan and Sechuen:—the races dwelling on the eastern borders of the Irawaddí are seeking an asylum under our protection, and the tide of emigration already begins to flow from east to west. The holy zeal of the missionary, tempered by a necessary discretion, has outstripped the advances of commerce, and the most extensive schemes have been planned for conferring upon their population the blessings of civilization and letters. In Assam, Manipúr, and Arracan, a propitious commencement has been already made; and standing on the neutral ground which separates Hindooism, on the one hand, from Boodhism on the other, the disseminators of sound knowledge, aided by all the influence and talent of the local authorities, are kindling an intellectual flame which, spreading east and west, will illumine the gloom of superstition and ignorance in which their benighted inhabitants now rest, and qualify them for higher destinies than they have ever yet fulfilled. If such be the result of the extension of British influence over the numerous tribes and nations which dwell on our eastern frontier, the recollection of the horrors of the Burmese war will fade before the glorious prospect of redeeming many millions of men from such mental debasement, and elevating them to that higher station, in the intellectual and moral world, upon which the favoured inhabitant of Europe now stands. We rescued them from a yoke which has bowed to the dust the energies of every people over whom it has been cast; and we may fearlessly refer those who doubt the ameliorating influence of our rule to Arracan, to Kachar, to Manipúr, and Assam, and abide the result of a comparison between their past and present condition—between the sufferings they formerly endured, and the peace they now enjoy."

The appendix contains a list of the seventeen principal tribes around the valley of Assam; and of the twenty chief passes from the borders of Assam into the Búthan territory, with a table of the geographical positions and elevations of 300 places along the eastern frontier.

The work is also accompanied by various maps, chiefly constructed from Captain Pemberton's own observations and surveys. Ten of these are special maps on a large scale of the chief passes through the frontier; besides a general map of the Manipúr territory; and also of the whole eastern frontier from Thibet and Bhútan on the north, to Rangún and Múlmein on the south; and from the meridian of Calcutta to the western frontier of Yunan in China.

VII.—*Etudes Grammaticales sur la Langue Euskarienne.* Par A. Th. D'Abbadie et T. Aug. Chaho. Communicated by J. Cowles Prichard, M.D., F.R.S., &c.

The Euskarian language, the native idiom of the aboriginal inhabitants of Western Europe, has been, until within a comparatively recent period, almost entirely unknown to philologists. Many old Spanish writers have devoted themselves to the illustration of their national antiquities, and the monuments of ancient Iberian art. The works of Velasquez, Zuaiga, and Flores, on the Celtiberian, and on the Turdetan, or Bætic coins and inscriptions, are well known. The same subject was treated by Jacob Barry, a Dutch consul at Seville. Of later date is the work of Iztueta on the ancient usages, dances, and games of Guipuzcoa. The idea of investigating the history of nations by a comparison of their languages is almost of modern invention. Scarcely any attempt had been made to illustrate the history of the primitive language of Spain till the early part of the last century, when a dissertation, written on that subject by Jezreel Jones, was published at the end of the collection by Chamberlayne and David Wilkins, containing specimens of the Lord's Prayer in different languages. In that paper little more is to be found than mere conjectures. The writer imagined the Basque or Biscayan language to be allied to the dialects of the Berbers and the Shúlúh, the aboriginal inhabitants of Mount Atlas. The earliest Spanish works on the Bascuence or Biscayan were not, as it may be supposed, written with any view to speculations on the history of languages, or with the design of illustrating through that medium the origin and affinities of the old Iberian people. The earliest Biscayan grammar was printed by a singular chance in a country where the nearest approximation to the structure of that curious idiom has been discovered, viz. in the Spanish dominions of the New World. The author was Balthazar de Etchabié; he was a native of Guipuzcoa, and published his work in Mexico, in 1607, for the use of the numerous Biscayans who were settled there. This work fell into oblivion after the more splendid achievement of Larramendi had appeared, which the author pompously entitled '*El impossible vencido.*' * To this indefatigable Jesuit we owe likewise the first Biscayan dictionary, the "prologo" of which is the earliest production that aspires to the character of a philosophical account of the Biscayan language. The grammar of Larramendi, and his dictionary, published in 1746,† have long been well known in Spain, and they have inspired some native writers with the desire to illustrate the language of their ancestors—a task which they have undertaken just in the manner that might be expected. The most celebrated attempts are by Astarloa and Juan Bautista de Erro. The former, in his '*Apologia de la Lengua Bascongada,*' devotes himself to a proof of the position that the Biscayan was one of the eight primitive languages which he supposes to have originated at the building of the tower of Babel. Erro's works are still more absurd. In his '*Alfabeto de la Lengua Primitiva,*' and his '*Mundo Primitivo,*' he even ventures more in favour of the Basque than did General Vallancey for his favourite Erse. The Biscayan is at least as old as the planting of Paradise: the Hebrew is but a late offspring from it.

German philologists were the first who investigated with any degree of success the history of the Iberian language. English writers, from the time of Edward Lluyd, though they had nearer access to the means of better information, had always supposed the Basque to be a Celtic dialect. That notion was commonly entertained till the publication of the Mithridates, in which the treatise on the Basque was written by Adelung. Several years afterwards there appeared an appendix to the same work by Baron Wilhelm von Humboldt, containing much valuable information, not only on the language, but on the literature of the ancient Iberians, of which the author had, during his abode in Spain, been indefatigable in collecting the remains. Fragments of an ancient national poem on the taking of Numantium by Scipio carry very far back indeed the commencement of this literature, or rather of its scanty relics. Humboldt's subsequent work, his '*Prüfung der Untersuchungen über die Urbewohner Hispaniens,*' contains a most successful attempt to investigate, by

* El impossible vencido. Arte de la Lengua Bascongada. Su autor el P. Manuel de Larramendi de la C. de Jesus, Maestro de Teologia de su real Colegio de Salamança. Salamanca, 1729.

† Diccionario trilingue del Castellano, Bascuence, y Latin, &c. San Sebastian. In folio, 1745.

topographical terms plainly derived from the Biscayan language, the history of the Iberian race in different parts of western Europe. But both Vater and Humboldt wanted the advantage of a thorough acquaintance with the idiom of ancient Spain; an incalculable advantage, if only it could be found united with the knowledge of the true principles of scientific philology and sound critical judgment. This rare combination has displayed itself for the first time, if we are not mistaken, in the writer whose name stands at the head of this article. M. D'Abbadie is an enterprising and highly talented man, from whose zealous exertions in the cause of geographical discovery, if his life be spared, much may be expected. He is now in Abyssinia. Previously to his departure, he left with us a proof of his possessing the qualifications which are most important for those who embark on investigations respecting the history and affinities of human races in his '*Etudes Grammaticales sur la langue Euskarienne.*'

This work contains, if we are not greatly mistaken, a far more complete analysis of the Biscayan language, and especially of the curious and elaborate construction of its parts of speech, than any that has before appeared. The multitude of inflections which the verb undergoes in its numerous conjugations are clearly and accurately analysed. Each of these modifies the sense in an infinitely more various manner than the conjugations of the Semitic verbs. The only real counterpart or parallel to the conjugation of the Biscayan is to be found in the native idioms of America. In the form of the verb itself, every modification of meaning is expressed, which we denote by inflections, or by adverbs and conjunctions, including even the sex, and number, and person, of the subjects and objects.

For an European of any other race to learn this language by study would be impracticable. The only individual who could by possibility achieve the task would be an Algonquin or Mexican, or some native of the New World, whose original speech, though totally different in vocables, is constructed on a model nearly similar, and indicates at least a direction of mind in the process of the development of language in a great measure analogous. This analogy had not escaped the observation of Vater and Du Ponceau. These writers, who first made known to the learned of Europe the very remarkable facts connected with the structure of the American languages, were at first struck forcibly by the resemblance which discovered itself between those languages and the Biscayan: and they would have been led by it to conjecture that the Basques are the last relic of an ancient population which may have found its way from the west of Europe across the Atlantic, had not a further investigation proved that, together with these marks of resemblance, there are others equally important of discrepancy. It is now the opinion of the best informed philologists that the analogy in structure belongs rather to a similar stage and mode of cultivation, than to the results either of accidental intercourse or of a common origin. Such is at least the supposition in which Du Ponceau and W. von Humboldt acquiesce. They appear to have resorted to it as affording a probable explanation of the facts, without observing that it involves much that is hypothetical. M. D'Abbadie has adopted this opinion. He says,—

"L'Eskuara, différent en cela de toutes les autres langues d'Europe, appartient à la famille des idiomes polysynthétiques, si riche et si variée chez les races primitives d'Amérique, et qui ne se trouve dans l'ancien monde que chez les aborigènes de l'Espagne et du midi des Gaules," &c. —p. 10, *et seq.*

M. D'Abbadie has more fully than any other writer compared the structure of the Basque with that of other languages in the Old World. He gives an account of its relations to the Finno-Hungarian idiom, the only European family of languages which bear such a comparison, to the Georgian in Asia, and to the Wolof or Jolof in Africa. The observations which follow will be read with interest:—

"Passant aux contrées transatlantiques, nous nous trouvons dans le dédale des langues Américaines, si nombreuses et si peu connues. On sait qu'en général ces idiomes ont une ressemblance commune dans la grammaire, mais qu'ils diffèrent singulièrement dans les racines. Des analogies qui se rapportent à ces deux ordres de phénomènes, le fond et la forme, suivent quelquefois une marche parallèle dans les langues que nous avons examinées ci-dessus, le Wolofe seul excepté. Ainsi les langues Laponne, Hongroise, Finnoise, Latine, Hébraïque, Arabe et Georgienne, nous ont offert plusieurs racines communes au Basque, malgré le peu de développement que nous avons donné à ce genre d'étude." "Dans les langues primitives de l'Amérique au contraire, la constitution de chaque mot a une physiognomie étrangère, et pour trouver des rapports avec le Basque, il faut se borner à la grammaire. Ici les analogies sont nombreuses et le seraient peut-être davantage, si la plupart des auteurs n'avaient suivi de trop près la marche de la grammaire Latine. On serait donc obligé de posséder ces langues à fond pour décider si elles ne se prêteraient pas plus aisément à une autre classification." "Le nom Mexicain n'a pas de genre et fait sa déclinaison par des post-positions. Il ne forme pas ces mots composés par une désinence que prendrait l'un des noms composans." "La langue Quichua, parlée par les aborigènes du Pérou, est l'une de celles qui ressemblent le plus à la nôtre," &c.—p. 24.

The author concludes by observing that by assuming as types of this small number of comparisons, the principal languages of the two hemispheres, he trusts to have said sufficient to encourage philologists and those who take an interest in ethnography as a branch of geography, to follow out the connexion, as there is little doubt their labours will not be in vain.

MISCELLANEOUS.

I.—*Recent Intelligence upon the Frozen Ground in Siberia.* By Prof. K. E. Von Baer, Imp. Ac. Sci., St. Petersburgh.

As the Royal Geographical Society of London has taken some interest in the communications it has received on the depth to which the ground at Yakutzk remains in a frozen state, it will, I conceive, be glad to learn the measures adopted by the Imperial Academy of Sciences at St. Petersburgh, to ascertain precisely, not only the law which regulates the temperature of the ground to the depth which is affected by the periodical changes of summer and winter, but also the influence of the external air penetrating into the well or shaft at Yakutzk, upon its sides; and, finally, to ascertain the depth which the summer heats, which in Yakutzk are very considerable in respect to its position, generally reach.

As soon as I had laid before the Academy the observations made by the merchant Schergin, which had been communicated to me by Admiral Wrangel, a commission was nominated for the purpose of pointing out the means which were to be applied to carry them on in the most desirable manner. Accordingly the commission recommended that a thermometer of a peculiar construction should be introduced into the side of the well, and placed horizontally, at the several depths of 1, 3, 5, 10, 20, 50, 100, 150, 200, 250, 300, 350 feet: two thermometers, indeed, at each of these depths, close to each other, one of which should be inserted a foot deep, and the other a fathom, so that the difference between the two would show the effect of the air in the shaft on the temperature of the ground. These thermometers are to remain a whole year in their places, and to be observed daily. A note is also to be taken of the depth at which the ground is thawed in the autumn, on the approach of winter, whether in forest ground, or under the dry steppes, in marshy countries, sandy places, &c. &c. Attention is also to be paid to the greatest degree of heat attained by the ground in these several districts, at a foot and a half beneath the surface. These directions were sent as early as February last, to Yakutzk, with orders to inquire whether any competent person could be found there to undertake these observations. But in all probability the academy will be obliged to send some one thither on purpose, properly qualified. They will also in due time, if the means at their disposal allow of it, direct deeper excavations to be made, to pursue still further the observations on the increase of temperature.

The members of the Royal Geographical Society have very justly observed that the observations of M. Schergin are already sufficient to demonstrate the gradual increase of the temperature, although it appears to be more rapid than has been found in other countries where the ground is not frozen. Nor did the mention I made of the doubts expressed by M. von Buch refer to this point; on the contrary, M. von Buch, in an address which he read to the academy of Berlin in the year 1825, distinctly expressed it as a matter of doubt, whether in a district covered with wood the ground can be in a frozen state at so great a depth as the first excavation in Yakutzk had reached, namely, 91 English feet.

Professor Erman, it appears, is not satisfied with the expression "ground ice" (bodeneis, Germ.) which I have proposed. I did so because it seemed to me to embrace all the requisite modifications, and is very concise. It is clear that ground which is totally free from moisture cannot be frozen; but the ground in high northern latitudes is never in this state. Even the sand, though in the arctic summer its surface may now and then be perfectly dry, is yet before the winter begins always saturated with wet. If we examine ground which contains only very little moisture in a frozen state, it is very difficult to detect the ice, as it forms an extremely thin partition between the single particles of the earth. Should the moisture be somewhat more considerable before the freezing comes on, we perceive in its frozen state small pieces of ice, wherever the spaces between the particles of the soil are large enough to admit them. These bits of ice, which look like small crystals, I have particularly noticed between the upper layer of soil which is thawed, and the lower layer in a frozen state. But in the flat marshy districts of the high northern latitudes, which in Russia are called Fundun (originally a Finnish word), there is so much water in the ground that the quantity of water frequently exceeds that of the soil mixed with it. If in the summer you drive a pole into the turf, which is here formed by the grass, or by the moss, a dirty water, mixed with soil, spirts up in a stream to a considerable height. In some places, too, and particularly those which are surrounded by hills, the ground is covered with pure ice. I have seen such a situation in Nóvaïa-Zemlïa, which was entirely covered with perfect fresh-water ice more than a foot thick. This was not a steep cliff covered over with a bed of ice, like that of which Captain Beechey speaks in Kotzebue Sound, but it lay horizontally upon the ground; nor was it the least doubtful how this layer of ice had been formed. It was almost entirely surrounded by hills: the water from the melted snow had collected in the hollow, and had been frozen in the winter. In the following spring or summer the ice, before it could be again melted, had been covered with warm soil from the neighbouring hills to the depth of three feet; and, as in Nóvaïa-Zemlïa, the increasing temperature of the earth during the summer does not thaw it so deep as three feet, the ice had remained unmelted. From the narrative of Admiral Wrangel's Journey, which I have read in MS., I observe that similar layers of ice are not rare in the valleys of Siberia. It sometimes indeed happens, as I have observed in Nóvaïa-Zemlïa, that the ground is penetrated by perpendicular clefts or shafts of ice. These clefts, which are, however, extremely thin, and which I have never found above four inches thick, occur most frequently in loamy soils. This ground in Nóvaïa-Zemlïa is penetrated by fissures in all directions, which are occasioned by contractions produced by the frost. They are usually from one to three inches in width. Water is collected in them in summer, which is frozen in the following winter; and if the fissures go to any depth, it is never thawed. This is especially the case if the spot be gradually overgrown with a layer of moss. All these modifications may be comprehended under the term of ground ice, which has also the advantage of allowing of the expression, "perpetual or permanent ground ice," as we speak of perpetual snow. Perpetual ground ice is, then, that which is found in the arctic regions, in that layer of earth which is immediately below that which is thawed by the summer heat, reaching, as this does, to the depth where the temperature of the earth is at freezing-point.

It seems to me very important for physical geography to ascertain the thickness of perpetually frozen ground in countries of which the mean temperature is considerably under the freezing-point. I will merely mention one approximation. If, as is the case at Yakutzk, the ground never thaws at the depth of 300 or 400 feet, all the small streams whose superficial waters only are kept in a liquid state in the summer, must be in winter entirely without water, and *vice versâ,* we may conclude that all rivers which do not come far from the south, and of which the course is entirely within those countries which preserve perpetual ground ice, and yet do not cease to flow in winter, must receive their waters from greater depths than those which remain in a frozen state. It is thus clear that these veins of water penetrate the perpetual ground ice. This circumstance strikes me as not devoid of interest with respect to the theory of the formation of springs, and it would be very desirable that some researches upon this subject should be set on foot in high northern latitudes. In the narrative of Admiral Wrangel's travels there occurs a remarkable instance of very considerable rivers in very cold countries being without water in winter, like our ditches and inconsiderable brooks. He was riding (to the north of Yakutzk, in about 65° N.) over the ice of a pretty considerable river, when the ice suddenly gave way, and his horse sunk: he was himself saved by being thrown on the ice, at the moment his horse fell. He was lamenting the loss of his horse to the Yakutzkers who accompanied him, as he knew not how to get another, when they laughed, and assured him they would soon get his horse back, and with a dry skin too. They got some poles, and broke away the ice, under which the bed of the river was perfectly dry, as well as the horse and his pack. The Yakutzkers were therefore aware that there was no water in the winter time at the bottom of rivers of this size; and in this case the water must have disappeared before the ice had gained sufficient thickness to bear a loaded horse and his rider. Similar accidents and similar results must doubtless have frequently taken place during the many journeys which the English have made in North America; and the agents of the Hudson's Bay Company must be well acquainted with the real state of the small rivers in winter in those high northern latitudes, *i. e.*, whether all of them are in a liquid state below the ice or not; *i. e.*, whether water flows in them under the ice or not.

I am collecting materials to ascertain the *southern* limit of the perpetual ground ice in the Old World. Those I have are not yet very complete; but I am already aware that this phenomenon extends much further in a southerly direction in Siberia than in Europe. I do not believe it has yet been observed in Lapland out of the mountainous districts. Nor did I ever hear of it at Archangel; though Herr Schrenk, a young botanist, who, during last summer travelled through the country of the Samojedes in the suite of the imperial botanist Sartens, and who reached the northernmost extremity of Ural and the island of Waigatz, assured me that on the Petchora the ground at a certain depth is never free from ice. It has long been known that further north the ground is at all times frozen, and a thaw is very rare.

The further we go east, the more southerly do we find the limit of perpetual ground ice. Humboldt found in Boguslowsk, in lat. $59\frac{3}{4}$° N., 60° E., at the foot of the Ural mountain, small pieces of ice at the depth of six feet in the summer; but Boguslowsk lies very high. No permanent ice has been found in Tobolsk in 58° N.; but at Berezov, in 64° N., where Erman found the temperature of the ground above + 1 R., at a depth of 23 feet, we learn from other observations, and particularly those of a Russian traveller of the name of Belawski, that the lower districts are never without ice in the ground; so that Berezov is probably very near the limit of perpetual ground ice; for it is clear that the peculiarities of soil must have considerable influence in countries which lie near this limit. Further east, this frozen soil extends much more to the southward. Already in the last century, travellers who visited Siberia, and especially Georgi, related that ice remained in the ground throughout the whole year, in the environs of the Lake Baïkal, and upon an island in lat. 52° N., 106° E. long. It was generally supposed that this circumstance was occasioned by the high mountains surrounding the lake. But Herr von Helmersen has communicated to me some ancient notices, from which it seems that the whole of this south-east angle of Siberia has perpetual ground ice. Captain Frehse, who, according to 'The Berlin Review,' went in search of gold-sand in 1836, states that so early as September, *i. e.*, very little before the beginning of the winter in those latitudes, he found the ground frozen at some distance below the surface, and that this frozen stratum was continued uninterruptedly quite to the under-lying rock, to the depth of from one to six Russian sashén, 42 English feet. But, as at that depth there was always rock, it was not possible to say how thick the layer of frozen earth would be under this latitude (52°). It thawed on the surface of the banks of the river to the depth of from 2½ to 6 English feet, and still deeper (6 to 9 feet) on the naked heights. But in the forests, where the light of the sun was intercepted, the thaw reached only from ¾ to 6 feet. If it be true that there are places in forests where the ground is never thawed a foot deep, it would demonstrate how little it is necessary for the ground to be thawed for trees to grow on it. The development of the leaves of the pines depends less on the temperature of the soil than on that of the air in the spring; it only requires that the ground should be so far thawed that the tree may be able to draw from it a sufficient quantity of moisture for its growth. I was convinced of this upon observing in Lapland large leaves already upon the shrubs when the thaw had scarcely reached a foot deep in the swampy soil on which they were growing; but the air was in consequence of the long days (in June) very warm. On one occasion, after having dug down six feet through the frozen ground in these trenches in the district of Nertchinsk, they came upon the pure ice, an archin, or 2¼ feet thick, and inclosing boulders (gerölle) of different kinds of rock. According to other observations, made equally in the circle of Nertchinsk by Colonel Tatarinoff, the frozen stratum was not more than six feet thick; and the annual thaw of the superficial stratum in each summer is rather less. The greater thickness of the frozen stratum, in the observations of Captain

Frehse, may be accounted for by the fact that they were made in higher latitudes; nor ought we to forget that the country of Nertchinsk is mountainous, and lies high.

Still further east, again, the perpetual ground ice is found at a less depth southwards, probably because the neighbourhood of the sea raises the temperature of the soil. Erman, at least, found no ice in the soil at Okhotzk.

It seems that Fort Churchill in America, in lat. 59° N., long. 94° W., lies exactly on the limit of perpetually frozen ice, as the mean temperature of this place is only a little below freezing point. (Signed) BAER.

II.—*Astronomical Positions in European Turkey, Mount Caucasus, and Asia Minor, fixed* by F. G. W. STRUVE, from the Observations by Officers on the Imperial Russian Staff, in the years 1828 to 1832. Communicated by Captain H. G. HAMILTON, Royal Navy.

[The following important Table of Geographical Positions is extracted from the *Bulletin de l'Académie Impériale des Sciences de St. Petersbourg*.]

DURING the campaigns of the Russian armies in the last Turkish war in Europe and Asia, a series of astronomical and geographical observations was carried on, which were as important as they were remarkable for their extent and accuracy. On the appointment of the Director of the Dépôt of Maps belonging to the Imperial Staff (Lieutenant-General von Schubert), some officers were named to make astronomical observations at the points occupied by the armies, and were for this purpose supplied with a complete set of instruments. These observations were made in European Turkey, from 1828 to 1832, by Captains Vrontschenko, Ortenberg, and von Essen. They made use of two portable transit instruments, two astronomical theodolites, pendulums and chronometers, and telescopes for the observation of occultations, sextants, and a magnetic dipping-needle. In the further Caucasus and in Asia Minor, Captain Birdin worked with similar means, but without theodolites or a dipping-needle. On this side of Caucasus, and at some points of Turkey in Europe, the observations were made by officers of the Topographic corps. These observers had previously followed a complete course of practical astronomy at the Dorpat Observatory. From this circumstance, their labours became an object of peculiar interest to me; and I had much pleasure in undertaking the office of comparing their occultations and lunar culminations with corresponding observations—of deducing the longitudes from them, of revising the latitudes, and azimuth and chronometrical longitudes of the observers, and of bringing out the final results, particularly of the longitudes, by a comparison of what had been done by the several methods. For this purpose the original journals, as well as the calculations of the observers, were all brought to Dorpat. In making these various calculations, I had the assistance of the officers, both of the staff and of the fleet, who were residing there for their astronomical studies, as well as of some other young men attached to this pursuit. Besides, the astronomer Captain Lemm, of the Imperial Staff, had been ordered to Dorpat for nine months, as well to assist in the calculations as also to prepare the manuscript for the press. When this was completed, nothing remained for me to do but the comparison I have already alluded to, and the bringing out the results, which I took in hand a short time ago; and I have thus been able to bring to a conclusion, after several years' labour, the working up of these rich materials.

The result of this operation is the fixing astronomically of 89 points in Turkey in Europe, of 14 points in Caucasus and in Asia Minor, in addition to 6 positions on Ararat ascertained by Fedorov—making in all 109 points.

Of these points, the latitudes of 85 are determined by observations made with the astronomical theodolite; and these are correct, at least, to two seconds. The sextant was made use of for those in Servia, Caucasus, and Asia. In Servia also Captain von Essen observed with this instrument the altitudes of the stars. The agreement of the results by the northern and southern stars, generally α Aquilæ and the polar star, ensures the certainty of the latitudes within ten seconds. In Asia Minor, Captain Birdin found the latitudes by altitudes of the sun, with the exception of Erẓ-rúm, where he made use of the circle in the prime vertical, the results of which agree very well with that obtained from the altitudes of the sun. Fedorov determined with great accuracy the latitude of Tiflís with the theodolite, as well as that of one of the points on Mount Ararat, from whence the others were trigonometrically deduced.

It is well known that longitudes are much more difficult to fix than latitudes. It is only of late years that the right ascensions of the moon and of neighbouring stars have been applied for fixing longitudes correctly; and they were first used for geographical purposes by Preuss, in fixing the position of the harbour of St. Peter and St. Paul in Kamtchatka, and of St. Francesco in California, during the second voyage of Kotzebue round the world. This was prior to their application to the same end by Foster, when with Parry in Port Bowen. If the Russian navy may claim the honour of having first used this method, where it is much wanted—namely, in fixing positions at a distance from observatories—it has been much more extensively adopted by the officers of the Russian army: inasmuch as 22 longitudes, fixed in this way by lunar culminations, served as a basis for all the rest. Of these 22 longitudes, 16 belong to Turkey in Europe, and 6 to the other continent. Ten positions—that is, 7 in Europe and 3 in Asia—were also observed by occultations of stars. Finally, the longitudes of these positions, Várnah, Burgház, and Adrianople, were correctly fixed by a new method proposed by me—that is, by observing the zenith distances of the moon, and of a neighbouring fixed star. The advantage of this method is greater in proportion as the moon is nearer to the meridian. The coincidence of the longitude of Várnah, as found by this method with that derived from lunar culminations, as well as that of the longitudes of Burgház and Adrianople with those which Captain Manganari ascertained by the chronometer on some neighbouring points of the coast, from the observatory of Nicolaye, have proved the utility and the correctness of this method.

From the positions thus fixed, the longitudes of the remaining points have been ascertained, by the chronometer or by instantaneous lights, and by latitudes and azimuths. Besides single chronometric observations, five journeys (comprehending a greater extent of country) were also made with chronometers; among which journeys the most important was that which extended from Tchernetz, on the frontiers of Servia, upon the Danube, as far as Shurshhah, and thence northwards as far as Butushan in Moldavia. Five chronometers were used on this journey—(Barraud No. 542 and 810, Parkinson No. 542, Breguet No. 4160, Arnold No. 293).

The materials thus obtained for ascertaining the longitudes of places in European Turkey, between the Servian frontier and the Black Sea, have been so important, that, after a comparison between the mean longitudes and the various differences, a very remarkable degree of certainty has been obtained for all the results.

The following table contains all the longitudes, latitudes, and magnetic variations. In those longitudes which are marked with one asterisk there may be an error of something more than 10 seconds of time; in those with two asterisks, the error may be something more than 20 seconds of time. All the other longitudes must be correct within from 2 to 4 seconds of time:—

No.	Name of the Place.	Locality.	Latitude, North.	Longitude East of Greenwich.	Variation. Date.	Variation. West.
		BESSERABIA.				
			° ′ ″	° ′ ″		° ′
1	Ismá'íl . . .	Sobor Cupola*	45 20 29·6	28 47 49		
2	Tuchkovah . .	Church Tower	45 20 35·6	28 49 29		
3	Soflyání . . .	Church Tower	45 24 16·0	28 51 50		
4	Kíliyá . . .	Sobor	45 26 2·6	29 15 56	1828·77	8 53·2
		MOLDAVIA.				
5	Yassi . . .	Church of St. Charalampia .	47 10 24·1	27 34 43	1828·56	11 50·7
6	Chitatsua Convt.	North Tower	47 8 13·9	27 34 45		
7	Skuléni . . .	Posthouse	47 19 13·1	27 36 44		
8	Butusháu . .	Church of St. Elias . . .	47 45 4·5	26 39 38	1831·24	9 43·5
9	Rómano . . .	Armenian Church . . .	46 55 21·8	26 55 19	1828·67	11 30·5
10	Fokshán . .	Church of St. Spiridion . .	46 13 49·7	27 54 17	1830·69	10 44·9
11	Berlát . . .	Church of St. John . . .	45 41 48·9	27 10 19	1829·30	11 19·0
12	Galátz, or Galách	Uspenski Church . . .	45 26 12·3	28 2 56	1828·70	11 1·9
		WALLACHIA AND BULGARIA.				
13	Ibráïl, or Bráïlov	Minaret of Láz-jámi' . .	45 16 11·3	27 58 11	1828·72	10 43·1
14	Ditto . . .	Church of St. Michael . .	45 16 19·8	27 58 10		
15	Buzeo . . .	Ch. of the Banu Monastery .	45 9 1·4	26 48 52	1829·36	12 4·8
16	Ditto . . .	Church of Episcopia . . .	45 9 31·9	26 48 32		
17	Ployeshti . .	Uspenski Church	44 56 20·7	26 16 9	1828·98	11 48·8
18	Bucharest, or Bukuresti	Metropolitan Church . . .	44 25 39·0	26 35 23	1829·05	9 14·4
19	Kalarash . .	Church	44 11 29·2	27 19 24	1829·54	11 13·8
20	Silisterah, or Silistria	Mesjid, *i.e.* Mosque . . .	44 7 9·8	27 14 41		
21	Shurshah . .	Church Tower of St. Nicholas in the Citadel	43 53 14·7	25 57 33	1831·39	9 7·3
22	Rushchúk . .	Tower	43 50 36·8	25 56 40		
23	Zimnitza . .	Church of St. Constantine and Helena	43 39 8·0	25 21 14	1831·40	10 15·2
24	Sistovah . .	Mosque	43 37 13·8	25 19 45		
25	Mogureni . .	Village	43 44 39·1	24 52 9	1831·40	11 1·7
26	Nicopolis, or Ní-gebólí	Eastern Mosque in the Fort	43 42 18·1	24 53 19		
27	Kalafát . . .	Village	43 59 33·9	22 55 36	1831·65	11 42·3
28	Viddín . . .	Tower	43 59 28·9	22 52 42		
29	Viddín . . .	Mosque in the Citadel . .	43 59 34·9	22 52 50		
30	Chernetz . .	St. Troitzki's Church . .	44 38 3·9	22 42 6	1831·65	14 51·0
31	Ditto . . .	St. Peter's Church . . .	44 38 1·5	22 41 13		
32	Ditto . . .	St. Nicolas' Church . . .	44 38 24·6	22 42 16		
33	Ditto . . .	Lazaretto	44 37 22·8	22 36 24		
34	Kladovah . .	Mosque	44 36 53·1	22 35 57		
35	Tirgoshil . .	Bisserika Domneska Church Tower	45 2 10·5	23 15 53		
36	Ditto . . .	Delya-Kimp Church-Tower .	45 2 0·7	23 16 16		
37	Kráyovah . .	St. Anne's Church-Tower .	44 19 23·5	23 47 39		
38	Ditto . . .	All Saints' Church-Tower .	44 19 8·4	23 47 43	1831·64	12 48·1
39	Ditto . . .	St. Nicolas' Church-Tower .	44 19 7·5	23 47 20		
40	Ditto . . .	Tower of the Light-house .	44 19 7·6	23 47 10		
41	Slatina . . .	St. Troitzki's Church-Tower .	44 25 55·8	24 20 56	1831·60	13 22·8
42	Piteshti . . .	St. Elias' Church-Tower . .	44 51 4·8	24 52 4		
43	Ditto . . .	Uspenski Church . . .	44 51 16·9	24 52 1		
44	Ditto . . .	Tower of St. George's Church	44 51 25·3	24 52 0	1831·58	12 47·0
45	Tirgovist . .	Church Tower of St. George	44 56 15·2	25 26 27		
46	Ditto . . .	Ch. Tower of St. Demetrius	44 56 6·2	25 26 35		
47	Ditto . . .	Tower of the Metropolitan Ch.	44 55 33·6	25 27 5	1831·57	12 48·4
48	Ditto . . .	Tower of the Mountain Convt.	44 57 29·4	25 28 22		
49	Bábá-tágh . .	Minaret	44 53 39·6	28 44 25	1828·82	12 9·8
50	Hirshóvah . .	Mosque	44 41 3·8	27 54 30	1828·94	11 48·8
51	Chernovodi . .	Mosque	44 20 23·2	28 0 26		
52	Zimíni-dindel .	Centre of the Village. . .	44 22 27·3	28 0 50	1828·93	11 40·4
53	Tarkutaï . .	Mosque	44 3 35·5	26 36 35	1829·50	11 36·0

* The Cathedral.

No.	Name of the Place.	Locality.	Latitude, North.	Longitude East of Greenwich.	Variation. Date.	Variation. West.
		WALLACHIA AND BULGARIA—CONTINUED.				
54	Kustenje, or Chiastenza	Mosque	44° 10′ 21″·2	28° 42′ ″2	1828·93	11° 32′·8
55	Mankaliyá, or Mangalia	Mosque in the Market-place	43 48 31·3	28 37 18	1830·48	12 13·1
56	Bázárjik . .	Mosque in the Market-place	43 34 16·8	27 54 3	1830·36	10 41·2
57	Kávarnah, or Egerneh	N. W. Mosque	43 25 49·8	28 22 50	1830·46	10 11·9
58	Várnah . . .	Mosque of Ḥasan Baïrákdár	43 12 3·3	27 57 33	1829·72	9 49 5
59	Parávádí, or Pravódí	Mosque	43 10 30·4	27 28 7	1830·44	14 41·1
60	Yeñí bázár . .	Mosque	43 20 32·1	27 13 24	1830·36	11 6·2
61	Shumlá . . .	Minaret	43 17 23·2	26 58 47		
62	Near Shumlá .	Point of Rock	43 14 55·4	26 56 53		
		BULGARIA AND ROUMELIA.				
63	Slivno . . .	Mosque, called Hájí Ibráhím jámi'-sí	42 40 45·0	26 19 48		
64	Karínábád, or Carnabat	Mosque, called Hájí Ades jámi'-sí	42 38 57·6	27 1 13	1829·70	11 20·2
65	Aïdós, (Aëtos)	Mosque, called Gornatá-jámí	42 42 17·5	24 48 28	1829·70	11 32·3
66	Misivrí (Mesembria)	Metropolitan Church . .	42 39 44·9	27 47 28	1829·68	10 47·6
67	Akhyólí (Anchialé)	Preobrazhenski Church . .	42 33 25·05	27 42 21	1829·69	11 19·3
68	Sozeh-bólí (Sozopolis)	Centre of the Hospital, on an insulated height in the City	42 26 46·3	27 45 25		
69	Burgház (Pyrgos)	Chief Mosque in the Bázár .	42 29 35·87	27 31 48	1829·70	115 ·4
70	Yáubólí (Joannopolis)	Old Jámi'	42 29 6·3	26 33 31	1829·84	11 35·3
71	Adrianople . .	Tower of the Eski-Saraï (old Seraglio)	41 41 26·5	26 35 41		
72	Kirk kilísá (40 Churches)	Mosque, called Mohammed Debender-jámi'	41 43 58·8	27 12 38		
73	Vízah (Byzia) .	Mosque, formerly the Ch. of St. Nicolas	41 34 26·9	27 45 28		
74	Saráï . . .	Mosque, called Ages Páshá-jámí'	41 26 27·0	27 56 23		
75	Chórlú . . .	Yení Jámí' (*i.e.* New Mosque)	41 9 46·3	27 48 2		
76	Lúleh Burgház .	Principal Mosque	41 24 25·1	27 21 21	1829·75	11 25·0
77	Dímotíkah (Didymotichos)	Tower of the Castle . . .	41 21 3·5	26 30 39	1829·74	11 41·4
		SERVIA.				
78	Porech . . .	Island in the centre of the City	44 30 36·3	22 2 51		
79	Gradeshti . .	End of the Great Road on the bank of the Danube . .	44 45 53·8	21 29 37		
80	Smedreva . .	Stone Church in the suburb towards Belgrade . . .	44 39 51·3	20 24 17		
81	Shabtza . . .	Suburb, not far from the new stone church	44 45 22·1	19 41 58		
82	Belgrade . .	Vracha, Prince Milosh's country house, outside of the fortress, not from the river Tópchíder	44 47 57·0	20 25 12*		
83	Krakoevatz .	Centre of the City . . .	44 0 29·4	20 55 32*		
84	Karanovatz .	Soshestwi Church . . .	43 43 26·2	20 39 17* *		
85	Chachak . .	Near the Church of All Saints, now a mosque—stone-house with a cupola	43 53 29·7	20 21 7* *		
86	Svoïdrug . .	Pot-house	44 2 15·3	19 16 45* *		
87	Tupriyeh . .	Lazaretto	43 56 3·3	21 19 59* *		
88	Poshegi . . .	Pot-house	43 50 49·6	19 46 44* *		
89	Yassika . . .	Centre of the City . . .	43 36 37·3	21 17 9* *		

No.	Name of the Place.	Locality.	Latitude, North.	Longitude, East of Greenwich.	Variation. Date.	West.
		CAUCASUS AND ASIA MINOR.				
			° ′ ″	° ′ ″		′ °
1	Tiflís . . .	Height in the Governor's Garden	41 41 4·0	44 50 39		
		A place near it, south of the Catholic Church . . .	41 41 27·4			
2	Piati-gorsk . .	Alexandrov's Spring . . .	44 2 39·3	43 5 8		
3	Kizlovodsk . .	Centre of the Fortress . .	43 54 21·0			
4	TzarskieKolodzi	. . .	41 27 57·7			
5	Jar, or Novaya Zakatali . .	. . .	41 37 40·6			
6	Nukhi . . .	. . .	41 11 46·3			
7	Akhaltzikeh .	Second Gate of the Citadel .	41 39 4·7			
8	Kárs	In the Fortress	40 37 1·7	43 9 2		
9	Ardagán . .	Near the Fortress, the House of the old Caravanseraï .	41 7 15·9	42 48 24		
10	Gúmrí . . .	Close to the Lazaretto Surgery	40 46 57·6	43 46 54		
11	Hasan Kal'eh .	. . .	39 58 47·4			
12	Erz-Rúm . .	. . .	39 55 16·1	41 18 31		
13	Báíbúrt . . .	. . .	40 15 36·5	40 8 48		
14	Kalkit chiftlik .	. . .	40 8 2·6	39 10 23		
		FEDEROW'S DETERMINATIONS ON ARARAT.				
15	. . .	Point of Little Ararat . .	39 39 10·68	44 24 34		
16	. . .	Foremost point of Great Ararat	39 42 24·17	44 17 53		
17	. . .	Hindermost point of Ditto .	39 42 21·94	44 17 38		
18	. . .	Convent of St. Jacob . . .	39 46 12·10	44 21 52		
19	. . .	Village of Bajád	39 52 38.78	44 31 5		
20	. . .	Point of the Hill of Alagetz .	40 31 35·65	44 11 23		

III.—*On the recent Russian Expeditions to Novaïa Zemlïa.* By Professor K. E. von Baer.

[Abridged from the *Bulletin de l'Académie Impériale de Sciences de St. Petersbourg.*]

[While our enterprising countrymen in the west are gradually prosecuting discovery on the northern shores of America within the arctic circle, it may not be uninteresting or uninstructive to turn our eyes to the east, and observe what our active rivals in the polar regions, the hardy Russians, are doing towards the improvement of the geography of the more northern lands that lie immediately adjacent to their own coasts;—and with this view, as also with the object of enriching our Journal with some valuable geographical information, an outline is subjoined of the recent expeditions to the islands of Novaïa Zemlïa, abridged from several communications made by Professor Baer, during the past year, to the Academy of Sciences at St. Petersburg, and accompanied by a chart showing the *actual* outline of its coasts, as traced by the pilot Ziwolka, from the latest examinations, by which it will be seen that more than the eastern half represented on our maps has no existence in reality.—Ed.]

The first discoverer of these islands, as far as is known, appears to have been Stephen Burrough, who in the year 1556 reached Waigatz Island; this being only separated from Novaïa Zemlïa by a strait thirty miles wide, there can be little doubt but that he saw its southern coast. Forty years later Barentz and the crew of a Dutch vessel wintered on the islands, and in the seventeenth century the Dutch are said to have pushed some distance to the eastward; but the eastern coast does not appear to have been approached before the time of Loshkin, a walrus-fisher, who in 1742 perished near the river, named after him the Sawina, in 71° 30′ N. lat. Twenty years later his countryman Rosmysloff penetrated through the Matotshkin-Shar, a strait so called extending in an east and west direction about forty-five miles long by three broad, which in the parallel of 73° 20′ divides the length of Novaïa Zemlïa into two nearly equal portions.

From 1762 till the year 1819 no expedition is on record, yet, doubtless, its shores were visited annually by seal, whale, and walrus-fishers; but between that and the year 1824, five successive exploring voyages were ordered by the Russian government. Four of these were under the command of Lieutenant, now Admiral, Lütké; yet all the skill and perseverance of this officer could not effect the desired object, namely, to explore the eastern shores of the islands.

The attempt, so often frustrated, to survey the shores of Novaïa Zemlïa, would probably have been abandoned but for the enterprise of Mr. Brant, a merchant of Archangel, who, desirous of restoring the former trade along the northern coast of the continent, from the White Sea to the gulf of Obi, and of establishing the walrus-fishery on the eastern shore of Novaïa Zemlïa, fitted out three vessels commanded by experienced officers. One vessel, commanded by the Master Pachtussoff, was to sail through the strait of Kara, and to examine the eastern shore of the islands. The second, under Lieutenant Krotoff, was to keep on the west coast as far as the entrance of the Matotshkin-Shar, then working through the strait, to steer for the gulf of Obi, or the mouth of the Yenisei. The third was merely to fish for walrus on the west coast, in which she was successful.

Quitting Archangel together on the 13th of August, 1832, Pachtussoff separated in a fog from Krotoff in the White Sea; they rejoined off Kanin Noss, and again parted for their respective stations, since which nothing has been heard of the latter, but the wreck of a large ship found in 1834 by Pachtussoff, at the western entrance of the Matotshkin-Shar, leaves no doubt as to the fate of Krotoff and his crew.

Pachtussoff was more fortunate; he reached the south shore of Novaïa Zemlïa, and, steering to the eastward, he met with so much ice that he was obliged to take up his winter-quarters on the 12th September in Felsen or Rocky Bay; and employed himself and his crew for the dreary months he remained here as usual with persons in similar circumstances. On the 6th May a heavy snow-storm occurred, and lasted three days, which M. Baer observes is remarkable for its great extent, having learnt from Mr. Helmersen, who was at that time travelling in the Ural mountains, that it was felt throughout the whole extent of that range as far as the parallel of 48°, a distance of nearly 1500 miles from the coasts of Novaïa Zemlïa.

On the 6th July Pachtussoff profited by a clear sea in the Strait of Kara to examine in his boat the eastern coast as far as a small river in 71° 30′ N., now named the Sawina, from the circumstance of the cross left by Ssawa Loshkin, who perished in 1742, having been found there.

On the 26th July, ten months from their arrival, the vessel was extricated from the ice, and left Felsen Bay, and they continued to the northward along the eastern coast, at one time detained in Lütké's Bay in 72½° N. for eighteen days. On leaving it they met with an iceberg, grounded in eleven fathoms, and on the following day, August 25, Pachtussoff entered the Matotshkin-Shar, having surveyed the eastern coast of the southern island.

At the western outlet of the strait he encountered a heavy gale of wind, which obliged him to run for shelter to Petchora on the Russian coast, where on the 15th September the vessel was wrecked, and he returned by land to St. Petersburg.

The following year, 1834, the Russian government sent two vessels under the command of Pachtussoff and the master Ziwolka to follow up these discoveries. They sailed from Archangel on the 5th August, pursued different courses along the shores of Novaïa Zemlïa, and met again at the western entrance of the Matotshkin-Shar. By the 26th September they contrived to effect their passage to the eastern outlet of the strait, but there the closely-packed ice compelled them to retreat to their winter-quarters, near the western end, which they occupied on the 20th October. In the following spring they constructed sledges; Pachtussoff completed the survey of the strait, while Ziwolka with great exertion pushed 100 miles to the northward along the east coast as far as Cape Flottoff, but want of provisions obliged him to return.

Pachtussoff, who had by this time completed a boat, started on the 12th July to the northward along the west coast, and met with the first ice off Admiralty Peninsula on the 8th July; two days after the ship was nipped and went to pieces; the crew were saved and proceeded to the northwards, dragging their boats over the fields of ice as far as Bucklige Island, in 75° 45′ N. The persevering Pachtussoff here set about surveying the neighbouring islands and coasts as far as 76° N. when a walrus-fisher approached the coast, rescued them from imminent peril, and carried them back to their winter-quarters. On the 26th August Pachtussoff started for the east coast, and explored as far as the island which bears his name, 74½° N., when the ice compelled him to retreat, and he returned to Archangel, where shortly after his arrival he died.

The ardour of M. Baer in natural history would not permit the examination of these islands to remain in this imperfect state. By his representations another expedition, being the eleventh to these shores, was fitted out under the pilot Ziwolka, accompanied by MM. Baer and Lehman as naturalists, and sailed from Archangel on the 1st July, 1837: touching on the coast of Lapland, they reached the western entrance of the Matotshkin-Shar on the 29th, where they found the sea in high latitudes clear of ice. With some difficulty they succeeded in penetrating to the eastern end of the strait, but stormy weather drove them back to their former anchorage. Excursions were made in various directions to explore the natural history of the country, &c. On the 16th August they quitted the strait and steered to the southward, examined Nameless Bay in lat. 73°, and proceeded round the western extreme of the land to the strait on the S.W. coast called Kostin Shar, which was found studded with small islands; here they ascended the small river Nechwatowa into the interior, but, the weather becoming very stormy, on the 12th September they quitted these shores for Lapland, and finally reached Archangel on the 23rd September.

Although but six weeks were spent on the coasts of Novaïa Zemlïa, and no opportunity was afforded for a further examination of its shores, still the voyage was far from being unproductive in scientific results, and especially in natural history. These may already be partly found, and will be more fully developed hereafter, in M. Baer's reports to the Academy of Sciences at St. Petersburg, and in which he has also discussed the latitudes and longitudes of the various points, the result of which is shown in the chart, and has entered fully into the meteorology and climate of Novaïa Zemlïa; but our object is entirely with the physical geography of the islands. On reference then to the chart, it appears from the recent examinations that Novaïa Zemlïa stretches

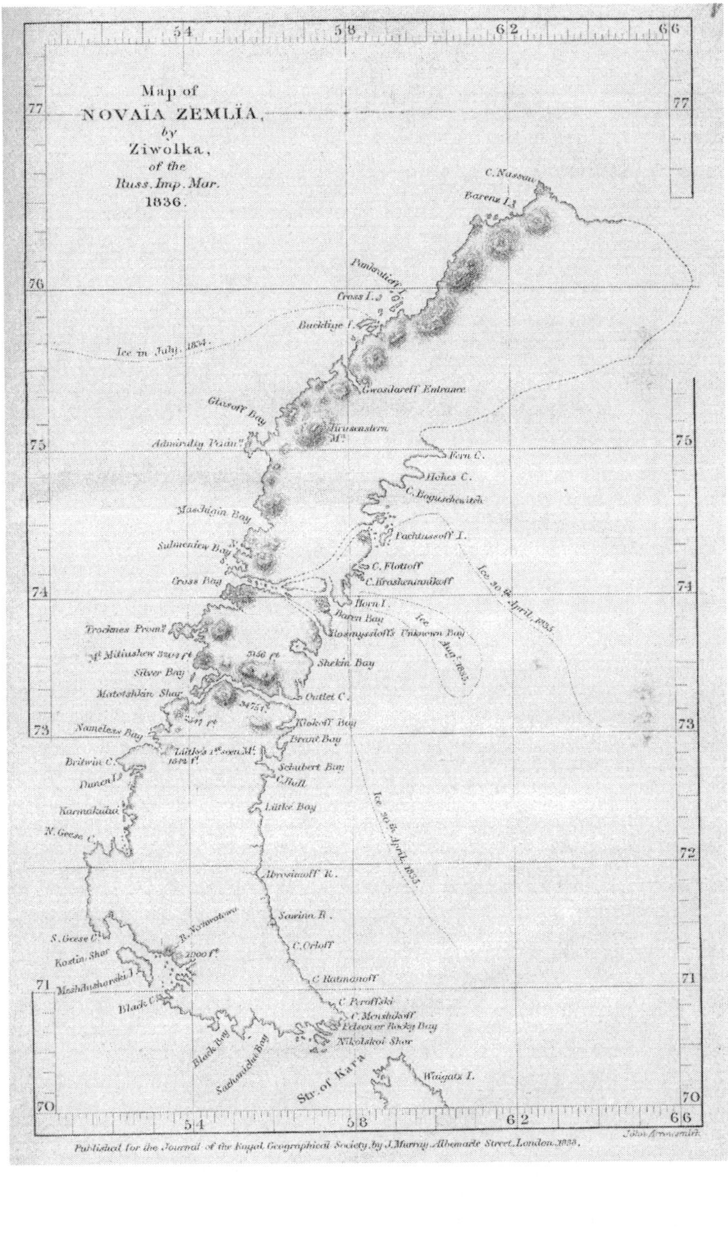

Published for the Journal of the Royal Geographical Society, by J. Murray, Albemarle Street, London, 1838.

in a N.N.E. and S.S.W. direction, slightly curving to the westward, nearly 400 geographical miles in extent, between the parallels of 70½° and 76½° N. and 52° and 66° long. E. of Greenwich, with an average breadth of 50 miles, or about one-half the size hitherto represented on all our maps. The land on the western side of the islands is mountainous, rising generally 2000 feet above the sea; the mountains overhanging the Matotshkin-Shar both on the north and south side reach respectively 3156, 2547, 3204, and that nearer the eastern entrance of the strait full 3475 feet above the level of the sea: they are chiefly composed of a very black clay-slate: in the more southern portions of the islands near the Kostin Shar, grey primitive limestone prevails, similar to that found in the northern parts of the Ural chain, of which Novaïa Zemlïa may be considered a northern prolongation.

The eastern shores are comparatively low and barren. The various portions of the coast are coloured on the original chart to show by whom they were explored; whence we learn that nearly all the west coast was copied from Lütké's chart; that all the east coast of the southern isle was explored by Pachtussoff in 1832-3; that the Matotshkin-Shar, and the east coast, from 73¼° to 75°, was examined by Pachtussoff and Ziwolka in 1834-5; from 72° to 73° on the west coast is the information derived from Bashmakoff, a walrus-hunter; and the rest of the dotted line, or the north-eastern portion of the coast, that obtained from other walrus-hunters.

About 100 miles of the north-eastern shore yet remain unexplored; but we learn from a letter from M. Baer to Dr. Berghaus,* that the master Ziwolka was to sail again this summer with two vessels, with the intention of wintering at Bucklige Island, in 75° 45′ N., on the west coast, and in the year 1839 was to use all his exertions to sail round the north-eastern point of Novaïa Zemlïa, and there complete the discoveries carried on in so persevering and praiseworthy a manner by the Russian government.

* Annalen, No. 154.

IV.—*Notes on Cloudy Bay and Harbour, and on some other Harbours in New Zealand.* Communicated by Captain W. C. Symonds, 74th Regiment.

So little is generally known of the coasts and harbours of New Zealand, that I am tempted to offer, for the perusal of any who may take an interest in its geography, the following notes gathered from several quarters, and originally put together in another form, for my own guidance, in the event of my ever carrying into effect a project long contemplated, of visiting that delightful country. If they have not the merit of being the result of personal observation, the correctness of their details will, I hope, entitle them to some consideration.

It is a subject for regret that, among the many who have lately visited New Zealand, none should have communicated the contents of his note-book to the Geographical Society, more particularly if his pursuits should have led him to the southward of Cook's Straits. All who have published their remarks have limited them to a description of the northern part of the northern island in its narrowest part, and very few have even there explored beyond much above a dozen miles from the coast, though New Zealand has been the resort of Englishmen for upwards of five-and twenty years. All that can be learned of the southern and largest island is chiefly culled from the crude remarks of whalers, who naturally take but little interest in the coast beyond its capabilities of affording them an easy means of speedily filling their butts and shelter from occasional storms; consequently our information respecting it is very meagre, and confined principally to accounts of two or three harbours, and the appearance of parts of the eastern shore.

Owing to the prevalence of westerly winds, the western coast of either island is little frequented.

In the northern island we have a knowledge of the existence of several very fine harbours on the western coast, namely, Wharo, Whangape, Hokianga, Kaipara, Manakou, &c., in all ten; but of these (Hokianga excepted, of which there exists a tolerable survey) very few have been explored. We are chiefly indebted to the missionaries, and also to Lieut. Mc Donnell, R.N.,* for what we do know of them. The following account of those most known is extracted from details for which the writer is, I believe, indebted to the Rev. William White, late chairman of the Wesleyan Mission in New Zealand.

"The harbour of Kaipara is distant, southwards, from Ho-

* The first vessel that entered this harbour was the *Túi*, of 25 tons, on the 23rd Nov. 1835, sent by Lieut. Mc Donnell, R.N., many years resident at Hokianga. This vessel also explored the river Wairoa. Mr. Mc Donnell has obligingly communicated to the Geographical Society a plan of the harbour of Kaipara.—Ed.

kianga about sixty miles. The distance between the two heads of the harbour is about five or six miles. There is a sand-bank about mid-channel, but on each side of it there is abundant water to carry in a vessel of any tonnage in all states of the tide: it is at least ten fathoms deep at low water. A false impression has been created against this harbour in consequence of the position of a very extensive sand-bank which runs out to sea six or eight miles, and extends from the northward of the north head to the southward of the south head, and in approaching from the westward appears to form a complete bar right across the entrance. But it is not so by any means. Inside of this bank there is a deep channel of at least two miles broad at its narrowest part, affording a safe passage for a ship of the largest class to work in. One of the first vessels which discovered this channel was the schooner Fanny, on the 6th January, 1836, at which time were written down the following directions for entering:—'Sailing into Kaipara, middle channel, go well to the southward of the south head; then steer E.N.E. by compass for a green patch on the sandy land, until you bring the middle green patch on the northward N. by E. ½ E.; steer in that course until you are clear of the north end of the inner sand-bank, then steer direct for the inside point of the north head.' After passing the heads the channel extends about five or six miles, when the vessel gains the bay and is completely sheltered from every wind. The expanse of the bay from north to south is not less than twenty-five or thirty miles. On the north there fall into the bay two large rivers, the Wairoa and Otamatea, and on the south the river Kaipara, which gives its name to the bay.

"The banks of these rivers abound in magnificent timber, particularly the banks of the Otamatea. The natives describe this as the largest pine or cowdie district in the island. Three or four Europeans have lately located here. Towards the head of the Wairoa there is a missionary station. Mr. White recently, at different times, ascended the Kaipara for about forty miles in two vessels, the Fanny, about 45 tons, and the Martha, about 200 tons. The same vessels ascended the Wairoa for about eighty miles, and the natives stated it to be equally navigable for thirty miles farther. From the top of a high mountain there were seen numerous and most extensive forests of timber, chiefly of cowdie. Great part of the shores of the inland bay are covered with forest trees: there are few natives. It is one of the districts nearly depopulated by the celebrated Houghi's wars several years ago."

The next harbour is Manakou, distant from the entrance to Kaipara about 30 miles. A tolerably detailed description of this fine harbour is contained in a work lately published;* but as no

* Present State of New Zealand. London, 1837.

correct survey of it has ever been made, comparatively nothing is known respecting it. More correct information may soon be expected of Manakou, as it is reported that H.M.S. Buffalo has gone there for spars, and will shortly return to England.

To the southward of Manakou is Waikato—distant from Manakou 25 miles. "It is a bar harbour, and no survey of the entrance has been obtained; but vessels of 250 tons have often been in for fresh provisions and water, and for flax. There are a few Europeans settled there, and the natives come down the rivers in their canoes, with flax, in great numbers. This harbour is formed by the Waikato and the Awaroa rivers: the Waikato is stated by the natives to have its source in an extensive inland lake, called Roturoa, but no account has been obtained of a passage from the lake to the sea. About 80 miles from the entrance of the harbour the river divides into two streams, the one called Horoteu the other Waipa."

The Horoteu flows in a N.W. direction, and is supposed to communicate with the lake Roturoa in the centre of the island. The Waipa is believed ~~to rise in the high snowy mountains~~ behind Tui, which are part of a range, being a spur from the great range or back-bone of this island.

The district about the lake Roturoa has the credit of being the most fertile portion of this part of the country, and in this report reliance may be placed, though we are chiefly indebted to natives for the information, assisted by the account given by that extraordinary man Rutherford, who was for several years in the interior.

The river Horoteu offers the easiest means of exploring in these parts; its stream is broad, deep, and unbroken by falls or torrents as far as has, up to the present, fallen under the eye of a traveller. And should it flow out of the Roturoa and prove navigable up to that lake, it is invaluable as the high road for the inland commerce to the coast. The inhabitants of the country through which it runs are few in number, peaceably disposed, and courting communication with Englishmen. The missionaries have not extended their labours beyond the neighbourhood of Waikato, but there they have been well received, and their exertions have been rewarded with success, equal to any which they have met in a more northerly part of the island.

Of the other harbours on this coast little or nothing is known beyond the report of their existence and excellence. Along the shore from Manakou to the river Mokou, and on the banks of the several rivers between these, great quantities of iron is found in the shape of dust or sand, for a distance estimated at 80 or 100 miles.

The principal harbours on the eastern side of the northern island are too generally known to require any notice here. Of the coast I shall offer a few remarks, for which I am indebted to my brother, Captain Symonds, R.N., who was on the coast of New Zealand in H.M.S. *Rattlesnake,* and has recently returned.

The Bay of Plenty is interesting as containing a volcanic island, like Stromboli, always burning, and affording a beautiful spectacle to the voyager by night in the columns of fire which issue from the crater at its summit. It is called White Island, is between 1000 and 1500 feet high, and would be visible from forty to forty-five miles from the deck of a ship.

Of Cape East, Captain Dumont D'Urville's account is generally correct, but he has omitted to call the attention of the navigator to the remarkable and high land which rises immediately behind it, and is estimated to be about 3000 feet above the level of the sea; it is a table land gradually shelving down towards the cape.

The *Rattlesnake,* in her passage from the Bay of Islands to Cloudy Bay, was driven by northerly gales so far to the southward of Cook's Straits that she sighted Banks' Peninsula, which my brother describes as the most remarkable land he ever saw. It is high table-land of about forty miles extent, rising abruptly from the sea, white, and visible from a great distance.

Of the south coast of Cook's Straits we are also indebted for an account to the French navigator M. D'Urville, who, as I before remarked, is considered very correct in his description, though he does not enter into all the details which the peculiarities of the coast seems to call for. He was prevented from visiting Cloudy Bay by a lack of wind; and being carried to the southward by a strong current, which has a set out of the straits in that direction, he anchored at a station six miles to the eastward of Cape Campbell, which he describes as of a moderate height, terminated to seaward by a low spit of land. Mount Tako rears itself behind Cape Campbell, and is a lofty peak covered with snow, seen at a great distance, and an excellent land-mark for making Cook's Straits on this side.

For the details of the following description of Cloudy Bay and Harbour I have to thank Mr. Rossiter, an Englishman, the sailing-master of the French whaler Mississippi, who gave me a copy of a survey* made by himself while lying there; and to my brother, whose notes, though few (having remained in the harbour but twelve hours), are valuable, as confirming the report of Mr. Rossiter. I copy the following from my brother's journal.

"*June* 14, 1837.—After much tossing about in gales all round the compass, from S.E. to N.W., which drove us considerably to the southward in sight of Banks' Peninsula, the most extraordinary promontory I ever saw, we this day made Cape Palliser, a fine, high, bold point, the entrance to Port Nicholson showing

* Presented by Captain Symonds to the library of the Geographical Society.—Ed.

itself very distinctly. A course, west by compass, of forty-five miles, brought us to the entrance of Cloudy Bay, off the White Cliff, its eastern point on entering. This is easily distinguished at some distance, and has behind it a high mountain, consisting of three peaks, the summits of which are covered with eternal snow. The bay affords good shelter from all but S.E. winds, which is the severest gale in these parts. The soundings are from seven to ten fathoms, and is a favourite resort of the black whale.

"While we lay here at anchor we saw several whales killed, and there were not less than thirty boats out, manned by natives, with many European ones also—English, American, and French. The difference in the temperature here and at the Bay of Islands is sensible, being here much colder. From Cloudy Bay we went into Cloudy Harbour, steering a north course; thermometer at freezing point. There is not a single rock or obstacle in the way of entering, and this perhaps is one of the finest natural basins in the world. There are creeks all over it on both sides, into which the whalers haul to boil down their fish. In them is deep water, and the natives supply ~~the ships with pigs, potatoes, fish~~, wood, and water, which are in great abundance and of excellent quality. The country is mountainous and the soil very rich. Queen Charlotte's Sound is a short distance from the head of the harbour, across a neck of land which divides them. Many of our countrymen have settled here, residing in log houses, chiefly on the west coast, and in Cook's Straits; their occupation—procuring oil from the black whale, which runs from forty to seventy feet in length, and gives from seven to ten tons of oil. This they supply to the ships which call here to take in their cargoes."

Mr. Rossiter lay in Cloudy Harbour for some time—in all its different bays on either hand. He gives it the highest character as a haven; its great draw-back however being, that in heavy S.E. gales the sea breaks across its mouth in from nine to seven fathoms water, which is open to the whole roll of the southern ocean; but in any of the bays, which are eight in number, the water is as still as in a mill-pond in the heaviest gales.

The southern point of Cloudy Bay is a white bluff cliff, called by the natives Buri-nui-hohiti, rising from 300 to 400 feet above the sea, and which, from being visible for a considerable distance, is an excellent land-mark. Its southern shore is high for some miles, as far as the mouth of the Wai-rau River, which falls into it at its south-western angle, and runs nearly parallel with the west shore, forming a long narrow peninsula, low and swampy at its extremity and covered at its higher and upper part with a forest of fine timber of three or four leagues in length. The Wai-rau has its source in the Snowy Mountains, and is navigable for several miles from its mouth: its banks swarm with ducks and other wild fowl of several species. Here also the apteryx abounds and offers capital sport to the hunter, to whom it is valuable for its feathers, which are held in high estimation.

Cloudy Harbour is in the northern shore of this bay, and extends in a N.N.E. direction for about 5 miles; the breadth from cape to cape, at its mouth, being about 1¼ mile; the eastern point of entrance lies in lat. 41° 20′ S., long. 174° 10′ E. of Greenwich, and from this the White Bluff at the southern entrance of the bay bears S. by E. ½ E. about twelve miles.

The head of the harbour is divided from Queen Charlotte's Sound by a narrow neck of land, over which the natives drag their canoes, and not exceeding, according to Mr. Rossiter, half a mile.

The five small bays on the western side of this harbour afford excellent anchorage, with a bottom of sand, and shells, and mud; deep water close in-shore. The best of these are "Europe Bay," "Tom Cary's Bay:" "Liln's Bay;" the country about them is beautiful; the soil fertile, and rendering a plentiful supply of provisions, wood, and water, to the many ships which frequent Cloudy Harbour. Several Englishmen have settled on their shores, and possess all things requisite for refitting ships of the largest dimensions. The natives are very few in number, having been nearly exterminated by the Kapiti tribe from the southern part of North Island and Entry Island. They live scattered on the banks of the Wai-rau and its neighbourhood, in miserable huts, in extreme poverty and wretchedness, whilst their foes, possessing the finest part of their territory, drive a lucrative trade with the whalers, and aid the English settlers in their establishments.

The three bays on the eastern side are larger than those on the western, affording equally good anchorage, and perhaps more perfect shelter from S.E. gales, but the country around them is steril for the most part, the inhabitants few, and the water of a bad quality. Their names are; "Solyan's Bay," "Martin's Bay," and "Poverty Bay."

"Point Underwood" divides the head of the harbour from "Poverty Bay," by a peninsula 2 miles long, by ½ a mile broad, and forms a beautiful, land-locked bay, 3 miles in length by about 1 in breadth, sheltered from all winds, with from 5 to 7 fathoms water, muddy bottom. This point received its name from being covered with low brush-wood, flax, &c.

"Queen Charlotte's Sound"—an excellent harbour in D'Urville's Island, Cook's Straits—"Akerva," in Banks' Peninsula—and "Dusky Bay," on the S.W. coast, are the only other havens in this island of which we have any account. All, except the harbour in D'Urville's Island, are the resort of sealers and whalers, and have been surveyed. Of the interior nothing is known. "The Southern Alps," which extend along the whole length of the island, *are said* to abound in coal and iron: on what authority this report is grounded I know not, but I conceive it highly probable that such may be the case.

On the S.E. coast of this island are several immense lagoons. into which flow rivers said to be navigable for vessels of heavy burthen for many miles, but I am not aware of our having the testimony of any one who has so navigated them.

The native inhabitants are very few in number, and much scattered all over the island: whether they, like the natives of Northern Island, are divided into different tribes is unknown, but it is probable that the high mountain range should cut off all communication between those on the eastern and those on the western coast, and that therefore they have become distinct, having separate interests and acknowledging different chieftains.

By the kindness of Captain Beaufort, R.N., hydrographer, we are enabled to add some extracts from the "Remark Book" of Mr. George Johnson, Master of H.M.S. *Conway*, Captain C. R. Drinkwater Bethune, who visited and partially surveyed Cloudy Bay and Harbour in October, 1837:—

"Approaching Cloudy Bay from Cook's Straits, a large grove of trees in the bottom of this deep inlet appears like an island, and it is not till well in that the low, sandy shore can be seen:—the north point of the bay has high, rocky cliffs, with an inlet a short distance off it, and this side generally is well wooded: the south projection of the bay is a bold white bluff, with numerous little green hills gradually rising to the interior, without either trees or verdure, and backed at a considerable distance by lofty mountains covered with snow.

"Cloudy Harbour bears from the white bluff N. by W. $\frac{3}{4}$ W. 10 or 12 miles: it is difficult to distinguish, but may be knowu by a high conical hill over its upper extremity: it is commodious, with moderate depth of water from 7 to 10 fathoms: the shore is bold to, by just keeping without the long kelp (peculiar to this island), which runs off most of the points. Wood, water, and pigs are plentiful."

The above remarks are accompanied by a sketch-plan of the harbour, on the scale of 4 inches to a mile, showing the soundings generally, drawn by Mr. C. R. Read, Midshipman of the *Conway*, from which it appears that the harbour is 4 miles long, by an average breadth of 1 mile: width of entrance $1\frac{1}{4}$ mile. The east point of entrance lies in lat. 41° 20′ 40″ S., long. 174° 9′ 50″ E., measured by five chronometers in a nine-days' run from Sydney. Variation 17° E.

The *Conway* also anchored for two days off Capati or Entry Island, Cook's Straits, where a sketch survey of the eastern shore of the island was made by Mr. Johnson, and the position of the

Brothers and Hummock Isles fixed. The observatory on the latter isle was found to be 40° 52′ 26″ S. lat., 174° 56′ 45″ E. long.; var. 11° 33′ E. in October, 1837. Entry Island is about 6 miles long in a N.E. and S.W. direction, by 1½ broad. The highest land rises 1800 feet above the sea. Water, wood, and stock plentiful. The tide rises from 4 to 6 feet; the flood-tide sets to the E.N.E. from 2 to 3 miles an hour; good anchorage in 10-fathom water; Hummock Island E., the Brothers S.W., about ¾ mile off shore.

V.—*Geographical Positions of the Principal Points of Eastern Greece, determined by the triangulation of* Monsieur Peytier, Capitaine d'Etat Major. Communicated by Captain H. G. Hamilton, Royal Navy.

[Extracted from the *Bulletin de la Société de Géographie de Paris.*]

The operations executed in Eastern Greece by M. Peytier are a continuation of those in the Morea, an account of which has already been published.* These observations have been made with the same instruments (Gambey's Theodolites), and with the same care. No new base has been measured, nor any further astronomical observations made. Several sides of the triangulation of the Morea have been made use of, as the basis for the calculations of the new triangles; and to determine the geographical positions, the extremities of these bases have been taken.

The heights above the level of the sea have been calculated from a massive ruin at the entrance of the Piræus, which has been connected with the triangulation, and of which the precise height above the level of the sea had been measured.

Starting from this point, the elevation of two mountains in the Isthmus of Corinth were calculated, the heights of which had been already obtained, starting from the Gulf of Napoli; and the agreement of the two results confirms what had already been said in the paper before alluded to, that the Gulfs of Napoli, Athens, Corinth, Marathonisi, and the sea near the Ionian Islands, were on the same level.

M. Peytier's recent operations extend over the Negropont, Attica, Bœotia, and Phocis, as far as the high mountains on the W. of Salona, and the frontier near Zeitún. They reach over a surface of more than 700 square leagues; and the number of positions fixed are nearly 600; the principal of which are contained in the following table:—

* Bull de Geog. vol. xix. p. 89.

Names of Places.	Designation of the Points observed.	Latitude North.	Longitude East from Greenwich.	Height above the level of the Sea in English ft.
		° ′ ″	° ′ ″	
Ambelaki	Windmill on the ruins of Salamis	37 56 6	23 33 37	
Ambelo-Nisi	Small island to the S.E. of the Bay of Aspraspitia.	38 17 34	22 43 14	
Andera	Cape to the N. of the village	38 51 0	22 43 3	
Arakhova	Church in the upper part of the village	58 28 43	22 36 16	
Argyro-Nisi	Small island towards the entrance of the Gulf of Volo.	39 0 27	23 5 41	
Atalanta	Summit of the island	38 40 15	23 7 00	426
Athens	Top of the W. pediment of the Partheum	37 58 8	23 44 51	571
Bay of Cuma	Small island to the E.	38 36 31	24 14 36	
Bisbardi	Bridge on the Cephissus in Bœotia, near to	38 30 33	22 53 00	371
Bodonista	The upper part of the highest tower of the fort of	38 45 1	22 38 12	1902
Bourzi	Fort on the Coast to the S. of Egripo	38 24 16	23 39 35	
Cape Colonna	The Temple (upper part)	37 38 51	24 2 46	269
Cape D'Oro	. . . . .	38 9 25	24 37 21	
Daphne	Summit of the mountain (Mount Ægialeus)	37 59 40	23 38 25	1535
Daulis	Church on the ruins of	38 30 16	22 45 10	
Delphi	Summit of the mountain (Eubœa)	38 37 26	23 51 44	5725
Disto	Tower of (Eubœa)	38 21 9	24 9 56	
Distomo	The southernmost house of the village of (Ambryssus).	38 25 29	22 41 6	
Dramesi	Ruined Tower near the village of (Delium?)	38 22 52	23 38 51	
Egripo	Fort Kara-Baba, upper part (Chalcis)	38 27 46	23 36 15	
Ekhinos	The Tower	38 53 31	22 44 50	
Elatea	Summit of the mountain (Cithæron)	38 10 53	23 16 14	4629
Eleusis	The Tower	38 2 25	23 33 13	
Eretria	Windmill of the Ipsariots, to the W. of (Eubœa).	38 23 18	23 47 48	
Gaidouro-Nisi	Patroclus Island (summit of the island)	37 38 49	23 58 15	836
Galaxidi	Windmill situate to the S.E., and near the town.	28 22 9	22 24 31	
Gatza	Summit of the island	38 31 3	23 24 29	
Gavaliani	Small summit near the S. cape of the island.	38 12 39	24 7 13	
Gerako-Vouni	The summit (Othrys)	39 0 55	22 43 43	5570
Granitza	Summit of the mountain of	38 24 8	22 55 11	2940
Guiona	Highest point of the mountain (the highest mountain in Greece).	38 38 40	22 16 24	8239
Gypto-Kastro	Highest tower (Acropolis of Eleuthera)	38 10 38	23 23 51	
Hagios Ioannis	Chapel to the S. of Topolias	38 28 59	23 11 5	321
Hagios Ioannis-Kinigos.	Monastery	38 0 1	23 51 29	
Hagios Meletios	Monastery	38 11 21	23 28 44	
Hagios Nikolaos	Mountain with a Palæokastro, near to Styra (Eubœa).	38 8 36	24 17 10	
Hagios Nikolaos Galatos.	Monastery (Eubœa)	38 42 50	23 23 33	
Hagios Nikolaos	Ruined tower in the island	38 29 11	23 31 33	
Haliartus	Small tower on the ruins of	38 22 10	23 6 55	
Hasseki	Large ruined Turkish Pyrgos, near Athens	37 58 50	23 43 35	
Hellada	Mouth of the river	38 50 14	22 36 54	
Jerusalem	Monastery on Parnassus	38 30 45	22 42 49	
Kakosio	The Tower (Thisbe)	38 15 13	22 59 35	
Kandili	Ruined Khan of (Isthmus of Corinth)	38 1 5	23 13 39	
Kaprena	Church to the N. of the village (Chersonese).	38 29 36	22 51 51	
Karysto	The highest part of the citadel (Eubœa)	38 1 57	24 27 9	
Katabothra	Highest point of the mountain (Œta)	38 47 22	22 16 34	7061
Kavasala	Tower on a height to the W. of the village	38 10 53	23 31 37	
Keratea	Highest point of the mountains of	87 47 18	23 59 28	2136
Kerato Pyrgos	Ruined Tower near the anchorage off Salamis.	37 57 27	23 37 16	

Names of Places.	Designation of the Points observed.	Latitude North.	Longitude East from Greenwich.	Height above the level of the Sea in English ft.
		° ′ ″	° ′ ″	
Khlomo of Talanta .	Summit of the mount	38 35 46	23 0 20	3547
Khlomo of Volo . .	Summit of the mount	39 5 2	22 57 57	2952
Kiphissia	White house in the shape of a tower . .	38 4 4	23 50 4	
Korombili . . .	Summit of the mount	38 11 43	23 5 41	2982
Koumi	The westernmost and most isolated of the windmills to the S. of.	38 37 23	24 7 29	
Kouroublia . . .	One of the summits of the chain called Kandili on the maps (Eubœa).	38 40 12	23 29 16	3967
Konveli	Small island in the bay of Dombrena . .	38 12 8	23 0 42	
Krevassara . . .	Bridge on the Cephissus of Bœotia, near to	38 34 24	22 49 3	
Ktypa	Summit of the mountain (Messapius) .	38 27 43	23 30 33	3363
Larymna	Small island, with chapel, in the gulf of .	38 34 27	23 19 16	
Lesko-Nisia . . .	The most eastern of the small islands .	38 57 16	23 28 28	
Likeri	Ruined Pyrgos on a cape in the lake (the soil).	38 24 29	23 16 37	426
Limni	Windmill on the coast	38 45 42	23 20 29	
Lithada	Summit of the mountain (Eubœa) . .	38 51 28	22 53 59	2221
Lithada	Island between Cape Lithada and the continent in the centre of the channel.	38 48 21	22 50 32	
Livkdia	Highest tower of the castle	38 25 40	22 53 40	
Livadostro . . .	Ruined tower of	38 12 17	23 8 17	
Livanotœs . . .	Ruined tower with Grecian ruins near cape of.	38 43 15	23 5 00	
Liakoura . . .	Highest summit of the mountain (Parnassus).	38 31 57	22 38 36	8068
Makro-Nisi . . .	Long Island, highest point of the island .	37 44 17	24 9 37	922
Mandoudi . . .	Ruined tower to the west of the village .	38 47 45	23 29 53	
Mandoudi . . .	Small island with two high rocks, near the coast of (Eubœa).	38 48 36	23 32 21	
Mantelo	English Island, south summit of the island	37 55 51	24 32 48	
Marathon . . .	Cape	38 7 9	24 4 43	
Marathon . . .	The tomb of	38 6 43	23 59 39	
Mariolatas . . .	The church (the ruins of Lilea are 3828 yards to the S.E.).	38 39 3	22 29 8	
Mazi	Small island near the coast N.W. of Mazi (Gulf of Corinth).	38 3 49	23 9 21	
Megara	Square tower in the upper part of the town	37 59 46	23 21 34	
Mendeli	Summit of mountain (Pentelicus) . . .	38 4 44	23 54 14	3642
Menidi	Church of Panagia to the west of the village of.	38 4 39	23 45 20	
Molo	Isolated new church to the S.E. of the village.	38 48 16	22 40 12	
Monomati . . .	Ruined tower of	38 5 16	23 47 49	
Moriki	Tower	38 25 13	23 22 40	
Moulki	White Turkish Pyrgos in ruins . . .	38 22 47	23 8 14	
Okthonia . . .	Summit of the mountain (Eubœa) . .	38 31 30	24 12 6	2403
Okthonia . . .	The smallest of the two islets off the cape	38 30 53	24 15 23	
Olympus	Summit of the mountain (Attica) . .	37 44 59	23 57 15	1501
Olympus	Summit of the mountain (Eubœa) . .	38 28 33	23 53 8	3848
Orchomenus . .	Highest ruin of the Acropolis	38 29 33	22 58 49	
Oreos	Chapel in the Acropolis	38 56 45	23 7 4	
Oropo	Large house in the port of	38 19 5	23 48 36	
Oxylithus . . .	Summit of the mountain (Eubœa) . .	38 35 7	24 7 45	1309
Oxea	Summit of the mountain (Parnes) . .	38 10 20	23 44 18	4636
Palæo-Vouna . .	Summit of the mountain (Helicon) . .	38 17 47	22 54 8	5738
Panagitsa . . .	Small island with chapel near the bay of Oreos.	38 56 27	23 4 37	
Panopea	Tree and chapel in the centre of the ruins of	38 29 35	22 49 1	
Patissia	House of Admiral Malcolm near Athens	38 0 7	23 45 31	
Petradgik . . .	The Tower	38 52 12	22 15 44	
Paximada . . .	Small island near Karysto	37 57 20	24 24 30	
Petali	Highest point of the largest of the islands	37 59 29	24 17 14	1214
Petra	Steep mountain with Greek tombs and ruins.	38 23 1	23 2 0	
Phaga	Summit of the mountain (Sphingius) .	38 22 10	23 12 21	1860
Phaneromem . .	Monastery (island of Salamis) . . .	37 58 51	23 27 26	
Piræus	Massive ruin at the entrance of the port	37 56 15	23 39 3	
Platæa	Chapel on the ruins of	38 13 10	23 17 42	

Names of Places.	Designation of the Points observed.	Latitude North.	Longitude East from Greenwich.	Height above the level of the Sea in English ft.
		° ′ ″	° ′ ″	
Politiki	The large tower of (Eubœa)	38 35 40	23 34 1	
Porto Thoriko . .	Or Porto Mandri (peak of)	37 44 18	24 4 37	479
Pyxaria	Summit of the mountain (Eubœa) . .	38 42 29	23 40 3	4436
Rafti	Statue on the largest island	37 52 48	24 3 57	
Rovies	The tower (Eubœa)	38 48 18	23 15 8	
Sagmata	Monastery	38 23 51	23 25 56	2457
Saint Elias . . .	White chapel on a summit to the south of Legrana.	37 41 29	24 0 2	1171
Saint Elias D'Oro .	Mount Ocha, summit of the mountain (Eubœa).	38 3 26	24 29 8	4607
Saint Elias of Salona	Highest summit of the mountain . . .	38 30 9	22 19 59	6277
Saint George . .	Chapel on a summit near Athens (Mount Lycabettus.	37 58 45	23 45 53	912
Salamis	Mavrovouni, highest summit of the island	37 55 42	23 31 3	1247
Sarandavli . . .	Woody summit on Parnassus, the Corycian cave is on the southern slope.	38 31 52	22 32 12	
Saromata . . .	Naked summit of the mountains . . .	38 44 6	22 35 29	4503
Skripou	Monastery of	38 29 28	22 59 51	
Salona	Cape of the harbour of	38 25 46	22 26 14	
Stephani	The northernmost of the two ruined mills near the sea.	38 1 37	23 37 7	
Stilida	The custom-house	38 54 34	22 38 12	
Stoura	Highest point of the island	38 10 2	24 10 58	
Talanta	Large house in the port of	38 40 8	23 5 40	
Thebes	The school (the ground)	38 19 5	23 20 17	676
Thebes	The tower	38 19 16	23 20 20	
Topolias	The church (Copais)	38 29 25	23 10 51	
Tower	To the south of Aliveri near the sea (Eubœa).	38 23 14	24 4 23	
Tower	S.W. of the lake Topolias (Coronea) . .	38 23 26	22 58 57	
Tourko-Khori . .	Bridge on the Cephissus in Bœotia, near to	38 37 2	22 43 39	518
Tlo-Vouni . . .	Summit of the mountain (Hymettus) .	37 56 37	23 50 7	3370
Trakones . . .	The tower on a small summit . . .	37 54 49	23 45 35	
Trikeri	Ruined windmill on the coast at the entrance of the Gulf of Volo.	39 5 19	23 4 51	
Vardoussia . . .	Highest rock of the mountain	38 40 42	22 9 54	8176
Vasilikos . . .	Tower of (Eubœa)	38 25 39	23 41 24	
Vatero	Ancient ruined tower to the S.E. of the ruins of Thespia.	38 16 52	23 11 41	
Velissa	Church in the eastern part of the village (Tithorea).	38 34 48	22 41 26	
Vraona	Tower near to	37 54 40	23 58 42	
Xero-Khori . . .	Large ruined house belonging to the Pacha	38 57 2	23 10 22	
Xero-Vouni . . .	Summit of (Mount Cirphis)	38 27 32	23 33 45	5128
Xero-Vouni . . .	Highest rock (Eubœa)	38 34 32	23 54 45	4689
Zagara	Highest point of the mountain (chain of the Helicon).	38 19 3	23 2 14	5010
Zeïtún	Minaret of the citadel	38 54 5	22 27 20	

VI.—*On the Site of Cusco, &c.* By J. B. PENTLAND, Esq., H.M. Consul in Bolivia. Communicated by Captain BEAUFORT, Royal Navy.

La Paz, March 28, 1838.

I HAVE just returned from a two months' tour into the southern provinces of ancient Peru, during which I have visited the capital of this ancient empire, Cusco, and the many interesting localities round that imperial city. An account of this journey, hitherto untrodden by scientific travellers, I hope to send to the Geographical Society; in the mean time, as a wish is expressed by the Secretary, in one of the late volumes of its Transactions, that the position of Cusco should be accurately determined, I beg you will tell him I have acted upon his suggestion, and that the ancient capital of the Peruvian empire is situated (the Temple of the Sun, now the Church of San Domingo) in lat. S. 13° 30′ 55″, long. W. of Greenwich, 72° 4′ 10″, and at an elevation above the sea of 11,380 feet.

I have determined, also, the position of all the principal places between La Paz and Cusco, and of the western shores of the great inter-alpine lake of Titicaca.

In latitude 14° 33′ S., on the mountains of *Vilcanota*, which tvansversely connect the two great east and west chains of the Cordillera, the limit of perpetual snow is at an elevation of 15,800 feet.

The eastern chain of the Cordillera of Upper Peru, from the parallel of *Sorata*, in 15° 50′ S., to that of *Salcantaï*, in 13° 10′ S., is composed of an almost uninterrupted series of snowy peaks; and the whole of this central chain is a compound of quartziferous porphyry, posterior to the transition state, and new red sandstone formation.

On the *Nevado de Guaracolta*, in 14° 30′ S. lat., a copious spring, which issues from the mountain about 250 feet below the limit of perpetual snow, has a temperature of 38½° Fahr. [+3°·6 cent.]

It would be very possible to measure an arc of the meridian extending from 13° 20′ to 20° S. lat. in the centre of the Andes. The ground towards each extremity of the arc is well adapted for the measurement of the bases.

VII.—*Brief Notice of the Gulfs of Kos and Symi.* By Lieutenants GRAVES and BROCK, R.N. Communicated by Captain BEAUFORT, Royal Navy, F.R.S., Corr. Inst. France.

AT Patmos, where we commenced this year's survey of the islands and coast of Asia Minor, we made a plan of Gumishlú, which I can with great confidence assert to be the ancient Myndus, from the extent of ruins and a colonnade of fifty-two columns, of which the pedestals all remain, besides foundations of temples and remains of ancient baths, tombs, &c.—the plan of them shall be forwarded as soon as we have time to copy it. From Patmos we proceeded to Kos, picking up a boat that had been detached to survey Zinari and Livata. Kos is a beautiful island, and it was quite a pleasure to behold it after the rugged and barren spots we have been so long accustomed to. Here our reports were confirmed of the wretched state of Kalymno, which, since our examination of it last year, has been depopulated by plague—the corn is uncut, and dead bodies are lying about in the streets. From Kos we steered for Symi. On reading over Mr. Brooke's communication, in the Geographical Journal (vol. viii. p. 133), respecting the Gulf of Symi, it appears that, after passing five islands, he entered a bay, the scenery of which was grand and imposing, and that on its further shore were the remains of a castle crowning the summit of a hill. This forms the northern coast of the harbour of Losetho. Again, his cruise ended at what he terms "Gothic Island." Now at about 2 miles to the northward of this, at the head of a narrow creek, on each side of which are high and precipitous cliffs, is, I believe, the narrow isthmus forming the ancient Triopium promontory. We levelled it across, and made a plan of this interesting locality, which agrees well with ancient authorities, and in no other place do the gulfs approach so near each other, although at Dahtchak, a bay on the north shore, nearer to Cape Krio, there is no great distance.

The boats being all occupied on the survey, I started for Mughlah (the ancient Alinda), then and now the capital of Caria; we took horse at the head of the gulf, at a place correctly called by Colonel Leake, Assereneh (modern): 18 hours of execrable road took us to our destination, the route leading us close to the head of the Gulf of Kos, but no ruins of any extent repaid us for our trouble, so I proceeded 6 hours farther to Eski Hisar (Stratonicea), the position of which I thought it would be satisfactory to determine. Here we were partly repaid for our toils, and found, as we almost invariably have done, that Chandler's description was perfectly correct.

The coast-line of the Gulf of Symi and the Island of Symi are finished, so that by the end of the season I hope to get as far as the harbour of Marmorice and Kara-kach.

No inscriptions worth mentioning have been found in the Gulf of Symi, but many ruins, chiefly buildings of the middle ages, built upon ancient foundations; from Kos they are very numerous; they shall be copied and sent home as soon as possible.

The plates published by the Society of Dilettanti have reached me safely. They are very beautiful, and, as far as I have yet examined, very faithful and correct.

Lieutenant Brock, describing the Gulf of Kos, states that the crew of his vessel suffered very severely from a fever caught on the unhealthy plains of Keramus, the town which formerly gave its name to the gulf. He proceeds to say that, in consequence of the various indentations and harbours on the south side of the gulf, they had much hard work, but still should have completed their survey, had it not been for this unfortunate sickness.

He then adds, "I was much disappointed in not finding any inscriptions among the numerous ruins, both ancient and modern, with which the gulf abounds. There are some buildings at Keramus which I shall narrowly examine, but am afraid shall only meet with the customary disappointment. The gulf is considerably deeper than any charts show, being 52 miles in length, of a very peculiar shape, taking that of a canal toward the bottom, bounded on the north side by precipitous mountains falling in a series of cliffs nearly to the water's edge, and on the south side by a constant succession of broken hills and deep ravines, with small patches of ground capable of cultivation—in some places well watered, and covered by most luxuriant vegetation. The rivers which water these small plains find an outlet in low, marshy ground in the harbours, but, owing to the flat nature of the ground, are salt at some distance from the sea, and can only with great difficulty be made use of for watering, and then only by carrying breakers up the valleys. There are no inhabitants near the coast, and it was with great difficulty we procured a few goats and fowls, and seldom met any one from whom we could procure information."

[In addition to this prospect of geographical information from the zealous officers intrusted with the survey of the coasts of Asia Minor, we have the satisfaction to add that Mr. Charles Fellows, just returned to England, made during the last summer an extensive journey through an almost unknown part of that country.

From Constantinople Mr. Fellows went to Kútáhiyah; not far from this he examined the ruins of Tóghánlú, which are misplaced on our maps; thence to Sandúklí, Búrdúr, Isbartah, Aghlásun, where were fine ruins, Súsah, and Antáliyah; thence eastward along the coast about 40 miles to Manavghát; returned to Antáliyah, then to Delik, Tegrovah, Cape Khelidonia, Phínika, Kákava [Andriace], Antiphellos, Pátara; thence up the valley of the Xanthus, the scenery of which is described as very beautiful, to the town of Xanthus; thence to Pinara, a town 4 miles in circuit, with many and splendid ruins; then westward to Makri, to 'Húlah? Mughlah, Mélasso, Palátia [Miletus], and to Ephesus; and Mr. Fellows has kindly offered to communicate to the Society the journal kept during this novel route.—Ed.]

VIII.—*Itinerary from Tehrán to Alamút and Khurrem-ábád in May,* 1837. By Lieut.-Col. JUSTUS SHIEL, serving in Persia.

MAY 19, 1837.—Marched from Tehrán to Kend, a large village about 10 miles N.W. of Tehrán, close to the range of El-burz, and near to which is a pass over the mountains into the district of Shehristánek.

May 20.—W. by S., 16 miles to Kerej or Suleïmáníyeh. The road was over an undulating, barren, stony plain, near to the foot of El-burz. At half a mile crossed the Kend stream, flowing from north; at 2 miles the peak of Demávend bore N. 73° E.;* the pass over El-burz was 2 miles to our right; at 12 miles Demávend still bore E. by N. ½ N., the hill above Reï E.S.E. At 16 miles crossed the Kerej stream, flowing from north, and which is afterwards lost in the Kevír or salt desert south of Tehrán.

May 21.—Left Suleïmáníyeh or Kerej, and proceeded to Fúshend, 27 miles. At Kerej the village of Hájí-ābád, in El-burz, bore N. 10° W., distant 3 miles; Senger-ābád, village on Kazvín road, N. 70° W., 12 miles; the mountains above Kinari-gird, the first day's march from Tehrán to Kum, bore from S. 11° E. to S. 16° E. The general direction of the road for 8 miles was N.W. ½ W.; then for 7 miles N. 30° W.; and the remainder N. 35° W. At the third mile from Kerej the road to Kazvín turned off N. 63° W.; at fourth mile, village of Imámzádeh was half a mile to our left; at thirteenth mile we passed close to the village of Kurdán on our right; and at 16 miles the village of Chendár was 1 mile on our left; at 27 miles, reached Fúshend, a large village close under the range of El-burz. The plain we travelled over to-day contained many villages.

May 22.—Left Fúshend for Mír, 27 miles; the first 4 miles the direction was N. 40° W.; then 3 miles N.; then 2 miles N. 45° W.; and the remaining 18 miles by a winding road, in the general direction of west. Immediately on leaving Fúshend we entered a defile and pass through El-burz, of 6 miles in length, and then descended into the district of Tálkán, through which the Sháh-rúd, rising at the eastern extremity, flows from E. to W.; at 10 miles we passed close to the village of Fúshendek, and then descended to the Sháh-rúd, down the left bank of which we proceeded until opposite to the village of Shehreser, when we crossed to right bank. At 21 miles passed the village of Leheren, and then, ascending a high hill, reached Mír. Tálkán is about 30 miles in length and 10 in breadth. It is fertile and populous. The western extremity of the valley is bounded by a high mountain called Yúzá, said to be 13 miles from Mír, and which bore S.W. The eastern boundary is a high mountain called Eserlik, where the Sháh-rúd rises. This river is here about 36 yards in width, with a rapid stream.

* All the bearings are *magnetic,* and were taken with a Kater's or prismatic compass. The variation at Tehrán in 1837 was 2° westerly.—ED.

May 23.—Left Mír, and proceeded 10 miles to the small village of Yerek, in the district of Alamút. For 3 miles the direction was west; then 4 miles N. 60° W.; and 3 miles N. 40° W. On leaving Mír there was a very bad descent of 2 miles to the right bank of the Sháh-rúd, down which we travelled for 2 miles; at the third mile passed the village of Kúshán; at the fourth mile commenced ascending the very fatiguing pass of Dú-derán, through a chain extending north and south, and left the Sháh-rúd flowing west; at 8 miles from summit of pass saw the hilly district of Rúd-bár to the west, with the Sháh-rúd flowing through it. It is called the Sháh-rúd of Ṭálḳán, to distinguish it from that of Alamút. Rúd-bár is bounded on the north by a high range which separates it from Gílán. From the top of the pass Yúzá mountain bore W. 17° N., and Eserlik mountain, at the eastern end of Ṭálḳán, S.E. ½ E. Having crossed Dú-derán, we were in the district of Alamút. Yerek is a small village at the southern foot of the Píshakúh range, which runs from east to west.

May 24.—Marched from Yerek to Gaser Kháni, in Alamút, 10 miles. For 2 miles the direction was N. 45° E.; then 4 miles north; then 4 miles N. 30° E. We commenced by an ascent of 2 miles over the Píshakúh range, at the top of which Yúzá mountain, in Ṭálḳán, bore S. 27° W., Yerek south; Selembár, a high peak at eastern extremity of Alamút, bore E. 6° N.; and Siyálán, a peak in the same quarter in which the Sháh-rúd of Alamút rises, bore N. 42° E. At the third mile we saw the village of Gaser Khání N. 29° E. At the fifth mile we passed the village of Díhek; at the seventh the village of Maḥmúd-abad, and then crossed to the right bank of the Sháh-rúd river, which flows east and west, and falls into the Sháh-rúd of Ṭálḳán. At 8 ~~miles we passed the village~~ of Shutur Khán. From Gaser Khání, El-burz, the name applied to a large high mountain at the extremity of Alamút, bore S. 35° E. From that village the rock of Alamút is distant 2 miles north, and said to be 8 farsakhs, or 32 miles, from Kazvín. The district of Alamút is enclosed by a high range to the north, which separates it from Gilán and Mázanderán; on the south is the Píshakúh range; on the east are El-burz and Siyálán; and on the west is Rúd-bár, and may be about 30 miles in length by 20 in breadth. The rock of Alamút stands alone; it is about two miles north of the village of Gaser Khání. The ridge is about 300 yards in length from east to west, and very narrow, not 20 yards at the top. The height is about 200 feet on all sides excepting the west, where it may be 100. It is a bare naked rock, exceedingly steep. There are no habitations in the vicinity nearer than Gaser Khaní, nor are there any traces of ruins. Within a short distance of the rock there is a burying-ground. There are several remains of walls belonging to apartments at the top of the rock, besides two or three cisterns or reservoirs for water. It must have been a place of great strength; and its capture would seem almost impossible by an army unprovided with the modern weapons of war. There are several excavations or pits in the side of the rock, which probably were used as store-rooms, as described by Von Hammer.* The vicinity of the rock is a most dreary solitude: excepting eagles and lizards, not a living thing is to be seen, nor even a single tree. The view from the summit of the rock is very fine, embracing nearly the whole of the valley of Alamút, and all the high mountains by which it is enclosed. Among them the most conspicuous is El-burz, which bore S.E. ½ S., and Siyálán E. ½ N., the pass through the Píshakúh range, S.W.

May 26.—Marched 12 miles to the village of Germ-rúd, in Alamút. For 2 miles the road was south; then the general direction for the remainder of the march was E. 15° S. We returned to the village of Shutur Khán, and then went up the banks of the Sháh-rúd, near to which Germ-rúd is situated; at fifth mile, Selembár mountain bore E. ½ S.; at sixth mile passed the hamlet of Chelmír; and at eighth mile, the village of Serráj Meḥelleh. We travelled through a great deal of cultivation, and near a number of villages. Germ-rúd is near the eastern end of the valley of Alamút, on right bank of Sháh-rúd, close under a steep bare range of mountains, which separate it from Mázanderán. El-burz mountain is immediately opposite to it, on the left bank of Shah-rúd.

May 27.—Left Germ-rúd, and proceeded 12 miles to Mírán, in the district of Ṭúnáḳabán,† province of Mázanderán; our courses were east 1 mile, N.W. ½ W. 3 miles, N. 20° E. 4 miles, north 2 miles, N. 25° W. 2 miles. After 1 mile we commenced ascending the very long and fatiguing pass of Germ-rúd into Mázanderán. We reached its summit at 8 miles, and found much snow. At 4½ miles El-burz bore S. 51° W. From the top of the pass we descended continually to Mírán; nothing but hills to be seen. The Sih-hezár‡ or Mírán river rises here.

May 29.—Left Mírán, and marched 14 miles to a spot in the thicket, where we encamped. The direction was very winding, but the general course was N. 20° W. The road was through the ravine in which the Mírán or Sih-hezár river flows, and which we crossed during to-day's march twenty-five times. It enters the Caspian one mile to the north-west of Khurrem-ábád. The road was a continued descent, and was often so bad and narrow that we were obliged to unload the baggage, and have it carried by porters. At the fourth mile we entered a thick jungle of oak, beech, elm, walnut, and brushwood; at the fifth mile we passed the village of Ḳáz Meḥelleh; at night we stretched our carpets in the thicket, and halted.

---

* History of the Assassins, p. 70.

† Túrná-ḳapán, *i.e.* Crane-catchers. (Turk.)

‡ Three thousand.—S.

May 30.—Left our camp, and proceeded 12 miles to the large village of Khurrem-ábád, in the district of Ṭúnikaban. The general direction was north, through the thicket, and among hills which gradually diminished in height. At the eighth mile we saw the Caspian, and then descended some low wooded hills into the flat plain which separates the hills from the sea, and which here was three and a half or four miles in breadth. This plain was covered with thick wood, intermixed with rice cultivation, and many villages. At 8 miles a peak and cape on the shore called Már Kúh* bore N.W. 6° W.; another cape called Zanes bore E. 13° S. Khurrem-ábád is a large village; the houses chiefly scattered through the thicket; the capital of the district of Ṭúnakábán.† Salmon is abundant here in the Mírán river, or Sihhezár, which falls into the Caspian one mile and a half from the village. The water of the sea here is not so brackish as to prevent horses from drinking it. From this spot Mount Selembár bore S.W. 6° S., and Siyalán S. $\frac{1}{2}$ W.

June 2.—Left Khurrem-ábád, and proceeded to Kelár-ábád, in Ṭúnikabán, 24 miles. Marched E.S.E. for 3 miles through the thicket, and then reached the sea, along the shores of which we travelled for 19 miles in the general direction of E. $\frac{1}{2}$ S. The thick wood reached nearly to the shore. At 6 miles Már-kúh peak bore N. 75° W.; at sixteenth mile we passed close to the village of 'Abbás-ábád (thirty houses); and at the twenty-second mile we turned south into the forest, and arrived at Kelár-ábád (fifty houses).

June 3.—Left Kelár-ábád, and proceeded to 'Alí-ábád, in the district of Keláristákh, 14 miles. We returned 2 miles nearly north to the shore, close to which we continued to travel; at the twelfth mile we turned south into the thicket, and, travelling through the latter, ~~mixed with cultivation,~~ reached 'Alí-ábád, the principal village of Keláristákh. Half a mile to the eastward flows the Chális river, which is the boundary of the district, and separates it from that of Kerán or Kujur.

June 4.—We left 'Alí-ábád and proceeded to Merzin-ábád, 24 miles. The road was a continued ascent, though not abrupt, through a forest of underwood, which now began to be rather less dense; its depth from north to south seems to be about 35 miles. Our path was very winding; the general direction a little to the west of south, and up the banks of the Chális river or torrent. We saw no villages excepting that of Tever, at the twenty-second mile, on the right bank of the river.

June 5.—Proceeded from Merzin-ábád and encamped in the thicket at seven miles. Our course was winding, its general direction being south, with a gradual ascent up the left bank of the Chális river. At 1 mile, village of Píshkhár, 1 mile on right

* Mount Serpent. † Steel Mount.

at 4 miles, village of Tewíl, 2 miles on right; cultivation more abundant, wheat and barley taking the place of rice.

June 6.—To I'nen, 16 miles. We first proceeded along the banks of the Chális, now a mere rivulet, nearly south, for 6 miles; here a high bare peak, called Fúlád Kúh, bore south; we then ascended a very fatiguing pass called Hezar Chesn, or "thousand twists;" in a general S.E. by S. direction for 6 miles farther, the 4 last miles were a little to the south of east; wood still abundant. The Chális stream proceeds from mountains east of I'nen; this latter village is in the district of Keláristákh.

June 7.—Journeyed 14 miles to Melik fáris, a village in the district of Laurá. We descended the mountain on which Enen is placed, and proceeding S.E., crossed the Chális, which rises to the east. At 4½ miles a remarkable rocky peak, 6 miles off, bore E. 16° S.; to the east of this is the district of Núr. After crossing the Chális we proceeded 4 miles a little to the east of south, and commenced ascending the high mountain and pass of Kildíwán; our course was south for 2 miles to the summit, and we then quitted the province of Mázanderán and re-entered that of 'Irák. We descended south into the district of Laurá to the foot of Kendíwán, and entered a ravine through which the Suleïmáníyeh or Kerej river was flowing from west to east, its source being higher up the ravine to the west; south of the ravine, and quite close to it, lay a high range with snow at the top, which separated us from the plains. We proceeded down the ravine, following the course of the stream, and at the twelfth mile reached the village of Gerchek on the right bank of the river; two miles farther on brought us to Melik fáris, situated on each side of the stream. Wood had disappeared altogether soon after quitting I'nen.

June 8.—To Ehár, 17 miles. For 6 miles the general direction was S.E. by S., then E. by S. 11 miles; at 5 miles the Kerej river, being joined by a small stream which came from N.E., turned off to W.S.W. There was considerable ascent and descent during this day's march; on our right was the ridge separating us from the plains. At the seventh mile we entered a long narrow valley covered with cultivation, and with a succession of small villages which seemed to form one large village; it was called Shehristánek or "little city."

June 9.—To Gúlchik,* 18 miles. We first proceeded south down the ravine, through which flows the Jáj-rúd river, and which, like that of Kerej, is lost in the salt desert to the southward. At 8 miles we ascended the last range of El-burz, from the top of which the peak of Demávend bore N. 74° E.; Jaj-rúd river E. 34° S.; Reï, or the ruins of Rhagæ, S. 6° W.; and Tehrán S. 25° W.; from this point we descended about 9 miles to the village of Gúlchik, 7 miles to the north of the city of Tehrán.

* Goljik (Turk.) "Little Lake." S.

IX.—*On the Use of Common Thermometers to determine heights.* By Lieut.-Col. W. H. Sykes, F.R.S.

Having been recently applied to by two gentlemen about to travel—the one in Africa and the other in Asia Minor—for a description of the thermometers and apparatus used by myself for some years in India for determining heights by the boiling temperature of water, I have ventured to believe that a brief account of a process which I found to produce results sufficiently near to the truth for most practical purposes, may not be unacceptable to some members of the Society, particularly as I carried on my barometrical observations contemporaneously, and thereby obtained data for fixing the value of certain points on the thermometric scale. To determine heights accurately, good barometers are necessary, which have been carefully compared with a standard barometer; the observations must be taken simultaneously at the upper and lower stations, and the temperature of the mercury and the air, and the hygrometric state of the latter, must be noted. Heights so determined, when tested again in the same or succeeding years, I have rarely found to vary more than 10 or 20 feet in 4000 or 5000. When barometers are used which have not been previously compared with a standard, when the observations are not simultaneous, and when the pressure and temperature at the level of the sea are *assumed*, the results may by accident be near to the truth, but they will usually be from 100 to 300 feet wrong,—at least such is the result of my experience within the tropics. But good barometers are very costly; they are troublesome to carry, are particularly exposed to accident on a journey, and get out of order by the escape of the mercury, which being frequently unobserved, the barometer continues to be used as if it were correct. The late Archdeacon Wollaston, aware of these facts, invented the thermometric barometer to supply the place of the ordinary barometer. This instrument is very sensible, but it is very fragile from the great weight of the bulb compared with the slenderness of the stem; moreover, there are some complex accompaniments, and the instrument is also expensive: in short, I found it not fit for *rough work* out-of-doors, having had three destroyed at the outset of my labours; and the same opinion is expressed by Mr. James Prinsep, of Calcutta, who is well known for the practical application of his scientific knowledge. I had then recourse to common thermometers, and, with certain precautions in their use, found them answer my purpose sufficiently well. A tin shaving-pot was my boiler; dry sticks and pure water were usually to be had, and by the time my barometers were settled I was ready to take the boiling temperature. The following is a sketch of the apparatus.

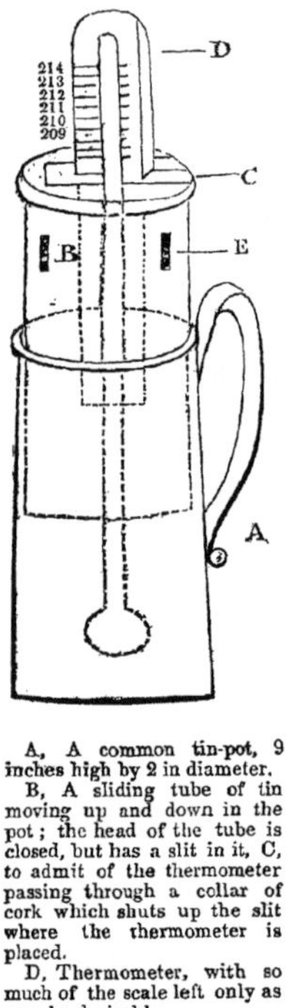

A, A common tin-pot, 9 inches high by 2 in diameter.
B, A sliding tube of tin moving up and down in the pot; the head of the tube is closed, but has a slit in it, C, to admit of the thermometer passing through a collar of cork which shuts up the slit where the thermometer is placed.
D, Thermometer, with so much of the scale left only as may be desirable.
E, Holes for the escape of steam.

It will be seen that the chief part of the scale usually attached to the thermometer is removed, only so much of it being left as may be desirable: I however permitted the brass scale of one of my thermometers to remain, and I did not discover that it was the cause of error. Previously to taking the thermometers inland, it is necessary to ascertain their boiling points at the level of the sea; for in many instances the scales are so carelessly applied, that a thermometer may indicate a boiling temperature of 213° 214° or 215° at the level of the sea; one of mine stood at 214.2 when water boiled. Nevertheless, by making a deduction of 2° 2′ in all observations, the indications rarely differed five-hundredths of a degree from the other thermometer, of which the boiling point was 212°: the temperature of the air and the height of the barometer at the time the *verification* of the thermometers is made must be noted. The following is the manner in which my observations were taken:—from 4 to 5 inches of *pure* water were put into the tin pot; the thermometer was fitted into the aperture in the lid of the sliding-tube by means of a collar of cork; the tin tube was then pushed up or down to admit of the bulb of the thermometer, being about *two inches*, above the bottom of the pot. Violent ebullition was continued for 10 minutes or a quarter of an hour, and the height of the mercury was repeatedly ascertained during that time, and the temperature of the air was noticed. Similar operations were repeated with a *second* thermometer, for it is never safe to rely upon *one* instrument. Having obtained the boiling points, it remains to determine the value of the indication of diminished pressure when the observations are taken above the level of the sea. The elastic tension of steam at different points on the thermometric scale has been determined by experiment, but not at regular intervals on the scale, nor with similar results, by different persons; tables, therefore, computed from the formulæ of the various experimenters do not accord; but, in three tables which I have in my possession, the heights computed by them, when compared with heights determined by corresponding barometrical observations, with previously compared barometers, (the only satisfactory way to ascertain heights not taken trigonometrically,) approximate sufficiently near for all practical purposes where great accuracy is not desired. These tables, however, differ slightly from each other.

The table which first came into my hands appeared anonymously in the Madras Gazette for 1824. In 1826 an able friend, Lieut. Robinson, of the Indian Navy, who entered warmly into my views to determine heights by common thermometers, thought he could improve upon the table I was using, and accordingly made a new computation; the third table came under my notice much more recently than the two former. It is computed by Mr. James Prinsep, of Calcutta, Secretary of the Asiatic Society of Bengal, a gentleman distinguished for his scientific research. He published it in the Journal of the Society. To admit of a just estimate being formed of the value of these tables,—of the value of corresponding barometrical observations, made with due precautions, although with different coadjutors and different instruments,—of the value of barometrical observations, with an assumed pressure and temperature, at the level of the sea,—of the value of thermometrical compared with barometrical observations,—out of many hundred heights determined in various ways, I have taken many at random, (the number it appears is eighty-eight,) and I have put them into juxtaposition in a tabular form. In thermometric heights the elements at the level of the sea were a boiling temperature of 212° Fahr. and a mean temperature of the air of 82°. The *assumed* pressure in heights determined barometrically, without corresponding observations, was 30 inches; mean temperature 82°. In looking over the tabulated results I was a good deal surprised to find that in no instance, by whatever method determined, do the barometric differences in height exceed 127 feet, and this only by comparing the highest indications with an assumed pressure with the lowest indications of corresponding observations. It will be seen that the various tables for determining heights thermometrically, with certain exceptions, do not differ very *materially* in their results from each other, nor from corresponding barometric observations; the formulæ on which they are founded may therefore be considered, on the whole, sufficiently accurate for the present state of our knowledge.

Lieut. Robinson's and Mr. Prinsep's tables give close approximations to each other in their results, but they are as much below the corresponding barometric observations, which I consider the true heights, as the results by the Madras table are above the true heights. Some of them curiously coincide within a foot or two of the heights determined by corresponding barometrical observations, but this coincidence must be the result of mere accident. Taking the mean of all the thermometric observations at a station calculated by the three tables, and the mean of all the corresponding barometric observations at the same place, the utmost difference is 107 feet in less than 600; and the least difference is 8 feet in about 3000; but, as the thermometric heights in which the difference of 107 feet occurs were single observations, made by a gentleman who had newly begun to use his thermometers, they may be looked upon as probably less accurate than subsequent trials would have made them. This is scarcely an unjust inference, as it will be seen that the next greatest difference made by the same gentleman was only 24 feet in 4490. It must be admitted however that this amount of error is just as likely to occur in heights of 100 feet as in those of 10,000. My thermometers were not graduated to less than half-degrees, and long practice enabled me to determine the height of the mercury in the stem to one-twentieth of a degree; but I would recommend thermometers being used in which the degrees are graduated to fifths or tenths of a degree. On the whole, I think the results of six years' experience justify me in saying that common thermometers may be satisfactorily used to supply the place of barometers in measuring heights where great accuracy is not required, and it will be recollected that what is usually looked upon as a difficult and troublesome operation with barometers, will be attainable by any person who carries with him a couple of thermometers, the requisite tin pot, and the tables, and who is master of the simplest rules of arithmetic.

Of the three tables in my possession I have chosen Mr. Prinsep's to submit to the Society, from their perspicuity and the facilities they offer for the conversion of boiling temperatures into heights with very little trouble; but a glance over the figures in my tables of altitudes will show that the tables are susceptible of considerable improvement, for, with two exceptions, all the heights deduced from Mr. Prinsep's and Lieut. Robinson's are much below those determined by simultaneous observations with good barometers; and I join with Mr. Prinsep in expressing a hope that every traveller boiling his thermometers will at the same time, if he possess a barometer, make a record of its indications, and thus render essential service to physics by fixing so many points on the scale of the elastic tension of steam at different temperatures.

Year.	Date.	Names of Places.	ALTITUDES DEDUCED FROM 1. Jones's Bar. No. 2, with assumed pressure of 30° in., and mean temp. 82° at the level of the sea.	2. Cary's Barometer, No. 2, ditto ditto.	3. Corresponding observations with Capt. Jervis's Gilbert's Bar. and Cary's No. 2.	4. Corresponding observations with Dr. Walker's Gilbert's Bar. and Cary's No. 2.	5. Corresponding observations with Capt. Jopp and Cary's Bar. No. 2.	6. Corresponding observations with Cary's Bars. Nos. 1 and 2.	7. Corresponding observations with Cary's Bar. No. 1, and Jones's No. 2.	8. Heights by boiling temperatures, therm. 1 by the Madras tables.	9. Heights by boiling temperatures, therm. 2 by the Madras tables.	10. Heights by boiling temp. by Lieutenant Robinson's tables, Indian Navy.	11. Heights by boiling temperatures, by tables of James Prinsep, Esq., Calcutta.	12. Difference between the means of *all* the boiling temp., and barometric corresponding observations.	13. Mean of corresponding observations by Barometers.	14. Mean of all the boiling temperatures.
1827	23 May....	Highest point, Hill Fort of Púrundhur..........	4588	4599	....	....	....	†4471	4528	4536	4553	4415	4427	—16	4499	4483
1827	10 May....	Singhur Hill Fort	4199	4180	....	....	....	†4211	4170	4341	4220	3927	3923	—86	4190	4104
1828	15 May....	Temple at Bíma Shunkur	....	....	....	....	†3090	....	....	3037	3037	2992	2991	—71	3090	3019
1825 1827	6 March 11 May..	Karleh, Cave Temple	2493	2652	....	....	....	†2530	....	2693 2526	2646 2526	2468	2473	+27	2530	2557
1827	23 May....	Highest point of Púrundhur above Púna	2697	2631	....	....	....	†2648	†2650	2661	....	2539	2566	—61	2649	2538
1828	9 Feb... 3 April.	Pait on the Yail River	....	....	†2478 †2493	....	†2470	....	....	2494	2494	2480	2484	+8	2480	2488
1828 1829	....	Temple in the Hill Fort of Huríchundurghur	3972	3931	3845	†3922	†3871	†3887	3935	3840 3887	3869 3887	3824	3788	—46	3892	3846
1829	11 to 17 Dec....	Source of Kristna River at Mahabuleshwur	....	....	....	†4496 †4503	....	....	....	*4498	*4556	*4422	*4425	—24	4499	4475
1828	27 April ...	Pokrí	....	....	....	....	†3197	....	....	3194	3194	3185	3141	—19	3197	3178
1828	6 April ...	Kullumb, on Goreh River	....	....	2043 2027	....	†1995	....	....	1971	2000	1988	1986	—36	2022	1986
1825 1826 1827 1828 1829	....	Púna, Hay Cottage	....	....	1810 1820	†1810	†1837	†1823	....	Means 1883	1897	1876	1861	+59	1820	1879
1828	16 Feb....	Downde, on the Bíma River..	....	....	....	....	1623	....	....	1591	1591	1567	1575	—41	1623	1582
1828	29 Oct.	Sasswur, above Púna	....	....	....	592	....	....	....	*514	*456	....	....	—107	592	485

† The heights most relied upon.

* Boiling temperatures determined by Dr. Walker.

TABLE 1.

To find the Barometric Pressure and Elevation corresponding to any observed Temperature of Boiling Water between 214° and 180°.

Boiling Point of Water.	Barometer Modified from Tredgold's Formula.	Logarithmic Differences or Fathoms.	Total Altitude from 30.00 in. or the Level of the Sea.	Value of each Degree in Feet of Altitude.	Proportional Part for One-tenth of a Degree.
°			Feet.	Feet.	Feet.
214	31.19		−1013		
213	30.59	00.84.3	507	−505	..
212	30.00	84.5	0	−507	..
211	29.42	84.9	+509	+509	..
210	28.85	85.2	1021	511	51
209	28.29	85.5	1534	513	..
208	27.73	85.8	2049	515	..
207	27.18	86.2	2566	517	..
206	26.64	86.6	3085	519	52
205	26.11	87.1	3607	522	..
204	25.59	87.5	4131	524	..
203	25.08	87.8	4657	526	..
202	24.58	88.1	5185	528	..
201	24.08	88.5	5716	531	53
200	23.59	88.9	6250	533	..
199	23.11	89.3	6786	536	..
198	22.64	89.7	7324	538	..
197	22.17	90.1	7864	541	54
196	21.71	90.5	8407	543	..
195	21.26	91.0	8953	546	..
194	20.82	91.4	9502	548	..
193	20.39	91.8	10053	551	55
192	19.96	92.2	10606	553	..
191	19.54	92.6	11161	556	..
190	19.13	93.0	11719	558	..
189	18.72	93.4	12280	560	56
188	18.32	93.8	12843	563	..
187	17.93	94.2	13408	565	..
186	17.54	94.8	13977	569	57
185	17.16	95.3	14548	572	..
184	16.79	95.9	15124	575	58
183	16.42	96.4	15702	578	..
182	16.06	96.9	16284	581	..
181	15.70	97.4	16868	584	..
180	15.35	97.9	17455	587	..
					59

The Fourth Column gives the Heights in Feet.

TABLE 2.

Table of Multipliers to correct the Approximate Height for the Temperature of the Air.

Temperature of the Air.	Multiplier.	Temperature of the Air.	Multiplier.	Temperature of the Air.	Multiplier.
°		°		°	
32	1.000	52	1.042	72	1.083
33	1.002	53	1.044	73	1.085
34	1.004	54	1.046	74	1.087
35	1.006	55	1.048	75	1.089
36	1.008	56	1.050	76	1.091
37	1.010	57	1.052	77	1.094
38	1.012	58	1.054	78	1.096
39	1.015	59	1.056	79	1.098
40	1.017	60	1.058	80	1.100
41	1.019	61	1.060	81	1.102
42	1.021	62	1.062	82	1.104
43	1.023	63	1.064	83	1.106
44	1.025	64	1.066	84	1.108
45	1.027	65	1.069	85	1.110
46	1.029	66	1.071	86	1.112
47	1.031	67	1.073	87	1.114
48	1.033	68	1.075	88	1.116
49	1.035	69	1.077	89	1.118
50	1.037	70	1.079	90	1.121
51	1.039	71	1.081	91	1.123

Enter with the mean temperature of the stratum of air traversed, and multiply the approximate height by the number opposite, for the true Altitude.

When the thermometer has been boiled at the foot and at the summit of a mountain, nothing more is necessary than to deduct the number in the column of feet opposite the boiling point below from the same of the boiling point above: this gives an approximate height, to be multiplied by the number opposite the *mean* temperature of the air in Table 2, for the correct altitude.

	°		feet.
Boiling point at summit of Hill Fort of Púrundhur, near Púna	204.2	=	4027
Boiling point at Hay Cottage, Púna . .	208.7	=	1690
Approximate height			2337

Temperature of the air above . . 75°
Ditto ditto below . . 83

Mean 79 = Multiplier 1.098

Correct altitude 2.566 feet.

When the boiling point at the upper station alone is observed, and for the lower the level of the sea, or the register of a distinct barometer is taken, then the barometric reading had better be converted into feet, by the usual method of subtracting its logarithm from 1·47712 (log. of 30 inches) and multiplying by ·0006, as the differences in the column of "*barometer*" vary more rapidly than those in the "*feet*" column.

		Feet.
Example.—Boiling point at upper station - -	185°=	14548
Barometer at Calcutta (at 32°) 29 in. 75°		
Logar. diff.=1·47712−1·47349=00363 × 0006=		218
Approximate height - -		14330
Temperature, upper station, 76° / Ditto lower, 84° } 80=multiplier		1·100
Correct altitude - - -		15763

Assuming 30·00 inches as the average height of the barometer at the level of the sea (which is however too much), the altitude of the upper station is at once obtained by inspection of Table 1, correcting for temperature of the stratum of air traversed by Table 2.

[Newman, Optician, 122, Regent Street, has been in the habit of making these instruments; he recommends the use of copper brazed, instead of tin, as more durable; and a free escape for the steam, or the results will be incorrect from the boiling taking place under pressure; a model may be seen at the apartments of the Royal Geographical Society.—ED.]

X.—*Proposed Exploring Expedition to the Asiatic Archipelago.* By James Brooke, Esq.

Shortly before leaving England for Singapúr, Mr. James Brooke, a member of the Royal Yacht Squadron, and to whom we are already indebted for an account of the island and gulf of Symi,* forwarded to the Society an outline of his intended expedition to the Asiatic Archipelago in his yacht the *Royalist*, a fine schooner of 140 tons burthen, on board which are embarked Mr. Murray, a master in the Royal Navy, a surgeon, and a crew of 20 men; the vessel being well equipped with all the requisite instruments for observation, including three chronometers, and the means for collecting and preserving specimens of natural history, sailed from Plymouth on the 10th November, to carry into execution the spirited enterprise of which an abridged account is contained in the following pages:—

"Touching at the Canaries and the Cape of Good Hope the *Royalist* will proceed direct to Singapúr, which may be considered as head-quarters for the necessary intervals of refreshment and repose, and for keeping open a certain communication with Europe. Here the best local information can be obtained, interpreters procured, the crew augmented for any particular service, and here, if needful, a small vessel of native construction may be added to the expedition, to facilitate the objects in view. An acquaintance may likewise be formed with the more respectable Bughis merchants, and their good-will conciliated in the usual mode,—viz., by civility and presents, so as to remove any misconceived jealousy on the score of trading rivalry, and to induce a favourable report of our friendly intentions in their own country, and at the places where they may touch. The *Royalist* will probably reach Singapúr in the month of March, 1839, at the latter end of the north-west, or rainy monsoon. The delay consequent on effecting the objects above mentioned, besides gaining a general acquaintance with the natural history and trade of the settlement, and some knowledge of the Malay language, will usefully occupy the time until the setting-in of the south-east, or dry monsoon. It may be incidentally mentioned, however, that in the vicinity of Singapúr there are many islands imperfectly known, and which, during the intervals of the rainy season, will afford interesting occupation. I allude, more especially, to the space between the straits of Rhio and those of Duryan,† and likewise to the island called Bintang, which, although laid down as one large island, is probably composed of small ones, divided by navigable straits; a better acquaintance with which might facilitate the voyage from Singapúr to the more eastern islands, by bringing to light other passages besides those of Rhio and Duryan, and, at any rate, would add something to our geographical knowledge in the immediate vicinity of our settlement. On the com-

* Journal, vol. viii. p. 129.

"† All the names of places mentioned in this paper will be found in Mr. John Arrowsmith's Map of the Asiatic Archipelago."

mencement of the healthy season I propose sailing from Singapúr, and proceeding without loss of time to Mallúdú Bay, at the south end of Borneo. This spot has been chosen for the first essay; and in a country every part of which is highly interesting, and almost unknown, the mere fact of its being a British possession, gives it a prior claim to attention.

"The objects in view may be briefly mentioned:—1. A general knowledge of the Bay, and the correct position of various points,—more especially the two principal headlands at its entrance, so as to determine its outline. The westernmost of these headlands, called Sampanmange, will likewise determine the extreme north point of Borneo. 2. Inquiries for the settlement of Cochin Chinese, reported on Earl's authority to be fixed in the vicinity of Bankoka: an intercourse will, if possible, be opened with this settlement, if in existence. 3. The rivers which flow into the Bay will be carefully and minutely explored, and an attempt will be made to penetrate into the interior as far as the lake of Kíní Ballú. 4. For the same purpose, every endeavour will be used to open a communication with the aboriginal inhabitants of the country, and every means employed to conciliate their good opinion; and (if the ceremony exists in this part of the island) to enter into the bonds of fraternity (described by Mr. Dalton) with some of the chiefs.

"I speak with great diffidence about penetrating into the interior of this country, for I am well aware of the insurmountable difficulties which the hard reality often presents, which are previously overlooked, and easily overcome in the smoothness of paper or the luxury of a drawing-room. The two points to be chiefly relied upon for this purpose, are a friendly intercourse with the natives, and the existence of navigable rivers. It is mentioned by Sir Stamford Raffles, on native authority, that a land communication, of not more than forty miles, exists between Mallúdú Bay and Lake Kíní Ballú; but neither this computation, nor any other derived from the natives, however intelligent otherwise, can be relied on, for the inhabitants of these countries are generally ignorant of any measure for distance, and their reckoning by time is so vague as to defy a moderately-certain conclusion. The fact, however, of the vicinity of the lake to the bay may be concluded; and it follows, as a reasonable inference, that the river, or rivers, flowing into the bay communicate with the lake. The existence of such rivers, which were from the locality to have been expected, is vouched for by Captain Forrest. 'Most of this north part of Borneo, (he says,) granted to the English East India Company by the Súlús, is watered by noble rivers: those that discharge themselves into Mallúdú Bay are not barred.' It is by one or other of these rivers that I should hope to penetrate as far as the lake and mountain of Kíní Ballú, and into the country of the Idaan. I have not been able to learn that any Malay towns of importance are situated in the bight of Mallúdú Bay, and their absence will render a friendly communication with the aborigines a matter of comparative ease. The advantages likely to result from such friendly relations are so evident, that I need not dwell upon them, though the mode of effecting such an intercourse must be left to the thousand contingencies which govern all, and act so capriciously on the tempers of the savage races. The utmost forbearance, and a liberality guided by prudence, so as not to excite too great a degree of cupidity, appear the fundamental rules for managing men in a low state of civilization. The results of an amicable understanding are as uncertain as its commencement, for they depend on the enterprise of the individual, and the power of the native tribe into whose hands he may have fallen. I will not, therefore, enter into a visionary field of discovery; but it appears to me certain, that without the assistance of the natives, no small party can expect to penetrate far into a country populous by report, and in many parts thickly covered with wood. Without entertaining any exaggerated expectation, I trust that something may be added to our geographical knowledge of the sea coast of this bay, its leading features, productions, rivers, anchorages, and inhabitants, the prospect of trade, and the means of navigation; and although my wishes lead me strongly to penetrate as far as the lake of Kíní Ballú, yet the obstacles which may be found to exist to the fulfilment of this desire will induce me to rest satisfied with the more moderate and reasonable results.

"It may not be superfluous to notice here, that a foregone conclusion appears to be spread abroad regarding the *aboriginal* (so called) inhabitants of Borneo, and that they are usually considered and mentioned under the somewhat vague appellation of Dyaks. They are likewise commonly pronounced as originating from the same stock as the Arafuras of Celebes and new Guinea, and radically identical with the Polynesian race. The conclusion is not in itself highly improbable, but certainly premature, as the facts upon which it is built are so scanty and doubtful as to authorize no such structure. On an island of the vast size of Borneo, races radically distinct might exist; and at any rate, the opposite conclusion is hardly justifiable from the specimens of language, or the physical appearance of the tribes of the southern portion of the country. We have Malay authority for believing that there are many large tribes in the interior, differing greatly in their degree of civilization, though all alike removed from the vicinity of a superior people. We have the Dyaks of the south; the Idaan of the north; the Kagins, and a race little better than monkeys, who live in trees, eat without cooking, are hunted by the other tribes, and would seem to exist in the lowest conceivable grade of humanity. If we may trust these accounts, these latter people resemble in many particulars the Orang benua, or aborigines of the peninsula; but the Dyaks and Idaans are far superior, living in villages, cultivating the ground, and possessing cattle. Besides these, likewise, we have the names of several other tribes or people, and, in all probability, many exist in the interior, with whom we are unacquainted.

"There are strong reasons for believing that the Hindú religion, which obtained so extensively in Java and Sumatra, and yet survives at Bali and Lombock, was likewise extended to Borneo; and some authors have conceived grounds for supposing a religion anterior even to this. If only a portion of these floating opinions should be true, and the truth can only be tested by inquiry, we may fairly look for the descendants of the Hindú dynasty as well as an aboriginal people. It never seems to have occurred to any one to compare the Dyaks with the people of Bali and Lombock. We know indeed but little of the former, but both races are fair, good looking, and gentle. Again, respecting the concluded identity of the Dyaks and the Arafuras: it is clear we have a very limited knowledge indeed of the former; and, I may ask, what do we know of the Arafuras?

"In short, I feel as reluctant to embrace any preconceived theory, as I am to adopt the prevailing notion on this subject, for it requires a mass of facts, of which we are wholly deficient, to arrive at any thing approaching a reasonable conclusion. To return, however, to the proceedings of the *Royalist*, I would remark that it depends greatly on the time passed in Mallúdú Bay, whether our next endeavour be prosecuted at Abai on the western, or Tusan Abai on the eastern coast. The object in visiting Abai would be chiefly to penetrate to the lake, which, on the authority of Dalrymple and Burton, is not far distant thence, by a water communication: but should any success have attended similar efforts from Mallúdú Bay, this project will be needless, as in that case the enterprise will have been prosecuted to the westward, and reach to the vicinity of Abai. At Kamínís is the limit of the British territory to the westward, so Point Kaní-úngan, situated to the southward of the bay of Sandakan, forms the eastern boundary; and a line drawn from coast to coast between these points, is represented as including our possessions. A reference to the chart will show the extent to be considerable, and the eastern coast from Mallúdú Bay to Point Kaní-úngan is so very little known, that it is highly desirable to become acquainted with its general features and conformation, and to seek thence the means of gaining an inlet into the interior, should it be denied at Mallúdú Bay.

"The reported proximity of Kíní Ballú to Mallúdú Bay, and likewise to Abai, would (supposing it is anything like the size it is affirmed to be) lead us to expect that it cannot be far distant from the eastern coast; and it is but reasonable to conclude that some rivers or streams discharge themselves into the sea in the numerous indentations that abound on this shore. However this may be, the coast, with its bays, and islands, and bold headlands, is one of great interest, and almost unknown; and the careful inspection of it as far as Point Kaní-úngan will, I trust, add something to our knowledge. The longitude of Point Kaní-úngan and Point Unsang will likewise determine the eastern extremity of Borneo.

"Much more might be added on this topic, especially of the reported communication by a line of lakes from Mallúdú Bay to Benjar Massín, which, if true, would in all probability place some of these lakes near particular points of the east coast, as the whole line, from the relative position of the two extremes, must be on the eastern side of the island. These reports, and the various surmises which arise from them, are rather matters for verification than discussion, and I will therefore only add, that, tempted by success, I shall not devote less than a year and a half to this object; but in case of finding a sickly climate, or meeting with a decidedly hostile population, I shall more easily abandon the field, and turn to others of not less interest, and perhaps of less risk.

"Equal to Borneo in riches, and superior in picturesque beauty to any part of the Archipelago, is the large and eccentric country of the Bughis, called Celebes. So deep are the indentations of its coasts, that the island may be pronounced as being composed of a succession of peninsulas, nearly uniting in a common centre in the district of Palos, and thus, by the proximity of every part to the sea, offering great facilities for brief and decisive interior excursions. The Dutch are in possession of Macassar, and had formerly settlements on the north-west coast, and in the bay of Sawa. Their power appears, however, never to have been very extensively acknowledged; and at present I have not been able to meet with any account of the condition of their factories. This information will probably be gained at Singapore. Avoiding the Dutch settlements, I propose limiting my inquiries to the northern and north-eastern portion of the island, more especially the great bay of Gúnong Tella. It is impossible to state here the direction of these inquiries, or any definite object to which they should be turned, as I am acquainted with no author who speaks of the country, save in a general and vague manner. It is reported as rich, fertile, mountainous, strikingly beautiful, and possessed of rivers; abounding in birds, and inhabited, like Borneo, by wild tribes in the interior, and by the Bughis on the sea shores and entrance of rivers. The character of the Bughis, though so variously represented, gives me strong hopes of rendering them, by care and kindness, useful instruments in the prosecution of these researches; for all writers agree that they are active, hardy, enterprising, and commercial; and it is seldom that a people, possessing such characteristics, are deaf to the suggestions of self-interest or kindly feeling. The arrogance, and especially the indolence, of the Malays, counteracts the influence of these strong incentives; and the impulse which governs such rude tribes, as the Dyaks and Arafuras, is a dangerous weapon, which cuts all ways, and often when least anticipated. The Bajús, or sea gipsies, are another race, on whom some dependence may be placed. Mr. Earl, who had a personal acquaintance with this tribe, and could speak their language, always expressed to me a degree of confidence in their good faith, which must have had some grounds.

"I may here conclude the first stage of the expedition, during the progress of which the head quarters will be fixed at Singapúr. During some of the intervals I hope to see Manila, and to acquire a cursory knowledge of the unexplored track at the southern extremity of Celebes, called in Norie's general chart the Tiger Islands.

"The time devoted to the objects above mentioned must, as I have before said, be regulated by the degree of fortune which attends them, for, cheered by success, I should not readily abandon the field; yet, if persecuted by climate, or other serious detriments, I shall frequently shift the ground, to remove myself beyond such evil influence. It is scarcely needful to continue a detail of projects so distant, having already carved out for myself a work which I should be proud to perform, and which is already as extended as the chances of human life and human resolves will warrant. The continuation of the voyage would lead me to take the *Royalist* to Timor or Port Essington, thence making excursions to the Arru Isles, Timor Laut, and the southern shores of New Guinea. That part of the coast contiguous to Torres Straits I am particularly desirous of visiting, as it has been suggested to me by Mr. Earl, and I think with reason, that a better channel than the one we are at present acquainted with may be found there. That such a channel exists, and will be discovered when the coast is surveyed, I entertain but little doubt; but the navigation is hazardous, and must, from the westward, be attempted with great caution.

"My own proceedings must, of course, be regulated by the discoveries previously made by Captain Wickham or others; and as this gentleman has orders to survey Torres Straits, the field may be well trodden before I reach it. The rest of the voyage I shall consider as one merely of pleasure, combining such utility as circumstances will permit. It is probable that I shall visit our Australian settlements, glance at the islands of the Pacific, and return to Europe round Cape Horn. Before concluding, I may observe that there are points of inquiry which may be useful to the studies of the learned, which (provided the process be moderately simple) I shall be willing to make, and I shall always be happy to receive any directions or suggestions regarding them. I allude to observations on the tides, to geology, to the branches of natural history, &c. &c., for the general inquirer often neglects or overlooks highly interesting facts, from his attention not having been called to them. The specimens of natural history will be forwarded home on every visit to Singapúr; and the information will be sent to the Geographical Society, and may always, if it be of any value, be used as freely as it is communicated. In like manner, the objects of natural history will be open to any person who is at all interested in such pursuits. I cannot but express my regret, that from pecuniary considerations as well as the small size of the vessel, and the limited quantity of provisions she carries, I am unable to take a naturalist and draughtsman; but I should always hail with pleasure any scientific person who joined me abroad, or who happened to be in the countries at the time; and I may venture to promise him every encouragement and facility in the prosecution of his pursuits. I embark upon the expedition with great cheerfulness, with a stout vessel, a good crew, and the ingredients of success as far as the limited scale of the undertaking will permit, and I cast myself upon the waters —like Mr. Southey's little book—but whether the world will know me after many days is a question which, hoping the best, I cannot answer with any positive degree of assurance."

XI.—*Some Account of Mohammedu-Siseï, a Mandingo, of Nyáni-Marú on the Gambia.* By Capt. Washington, Royal Navy.

Although the special object of our inquiries, as geographers, is the surface of the earth which we inhabit, yet, as has been well remarked, as that earth is only interesting to us as the abode of man "for whom it was created," it may be permitted to pause for a moment in our more ordinary researches, for the purpose of contemplating a native of one of the least known regions of the globe —and to mark the vicissitudes in the life of a Mandingo, born on the banks of the river Gambia, who in his native village had seen and been in company with Mungo Park, one of the first and best of our African travellers—and successively to notice him as a slave—a soldier in the British army—a freeman—and finally as about to return to the home of his fathers, and to impart to his countrymen some few of the blessings of civilization which he may have acquired during an absence of more than a quarter of a century from the land of his birth.

Mohammedu-Siseï, was born at Nyáni-Marú, a village of about 200 houses on the northern bank of the Gambia,* of Mandingo and Muselmán parents; his father's name was Abú-Bekr, his mother's Aiseta [Ayishah]. At eight years old he was sent to school, at a place called Där Salámi (House of Peace), about 2 hours N.W. of Fatatenda, on the limits of the Wúlí country, where he learned to read the Korán and to write.

After about eight years at school, he underwent the rite of Súnnah or circumcision when he returned home. At the age of seventeen years old, he distinctly remembers Mungo Park passing through Nyáni-Marú on his way from Joka-kúnda, where he had landed, to Pisaniáre; this we know from Park's travels occurred in the year 1805, when he started on his second journey; and Mohammed says, among other circumstances that cause him to remember his arrival, was the quantity of rum given to the governor in exchange for his horses. This date also enables us to fix his age at from forty-eight to fifty.

In this same year Mohammed went by sea in a vessel to the French settlement at Gorée, to purchase trinkets apparently for his approaching marriage. He also made a journey by land as far as the Bundú country, and Bulibane, about seven days' journey to the N.E., a mountainous country which divides the waters of the Gambia and the Falémé, for the same purpose; in both these routes he names the places marked on our maps correctly, and many others which do not there appear.

On his return he married his cousin Aïseta, and then kept a school for five years in his native village.

At this time, it would seem, that the king of Wúli, Mansa Koï (or White King), and the king of Janjan-búre, Salám Mansa (or King of Peace), went to war, and the King of Wúli proving the more powerful, Salám Mansa retreated to the Kabú country for more troops; on his return he attacked Nyáni-Marú, and after some days' fighting, Mohammed, among many others, was made prisoner, and sent to the village of Kánsalá [Kansorály] in the Kabú country, (whose king was called Mansa Wal,) some days' march to the southward, where he was kept for five months.

At the expiration of this period, Mohammed, with many others, was marched to Sikka, near the mouth of the Gambia, sold to a French slaver, at once embarked, and sailed from the coast.

---

* In lat. 13° 42′ N., long. 14° 58′ W., exactly 100 geographical miles from the mouth of the river *Gambi*, as Mohammed always called it.

Thus, in a few months, was this poor African, living peaceably in his native village, torn from his parents and his wife, sold into slavery, and quitted, apparently for ever, the land of his birth.

Would to God that we could hope that even now a similar fate may not await him! but the recent capture of some liberated Africans, who returned from Demerara to Sierra Leone, and the inefficient measures adopted by other nations to suppress the slave-trade, prove too clearly the state of insecurity to which the European traffic in human flesh has reduced the region of western Africa.

But the hour of Mohammed's freedom was nearer at hand than he supposed; five days after the vessel sailed from the Gambia, she was captured by a British frigate, commanded by Sir Thomas Cochrane, and carried into Antigua. Of course, instantly on landing, he became free, and was put into the third West India regiment, in which he served from 1811 to 1825, in the grenadier company; he was present at the capture of Guadaloupe and the Saintes; and, after being stationed at Barbadoes, Dominica, &c., in 1816 he was sent to Trinidad, where he remained as a soldier till the reduction of the regiment in 1825, when he was discharged with a good character.

Land, at Manzanilla, on the eastern side of Trinidad, was given, in lieu of a pension, to the greater part of the soldiers when disbanded, but Mohammed appears to have lived chiefly at Port of Spain, the capital. In 1831 he married a creole of Grenada, who, with one child, accompanied him to England, whither he came in the hope of obtaining a pension for his fourteen years' service, and then of returning to his native country.

During the twenty years he had lived at Trinidad, Mohammed was a member of the Mandingo society of Mohammedans, who voluntarily subscribed funds among themselves to rescue their brother Mohammedans from slavery. He remembers more than twenty being thus ransomed at an expense of from 300 to 700 dollars each (70*l.* to 150*l.*), which the society voluntarily paid; and for several years, he says, no Muselmán has existed in slavery in Trinidad.

Very little intercourse with Mohammed will suffice to show that he has a good share of the intelligence generally found by travellers among the Mandingos. He is a strict Mohammedan; is well acquainted with the Korán, certain texts of which he always carries about him. He writes Mandingo indifferently in the Arabic character. He speaks English tolerably well, and, in forming a vocabulary of his language, he has often surprised me by naming the correct English word for a circuitous sentence, by which I had endeavoured to arrive at the Mandingo term.

While passing a few days with a kind friend, to whom we are indebted for the excellent account of Abú Bekr,* the com-

* Journal, vol. vi. p. 100.

panion of Davidson, Mohammed was introduced to Mr. Park, brother of the well-known traveller, when he mentioned circumstances and described persons with whom Mr. Park was well acquainted from his brother's accounts.

When taken to the Zoological Gardens and to the British Museum, Mohammed at once recognised and gave the Mandingo name to several animals and plants, natives of his country. At the Museum he pointed out the conical hat, the Fúlah cloak and mandolin, brought home from Africa by Clapperton; but at the sight of the elephant walking about at liberty in the Zoological Gardens, he clapped his hands with delight, and exclaimed, "O *Sama! Sama!* now I really see my own country again."

As has already been correctly observed by Golberry and Laing of the Mandingos, Mohammed resembles, in his complexion and character of face the Hindús or blacks of India, more than those of Africa in general. His features are regular and open, his person well formed, full six feet in height, his nose Roman, with the nostrils rather flattened, not thick lips, beautiful teeth, hair woolly, colour a good clear black, but not jet.*

The country of the Mandingos, as stated by Ritter in his well-known *Erdkunde*, and derived from the accounts of travellers, as Park, Mollien, Durand, &c., is the northern slope of the high table-land of the Senegambia. Park, on his return from the interior, first heard the Mandingo language spoken to the west of Taffara and Jabbi, from whence a high tract of fertile land extends as far west as Worombáka, between the head waters of the Niger and the Senegal; to the north is the Ṣahrá or great desert; to the south and west the high mountains of Jallonkadú, traversed by numerous rivers which descend through parallel ravines to the north. This, according to historical tradition, is the proper and immemorial abode of the Mandingo race.

The Mandingo language, we learn from Balbi and others, is one of the most extended of the 36 families of languages into which that author has divided the 115 languages of Africa, and is, on several accounts, the most important of the 21 families of languages of Nigritia or of the Negroes.

With the assistance of Mr. Renouard, I set to work diligently to obtain from Mohammed a vocabulary of his native language, not being aware that more of it was known than a list of about 200 words, collected in 1730 by Francis Moore—about 300 words given by Mungo Park in 1796—200 words by Caillié in 1829, and the small collection by Hannah Killam, in her specimens of African languages; after having written down about 1000 words and phrases, I was much surprised to find that the Gospel of St. Matthew in Mandingo had just been printed by the British and Foreign Bible Society, from a translation by Mr.

* Mr. W. Carpenter has obligingly painted an excellent portrait of Mohammed.

R. Maxwell Macbrair, a zealous and active agent of the Wesleyan Missionary Society, a copy of which was very civilly sent to me by Dr. Bunting. I also learnt that Mr. Macbrair had lately printed a grammar and vocabulary of the language, compiled during a residence of some years at the Gambia: finding this to be the case, and having very little leisure for so tedious an undertaking as compiling a vocabulary, (if done conscientiously,) I at once abandoned it. Yet while on this subject it would be unjust to omit Mohammed's spontaneous testimony to the accuracy of the vocabulary given by that excellent traveller Mungo Park. It was my custom while writing to have the three lists of Park, Moore, and Caillié open before me, and to refer to them as a new word occurred. Mohammed noticed this, and became curious to know what version each author gave, which I always read to him, and at length said in his negro English, pointing successively to the three books, "This man speak real Mandingo"—"That speak Mandingo, but not Mandingo of Nyáni Marú"—"Him never speak Mandingo at all."

Of the origin, progressive extension, and change of territory of the Mandingo race, as well as of their distribution into tribes, Mohammed could tell me nothing, although strictly questioned in compliance with the wish expressed by one intimately acquainted with African history and geography.

At my request he wrote down the following list of 25 places where Mandingo is spoken; many of them are probably only villages near the Gambia, at least the names are not familiar to me, and I have not leisure to search them out: they are given in the order in which he wrote them, as they may be useful to others:—Saman-galla, Gimbara-kunda, Dide, Welingara, Kanáko, Delafing, Kusang, Bokan-dandi, Hojuliri, Talibaji, Jalakoto, Jikir-illahi, Dar Salámi, Falamah, Palingo, Kanja, Tinkidah, Nyalen-kunda, Puró-pana, Sotoma, Paisori, Faraba, Garja, Lalemulé, Marená.

The 3rd West-Indian regiment, to which Mohammed belonged for fourteen years, was composed of natives of Central Africa; among them he enumerated some of Bambara,* Súsú, Fúlah, Bassarí, Jolah, Serawúlli, Yáriba, Hausa, Futa-toro,* Futa-jallo,* Krú-man,* Búndú, Jenné,* Ibú, Karamanti, Moko, Wamvi, Kanjá, Sereri, and Kalbé (as he always called Senegal); I also find Tumbúkatú* (as he pronounced it) in my notes; but I am not positive that he said there was a native of that city in his regiment,—my impression is that he did. These men always conversed with each other in English, although, said Mohammed, those marked * could speak 'half Mandingo.' Jonas Bath, a well-known, intelligent, and highly respected man in Trinidad, was a Súsú by birth, and was considered chief priest and patriarch of the Mohammedans; he is lately dead. The greater part of these men are now about to return to their native country, her Majesty's government having granted their petition to send a vessel with them to Africa. Should they succeed in reaching their several homes, which is much to be feared, we may hope that the quarter of a century passed in the midst of civilization will not have been entirely lost upon them, but that they may carry with them some useful arts, at least, which may tend to raise their countrymen from their present degraded state.

Mohammed has just left England for Bathurst, at the mouth of the Gambia, where it is hoped, through the kindness of the governor, Major Mackie, he may be usefully employed; as an interpreter with his nation he would be invaluable, as he is intelligent, and speaks English very fairly—and should any traveller be disposed to attempt to penetrate Africa, by following the route of Mungo Park, he would find a useful companion in Mohammed, who, before he left London, assured the writer that he was ready and willing to travel to any part of the interior.

It may not be generally known that there are several establishments besides Bathurst on the river Gambia belonging to Great Britain. Fort James, near Sikka, situated on an island about 16 miles up the river; opposite this, on the north side, is Jilifrí, in a healthy situation, and surrounded by a fertile district. On the south bank are Bintang; Tankoral, at 37 miles; and Jokakúnda, at 95 miles (?) from the entrance of the river.

The import trade to the Gambia, in 1836, was valued at 114,772*l.*; ships, 275; tonnage, 14,522; the exports were valued at 147,732*l.*; ships, 284; tonnage, 14,081; men, 2273.

The most valuable part of our traffic is carried on high up this extensive river, which, in a course of upwards of 400 miles from its hitherto unexplored sources to its mouth, fertilises a productive and highly populous country; and had we but stations to protect our trade, there is little doubt but that it might become the most valuable of our possessions in this quarter of the world.

But by far the most important station, and one that we cannot contemplate without every wish and hope that it is destined to form the nucleus of civilization in this portion of Africa, is MacCarthy's Island, as we term it, the *Janjan-Búré* of the natives, an island with an area of about 3 square miles, situated 127 miles* from the entrance of the river, in the midst of a populous country, and 60 miles below the falls of Barra-kunda, up to which spot the river is navigable for vessels of fifty tons burthen.

It is on this island that we have now an establishment for liberated slaves, and by the last returns for 1836 it had a popula-

* Measured on Owen's chart of the Gambia, surveyed in 1826. Fort George, MacCarthy's Island, is in 13° 33′ N., 14° 45′ W. long.

tion of 1600 persons; of these seven only were white; and 350 were Mandingos, who inhabited a village before the British took possession of it. Thanks to the active zeal of the Wesleyan missionaries, here are now two chapels, two schools for children, and one for adults, and the natives seem cheerful, orderly, and well-disposed to cultivate the ground, and to apply themselves to learn the various arts in which they are instructed.

When we consider the dreadful state of the slave-trade, which recent inquiries have brought to light, and the absolute necessity that exists, if we really wish to put a stop to it, of inducing the negroes to turn their attention either to cultivating the ground, or to some peaceful occupation, in their own country, we are sanguine enough to look upon this establishment at *Janjan-Búré* as a normal school, where people of nearly every tribe of Western Africa are receiving instruction in arts and civilization, if not in Christianity, which they may eventually carry with them to their several homes, and thereby be the means of diffusing civilization in places the most remote and inaccessible, which for the next century neither missionary nor traveller can hope to visit.

XII.—*A brief outline of the recent Expedition to the North-west Coast of Australia; under Lieutenants* GREY *and* LUSHINGTON. Principally from documents in the Colonial Office.

[IN offering to the readers of the 'Geographical Journal' a brief outline of the recent expedition to the north-western coast of Australia, extracted chiefly from documents liberally furnished to the Society by the Colonial Office and the Admiralty, it is not with the slightest hope of satisfying curiosity, or to anticipate the interest which the public in general, and geographers especially, always feel in enterprizes of this nature, but merely to give such a sketch of the principal features of the expedition as may serve to direct those who are desirous of obtaining information respecting a portion of this remarkable country—hitherto only visited by Tasman, Dampier, Baudin, and King, and never before, we believe, penetrated by any European—to look forward to the detailed journals of the spirited officers who had the conduct of the expedition, assured that they will contain much to interest them, not only in geography and in the light thrown upon the origin of the races who now inhabit this vast island, but in natural history, geology, and other branches of science.]

Her Majesty's Ship *Beagle,* Captain Wickham, destined for the nautical survey of the north-western and other parts of the coast of Australia, on board which vessel were embarked Lieutenants Grey and Lushington, and their party for the land expedition, left England on the 14th of July, and touching at Teneriffe and Bahia in the Brazils, reached the Cape of Good Hope on the 21st of September, 1837. Here the parties separated, as the *Beagle's* orders directed her to Swan River, while the leaders of the land expedition adopted the spirited but hazardous determination of proceeding direct to their destination on the N.W. coast of Australia.

Having freighted the *Lynher*, a schooner of 160 tons, taken on board 50 sheep and goats, and made all the requisite arrangements for the thorough equipment of their party, Messrs. Grey and Lushington quitted the Cape on the 20th of October, and on the 3rd of December reached Hanover Bay, at the outlet of Prince Regent's River, in lat. 15° 20′ south, long. 124° 40′ east.

Here having landed and pitched their tents in a beautiful valley now for the first time trodden by European feet, and having formally taken possession of this part of the country in the name of her Majesty, the schooner, under charge of Mr. Lushington, was despatched to Coepang, in the island of Timor, distant about 300 miles to the north, to embark ponies, of which six-and-twenty were obtained, at the rate of about two pounds each, chiefly in exchange for muskets and powder.

During the vessel's absence Mr. Grey and his party examined the country in the immediate vicinity of their camp—and explored a small stream which watered the valley, for about five miles to the southward, where it separated into two branches; one of which is little more than a mountain torrent; the other a much more considerable stream, winding through a deep and narrow valley: the soil brought down by it was rich and good, and therefore it was probably the drain of a considerable tract of fertile country.

The country generally as viewed from the sea, promises well, but upon landing the first appearance is monotonous and steril; being composed of rocky hills of sandstone about 300 feet in height, covered with brushwood and prickly grass; but between these hills are beautiful vallies, the soil of which is fertile, and where fresh water may always be found. Although at this time, from prudential motives, no communication was held with the natives, still their huts were visited, and found to consist of a conical frame-work of wood, about 4 feet high and 10 feet in diameter at the base, and were upon the whole well and neatly made. The natives were evidently in possession of sharp-cutting instruments, and many large trees were seen with notches or steps cut in them, by means of which they ascend for the purpose of stripping off the bark, which furnishes them with all the clothes they need in this fine climate—the hatchets may probably be of stone. The principal food of the natives would seem to be here, as in other parts of this land, kangaroo and shell-fish, as the remains of them were found near their fires, and both of them are abundant.

On the return of Lieutenant Lushington from Timor with the ponies and other necessaries, on the 20th of January, 1838, the party commenced their preparations for their immediate departure; and on the 1st of February quitted their encampment at Hanover Bay for the interior. The party consisted of 12 persons, including besides the leaders of the expedition, Mr. Walker, as surgeon and naturalist, two sailors, three men of the corps of sappers and miners, &c., with their ponies laden with instruments, provisions, and baggage, and followed by part of their flock of sheep and goats. They proceeded first about 15 miles in a nearly due south direction, until they had reached the parallel of 15° 29′ S. lat. The whole of the country lying between this point and Hanover Bay was composed of ridges of sandstone, of no great elevation, but intersected by deep ravines; their progress was consequently slow and toilsome, for they had to construct paths for the horses to travel upon before they were able to move from one encampment to the next spot where they intended to halt. In this first part of the journey they also lost many horses; indeed all of them suffered more or less from it.

After passing the parallel of 15° 29′ S. they entered upon a very rich tract of country, that even surpassed in fertility that small portion of the Brazils which they had had an opportunity of seeing. A large expanse of water having been seen a little to the west of south, they were induced to pursue that direction, and still found the country to be of the same rich and luxuriant character. Upon attaining the parallel of 15° 43′ S. lat., and 124° 44′ E. long., they found themselves upon the banks of a very considerable river, which Lieutenant Grey named Glenelg River, in testimony of the obligations which he and the whole expedition were under to the Principal Secretary of State for the Colonies.

The river was at this point salt; and as, from its magnitude and rapid current, no possibility existed of crossing it, they were obliged to travel up its banks in a north-easterly direction—their progress being much impeded by the number of tributary streams which poured into it, many of these being of such a size and depth, owing to their running through a low country, that they were in the same manner obliged to travel for several miles up their banks, before they could get an opportunity of crossing them.

The luxuriance of the vegetation also much impeded their progress; so that, although their horses again began to thrive from the goodness of the grass, their advance was still slow and difficult.

The river first became perfectly fresh at a point situated in 15° 41′ S. lat., and 124° 53′ E. long. Beyond this point it would not be navigable for large vessels; for a series of rapids occur here, so that a portage must necessarily be established: immediately below this point a shoal of porpoises was seen.

From hence they still continued their course to the eastward, in the direction of the river, but at some distance from it, until they arrived at a point situated in 15° 41′ 50″ S. lat., and 124° 59′ E. long. The river was here quite fresh, running at the rate of rather more than 5 knots an hour. Its banks were composed of fine white sand; and even close to the bank it was 2½ fathoms in depth.

On the other side of the river the country hereabouts appeared to be low and marshy; and a remarkable circumstance was observed, namely, that drift-wood, weeds, &c. were lodged in the forked branches of trees at least 15 feet above their heads when they stood upon the banks of the river: extensive inundations must therefore sometimes take place, and at these periods there is little doubt that the whole of the low country to the south is flooded.

A very large tributary stream here joined the river, which they could not cross, as it ran through a low and marshy country; and they were again obliged to turn off to the northward. After following its course for about eight miles through a marshy and almost impassable country, they succeeded in crossing it, but still experienced great difficulty in travelling upon the other side of it; for owing to the heavy rains which had lately fallen, the marshes were almost impassable.

Lieut. Grey here remarks that these heavy rains, which were of several days' continuance, had but very little effect upon the main river, and would by no means account for the signs of inundation which they there saw. This is a remarkable fact, and, viewed in connexion with some others, bears materially upon the physical geography of the country.

In 15° 49′ S. lat., and 125° 6′ E. long., they crossed another very considerable stream, which ran in the direction of Glenelg river. This was the largest tributary stream seen flowing down to it.

The next point at which they saw the river was in 15° 56′ S. lat., and in 125° 8′ E. long.; it was here 250 yards wide, but again formed a series of rapids; and they found a ford just above these, the average depth of which was not more than 3 feet. The soil on its banks was still good: the bed of the river at this point trended away to the eastward.

Only 12 ponies now surviving out of the 26 which had arrived from Timor, it was deemed prudent no longer to follow the course of this river; but Lieut. Grey resolved to push on with as much dispatch as possible in the direction of the great opening behind Dampier's Land; for although it was evident that from want of provisions they would soon be compelled to return to Hanover

Bay, he conceived it absolutely necessary that the part of the country above alluded to should be examined.

After crossing the river, the country again changed its character, the soil becoming sandy, and they began the ascent of a chain of hills running from S.E. to N.W. They continued this gradual ascent for three days, and then found themselves at a point situated in 16° 3′ S. lat. and 125° 13′ 30″ E. long.: they were now nearly at the summit of the part of the range that they were on, but to the S.E., south, and S.W. they could see nothing but inaccessible precipices.

Several days were occupied in trying to find a pass by which the horses might proceed over the mountains, but none such could be found; they, therefore, resolved to proceed with a small party to examine the country to the south of their present position.

At this time Mr. Grey's wound in the hip, which he received from the spear of a native at the outset of his journey, became so painful, that the surgeon positively forbade his proceeding a step farther. A detached party, therefore, started under the command of Mr. Lushington, and attained a point situated in about 16° 30′ S. lat. and 125° 15′ E. long.; the result of their *reconnaissance* was, that in all probability, no large river could exist immediately to the south.

As any further attempt to proceed in that direction appeared then to be useless, and as the party were compelled to return to Hanover Bay from want of provisions, they resolved to make for this point by a route different to that which they had already traversed, and thus fully to complete the exploration of this portion of the country.

On arriving at Hanover Bay, on the 15th March, they had the unexpected pleasure of meeting with Captain Wickham, R.N., in command of H.M.S. *Beagle,* who had, after a careful examination of the coast, arrived at the same conclusion, viz., that no large river could exist between the one that they had discovered, and *Fitzroy River,* which he had discovered at the south part of the great opening behind Dampier's Land.

These rivers, although of considerable magnitude, are still utterly insufficient to account for the drainage of this vast continent, and this interesting question, instead of being at all placed in a clearer point of view by the united exertions of these two expeditions is, if possible, at this moment involved in deeper obscurity and mystery than ever.

In the course of the journey Messrs. Grey and Lushington found a great many curious native paintings in caves, executed in a surprising way for a savage race. In these caves were some drawings of the human hand, which showed great knowledge of the art of producing effect: they selected a rock in the most gloomy part of the cave, and the hand must have been placed upon this rock and some white powder dashed against it. When the hand was removed, a sort of stamp was left upon the rock; the hand was then painted black and the rock about it quite white, so that on entering that part of the cave, it appeared as if a human hand and arm were projecting through a crevice, admitting light. Many of the figures in these drawings were clothed, though the natives themselves were in a perfect state of nature. These and other circumstances would countenance the belief that they are a race of Asiatic origin; indeed, Lieut. Grey expresses his conviction that the mass of evidence he has collected on this point will enable him to prove it.*

These caves and paintings were all far inland, and nothing of the kind was near the coast. Copies were also obtained of some of the drawings by the natives living on the coast, but these are said to be the productions of a quite distinct race.

Before quitting Hanover Bay the party had the gratification of seeing the bread-fruit and cocoa-nut trees which they had brought from Timor, and planted in the valley, as well as numerous seeds from the Brazils and the Cape of Good Hope, in a most flourishing state. They had also introduced and left there several animals, as ponies, goats, and sheep, and in short done everything in their power to make their visit a blessing to the natives and to the country.

On the 30th April the *Lynher* sailed for the Mauritius, whence Mr. Grey's letter is dated on the 2nd June, at which time, it is highly satisfactory to add, he had quite recovered the effects of his wound, and contemplated returning to a more southern portion of the same coast of Australia. Lieutenant, now, we are happy to say, Captain Lushington, has arrived in England, and it is gratifying to hear the distinct testimony he bears to the zeal and energy displayed by Lieutenant Grey, even when suffering from the effects of a wound and a tropical temperature, in his anxiety to carry into execution the expedition entrusted to his charge.

* Colonel Fox's letter.

XIII.—*Outline of the Survey of part of the N.W. Coast of Australia, in H.M.S.* Beagle, *in* 1838. By Captain Wickham, R.N. Communicated by Captain Beaufort, Royal Navy.

H.M.S. *Beagle*, 17th April, 1838.
Port-George IV., N.W. Coast of Australia.

Sir,—I have the honour to acquaint you that H.M. surveying vessel, under my command, sailed from Swan River on the 4th January, 1838, and proceeded direct to the N. W. coast of Australia.

We reached the shoal soundings off Cape Villaret on the evening of the 15th without being able to make out the land before dark; but continued to stand on until the water shoaled to 14 fathoms, when an anchor was dropped to enable us to retain our position, with a view to commence the examination of Roebuck Bay in the morning; consequently at daylight we were under way and standing into the bay, which we soon found to be very shoal, and were obliged to anchor with Cape Villaret bearing S. 15° W. 6 or 7 miles, and about $3\frac{1}{2}$ miles off shore, in 6 fathoms at high-water, and a rise and fall of 18 feet at the neaps.

Several attempts were made to get the ship farther into the bay, but we only succeeded in getting her about 5 or 6 miles above our first anchorage. The examination of the bay was consequently commenced in the boats, and was soon found to be of small extent, terminating in flats of muddy sand, which dried from 1 to 2 miles off at low water.

The shores of the bay were carefully traced without our being able to discover any opening whatever, thereby setting at rest the question respecting Dampier Land being an island.

From Roebuck Bay we proceeded to the northward, where the coast was found to differ a little from that laid down in the charts; and, thinking there might be some opening unobserved by Captain King, owing to his distance off shore, we narrowly examined every part of the coast as far as Point Swan, never being more than from 2 to 3 miles off shore, and at times considerably within that distance, anchoring every night, which we were fortunately enabled to do, owing to the favourable state of the weather.

The bay round Sandy Point, which has been named Beagle Bay, in lat. 16° 50′, affords the best anchorage on the coast, but it is exposed to westerly winds.

The coast from Roebuck Bay to Point Swan having been closely examined, without the slightest appearance of even a stream of fresh water running into the sea from any part of it, and the weather being evidently on the eve of a change, and the westerly monsoon setting in upon the coast, the ship was moored off the first sandy beach round Point Swan, which place appeared to offer the best anchorage we were likely to find. I therefore resolved to wait a few days for the purpose of rating the chronometers, and making the other necessary observations, and, if possible, to complete our stock of water: wood we saw was plentiful everywhere.

We remained at this anchorage until February 9th, during which time we had constant bad weather, blowing hard from the westward, between N.W. and S.W., with very heavy rain, and much thunder and lightning.

This bad weather, although very much against our surveying operations, was favourable to us, inasmuch as it enabled us to fill up our water, which was caught by the awning, and bailed from the hollows of the rocks. A well was sunk 16 feet deep, and some feet below the level of the sea at high water, but there the soil (still of a sandy nature) was perfectly dry.

From Point Swan we proceeded towards Sunday Strait, having previously ascertained that there was no passage fit for a vessel of any description between that and Point Swan, the whole space being thickly strewed with rocky islets and reefs, nearly all dry at low water, and causing heavy races and overfalls as they became covered at half-tide.

Owing to the unsettled state of the weather, we were obliged to anchor at three different times under the large island on the west side of the strait, in exposed anchorages; but this trifling delay enabled us to complete Captain King's plan of the place.

From these islands, which we have named Roe's Islands (in compliment to Lieut. Roe, who accompanied Captain King), we proceeded to Cygnet Bay, and continued a close examination of the coast to the southward. Many good anchorages were found, but no appearance of streams of fresh water.

Having reached as far as Foul Point, I sent Lieutenant Stokes with two boats to trace the shore farther to the southward, and to gain some information as to the probable extent of the opening. He returned on the eighth day, having succeeded in discovering that the southern part of King's Sound terminated in the mouth of a river, or of an extensive lake, as at low-tide (the fall being 36 feet) the water was quite fresh alongside the boat, and running in small streams from the southward, between the extensive flats that were left dry by the tide, and reached from shore to shore, a distance of 5 or 6 miles, without leaving a passage for a boat.

In consequence of this information the ship was moved to the

S.E., in which direction Captain King had seen land raised by refraction, but which was found to be only 18 miles distant, and very low. From this point the examination of the river was continued by Lieut. Stokes and myself.

We found it to be of trifling extent; but, from the quantities of drift-wood and weeds suspended to the trees from 12 to 15 feet over our heads, it must be subject to immense inundations at certain seasons of the year; and from the lowness of the land on each bank, as far as we could see, the whole country must then be under water for an extent of many miles.

We proceeded about 15 miles in a straight line, S. by E., at which place our progress was impeded by the numbers of drift-trees that completely blocked the passage. At this point the river was rapid, and ran in three or four small streams, occasioned by some islets, and a number of fallen trees.

The banks were nowhere above 12 feet high, and the land on each side perfectly level, as far as could be seen from the top of the highest tree. Quantities of rich grass covered both banks; and the country appeared in places to be thickly wooded.

The entrance to this river, which has been named *FitzRoy River*, in compliment to Captain R. FitzRoy, R.N., is in lat. 17° 34′ S., and long, 123° 38′ E. (nearly).

During the examination of the river, a boat was employed in tracing the shore to the eastward, which appeared to be a deep opening, as no land was visible from the mast-head; and I thought it not improbable that a passage might be found communicating with Collier Bay; but, like all the other openings we had examined, it terminated in low flat land, thickly studded with mangroves, amongst which the tide flows for a considerable distance; and at low water the whole shore is fronted by extensive flats of soft mud.

From this point the shore of the main land has been carefully traced as far as Port-George IV., chiefly by Lieut. Stokes, in the boats, who is perfectly satisfied that there is no stream of any consequence running into the sea from any part of it.

On the passage from King's Sound to Port George IV. we discovered a dry sand-bank, 10 or 12 feet above water, and which is not laid down in Captain King's chart. It lies in lat. 15° 19′ 20″ S., and long. 123° 35′ E., being about 26 miles N.E. by E. of Adele Island.

It lies in a direction about N.W. and S.E.; the dry part about a third of a mile; but the whole extent of the breakers occupying a space of about 4 or 5 miles in that direction, which appears to be the direction of all the islands, reefs, sand-banks, and tide-ripplings on this part of the coast, there is much uneven ground between Adele Island and this bank, which has been named Beagle Bank.

I take this opportunity also of making known the discovery of two dangers seen by Mr. Browse in the *Lynher* (the vessel in which Mr. Grey and party arrived on the coast).

One is situated in lat. 15° 26′ 30″ S., and long. 121° 55′ E., and is a reef about 2 feet under water; the other is in lat. 14° 4′ S., and long. 123° 30′ E. This last was fallen in with on the second night after sailing from Timor; therefore cannot be much out of position; besides, the vessel was furnished with a good chronometer. It is an island about a mile in extent, but very low, and was not discovered until about 2 cables distant, when Mr. Browse sounded in 10 fathoms. He afterwards sounded, when he thought himself about a mile to the westward of it, but had no bottom with 40 fathoms.

As we have been fortunate enough to meet Lieut. Grey and his party, who returned from his expedition three days after our arrival here, and who has kindly offered to take charge of any letters we may wish to send, I take the opportunity of sending this short account of our proceedings, and merely a sketch of our work in continuation of Captain King's charts, as there has not been time to lay down the work that all have been busy in collecting, and which will require at least two quiet months to put together.

As we have now only three weeks' provisions on board, and I am given to understand that no supplies can be procured at Timor, excepting for present use, and those with some difficulty, in exchange for gunpowder only, we shall make the best of our way to Swan River, where it will be necessary to refresh the crew, who, although we have had no sickness occasioned by the climate, which I consider equal to any in the world, will be much invigorated by a change of temperature, which, since our arrival on the coast, has rarely been below 83°, and frequently as high as 95° 5′ in the shade.

After receiving supplies and refreshments at Swan River, I shall proceed to Sidney for stores and to refit; likewise to get the work of this cruise completed, before going to Bass' Strait, the examination of which place will afford ample employment for the summer months. I have the honour to be, &c.

Jno. C. Wickham, Commander.

Extract from a private letter from Captain Wickham to Captain Beaufort:—

"I was afraid at the commencement of our survey that Captain King's luck was to attend us; as on the night of our anchoring off Cape Villaret, one of those furious squalls he so well describes came off the land at E.S.E., and, although we had nearly 80 fathoms of cable out, our anchor broke like a piece of glass, and, having previously broken one at Swan River, it was no very pleasing event; however, since that everything has favoured us. We have had frequent communications with the natives at different places, and all have been of a most friendly nature; indeed, they have invariably sought our acquaintance, by coming to the beach and beckoning to us to land; and, whenever our parties have landed, although at times in great numbers, for the purpose of fishing, wooding, and watering, they have never drawn back, but have put themselves (unarmed) entirely into our power, and have never once expressed a wish that our arms should be laid down. Almost all that we have seen have had the two front teeth of the upper-jaw extracted, and all perfectly naked, with the exception of a small grass apron.

"This appears to be a very extraordinary part of the world; the whole coast on the western side of King's Sound, also the entire coast between Roebuck Bay and Point Swan, being entirely of sand-stone, whereas, the islands and all the high land on the eastern side are entirely of quartz, and of so rugged an outline, that it is a most difficult matter to proceed in any direction. The islands are almost void of vegetation, and the whole seem to be thrown up into such fantastical shapes, as to lead one to suppose this part of the world to be in the last stage of utter ruin and confusion. From Valentine Island (where the cliffs end) the land is very low, and continues so to the banks of FitzRoy river. I am afraid that this river is never likely to become of any service, owing to its distance from the sea, and the risk in approaching it, occasioned by the strong tides amongst the islands of Buccaneer's Archipelago; but I am of opinion that, if a party is ever to reach the interior of Australia, it must be on its banks; the wood is nowhere so thick as to offer any impediment to such an undertaking, and there would be a certain supply of good grass. It appears to me that the inundations that evidently take place at times are in no way connected with the rainy season on the coast, as at our visit, which was immediately after the rains, there were no recent marks of the country having been flooded; these overflowings might prove a serious evil, as in many places there would be no escape but in the trees; the bed of the river is entirely of sand and gravel, therefore, no sickness would be likely to arise from keeping on its banks.

I find that Mr. Grey has discovered another river at no great distance from Prince Regent's River, and which he thinks discharges itself into the sea, somewhere between this port and Collier Bay; but, as Lieut. Stokes has just returned from the examination of that piece of coast, without having seen any opening, or anything to lead him to suppose the existence of a river, I think it more than probable that Glenelg River loses itself in the low land, and escapes by numerous creeks through the many mangrove-flats by which each small bay is bounded; and it is not improbable that it may lose itself in the boundless mangrove-flats at the head of Stokes' Bay. From the account of Lieut. Stokes, it appears very unlikely that any river of consequence empties itself into the sea on this side of Collier Bay, as no drift-trees or timber was seen by him; the tides were regular, and in the depth of Collier Bay there was scarcely any stream of tide. This was not the case in the entrance to FitzRoy River, as there the tides were very rapid; the flood ran only 4 hours, and the ebb 8, and in King's Sound there was a great deal of drift timber, which increased very much as we approached the river, and the water was quite muddy; between this and Collier Bay the water was not even discoloured. It appears strange where this river runs into the sea, but I am quite of opinion that it may be lost in numerous creeks which we generally find in all mangrove-flats, and into which it is next to impossible to penetrate in the boats. I am very glad that we have met Grey's party, and that all has gone on well, as I should have been reluctant to leave the coast without having heard of them; it is a satisfactory wind-up to our first campaign.

"I cannot conclude this without expressing to you the entire satisfaction I feel with respect to the manner in which *all* have performed the different duties allotted to them. Stokes has been most indefatigable. Natural history has not been forgotten by those who have had time to attend to it."

I. C. W.

[When that clever old navigator Dampier was off this part of the coast in the year 1699, he noticed the extraordinary rapidity of the tide, which had not been the case in the higher latitudes of this western coast, and he makes this observation, that, "by the great tides I met with a while afterwards, more to the N.E., I had a strong suspicion that here might be a kind of archipelago of islands, and a passage possibly to the south of New Holland and New Guinea into the great sea eastward, which I had thoughts also of attempting on my return from New Guinea." He did not, however, return. The examination made by Capt. Wickham sets the question of a *passage* at rest; and I think the great occasional rise of the two rivers, that of Prince Regent, observed by Grey and Lushington, and that of FitzRoy River, described by Wickham, both stating it to be about 15 feet, will fully explain the rapidity and irregularity of the tides mentioned by Dampier, when these rivers are in a state of flood, and the whole country adjacent inundated.

These rivers now discovered are not of sufficient magnitude to explain the discrepancy which seems to exist between the great land of Australia and all other continents and large islands on the face of the globe. The drainage of these, and the rest already known, are quite insufficient in ordinary cases to carry off the waters of so vast a country. May not the extraordinary occasional rise point to the old conjecture of some great Mediterranean sea? One spot only now remains where a large river can be looked for, and that is in the bottom of the deep and wide gulf of Carpentaria; and if not there, we must either conclude that there is some peculiarity in the construction and material of Australia, or that the waters are collected in numerous lakes or one great sea, and evaporated or absorbed before they reach the coast.—JOHN BARROW.]

LIST OF GEOGRAPHICAL WORKS RECENTLY PUBLISHED.

EUROPE.

AUSTRIA.—Travels in Austria, Russia, and Turkey, by the Rev. C. B. Elliott. 2 vols. 8vo. 1838.

BRITISH ISLES.—Ordnance Survey of the County of Londonderry. 4to. 1838.

Statistique de la Grande Bretagne et de l'Irlande, par A. Moreau de Jonnès. 2 vols. 8vo. Paris, 1837.

DENMARK, Sweden, and Norway. Hand-Book. Murray, 1838.

GREECE.—Residence in Greece, and Journey through Bulgaria, by F. Hervé. 2 vols. 8vo. 1838.

ICELAND.—Voyage en Islande et au Gröenland en 1835-6, publié sous la direction de M. Gaimard, 6 vols. 8vo. 7 livraisons du texte, et 12 livraisons des planches en vente. Paris, 1838.

ITALY.—Dizionario Geografico della Toscana, da E. Repetti. Firenze, 1838. Letter L.

Memorie sul beneficamento delle Maremme Toscane. 4to. Firenze, 1838.

Statistica d'Italia, dal Conte Serristori. 6 Nos. Firenze, 1838.

PORTUGAL.—Meine Reise durch Portugal in 1836, von Gust. v. Heeringen. Leipsig, 1838.

RUSSIA.—Narrative of a Tour in the North of Europe, by Lord Londonderry. 8vo. 1838.

SAXONY.—Mittheilungen des Statischen Vereins des Königreich Sachsen. 9te Lieferung. Dresden, 1838.

SCANDINAVIA, Ancient and Modern, by Dr. A. Crichton and H. Wheaton. 2 vols. 8vo. Edinburgh, 1838.

SPAIN.—Sketches in the Mountains of Ronda and Granada, by Capt. C. R. Scott. 2 vols. 8vo.

SWITZERLAND, Savoy, and Piedmont. Hand-Book. Murray, 1838.

TURKEY.—Journal of a Tour through Rumelia, by D. Urquhart. 2 vols. 8vo. 1838.

ASIA.

ASIA.—Die Erdkunde von Asien, von Carl Ritter, Vol. IV., Part I. Berlin, 1838.

Asia, Von Dr. H. Berghaus. 4to. Part IV. Gotha, 1838.

Reise um die Erde, durch Nord Asien, &c., in 1828-30. Von A. Erman. 8vo. 2d vol. Berlin, 1838.

ARABIA.—Travels in Arabia—Omán, Mount Sinai, and Red Sea, by Lieut. Wellsted, I.N. 2 vols. 8vo. 1838.

ASSYRIA.—Researches in Assyria, Babylonia, and Chaldea, by W. Ainsworth. 8vo. 1838.

CAUCASUS.—Alte Geographie des Kaspishen Meeres des Kaukasus und des südlichen Russlands, von Dr. E. Eichwald. 8vo. Berlin, 1838.

Travels in the Western Caucasus by E. Spencer. 2 vols. 8vo. 1838.

EASTERN INDIA.—Topography and Statistics of Behar, Shahabad, Bhagulpúr, Assam, &c., by Montgomery Martin. 3 vols. 8vo. 1838.

Topography of Assam, by Dr. M'Cosh. Calcutta, 1838.

Report on the Eastern Frontier of India, with an Appendix and Maps, by Captain R. B. Pemberton. 1 vol. 8vo. Calcutta, 1835.

Instructions Nautiques sur les mers de l'Inde, traduites de Horsburgh, par M. le Capitaine Prédour. 3me. tom. 8vo. Paris, 1838.

HINDUSTAN.—Travels in the Panjab, Ladakh, Kashmír, &c., by Messrs. Moorcroft and Trebeck. 2 vols. 8vo.

Voyage dans l'Inde, par Victor Jacquemont. 4 vols. 4to. 24 livraisons en vente. Paris, 1838.

Personal Narrative of a Tour through the Western State of Rajwara in 1835, comprising Bíkaner, Jesulmer, and Jadhpúr, with the passage of the Great Desert, by Lieut. Boileau, Bengal Engineers. 4to. Calcutta, 1837.

Observations on the Nilgherry Hills, with an account of their Topography, &c., by Dr. Baikie. 8vo. Calcutta.

JAPAN.—Nippon Archiv zur Beschreibung von Japan; von P. F. Siebold. Folio. Leyden, 1832-8.

Expedition from Singapore to Japan, with a visit to Loochoo, by Dr. Parker. 12mo. 1838.

JAVA.—Ueber die Kawi-Sprache auf der Insel Java, von Wilhelm von Humboldt. 4to. Berlin, 1837.

MALACCA.—On Penang and Province Wellesley, by Captain Lowe. Singapore, 1838.

Notices of the Indian Archipelago, by J. H. Moor. Singapore, 1837.

PALESTINE.—Palästina, von Carl von Raumer. 8vo. Leipzig. 1838.

Reisen durch Syrien und Palästina nach Cypern und durch Kleinasien, in 1766, von Carsten Niebuhr. 4to. Hamburgh, 1837.

PERSIA.—A Winter's Journey from Constantinople to Tehrán, with Travels through various parts of Persia, Khorasán, and the Caspian Sea, to Tabríz, by J. Baillie Fraser, Esq. 2 vols. 8vo. 1838.

AFRICA.

ABYSSINIA.—Voyage en Abyssinie, &c., par MM. Combes et Tamisier. 2 vols. 8vo. Paris, 1838.

AFRICA, SOUTH.—An Expedition of Discovery into the Interior, through the countries of the Great Namáquas, Boschmans, and Hill Dámaras, by Captain Sir Jas. Alexander. 2 vols. 8vo. 1838.

Excursions dans l'Afrique Septentrionale, par MM. Falbe et Sir Grenville Temple. 8vo. Paris, 1838.

ALGIERS.—Tableau de la Situation des Etablissements Français dans l'Algérie. 4to. Paris, 1838.

CANARY ISLES.—Histoire Naturelle des Isles Canaries, par MM. Webb et Berthelot. 4to. Paris, 1838. 32 livraisons en vente.

EGYPT.—Letters from Egypt and Edom, by Lord Lindsay. 2 vols. 8vo. 1838.

MADAGASCAR.—History of, compiled from Original Documents, by the Rev. Wm. Ellis. 2 vols. 8vo.

AMERICA.

AMERICA.—Antiquitates Americanæ, sive scriptores septentrionales rerum ante Columbianarum in America, by the Northern Society of Antiquaries. Copenhagen. 1 vol. folio, 1837.

Examen Critique de la Géographie du Nouveau Continent, au 15me. et 16me. siècles, par le Baron A. von Humboldt. Imp. fol. Edition en 8vo. 4 tomes, en vente.

Reise durch Nord-America, von Maximilien Fürst von Wied. Coblenz. 1838.

Narrative of the Voyage of H. M. Ship *Terror* to Hudson's Strait, in 1837, by Capt. Sir George Back, R.N. 1 vol. 8vo. 1838.

Exploring Visits to the sources of the Hudson, in 1836-7, by W. C. Redfield. Pamphlet. 1838.

Reise nach Saint Louis am Mississippi. Weimar, 1838.

Journal of an Exploring Tour beyond the Rocky Mountains, by the Rev. J. Parker. Ithaca, U.S. 1837.

AMERICA, SOUTH.—Voyage dans l'Amérique Meridionale en 1824-33, par Alcide d'Orbigny, 7 vols. 4to. en 75 livraisons, dont 35 en vente. Paris, 1838.

PATAGONIA.—Surveying Voyages of H. M. Ships *Adventure* and *Beagle*, 1833-6, (Patagonia, Chile, and Peru,) by Captains P. P. King and R. FitzRoy, R.N., with extracts from the Journal of C. Darwin, Esq.

LA PLATA.—Coleccion de obras y documentos relativos à la Historia Antigua y Moderna de las Provincias del Rio de la Plata, por P. de Angelis. Vols. V. and VI. Buenos Ayres, folio, 1837.

A Journey across the Pampas and the Andes, from Buenos Ayres to Valparaiso, Lima, and Panama, with Remarks upon the Isthmus, by the Hon. P. C. Scarlett. London, 2 vols. 8vo. 1838.

Buenos Ayres and La Plata, by Sir Woodbine Parish. 8vo. 1838.

Letters on Paraguay, by J. P. and W. P. Robertson. 2 vols. 8vo. 1838.

Voyage à Buenos Ayres et à Porto Alegre, &c., par Arsène Isabelle. 8vo., Havre.

POLYNESIA.

AUSTRALIA.—Three Expeditions into the Interior of Eastern Australia, with descriptions of the recently-explored region of Australia Felix, and of the present Colony of New South Wales. By Major T. L. Mitchell. 2 vols. 8vo. 1838.

ARAFURA SEA.—Sailing directions for, by G. W. Earl. 8vo. 1837.

NEW ZEALAND.—Travels and Adventures in, between 1831-7, by J. S. Polack. 2 vols. 8vo. 1838.

MISCELLANEOUS.

AN Attempt to develope the Law of Storms, by Lieut.-Col. W. Reid. 8vo. 1838.

ANNALEN der Erd-Völker und Staatenkunde, von Dr. H. Berghaus. 8vo. Berlin.

ANNALES des Voyages, par MM. Eyries, Humboldt, et Larenaudière. 8vo. Paris.

——— Maritimes, par MM. Bajot et Poirré. 8vo. Paris.

ATLANTIC Ocean, Sailing Directory for the South, by John Purdy. 1837.

GEOGRAPHIE d'Edrisi, tome V., du Recueil des Voyages et des Mémoires, publié par la Société de Géographie de Paris. 4to. 1837.

JOURNAL de la Navigation autour du Globe de la frégate la *Thétis* et la corvette *l'Espérance*, en 1824-6, par M. le Baron de Bougainville. 2 vols. 4to. Paris, 1837.

NAUTICAL Magazine. 8vo. 1838.

PHYSIKALISCHE Geographie, von F. Hoffman. 2 vols. 8vo. Berlin, 1838.

POSITIONS Géographiques des Principaux Lieux du Globe, par M. Daussy. 8vo. Paris, 1838.

RESEARCHES into the Physical History of Mankind, by Dr. J. C. Prichard. 2 vols. 8vo. 1837.

STATISTIQUES des Colonies Françaises. 2 vols. 8vo. Paris. 1838.

VOYAGE Round the World in 1836-7, in the U.S. Ship *Potomac*, by C. J. Ruschenberger, M.D. New York, 1838.

ZEITSCHRIFT für die Kunde des Morgenlandes, von H. Ewald, C. N. D. Gabelentz, und R. Rödiger. 8vo. Gottingen, 1837.

MAPS RECENTLY PUBLISHED.

EUROPE.

AUSTRIA.—The Arch-duchy of Austria, in 47 sheets, scale 1-144,000th. Quart.-Mas.-Gen. Off., Vienna, 1838. 30 sheets published.

The environs of Vienna in 28 sheets, scale 1-14,000th. Quart.-Mas.-Gen. Off., Vienna, 1838. 9 sheets published.

BRITISH ISLES.—England and Wales, the ordnance survey of, in 120 sheets, scale 1 inch to a statute mile. 75 sheets published.

——— Plan of Teignmouth, Devon, by Mr. Spratt, R.N., scale 6 inches to a mile, Hyd. Off., 1838.

Wales.—Porth-dyn-llaen Harbour, by Lieut. Sheringham, R.N., scale 4 inches to a mile. Hyd. Off., 1838.

Ireland, the townland survey of, scale 6 inches to a mile. 15 counties published. Dublin, 1838.

——— Map in 4 sheets, scale 4 miles to an inch. Dublin, 1838; drawn by C. and J. Walker.

——— Map in 2 sheets, scale 10 miles to an inch, by the Diff. Use. Know. Soc., London, 1837.

——— Ballyshannon Harbour, by Capt. W. Mudge, R.N., Hyd. Off., 1838.

Scotland. The Shetland Isles, by Mr. G. Thomas, R.N., scale 2 miles to an inch. Hyd. Off., 1837.

FRANCE.—Carte Topographique de la France, in 258 sheets. 60 published. Scale 1-80,000th. Dépôt de la Guerre, Paris, 1838.

Côte de France, entre Fontenailles et Langrune, Dép. de la Mar., Paris, 1838.

——— entre Grandchamp et Fontenailles; pointe et raz de la Percée. Dép. de la Mar., Paris, 1838.

Plan du Port de Courseuilles. Dép. de la Mar., Paris, 1838.

Corsica, by M. le Capitaine Hell. Hyd. Off., London, 1838.

GERMANY.—Topographische Karte der umgegend von Frankfort. Von A. Ravenstein. Scale 1-75,000th. 1837.

HUNGARY.—Map of Hungary, in 9 sheets, by Professor Schedius. Pesth, 1836-8.

ITALY.—Map of the Lombard-Venetian Kingdom, in 43 sheets, scale 1-86,400th. Dépôt Topog., Milan. 36 sheets published.

Sardinia. Cagliari, by Capt. Smyth, R.N., Hyd. Off., 1838.

——— Channel of San Pietro, by Capt. Smyth, R.N., Hyd. Off., 1838.

——— Ports Conte and Alghero ,, ,, ,,

——— Graham's shoal, by Mr. Shadwell, R.N. ,,

Naples. Plan de la Baie de Naples. Dép. de la Mar., Paris, 1838.

RUSSIA.—Military Map of Russia, in 16 sheets, scale 1-680,000th. Dépôt Topog., St. Petersburg, 1837.

SAXONY.—Topograghical Map of the Kingdom of Saxony, in 20 sheets, scale 1-57,600th. Dépôt Topog., Dresden, 1836-8. 5 sheets published.

ASIA.

ARRACAN.—Route from Aeng in Arracan to the Iráwádí, by Capt. B. Pemberton. Calcutta, 1837.

CAUCASUS.—Karte des Kaukasus, scale 2½ inches to a degree; by F. G. Morin. Berlin, 1838.

CHINA.—La Mer de Chine, dressée par M. Daussy, en 1838; scale 1 inch to a degree. Dép. de la Mar., Paris.

INDIA.—Rajamandry.—Sheet 95 of Indian Atlas, scale 4 inches to a mile. E. I. Company, 1838.

Cochin.—Sheet 62 of Indian Atlas. E.I. Company, 1838.

La Mer des Indes, par MM. Daussy et P. E. Wissocq. Dép. de la Mar., Paris, 1837. Scale 3-10ths of an inch to a degree.

Les Entrées de la Rivière d'Hoogly, &c. Dépôt de la Mar., Paris, 1838.

JAVA.—Straits of Madura and Bali, scale 6 miles to an inch, by J. Schwart. Amsterdam, 1837.

MALDIVAS.—Trigonometrical Survey of the Maldivas, by Captain R. Moresby and Lieut. F. T. Paul, Indian Navy; scale 4 miles to an inch. London, 1838.

MANIPUR.—Map of the Manipúr Territory, with part of Kubo Valley and the Burmese Frontier, by Captain Pemberton. Calcutta, 1837.

PERSIAN GULF.—Bushire Port and Karak Island, by Lieuts. Brucks and Cogan, I.N. Hyd. Off., 1838.

SUMATRA.—Karte von Sumatra, von H. Berghaus. Scale 2 inches to a degree. Gotha, 1837.

SUNDA.—Chart of the Straits of Sunda, by Mr. J. Schwart, scale 4 miles to an inch. Amsterdam, 1838.

Sunda oder Borneo-Meere, von H. Berghaus. Scale 2 inches to a degree.

AFRICA.

ALGIERS.—L'Algerie en 3 feuilles Dép. de la Guerre. Paris, 1838.

Province d'Alger,—'Echelle, 1-400,000th „

„ de Constantine „ „

„ d'Oran „ „

ALEXANDRIA.—Plan du port et des mouillages d'Alexandrie, Dép. de la Mar. Paris, 1838.

SOUTH AFRICA.—Les côtes méridionales d'Afrique et de l'entrée du canal de Mozambique, Plan du Port Natal et de la Baie Delagoa. Dép. de la Mar. Paris, 1838.

Port Natal, by W. T. Haddon. Hyd. Off., 1838.

MADAGASCAR.—Le Canal de Mozambique et l'île de Madagascar. Plan du Port de Mozambique, du mouillage à la côte S. O. de Zanzibar. Dép. de la Mar. Paris, 1838.

Les îles situées à l'est et au N.E. de Madagascar. Dép. de la Mar. Paris, 1838.

AMERICA.

CANADA.—Gulf of St. Lawrence, by Captain Bayfield, R.N., Hyd. Off., 1838.

—— Mount Louis Bay and Magdalen River,„ „

—— Cawee and Egg islands, „ „ „

—— Gaspé and Mal Bay, „ „ „

—— Chateau Bay (Belleisle Strait), „ „

—— Little Miccatino Sound, „ „ „

—— Red Bay—Bradore Bay—Kegashka Bay, „

—— Bonne Esperance—Little Natashquan, „ „

—— St. Nicholas Harbour—Coacoacho Bay, „ „

—— Amherst Harbour—Grand Entry Harbour, „

—— Belles Amours—Manicouagan Rivers, „ „

CANADA.—Gulf of, S.W. Part of Anticosti—Wapitayum, by Captain Bayfield, R.N., Hyd. Off., 1838.
—— E. Part of Anticosti—Bear Bay, " "
—— Mingan Islands—St. Paul's Island, " "
—— Magdalen Islands, by Lieut. P. E. Collins, R.N.
River St. Lawrence, by Captain Bayfield, R.N., Hyd. Off., 1838.
—— Traverses, " " "
—— Above Quebec, sheets 1, 2, 3, 4, " "
—— Montreal Harbour, " " "

NEWFOUNDLAND.—Les attérages de l'île de Terre Neuve, entre Cap Raze et les îles St. Pierre et Miquelon. Dépôt de la Mar., Paris, 1838.

MEXICO.—Apalachicola Bay. Captain Reals. ½ inch to a mile. Hyd. Off., 1838.

WEST INDIES.—Bahama—Western Elbow of Cay Sal Bank. Capt. R. Owen, R.N., Hyd. Off., 1838.
Cuba.—Colorados and Sancho Pardo Shoals, from Span. MSS., and Capt. R. Owen, R.N. 5 inches to a mile. Hyd. Off., 1838.
Trinidad.—Bocas de Dragos, by Capt. Columbine, R.N. 2 inches to a mile. Hyd. Off., 1838.

SOUTH AMERICA.—General Map of, comprising the whole of the late Survey of the Coast, in one sheet, by John Arrowsmith. Nov., 1838.

GUAYANA.—Entrance of the river Corentyn, by R. H. Schomburgk, scale 1 inch to a mile. Hyd. Off., 1838.

AUSTRALIA.

AUSTRALIA.—General Map of Australia, scale 9-10ths of an inch to a degree, by John Arrowsmith. 1838.
Map of the South-Eastern portion, scale 7-10ths of an inch to a degree, by John Arrowsmith. 1838.
Spencer and St. Vincent's Gulf, &c., scale 3 inches to a degree, by John Arrowsmith, 1838.
Port Phillip, by Lieuts. Symonds and Henry, and Mr. Shortland, R.N. ½ mile to an inch. Hyd. Off., 1838.
Eastern Entrance to Torres Straits, by Capt. P. P. King, R.N. Hyd. Off., 1838.
Arafura Sea, by G. W. Earl. Hyd. Off., 1838.

MISCELLANEOUS.

ATLANTIC.—Carte reduite des mers du nord, depuis 48° jusqu'à 75° de lat., par M. P. Daussy. Dép. de la Mar., Paris, 1838. Scale 4-10ths inch to a degree.

PACIFIC.—L'Océan Pacifique, par MM. D'Urville et Lottin; revue en 1834. Dép. de la Mar, 1838. Scale 5 degrees to an inch.
Actæon Group, by Mr. G. Biddlecomb, R.N. Hyd. Off., 1838.
Atlas du voyage de la *Thétis* et de *l'Espérance*, en 1824, 5, 6. Dép. de la Mar., Paris, 1838.
Physikalischer Atlas, von Dr. H. Berghaus. Parts 1 and 2. Gotha, 1838.

WORLD.—Map of the World, coloured according to the different religious creeds, &c., by James Wyld. London, 1838.

INDEX.

END OF VOL. VIII.

London: Printed by W. Clowes and Sons, Stamford Street.

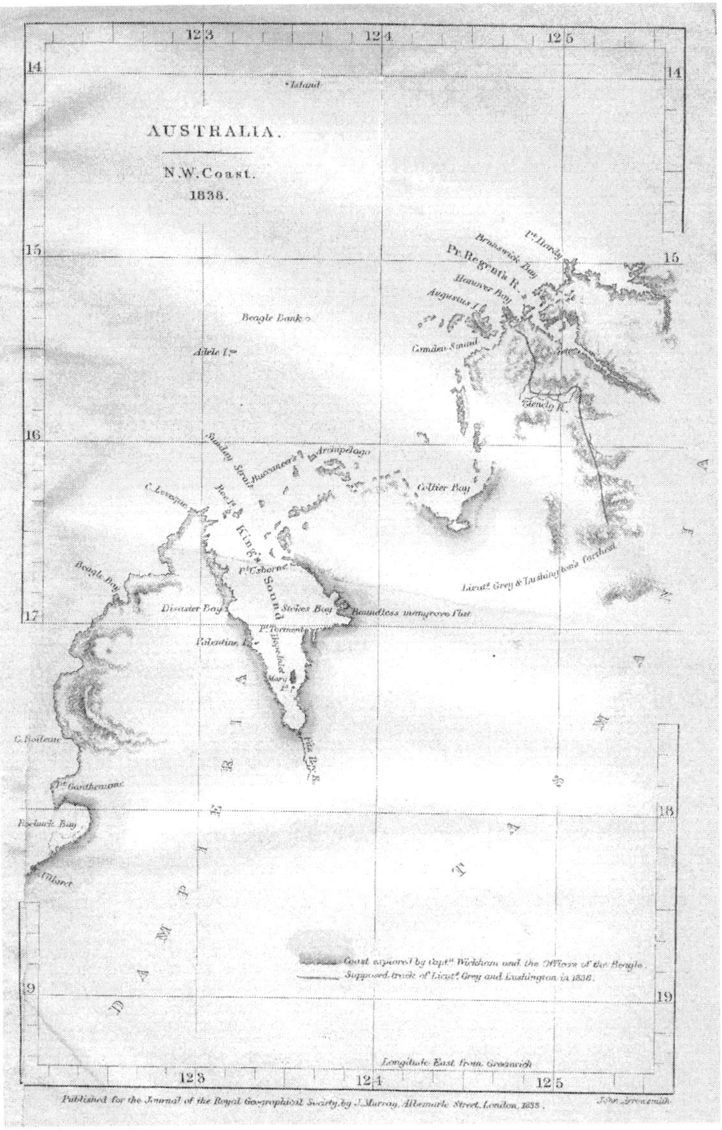

AUSTRALIA.
N.W. Coast.
1838.
Island
Beagle Bank
Adele I.
Brunswick Bay
Pt. Hardy
Pr. Regent's R.
Hanover Bay
Augustus I.
Camden Sound
Glenelg R.
Sunday Strait
Buccaneer's Archipelago
Collier Bay
C. Leveque
King's Sound
Pt. Usborne
Beagle Bay
Disaster Bay
Stokes Bay
Boundless mangrove flat
Lieut. Grey & Lushington's farthest
Pt. Torment
Valentine I.
C. Boileau
Pt. Gantheaume
DAMPIER LAND
Coast explored by Captn. Wickham and the Officers of the Beagle.
Supposed track of Lieut. Grey and Lushington in 1838.
Longitude East from Greenwich
Published for the Journal of the Royal Geographical Society, by J. Murray, Albemarle Street, London, 1838.
John Arrowsmith

839027

Printed in Great Britain by
Amazon.co.uk, Ltd.,
Marston Gate.